AF361896

Advanced Polymeric Scaffolds for Stem Cell Engineering and Regenerative Medicine

Advanced Polymeric Scaffolds for Stem Cell Engineering and Regenerative Medicine

João Carlos Silva
Frederico Castelo Ferreira

Basel • Beijing • Wuhan • Barcelona • Belgrade • Novi Sad • Cluj • Manchester

Editors
João Carlos Silva
iBB-Institute for Bioengineering
and Biosciences
Universidade de Lisboa
Lisboa
Portugal

Frederico Castelo Ferreira
iBB-Institute for Bioengineering
and Biosciences
Universidade de Lisboa
Lisboa
Portugal

Editorial Office
MDPI AG
Grosspeteranlage 5
4052 Basel, Switzerland

This is a reprint of articles from the Special Issue published online in the open access journal *Polymers* (ISSN 2073-4360) (available at: www.mdpi.com/journal/polymers/special_issues/polymer_scaffold_cell).

For citation purposes, cite each article independently as indicated on the article page online and as indicated below:

Lastname, A.A.; Lastname, B.B. Article Title. *Journal Name* **Year**, *Volume Number*, Page Range.

ISBN 978-3-7258-2592-9 (Hbk)
ISBN 978-3-7258-2591-2 (PDF)
doi.org/10.3390/books978-3-7258-2591-2

Contents

About the Editors

João Carlos Silva

João Carlos Silva (J.C. Silva) is currently a Marie Sklodowska-Curie Action (MSCA) Postdoctoral Fellow at Politecnico di Torino in Italy and a collaborating researcher in the Stem Cell Engineering Research Group from the Institute for Bioengineering and Biosciences (iBB), in which he is the PI of Biomimetic and Functional Regenerative Biomaterials Lab. He completed his PhD in Bioengineering—Cell Therapies and Regenerative Medicine in 2019 under a collaborative international project between iBB and the Rensselaer Polytechnic Institute (RPI, Troy New York, USA). He also holds an Integrated Masters in Biomedical Engineering from Instituto Superior Técnico—Universidade de Lisboa (2013).

J.C. Silva's main research includes i) tailoring biomaterials structure and composition to mimic stem cell and tissue microenvironments, particularly focused on the development of advanced 3D disease models for osteoarthritis and bone cancer research; ii) the development of novel bio-inspired, biomimetic, and functional ("smart") biomaterial scaffolds for regenerative medicine and cellular agriculture applications; and iii) the design and prototyping of bioreactor devices for the fabrication of tissue-engineered constructs under controlled "in vivo-like" biophysical (mechanical and electrical) stimulation.

J.C. Silva is the author of 44 original articles and 17 indexed conference abstracts published in international peer-reviewed scientific journals, 1 book chapter, 1 encyclopedia entry, 1 granted patent, and 31 oral and 30 poster communications presented in relevant international conference meetings. He also supervises a team of 6 PhDs, 1 research fellow, and 6 MSc students working toward the development of novel bioengineering and tissue engineering strategies for applications in orthopedic and dental regeneration.

Frederico Castelo Ferreira

At the Institute for Bioengineering and Bioscience (iBB), Frederico Castelo Ferreira (FFerreira), an Associate Professor at the Department of Bioengineering (IST), leads the Focus Areas in Process/Devices and Materials Engineering for Health and Sustainability (pMESH and dMESH), and he is the PI of the Laboratories on Sustainable Tissue Engineering, Materials and Processes and on Processes for Sustainable Food and Pharma. He obtained his Ph.D. in 2004 from Imperial College London (IC) and did a post doc with IC and GlaxoSmithKline (2004–2006). He also holds an MBA (2006–2008) and a BSc in Applied Chemistry (5-year degree with a 2-year specialization in Biotechnology), both from Nova University of Lisbon.

He has published c.a. 150 SCI papers, with c.a. 3000 citations and an h-index of 30. He has co-authored 7 patent applications and 7 book chapters. Ferreira is co-author of more than 200 presentations and participated on 25 R&D projects.

Ferreira's current research interests balance between fundamental and applied research, with potential translation into the market of sustainable products and processes. Primarily, he is focused on the following:

- Tissue engineering, biomaterials, and bioreactors, namely the development of electrical and magnetic stimuli-responsive nanotechnologies, scaffolds and bio-inks for developing stem-cell-based tissue constructs and/or targeting cancer therapies;

- Sustainable separations and biorefineries exploring unconventional yeasts for the production of fuels, biosurfactants, and enzymes from residues and wastes and using membrane technology and molecularly imprinted polymers for difficult separations;

- Cellular Agriculture, with a focus on the production of seafood with low ecological impact; the development of novel edible bio-inks and scaffolds made from algae and vegan materials; and fostering tissue growth, maturation, and organization.

Editorial

Advanced Polymeric Scaffolds for Stem Cell Engineering and Regenerative Medicine

João Carlos Silva [1,2,*] and Frederico Castelo Ferreira [1,2]

1. Department of Bioengineering and iBB-Institute for Bioengineering and Biosciences, Instituto Superior Técnico, Universidade de Lisboa, Av. Rovisco Pais, 1049-001 Lisboa, Portugal; frederico.ferreira@tecnico.ulisboa.pt
2. Associate Laboratory i4HB—Institute for Health and Bioeconomy, Instituto Superior Técnico, Universidade de Lisboa, Av. Rovisco Pais, 1049-001 Lisboa, Portugal
* Correspondence: joao.f.da.silva@tecnico.ulisboa.pt

check for
updates

Citation: Silva, J.C.; Ferreira, F.C. Advanced Polymeric Scaffolds for Stem Cell Engineering and Regenerative Medicine. *Polymers* **2024**, *16*, 2667.

https://doi.org/10.3390/polym16182667

Received: 2 August 2024

Accepted: 8 September 2024

Published: 22 September 2024

Polymeric scaffolds play a pivotal role in tissue engineering (TE) and regenerative medicine strategies, as they offer the possibility to closely mimic the architectural features of the native tissues' extracellular matrix (ECM) and support cell performance both in vitro and in vivo, creating a favourable regenerative microenvironment [1]. Natural and synthetic polymers have been widely used in TE to fabricate scaffolds with defined structural, biological, mechanical and degradation properties to guide specific cellular behaviours (e.g., adhesion, migration, proliferation and differentiation) and meet the particular requirements for the regeneration of different target tissues [2,3]. Recent multidisciplinary developments in the design and engineering of new stimuli-responsive polymers and hybrid bioactive composites, as well as significant advances in scaffold biofabrication technologies (e.g., additive manufacturing—3D bioprinting and electrospinning), have increased the structural complexity and regenerative potential of polymer-based scaffolds [4,5]. This enhanced functionality achieved by state-of-the-art polymeric scaffolds strongly encourages their use in combination with stem cells towards the development of disruptive TE strategies and reliable in vitro disease models with high potential for clinical translation.

This Special Issue focuses on bringing together contributions covering the most recent and exciting approaches to the development of polymeric biomaterial scaffolds for regenerative medicine and disease modelling applications. This multidisciplinary publication comprises sixteen manuscripts, of which eleven are original research papers and five are review papers, from research groups all over the world. These contributions, which are briefly summarized below, cover a wide range of topics related to polymeric scaffolds, from novel bioinks for 3D bioprinting, engineered hydrogels for soft tissue repair, and drug delivery systems for anticancer therapies to the optimization of scaffolds' structural and mechanical properties through the use of additives or following mathematical design and in silico modelling approaches.

Carvalho and colleagues [6] reviewed and discussed the potential of bone ECM non-collagenous proteins for bone and dental TE strategies. More specifically, the authors presented an overview of the different non-collagenous proteins found in bone ECM, highlighting their important role in regulating cellular processes such as adhesion, proliferation, osteogenic differentiation and angiogenic potential, all of which are crucial for successful bone TE strategies.

Hybrid composite scaffolds combining biological materials with synthetic polymers to achieve constructs with high bioactivity and appropriate mechanical properties hold great promise for regenerative medicine. Accordingly, Sawyer et al. [7] reviewed the use of a commercially available reinforced tissue matrix (RTM) device (OviTex), developed by embedding a polymer within a decellularized ovine-derived ECM, in several types of hernia repair treatments. The authors observed that the use of the RTM in hernia repair promoted wound healing, with no evidence of foreign body response, and provided structural tissue reinforcement, lowering the risk of hernia recurrence.

Electrically conductive polymeric scaffolds are attracting considerable interest for TE applications, particularly in strategies targeting the regeneration of electroactive tissues such as bone, skeletal muscle, neural, and cardiac tissue [4]. In this context, Bartoli et al. [8] presented an overview of the use of nanostructured carbon fillers (e.g., carbon nanotubes (CNTs) and graphene-related materials) to produce biopolymer composites with enhanced mechanical properties and electrical conductivity, particularly targeting TE applications. Accordingly, Sanjuan-Alberte and colleagues [9] developed conductive photocurable inks for two-photon polymerization (2PP) 3D printing by dispersing multi-walled CNTs (MWC-NTs) within a gelatin methacrylate (GelMa) solution. The enhanced electrical properties of the formulated hydrogel inks were demonstrated through impedance spectroscopy and cyclic voltammetry. Moreover, the conductive material was shown to support the viability and growth of human induced pluripotent stem cell (iPSC)-derived cardiomyocytes. Finally, the authors were able to manufacture, for the first time, micron-sized conductive hydrogels via 2PP, highlighting the potential of their strategy for applications in TE and bioelectronics. Considering the current need to develop appropriate scaffolds and setups for the electrical stimulation of cells, Garrudo et al. [10] reviewed the existing systems based on poly(aniline):camphorsulfonic acid (PANI:CSA), one of the most well-studied electroconductive polymers, with a particular focus on applications for neural cell culture and differentiation.

The application of innovative biomaterials, namely electrospun nanofibers, nanoparticles (NPs), hydrogels and 3D printed scaffolds, in anticancer therapies and regenerative strategies after tumour resection is comprehensively discussed in a review paper by Zaszczyńska and colleagues [11]. Considering this, Ferrentino et al. [12] synthetized and characterized amphiphilic poly(ε-caprolactone)-poly(ethylene glycol) (PCL-PEG-PCL) triblock copolymers, which self-assembled in the form of core–shell micelles. The PCL-PEG-PCL core–shell NPs were characterized using dynamic light scattering and nuclear magnetic resonance, as well as in terms of their capacity for uptake by human colorectal carcinoma cells. Finally, the developed micelles were loaded with the hydrophobic molecule quercetin, and the anticancer properties of the quercetin-loaded NPs were demonstrated using a human colon cancer cell line (HCT 116 cells). In a different study addressing cancer, Banda Sánchez and colleagues [13] describe the selection and optimization of a bioink and bioprinting process for pancreatic tumour modelling. The ink, composed of plasma-loaded alginate/methylcellulose (Alg/MC), was characterized in terms of its viscoelastic behaviour, gelation kinetics and degree of recovery, showing suitability for 3D bioprinting. When combined with pancreatic tumour cells (PANC-1), the cell-laden bioink was able to generate 3D-bioprinted models featuring proper geometrical fidelity while also supporting the PANC-1 cells' viability, typical phenotype and proliferation capacity.

The development of tissue-engineered vascular grafts (TEVGs) for the regeneration of small arteries remains an unmet clinical need. Nevertheless, recent approaches using polymeric scaffolds in vascular TE strategies have achieved highly promising results both in vivo and in vitro. Antonova et al. [14] developed a TEVG based on poly(3-hydroxybutyrate-co-3-hydroxyvalerate)/poly(ε-caprolactone) loaded with vascular endothelial growth factor (VEGF), basic fibroblast growth factor (bFGF) and stromal cell-derived factor 1α (SDF-1α) and surface-coated with heparin and illoprost (PHBV/PCL[VEGF-bFGF-SDF]$^{\text{Hep/Ilo}}$). The developed TEVG was evaluated in a sheep carotid artery interposition model using biostable vascular prostheses of expanded poly(tetrafluoroethylene) as a control. The authors observed that the PHBV/PCL[VEGF-bFGF-SDF]$^{\text{Hep/Ilo}}$ grafts showed a promising primary patency rate in comparison to the control, and the regenerated arteries presented high vascularization, full endothelialization and a multilayer hierarchical structure similar to that of the native blood vessels. Wang et al. [15] synthesized biodegradable and electrospinnable α-amino acid-substituted poly(organophosphazene) (PαAPz) polymers and subsequently fabricated and characterized PαAPz electrospun nanofibrous scaffolds for vascular TE applications. The authors were able to produce bead-free fibrous scaffolds, with average diameters of 100 to 300 nm, that supported the adhesion and spread of human smooth muscle cells (SMCs), bone marrow-derived mesenchymal stem/stromal cells (BM-MSCs) and iPSC-derived MSCs (iMSCs). Notably, in vitro studies evidenced that PαAPz nanofibers were able to promote the differentiation of iMSCs towards an SMC lineage, suggesting their potential for the fabrication of engineered functional vascular tissues.

Collagen membranes are the most widely used biomaterial for guided bone regeneration procedures in dentistry and periodontology. However, their fast degradation kinetics limits the success of these therapeutic approaches. To understand whether the collagen's origin might have an important effect on the biodegradation process, Vallecillo-Rivas et al. [16] studied the degradation patterns of five commercially available collagen membranes (Biocollagen, Heart, Evolution X-fine, CopiOs and Parasorb Resodont) of different origins in three different solutions (phosphate-buffered saline (PBS), bacterial collagenase from *Clostridium histolyticum* and porcine trypsin). Overall, the Evolution X-fine collagen membrane derived from porcine pericardium was the least affected when exposed to the three different solutions, while Biocollagen and Parasorb Resodont (both of equine origin) presented the highest degradation rates. Moreover, the bacterial collagenase solution was the most aggressive medium, with four out of the five membranes (with the exception of Evolution X-fine) being totally degraded after 7 days of exposure.

Focusing on the search for an appropriate biomaterial for the repair of dura mater (DM) defects, Lipovka and colleagues [17] compared the mechanical properties of bacterial nanocellulose (BNC) impregnated with chitosan nanoparticles (Novochizol^{TM}) and the antibiotic vancomycin with those of native BNC and with those of cadaveric DM (preserved in formalin) and human native DM. The authors reported that the BNC + Novochizol^{TM} + vancomycin achieved strength properties similar to those of DM (cadaveric and fresh). Moreover, the strength features were significantly improved by the addition of Novochizol^{TM} to the BNC and could be modulated by varying the polymer thickness.

The management and functional regeneration of articular cartilage and osteochondral (OC) defects are still among the most challenging clinical issues in orthopaedics. Biodegradable and non-biodegradable polymeric scaffolds have been widely used in cartilage regeneration and replacement strategies, respectively. Considering a cartilage replacement approach, Duque-Ossa et al. [18] developed a solid/liquid triborheological system to recreate the conditions in human synovial joints. The system was used to evaluate the potential of polyvinyl alcohol (PVA) hydrogels for articular cartilage replacement under physiological conditions and using hyaluronic acid (HA) as a lubricant. The pressures applied during the tests simulated the impact experienced daily in these joints. The PVA hydrogel characterization results showed an equilibrium water content of around 83%, which is very similar to that of native articular cartilage (80%), and a compression Young's modulus of 2.26 MPa. Triborheological tests confirmed a coefficient of friction in the range of 0.1–0.4, which is compatible to that of natural cartilage (0.1) and supports the hypothesis that PVA hydrogels lubricated with HA might represent a promising mechanical alternative for future tissue replacement strategies. In another study focused on cartilage repair, Fu and colleagues [19] optimized the formulations of injectable hydrogels based on tyramine (TA)-functionalized HA and dextran (Dex) towards improved chondrocyte growth and matrix synthesis. Based on the high biocompatibility, improved mechanical properties, and enhanced glycosaminoglycan and collagen production, the authors observed that the 5% *w/v* hybrid hydrogels produced with a 25%/75% HA/Dex ratio were the most promising constructs for cartilage TE. Another study performed by Cordeiro et al. [20] combined PCL with cellulose from different sources (commercially available: microcrystalline and methylcellulose, and cellulose obtained from agro-industrial residues) to fabricate scaffolds for cartilage TE using Fused Deposition Modelling (FDM). The produced composite scaffolds were characterized in comparison to pure PCL in terms of their morphology, chemical and mechanical properties, enzymatic degradation profile and cytocompatibility. Overall, the results suggest that cellulose (particularly methylcellulose) can be incorporated into PCL scaffolds to produce scaffolds with compressive properties closer to those of native cartilage and enhance the proliferation of human dental pulp stem/stromal cells. In a different study focused on OC repair, Marcelino et al. [21] presented a novel method for designing and manufacturing scaffolds with a mathematically defined curvature (based on the geometry of a sphere) able to mimic the native OC tissue shape. The printability of the scaffolds using poly lactic acid (PLA) and FDM equipment was evaluated, and a limit sphere radius was defined to ensure high printing fidelity without defects, as confirmed via scanning electron microscopy and micro-computed tomography. The mechanical behaviour of the produced curved scaffolds was evaluated experimentally under compressive tests and by means of finite element modelling, which allowed for the identification of the scaffold regions subjected to higher loads. Finally, the authors highlighted the potential of combining numerical modelling with experimental approaches towards the development of improved biomimetic scaffolds for OC TE strategies.

In conclusion, this Special Issue provides the readers of *Polymers* with an outstanding and multidisciplinary array of exciting recent research and literature reviews on the development of advanced polymer scaffolds for regenerative medicine applications, highlighting their potential for the development of more affordable and sustainable personalized therapies.

Author Contributions: Conceptualization: J.C.S. and F.C.F.; Writing—Original Draft: J.C.S.; Writing—Review and Editing: J.C.S. and F.C.F. All authors have read and agreed to the published version of the manuscript.

Funding: This research was funded by FCT—Portuguese Foundation for Science and Technology (FCT/MCTES) through the project InSilico4OCReg (PTDC/EME-SIS/0838/2021) and through funding provided to the research unit iBB (UID/BIO/04565/2022) and to the Associate Laboratory I4HB (LA/P/0140/2020).

Acknowledgments: The Guest Editors would like to acknowledge all contributing authors for their submissions to this Special Issue. A particular gratitude is expressed to the Editors-in-Chief for their invitation to organize this Special Issue of *Polymers* and to the Section Managing Editor, for his efficient and pro-active management of the editorial process.

Conflicts of Interest: The authors declare no conflicts of interest.

References

1. Khan, M.U.A.; Aslam, M.A.; Adbullah, M.F.B.; Hassan, A.; Shah, S.A.; Stojanović, G.M. Recent perspective of polymeric biomaterial in tissue engineering—A review. *Mater. Today Chem.* **2023**, *34*, 101818. [CrossRef]
2. Jafari, M.; Paknejad, Z.; Rezai Rad, M.; Motamedian, S.R.; Eghbal, M.J.; Nadjmi, N.; Khojasteh, A. Polymeric scaffolds in tissue engineering: A literature review. *J. Biomed. Mater. Res. Part B* **2017**, *105B*, 431–459. [CrossRef] [PubMed]
3. Molina, M.I.E.; Malollari, K.G.; Komvopoulos, K. Design Challenges in Polymeric Scaffolds for Tissue Engineering. *Front. Bioeng. Biotechnol.* **2021**, *9*, 617141.
4. Municoy, S.; Álvarez Echazú, M.I.; Antezana, P.E.; Galdopórpora, J.M.; Olivetti, C.; Mebert, A.M.; Foglia, M.L.; Tuttolomondo, M.V.; Alvarez, G.S.; Hardy, J.G.; et al. Stimuli-Responsive Materials for Tissue Engineering and Drug Delivery. *Int. J. Mol. Sci.* **2020**, *21*, 4724. [CrossRef]
5. Castilho, M.; de Ruijter, M.; Beirne, S.; Villette, C.C.; Ito, K.; Wallace, G.G.; Malda, J. Multitechnology Biofabrication: A New Approach for the Manufacturing of Functional Tissue Structures? *Trends Biotechnol.* **2020**, *38*, 1316–1328. [CrossRef]
6. Carvalho, M.S.; Cabral, J.M.S.; da Silva, C.L.; Vashishth, D. Bone Matrix Non-Collagenous Proteins in Tissue Engineering: Creating New Bone by Mimicking the Extracellular Matrix. *Polymers* **2021**, *13*, 1095. [CrossRef]
7. Sawyer, M.; Ferzoco, S.; DeNoto, G., III. A Polymer-Biologic Hybrid Hernia Construct: Review of Data and Early Experiences. *Polymers* **2021**, *13*, 1928. [CrossRef]
8. Bartoli, M.; Piatti, E.; Tagliaferro, A. A Short Review on Nanostructured Carbon Containing Biopolymer Derived Composites for Tissue Engineering Applications. *Polymers* **2023**, *15*, 1567. [CrossRef]
9. Sanjuan-Alberte, P.; Vaithilingam, J.; Moore, J.C.; Wildman, R.D.; Tuck, C.J.; Alexander, M.R.; Hague, R.J.M.; Rawson, F.J. Development of Conductive Gelatine-Methacrylate Inks for Two-Photon Polymerisation. *Polymers* **2021**, *13*, 1038. [CrossRef]
10. Garrudo, F.F.F.; Linhardt, R.J.; Ferreira, F.C.; Morgado, J. Designing Electrical Stimulation Platforms for Neural Cell Cultivation Using Poly(aniline): Camphorsulfonic Acid. *Polymers* **2023**, *15*, 2674. [CrossRef]
11. Zaszczyńska, A.; Niemczyk-Soczynska, B.; Sajkiewicz, P. A Comprehensive Review of Electrospun Fibers, 3D-Printed Scaffolds, and Hydrogels for Cancer Therapies. *Polymers* **2022**, *14*, 5278. [CrossRef] [PubMed]

12. Ferrentino, N.; Romano, M.P.; Zappavigna, S.; Abate, M.; Del Vecchio, V.; Romano, D.; Germinario, C.; Grifa, C.; Filosa, R.; Pappalardo, D. Poly(ε-caprolactone)-poly(ethylene glycol) Tri-Block Copolymer as Quercetin Delivery System for Human Colorectal Carcinoma Cells: Synthesis, Characterization and In Vitro Study. *Polymers* **2023**, *15*, 1179. [CrossRef] [PubMed]
13. Banda Sánchez, C.; Cubo Mateo, N.; Saldaña, L.; Valdivieso, A.; Earl, J.; González Gómez, I.; Rodríguez-Lorenzo, L.M. Selection and Optimization of a Bioink Based on PANC-1- Plasma/Alginate/Methylcellulose for Pancreatic Tumour Modelling. *Polymers* **2023**, *15*, 3196. [CrossRef]
14. Antonova, L.V.; Krivkina, E.O.; Sevostianova, V.V.; Mironov, A.V.; Rezvova, M.A.; Shabaev, A.R.; Tkachenko, V.O.; Krutitskiy, S.S.; Khanova, M.Y.; Sergeeva, T.Y.; et al. Tissue-Engineered Carotid Artery Interposition Grafts Demonstrate High Primary Patency and Promote Vascular Tissue Regeneration in the Ovine Model. *Polymers* **2021**, *13*, 2637. [CrossRef]
15. Wang, M.; Lin, S.; Mequanint, K. Electrospun Biodegradable α-Amino Acid-Substituted Poly(organophosphazene) Fiber Mats for Stem Cell Differentiation towards Vascular Smooth Muscle Cells. *Polymers* **2022**, *14*, 1555. [CrossRef]
16. Vallecillo-Rivas, M.; Toledano-Osorio, M.; Vallecillo, C.; Toledano, M.; Osorio, R. The Collagen Origin Influences the Degradation Kinetics of Guided Bone Regeneration Membranes. *Polymers* **2021**, *13*, 3007. [CrossRef] [PubMed]
17. Lipovka, A.; Kharchenko, A.; Dubovoy, A.; Filipenko, M.; Stupak, V.; Mayorov, A.; Fomenko, V.; Geydt, P.; Parshin, D. The Effect of Adding Modified Chitosan on the Strength Properties of Bacterial Cellulose for Clinical Applications. *Polymers* **2021**, *13*, 1995. [CrossRef]
18. Duque-Ossa, L.C.; Ruiz-Pulido, G.; Medina, D.I. Triborheological Study under Physiological Conditions of PVA Hydrogel/HA Lubricant as Synthetic System for Soft Tissue Replacement. *Polymers* **2021**, *13*, 746. [CrossRef]
19. Fu, Y.; Zoetebier, B.; Both, S.; Dijkstra, P.J.; Karperien, M. Engineering of Optimized Hydrogel Formulations for Cartilage Repair. *Polymers* **2021**, *13*, 1526. [CrossRef]
20. Cordeiro, R.; Alvites, R.D.; Sousa, A.C.; Lopes, B.; Sousa, P.; Maurício, A.C.; Alves, N.; Moura, C. Cellulose-Based Scaffolds: A Comparative Study for Potential Application in Articular Cartilage. *Polymers* **2023**, *15*, 781. [CrossRef]
21. Marcelino, P.; Silva, J.C.; Moura, C.S.; Meneses, J.; Cordeiro, R.; Alves, N.; Pascoal-Faria, P.; Ferreira, F.C. A Novel Approach for Design and Manufacturing of Curvature-Featuring Scaffolds for Osteochondral Repair. *Polymers* **2023**, *15*, 2129. [CrossRef] [PubMed]

Review

Bone Matrix Non-Collagenous Proteins in Tissue Engineering: Creating New Bone by Mimicking the Extracellular Matrix

Marta S. Carvalho [1,2,3,*], Joaquim M. S. Cabral [2,3], Cláudia L. da Silva [2,3] and Deepak Vashishth [1,*]

1   Center for Biotechnology and Interdisciplinary Studies, Department of Biomedical Engineering, Rensselaer Polytechnic Institute, Troy, NY 12180, USA

2   Department of Bioengineering and iBB—Institute for Bioengineering and Biosciences, Instituto Superior Técnico, Universidade de Lisboa, 1049-001 Lisboa, Portugal; joaquim.cabral@tecnico.ulisboa.pt (J.M.S.C.); claudia_lobato@tecnico.ulisboa.pt (C.L.d.S.)

3   Associate Laboratory i4HB—Institute for Health and Bioeconomy, Instituto Superior Técnico, Universidade de Lisboa, 1049-001 Lisboa, Portugal

*   Correspondence: martacarvalho@tecnico.ulisboa.pt (M.S.C.); vashid@rpi.edu (D.V.)

Abstract: Engineering biomaterials that mimic the extracellular matrix (ECM) of bone is of significant importance since most of the outstanding properties of the bone are due to matrix constitution. Bone ECM is composed of a mineral part comprising hydroxyapatite and of an organic part of primarily collagen with the rest consisting on non-collagenous proteins. Collagen has already been described as critical for bone tissue regeneration; however, little is known about the potential effect of non-collagenous proteins on osteogenic differentiation, even though these proteins were identified some decades ago. Aiming to engineer new bone tissue, peptide-incorporated biomimetic materials have been developed, presenting improved biomaterial performance. These promising results led to ongoing research focused on incorporating non-collagenous proteins from bone matrix to enhance the properties of the scaffolds namely in what concerns cell migration, proliferation, and differentiation, with the ultimate goal of designing novel strategies that mimic the native bone ECM for bone tissue engineering applications. Overall, this review will provide an overview of the several non-collagenous proteins present in bone ECM, their functionality and their recent applications in the bone tissue (including dental) engineering field.

Keywords: non-collagenous proteins; extracellular matrix; bone tissue engineering; biomimetic scaffolds

Citation: Carvalho, M.S.; Cabral, J.M.S.; da Silva, C.L.; Vashishth, D. Bone Matrix Non-Collagenous Proteins in Tissue Engineering: Creating New Bone by Mimicking the Extracellular Matrix. *Polymers* **2021**, *13*, 1095. https://doi.org/10.3390/polym13071095

Academic Editor: Alessandro Pistone

Received: 28 February 2021
Accepted: 20 March 2021
Published: 30 March 2021

1. Introduction

New promising solutions for bone tissue engineering have been developed over the last years following the dramatic increase of the number of bone-related medical conditions that require clinical interventions. In fact, each year, more than one million non-union fractures are treated in the United States [1]. Moreover, 5–10% of the bone fractures that occur worldwide do not heal [1]. Besides bone fractures, bone tissue can also be damaged by traumas, tumors, infections, or bone diseases. Furthermore, new strategies to engineer bone tissue are required as an alternative to the use of bone grafts, addressing the increasing worldwide incidence of bone disorders in an aging society severely impacted by obesity, lack of exercise, and with reducing healing capacity [2]. Even though bone tissue engineering appears as a promising alternative, to date the gold-standard treatment for bone regeneration still relies on bone grafts, autologous, allogenic, and xenogeneic grafts [1] (Figure 1). These approaches have some limitations and are not ideal for bone regeneration. Autografts have been applied since they can provide a matrix with osteogenic cells and osteoinductive factors to support new bone growth, without having immunological rejection and promote better osseointegration. Nonetheless, the availability of quality graft material is limited and possible complications may occur, such as pain, infections, scarring, and weakening of the donor bone. Moreover, there is a high morbidity associated with this procedure, since more than one surgery is needed [1,3]. Allografts, usually harvested from cadavers, have also some limitations, namely the higher risk of immunologic rejection and infection though it requires less procedures than autografts, minimizing the surgical time and accelerating the patient recovery [3]. As an alternative to allografts, xenografts consist of transplantation of bone tissue across species. The most common xenograft used in orthopedic surgery is bovine derived. Xenografts have some advantages compared to other grafts, such as being readily available due to the abundance of donor bone tissue and being less expensive than allografts [2,3]. In fact, commercially-available xenografts are approximately one-tenth the price of commercially-available allografts. Also, because of the extensive sterilization processes, their shelf life is generally long. However, xenografts present several challenges such as the risk of disease transmission and a higher risk of immune response of the host tissue compared to allografts [3]. Moreover, xenografts require intensive sterile processing, which can decrease their osteoinductive properties.

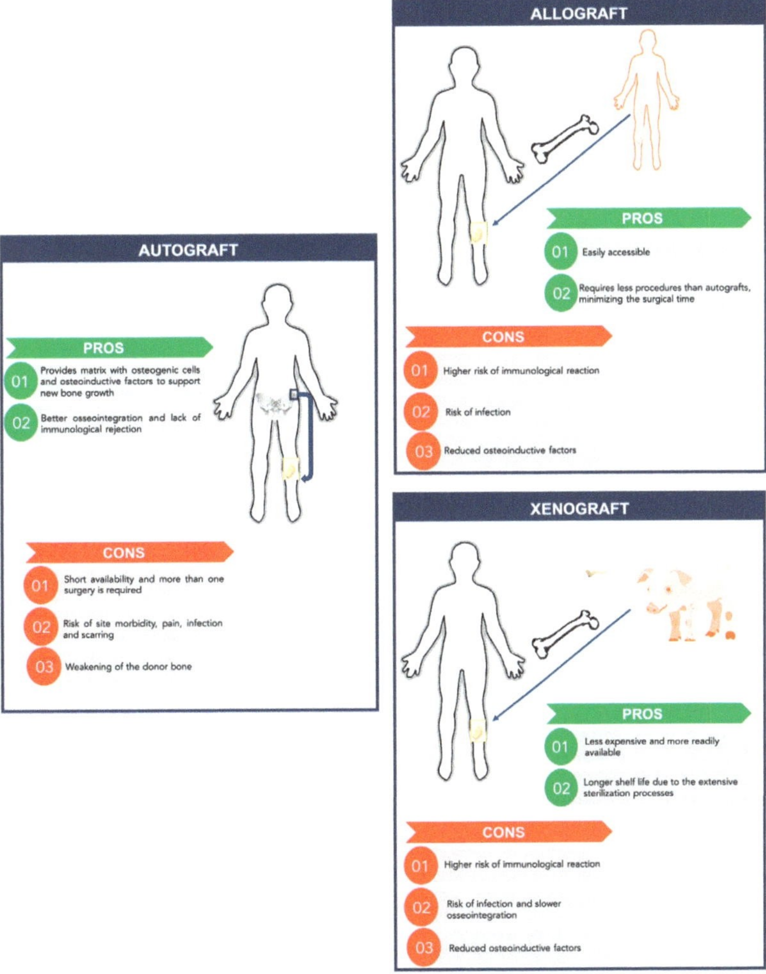

Figure 1. Comparison between autografts, allografts, and xenografts: Advantages and disadvantages.

Bone tissue engineering has the potential for solving these problems by combining different elements such as cells, molecules, and scaffolds. The standard tissue engineering approach uses a combination of growth factors, scaffolds and osteogenic cells (triangular concept). However, Giannoudis and colleagues developed and discussed a new concept, the diamond concept, in which a fourth element, vascularization, should be also considered as a contributor to bone healing. Thus, the diamond concept in bone tissue engineering combines four basic elements [4]: (i) A biomaterial with osteogenic ability for bone formation that acts as a scaffold for the other elements; (ii) osteogenic cells capable to creating or inducing new bone formation at the defect site; (iii) osteoinductive molecules that trigger cells and recruit resident cells to form new functional bone tissue; and (iv) vascularization to support the viability of the defect site thus allowing the diffusion of oxygen and nutrients to the defect region (Figure 2).

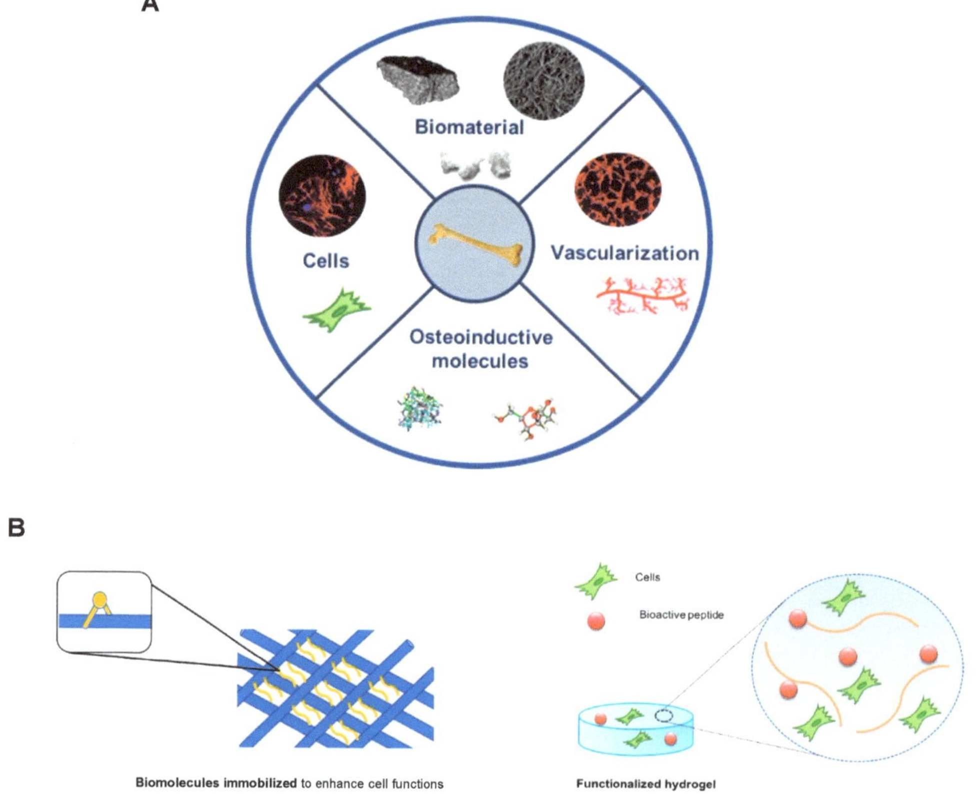

Figure 2. Bone tissue engineering strategies. (**A**) The bone tissue engineering paradigm highlights (1) Biomimetic scaffold, (2) osteogenic cells, (3) osteoinductive molecules, and (4) vascularization. (**B**) Schematic representation of biomolecules immobilized into a porous scaffold (left) and a functionalized hydrogel with bioactive peptides and cells incorporated (right).

Several scaffolds have been developed for bone tissue engineering, including natural biomaterials (such as collagen, gelatin, and chitosan), ceramic implants (such as hydroxyapatite), polymeric synthetic materials (such as polylactic acid (PLA) and polyglycolic acid (PGA)) and composite scaffolds [5]. Even though materials science technology has resulted in clear improvements for bone regeneration, challenges to achieve functional and mechanically competent bone growth still remain. One approach in the design of bone scaffolds involves the production of biomimetic bone matrices, in which the bone replacement material should interact with the surrounding tissues by biomolecular recognitions [6]. Among the molecular signals used for bone tissue engineering applications, the number of alternatives is yet smaller compared to the number of different scaffolds materials that can be used. Bone morphogenetic proteins (BMPs), a group of growth factors, play the leading role in the field, being able to promote proliferation and differentiation of osteogenic cells [7]. BMP-derived peptides have been widely used for bone tissue engineering applications, in particular derived from BMP-2 and BMP-7 [6]. BMP-2 peptides have been shown to promote osteogenic differentiation of MSCs [7,8]. Lin and co-workers developed a copolymer membrane loaded with a novel synthetic BMP-2 derived peptide, P24, and observed enhanced osteogenic differentiation of MSCs in vitro and bone regeneration in vivo [8]. Furthermore, BMP-2 and -7 received approval from the FDA and the EMA to be used in combination with type I collagen for the treatment of severe tibial fractures and posterolateral spinal fusions [9–11]. Although being widely used, several side effects of BMPs have been reported, such as postoperative inflammation and associated adverse effects, ectopic bone formation, osteoclast-mediated bone resorption, and inappropriate adipogenesis [12]. BMPs have other drawbacks, such as the high costs of production and the high doses required, raising questions about their cost-effectiveness [13]. In addition to above, calcitonin gene-related peptide (CGRP) has been widely applied due to its bone regeneration potential. CGRP might play a crucial role in promoting osteoblast proliferation and differentiation by bonding with functional receptors and transporters on the osteogenic cells and by stimulating growth factors production, such as BMP-2 [14]. In fact, Mi and colleagues demonstrated that CGRP administration increased new bone formation by promoting MSCs migration and differentiation [14]. A recent study from Lai and co-workers presented a new strategy to immobilize CGRP onto TiO_2 nanotubes through polydopamine [15]

Recently, the field has shifted towards investigating the interaction between extracellular matrix (ECM) proteins and cell membrane receptors [5]. This approach avoids the use of growth factors and better mimics the bone ECM, reducing the side effects and increasing the efficiency of bone healing process. ECM proteins can be used intact or reduced to peptides with specific sequences that will trigger the action required. Therefore, these new osteoinductive peptides are easy and less expensive to manufacture, more unlikely to elicit immune responses due to their small size and stable in physiological conditions [16]. An attractive strategy consists of combining different ECM peptides to enhance cellular processes, such as adhesion and proliferation but also to promote osteogenic differentiation and angiogenesis. Combining peptides with important functions/properties increases the effectiveness and versatility of the final product to be used in clinical applications. In spite of the great improvements on incorporating ECM peptides on biomimetic materials, the influence of non-collagenous bone ECM proteins on osteogenic differentiation remains to be evaluated. This review aims to present a summary of the different non-collagenous proteins found in bone ECM and their important functions in context of bone tissue engineering applications, specifically on their role in cell adhesion, proliferation, osteogenic differentiation, and angiogenic capacity.

2. Bone Extracellular Matrix: Characterization, Properties, and Quality

Most of the outstanding properties of the bone are related to its matrix constitution [17]. Bone ECM has two components: A mineral part comprising of hydroxyapatite (70–90%) and an organic part (10–30%) composed primarily of collagen (approx. 90% of organic matrix) with the rest being non-collagenous proteins (~10%) [17,18]. The organic matrix of bone is mainly composed of collagen, however, Herring and co-workers identified the presence of other non-collagenous proteins [19] (Figure 3). Type I collagen is the most prevalent protein in the body and can be found not only in mineralized but also in non-mineralized tissues, playing a critical role in the structure and function of different skeletal tissues [18]. However, type I collagen is not the only protein involved in mineralization. Improved technologies have led to the isolation of a large number of non-collagenous matrix proteins. It is known that some matrix proteins bind to collagen forming fibrils. Thus, collagen serves as a scaffold upon which nucleators of hydroxyapatite, such as non-collagenous proteins, are present (Figure 4) [18–20]. Although some studies have already described the potential role of these non-collagenous proteins, their contributions and role in bone tissue engineering applications remain to be well investigated. Moreover, bone ECM quality may be determined not only by the nature of collagen type I, but also by mineral and non-collagenous proteins composition [21,22]. Using different characterization techniques and diseased mice models, it has been demonstrated that the nano-structural organization influences bone properties. In fact, several diseases related with deregulation of type I collagen and mineralization showed impairment of bone quality and other bone properties, such as bone fragility and strength. Mice with osteogenesis imperfecta, a condition derived from mutation in type I collagen, presented bone fragility and reduction in strength [23,24]. Osteopetrosis is a condition responsible for hypermineralization of bone that increases bone fragility [25] and involves altered interactions between collagen and mineral component that modify the nature of organization in bone at the nanometer scale. Non-collagenous proteins have also been suggested to influence the mechanical quality of bone matrix. Studies on osteopontin (OPN) showed that it behaves like "glue" in bone [26]. In the presence of calcium ions, OPN is capable of sacrificial bonding, a nanoscale mechanism that dissipates energy and inhibits crack growth. Osteocalcin (OC), the most abundant bone specific non-collagenous protein, complexes with OPN [27] and regulates bone mineralization through its strong affinity to hydroxyapatite. Previous works from our group found that fracture in bone initiates as dilatational bands that form as a result of OC-OPN interaction. In the absence of either protein, the complex is disrupted, resulting in a dramatic loss of toughness [28].

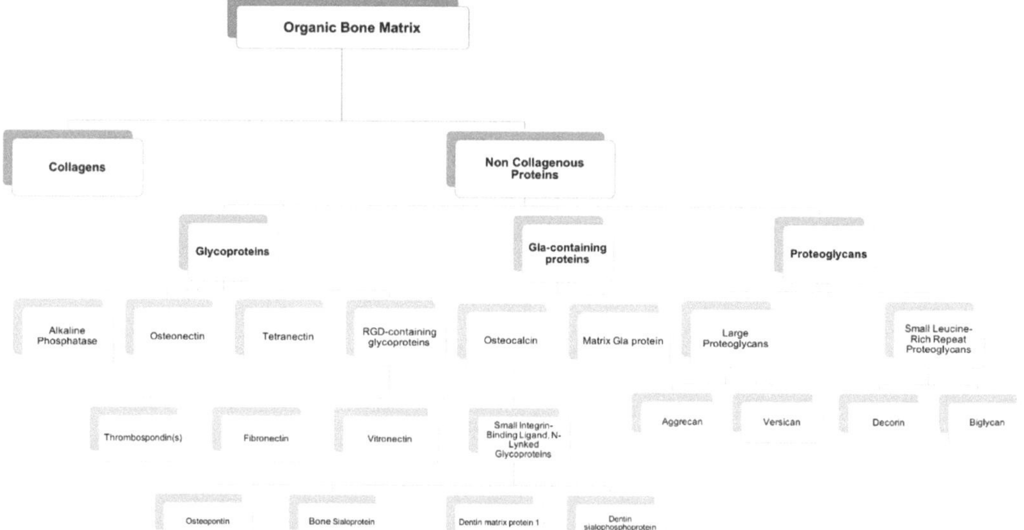

Figure 3. Organic components of the bone extracellular matrix.

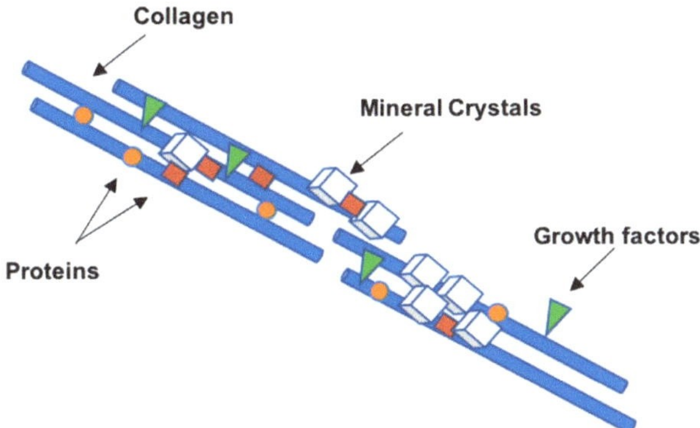

Figure 4. Schematic diagram representing the organization of the collagen molecules reinforced with calcium phosphate nanocrystals, proteins, and growth factors arranged in a semi-regular pattern.

3. Non-Collagenous Bone Matrix Proteins

Non-collagenous proteins have been isolated from bone and have been found to be biologically active, even though their functions are not yet completely understood. Based on their localization patterns, each of these proteins may perform different functions. Therefore, it is extremely important to better understand the properties and functions of these proteins, aiming to design innovative strategies for bone tissue engineering applications. It has been speculated that non-collagenous proteins might have an important role in cell attachment, cell differentiation, and regulation of hydroxyapatite minerals deposition [29]. Some of these proteins may be multifunctional, playing different roles in the bone, thus defining a single function may not be sufficient. Also, some of these proteins might act together, having a synergistic effect on cellular behavior and mechanical properties of bone, or they can compensate some effects resulting from deregulation of the levels of other non-collagenous proteins present in bone matrix (Figure 5). However, non-collagenous proteins may also present some drawbacks such as limited information about their mechanism of action, high-water solubility, and limited availability.

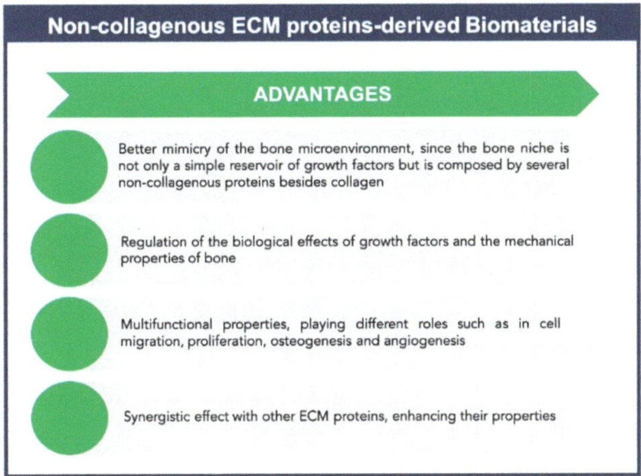

Figure 5. Advantages of non-collagenous extracellular matrix (ECM) proteins-derived biomaterials for Bone Tissue Engineering applications.

Interestingly, not all types of bones contain the same amount of non-collagenous bone proteins. In humans, for example, cortical bone contains 30× more OC than trabecular bone, but trabecular bone contains 21× more osteonectin (ON) [30]. Moreover, it is possible to find non-collagenous proteins in some other tissues besides bone, specifically OPN and ON present a general tissue distribution. Bone sialoprotein (BSP) and OC are also found in other mineralizing tissues, such as dentin. Therefore, importance of these proteins in bone physiology cannot be underestimated. Indeed some studies have reported that mutations in some of these proteins may result in abnormal bone [29].

The multifunctional properties of these non-collagenous proteins make them attractive agents for incorporation within an appropriate scaffold to enable stem cell-based bone tissue engineering (Figure 5). These proteins can be used successfully as signaling molecules to direct stem cell recruitment, attachment, and differentiation and create a mature and mineralized extracellular matrix.

In bone, non-collagenous proteins are mainly composed of two major types: Glycoproteins and gamma-carboxyglutamic acid (Gla)-containing proteins; however some proteoglycans can also be found in smaller content [31,32]. The most relevant and abundant glycoproteins are represented by alkaline phosphatase (ALP), ON, and the Arginine-Glycine-Aspartic acid (RGD)-containing proteins, which include, but are not limited to, OPN and sialoproteins. Of the Gla-containing proteins, OC is the major component.

4. Exploiting Non-Collagenous Proteins in Bone Tissue Engineering Applications: An ECM Mimicking Approach

For regeneration of soft and hard tissues, several synthetic and natural materials have been used to create "ideal" scaffolds that mimic and function as bone tissue. These scaffolds should be porous, biocompatible and biodegradable in order to promote cell recruitment, migration and proliferation. Additionally, these scaffolds should also stimulate neovascularization into the graft, incorporate into the host microenvironment and form new healthy tissue [33]. Such new designs have indeed led to great breakthroughs and improvements in bone tissue engineering, however their development has also highlighted several shortcomings over the years. Although fabrication of synthetic materials can be easily reproducible, they often lack the biological cues required for engineering new tissue [33]. In contrast, scaffolds derived from natural materials, such as type I collagen, possess bioactive motifs that can guide cell adhesion, proliferation, differentiation, and tissue regeneration and have demonstrated better osteoinductive properties. However, these natural scaffolds suffer from lack of reliable and reproducible quality standards and greater batch-to-batch variation [34,35]. To combine the advantages of natural and synthetic materials and minimize the drawbacks of each, hybrid scaffolds have been designed for bone tissue engineering. Thus, researchers have been recently focused on designing and engineering materials with structure, composition and functions similar to the bone ECM. Moreover, the defined structure of biomaterials can also provide mechanical cues and modulate the microenvironment, controlling cell behavior. Thus, when fabricating new scaffolds for bone tissue engineering applications, it is important to consider the influence of physical cues, such as porosity, pore structure, roughness, and stiffness. Aiming to tune the microstructure of the biomaterial, different methods can be applied, such as 3D printing and electric field-assisted techniques, which improve scalability and control of scaffold microstructure [2,36].

The effort to functionalize synthetic scaffolds with biological cues, such as growth factors and ECM peptides, to form new tissue involves elicitation of specific cellular responses that may be absent in the native tissue (Figure 2) [37,38]. In fact, ECM can provide an adhesive substrate to which integrin and other adhesive cell receptors can bind, modulating several signaling cascades. ECM peptides can interact with cell-surface receptors, regulate signaling cascades that control cell development and determine gene expression [39]. Therefore, the properties of materials can be enhanced, specifically by promoting osteogenic differentiation and by inducing expression of osteogenic marker genes in osteoblasts [33]. Among the several signaling molecules, BMPs have been used clinically. Instead of the use of growth factors, another common strategy is the incorporation of cell-binding peptides into biomaterials, mimicking naturally occurring processes such as cell-ECM signaling, cell proliferation and differentiation [40]. The most commonly used peptide for surface modification is RGD [41,42], the signaling domain derived from fibronectin (FN) and laminin and also found in collagen. These peptides can be chemically attached to polymers to facilitate cellular interactions at an injury site. Specifically, RGD peptides have been shown to enhance proliferation, differentiation and mineralization when attached to the surface of various biodegradable materials [37,42,43]. Biomimetic PLA scaffolds modified with RGD peptides have been fabricated to promote the attachment and proliferation of osteoblasts [42]. Moreover, it has been reported that, by controlling the distribution of RGD on hydrogels by nanopatterning, it is possible to maximize its beneficial effects on adhesion, viability, and differentiation of mesenchymal stem/stromal cells (MSCs) [44,45]. Additionally, other peptide sequences have been immobilized on different scaffolds, such as Tyr-Ile-Gly-Ser-Arg (YIGSR) and Ile-Lys-Val-Ala-Val (IKVAV) in laminin [46], as well as Arg-Glu-Asp-Val (REDV) and Leu-Asp-Val (LDV) in FN [47]. Peptide fragments from collagen have been used for surface modification in numerous studies. For example, Bhatnagar and colleagues identified a cell-binding domain P15 (GTPGPQGIAGQRGVV) from type I collagen that supported ECM synthesis [48]. Similarly, another type I collagen peptide-GFOGER was used to functionalize surfaces and shown to support the expression of osteogenic genes and to induce matrix mineralization, similar to type I collagen [49,50]. Recently, Aziz and co-workers have developed a 3D matrix metalloproteinase (MMP)-sensitive hydrogel to promote osteocyte differentiation. These poly(ethylene glycol) hydrogels contained the peptide crosslink GCGPLG-LWARCG (MMP-sensitive peptide) and RGD to promote cell attachment. This hydrogel was capable of cell-mediated degradation and enhanced mineralized collagen matrix and osteocyte differentiation [51].

Collagen has been described to be extremely important as a template to create bone tissue; however, it is unknown which of the other components from bone matrix are essential to engineer new bone tissue. Recently, investigators have been focusing on using the native non-collagenous proteins from bone matrix, instead of collagen-binding motifs, to enhance the properties of the scaffolds, such as cell migration, proliferation and osteogenesis. Although great improvements in developing peptide-incorporated biomimetic materials have been achieved, the influence of the non-collagenous bone matrix proteins on osteogenic differentiation remains to be evaluated, even though they have been identified and described some decades ago [19]. Thus, in vitro and in vivo studies must be conducted to better understand the functions of the bone matrix constituents. In fact very few studies have used components of the organic bone matrix other than collagen to create bone substitutes. Johnson and colleagues used a composite alloimplant of human bone morphogenetic protein and autolyzed allogeneic bone containing a mixture of ECM proteins. In a clinical trial, bone union was possible in 24 of 25 cases identified as non-unions. However, the addition of BMPs as osteoinductive factors did not allow for any conclusions on the performance of the non-collagenous bone matrix proteins alone [52]. Sun and co-workers integrated non-collagenous proteins from bone ECM into gelatin scaffolds to form an artificial matrix that mimicked natural ECM while enhancing osteogenesis and mineralization [53].

In the next section, we will introduce and briefly summarize the several non-collagenous proteins present in bone ECM, as well as describe their key characteristics in context of bone tissue engineering (Table 1).

Table 1. Non-collagenous protein-modified biomaterials for bone tissue engineering applications.

Non-Collagenous Protein	Modified Biomaterial	Outcomes	References
Alkaline Phosphatase	ALP-immobilized on microporous nanofibrous fibrin scaffolds by 1-ethyl-3-(3-dimethylaminopropyl)carbodiimide hydrochloride/N-hydroxysuccinimide (EDC/NHS) method.	Supported cell proliferation and osteogenic differentiation in vitro. In vivo, these scaffolds promoted bone formation.	[54]
Osteonectin	Oxidized alginate hydrogels with the GHK peptide, a fragment of osteonectin. The free aldehyde groups present in the oxidized alginate can form covalent bonds with molecules that contain amino groups, such as GHK (self-crosslinking).	Improved osteogenic differentiation of MSCs, demonstrated by enhanced gene expression, alkaline phosphatase activity and bone extracellular matrix deposition.	[55]
Fibronectin	Fibronectin-immobilized nanobioactive glass/polycaprolactone scaffolds by EDC/NHS treatment.	Improved cellular adhesion and proliferation.	[56]
Vitronectin	Vitronectin-derived peptide covalently grafted onto titanium scaffolds. Pretreated (oxidized and silanized) constructs were peptide-grafted by immersion overnight into a 1 mg/mL peptide solution.	The presence of the vitronectin-peptide bound to the titanium constructs improved the osteogenic activity immediately after implantation, accelerating bone ongrowth.	[57]
Osteopontin	Oligo(poly(ethylene glycol)) fumarate hydrogels modified with OPN-derived peptide. Peptides were coupled to acrylated-PEG by NHS treatment.	Improved osteoblast proliferation and migration.	[58]
Osteopontin	CO_3 apatite-collagen sponges containing the SVVYGLR motif (amino acids residues 12–18 of OPN). CO_3 apatite-collagen sponges were immersed in 10 ng/mL of SVVYGLR peptide solution.	In vivo studies presented improved angiogenesis.	[59]
Bone Sialoprotein	Bone sialoprotein coated 3D printed calcium phosphate scaffolds. 3D printed calcium phosphate scaffolds were coated with BSP via physisorption. Incubation was performed with different concentrations of BSP solution (50 and 200 μg/mL) under mechanical stirring at 8 °C.	Improved osteoblast viability and in vivo studies showed that BSP coated 3D printed calcium phosphate scaffolds promoted increased bone formation in comparison to uncoated scaffolds.	[60,61]

4.1. Proteoglycans

This class of molecules is characterized by the covalent attachment of long chain polysaccharides (glycosaminoglycans, GAGs) to core protein molecule. GAGs are composed of repeating carbohydrate units that are sulfated to varying degrees and include chondroitin sulfate (CS), dermatan sulfate (DS), keratan sulfate (KS), heparin sulfate (HS), and hyaluronan (HA, unsulfated) [31]. These proteins are commonly found in cartilage matrix. Moreover, since endochondral bone formation is mediated by a cartilage matrix, chondrogenic proteins can be incorporated into the initial matrix to promote bone formation [31,32]. Proteoglycans have been suggested to be responsible for matrix maintenance, organization and regulation of cartilage calcification, through interactions with the GAG chain of type IX and the type II collagen fibrils [62]. Proteoglycans and GAGs can inhibit hydroxyapatite formation and growth [63] and they can also chelate. Several proteoglycans have been identified in cartilage matrix and bone matrix, such as the large proteoglycans, aggrecan and versican, and small leucine-rich repeat proteoglycans, such as decorin and biglycan [64].

For tissue engineering applications, synthetic peptido-GAGs have shown promising results for biomedical applications [65]. These peptido-GAGs could replicate many biological functions of decorin, modulating fibril formation and stiffness of the new tissue and promoting cellular adhesion [66,67]. Also, aggrecan peptido-GAG have been used to enhance the properties of some scaffolds for cartilage regenerative applications [68].

Moreover, few studies have incorporated collagen and GAGs within scaffolds [68,69], providing a suitable 3D environment for inducing osteogenic differentiation of MSCs [70,71] and to enhance osteoblast activity [72]. In fact, incorporation of GAGs within collagen fibrils improved cell proliferation [73]. Moreover, the addition of GAGs reduced the degradation rate of the scaffold and improved the structural stability of the collagen/GAG matrix [73,74]. Interestingly, these scaffolds have demonstrated favorable results for bone tissue engineering applications, even without adding stem cells or growth factors to the system [70]. Therefore, approaches to mimic the native structure and composition of bone tissue with novel scaffolds comprising collagen, GAGs and calcium phosphate crystals have been successful [75,76].

The glycosaminoglycan CS has also been used for bone tissue engineering applications as CS supports osteogenic differentiation of MSCs and increases the regeneration ability of injured bone. A CS-bioglass composite encapsulating MSCs was reported to induce bone regeneration in vivo when incorporated with BMPs [77]. Also, a CS-collagen biomaterial, fabricated as a BMP delivery system, showed high biocompatibility and osteogenic stimulation [78]. Hyaluronic acid has been used in several medical fields. This GAG is also used as a carrier for regenerative growth factors [79]. Proteoglycans can be further divided into large proteoglycans (aggecan and versican) and small leucine-rich repeat proteoglycans (decorin and biglycan) (Tables 1 and 2).

Table 2. Proteoglycans in Bone Matrix: Protein functions and in vivo studies.

Proteoglycans	In Vivo Studies	Functions	References
Aggrecan	Aggrecan deficient mice presented cartilage matrix deficiency and were characterized by perinatal lethal dwarfism and craniofacial abnormalities.	Can have an important role in preventing cartilage calcification.	[80]
Versican	Versican deficient mice have presented an early lethality.	Can have an important role in preventing cartilage calcification.	[81–83]
Decorin	Decorin-knockout mice showed skin laxity and fragility and their bones did not demonstrate any visible bone phenotype. However, their teeth sowed alteration in matrix properties, presenting a hypomineralized dentin.	Binds to collagen and can regulate fibril diameter and orientation. Can prevent premature osteoid calcification and regulate the collagen-matrix interactions.	[84–88]
Biglycan	The biglycan-knockout mice presented reduced skeletal growth, having shorter femora and decreased bone mass.	Binds to collagen and can regulate fibril diameter and orientation. Can prevent premature osteoid calcification and regulate the collagen-matrix interactions.	[84,86–89]

4.1.1. Large Proteoglycans

Aggrecan and Versican are two large CS-proteoglycans that are found in bone matrix and can bind to HA, forming large aggregates.

Aggrecan

Aggrecan has been suggested to play an important role in skeletal development, having a molecular weight of approximately 205 kDa. First studies revealed that mice with a mutation of aggrecan gene presented cartilage matrix deficiency, perinatal lethal dwarfism and craniofacial abnormalities [80]. Since the amount of aggrecan present in bone is much lower than in cartilage tissue, it is not completely understood if its presence in bone represents residual calcified cartilage. Although aggrecan has been reported to have an important role in preventing cartilage calcification, its function in bone remains unknown.

Versican

Versican is another CS proteoglycan that is found in relatively lower levels compared to aggrecan in cartilage and bone. It has a molecular weight of approximately 360 kDa and is reported to be expressed during osteogenesis and in bone development [81]. Although versican stimulates chondrocyte proliferation [82], its function in cartilage and bone are still unknown. Potentially, it can serve as a bridge between cells and the ECM, allowing for cell binding to HA [83].

4.1.2. Small Leucine-Rich Repeat Proteoglycans

Small leucine-rich repeat proteoglycans are another family of proteoglycans with a protein core and leucine-rich repeat sequence. In cartilage and bone, several members of this family, such as decorin and biglycan, are present. They exhibit different patterns of expression and tissue localization, which might be indicative of different functions.

Decorin

Decorin has an apparent molecular weight of approximately 130 kDa and it has been shown to bind to and regulate the fibrillogenesis of type I, II, and VI collagens and collagen-matrix interactions [90]. In bone, decorin may function as regulator of collagen fibril diameter and fibril orientation and can prevent premature osteoid calcification due to its low affinity for calcium [84,90], in contrast to a high affinity to type I collagen. Moreover, studies suggest a role of decorin in matrix mineralization, since proteoglycans with low molecular weight are present in type I collagen fibrils but then disappear when mineralization occurs [85]. Studies with decorin-knockout mice showed skin laxity and fragility without any visible changes in bone phenotype. However teeth from decorin-knockout mice showed alteration in matrix properties, presenting a hypomineralized dentin [86].

Biglycan

Biglycan is another small proteoglycan present in both cartilage and bone with a molecular weight of approximately 45 kDa. Although its functions remain to be investigated, biglycan demonstrates different effects in solution depending on its concentration. Low concentrations promote apatite formation whereas in higher concentration biglycan inhibits the growth and proliferation of mineral crystals [84].

The biglycan-knockout mice present reduced skeletal growth with short femora and decreased bone mass [87]. Moreover, decorin and biglycan-double knockout mice have additive deficiency in dermis and synergistic effects in bone. Ultrastructural analysis of these mice reveals loss of fibril geometry [88]. The mineral within these bones has increased crystal size compared to wild-type controls [89]. However, the low amount of biglycan in bone matrix as well as its absence within bone collagen fibrils suggest that its principal function may not be directly related to mineralization.

4.2. Glycoproteins

This class of proteins is characterized by the covalent linkage of sugar moieties attached via asparaginyl or serinyl residues [32]. These glycoproteins may be further modified by post-translational sulfation and phosphorylation (Tables 1 and 3). Glycoproteins derived from bone matrix have been reported to enhance cell binding, being recently used in association with a variety of different scaffolds for tissue engineering applications.

Table 3. Glycoproteins in Bone Matrix: Protein functions and in vivo studies.

Glycoproteins	In Vivo Studies	Functions	References
Alkaline Phosphatase	Mice with null mutations for the tissue –nonspecific alkaline phosphatase showed increased osteoid and defective growth plate development.	Possible role in mineralization. ALP can act as a potential Ca^{2+} carrier and hydrolyzes inhibitors of mineralization such as pyrophosphates.	[91–95]
Osteonectin	Osteonectin deficient mice have presented a poor bone status, developing osteopenia.	Can promote mineral deposition and regulate growth and proliferation of mineral crystals, supporting bone remodeling. May influence cell functions, binding to growth factors and through cell-matrix interactions.	[96–103]
Tetranectin	Tetranectin deficient mice have presented a delayed fracture healing.	Can regulate matrix mineralization, playing a role in tissue formation and remodeling.	[104–106]
Thrombospondin	Thrombospondin deficient mice presented disordered collagen in their soft tissues, increased cortical bone thickness and density and altered fibroblast attachment.	Role in cell attachment. It binds to several ECM proteins. Role in bone development and remodeling, collagen fibrillogenesis and ECM organization.	[107–111]
Fibronectin	Elimination of fibronectin gene in transgenic animals is lethal in utero, since connective tissues do not form.	Role in cell attachment. It binds to several matrix proteins and cell surface proteins, like collagen.	[112–114]
Vitronectin	Vitronectin deficient mice have been shown to have a thrombolytic phenotype, but skeletal defects were not apparent in these mice.	Role in cell attachment. It can bind to collagen.	[113,115,116]
Osteopontin	Osteopontin deficient mice presented larger crystal size and an increased mineral content.	Role in cell attachment. It binds with other molecules present in bone matrix. Can regulate mineralization by regulating the nucleation of mineral crystals. Can regulate bone resorption through osteoclasts attachment and migration. Play a specific role in angiogenesis.	[117–130]
Bone Sialoprotein	Bone sialoprotein deficient mice presented shorter, hypomineralized bones with higher trabecular bone mass and with lower bone formation rate.	Role in cell attachment and matrix mineralization induction. It acts as a hydroxyapatite nucleator since it has high affinity for calcium. Can have an important role in osteoclasts formation and bone resorption.	[131–138]

Table 3. *Cont.*

Glycoproteins	In Vivo Studies	Functions	References
Dentin matrix protein-1	Dentin matrix protein-1 deficient mice have significantly lower mineral content when compared with their controls.	Role in cell attachment. It binds to collagen. If phosphorylated, may inhibit the formation and growth of hydroxyapatite, if dephosphorylated it facilitates nucleation of hydroxyapatite crystals, inducing mineralization. Can play a role in angiogenesis.	[139–148]
Dentin sialophosphoprotein	Dentin sialophosphoprotein deficient mice have shown decreased mineral content.	Can regulate type I collagen fibrillogenesis and acts as nucleator of hydroxyapatite formation at lower concentrations and inhibitor at higher concentrations.	[149–152]

4.2.1. Alkaline Phosphatase

Although ALP is not typically thought of as a matrix protein, several studies demonstrated that ALP can be released from the surface of osteogenic cells or in a membrane-bound form (matrix vesicles) [91,92].

Developmental studies, in vivo and in vitro, have suggested an important role of ALP in mineralization, since its expression precedes mineralization and it is maintained during early stages of hydroxyapatite deposition [93]. Moreover, it is reported that hypophosphatasia is characterized by mutations in ALP gene and results in improper mineral deposition. Indeed, mice with null mutations for the tissue-nonspecific ALP showed increased osteoid and defective growth plate development, reinforcing the importance of ALP in mineralization [94]. Furthermore, it was observed that cells that normally do not mineralize promote formation of a mineralized matrix when transfected with the ALP gene [95].

For bone tissue engineering applications, ALP has been immobilized on microporous nanofibrous fibrin scaffolds [54]. These scaffolds are nontoxic, biodegradable, and support cell proliferation and osteogenic differentiation in vitro. Furthermore, the immobilized ALP fibrin scaffolds were shown to support bone formation in a mouse calvarial defect model. ALP has also been coated onto titanium scaffolds exhibiting increased hydroxyapatite formation while enhancing the bioactivity of titanium scaffolds [153]. Additionally, ALP immobilized on collagen and alginate scaffolds have been shown to be a good candidate for improving osteogenesis [154,155].

In a different study, ALP was successfully incorporated into halloysite (HAL) nanotubes. Immobilized ALP effectively induced biomineralization processes, thus, ALP-HAL nanocomposite material may be an attractive bioactive scaffold for bone regeneration [156].

4.2.2. Osteonectin

Osteonectin (ON), a secreted phosphoprotein acidic and rich in cysteine (SPARC), was the first matrix protein to be isolated from bone with a molecular weight approximately between 35–45 kDa [96]. Although it is synthesized by osteoblasts, it can also be synthesized by fibroblasts from skin, tendon, sclera, and periodontal ligaments; however, most of ON found in circulation is derived from platelets [97].

ON binds to type I, type III, and type V collagen, thrombospondin and to hydroxyapatite through high-affinity calcium-binding sites [98–101]. Its affinity for calcium and phosphate ions may suggest that ON can promote mineral deposition; however, ON accumulates only within mineralized matrix, suggesting that it is not involved in the induction step of mineralization, but may have an important function in regulating growth and proliferation of mineral crystals. ON-deficient mice present poor bone condition and develop osteopenia with significant loss of trabecular bone associated with a decreased rate in bone formation [101,102]. These results reinforce that ON might support bone remodeling and maintenance of bone mass [101]. In vitro studies use both intact molecule and peptides derived from different regions of ON. Results suggest that ON is an important regulator of cell–matrix interactions [103]. However, many of these properties have not been evaluated in osteoblast cell cultures.

For bone tissue engineering perspective, few studies have incorporated ON into scaffolds to regulate the mineralization process [157]. A nano-hydroxyapatite/collagen/ON complex was developed to mimic the hierarchical structure of native bone from nanoscale to microscale, which surpasses the limitation of mineralized pure collagen synthesized in vitro [158]. Interestingly, the formation of the mineralized collagen nanofibers was influenced by the presence of ON [159]. Glycine–histidine–lysine (GHK) peptide, a fragment of ON, has been studied for tissue engineering applications. GHK was incorporated in alginate hydrogels and found to promote production of vascular endothelial growth factor (VEGF) and basic fibroblast growth factor (bFGF) from MSCs, increasing MSC proliferation [160]. Recently, biomaterial functionalization with GHK has been suggested to be beneficial for bone tissue engineering applications [55]. Klontzas and colleagues developed novel oxidized alginate hydrogels with the GHK peptide and demonstrated enhanced MSC osteogenic differentiation, suggesting a mechanism of GHK action related to integrin β1 and mediated by integrin linked kinase [55].

4.2.3. Tetranectin

Tetranectin has an approximate molecular weight of 21 kDa [104]. This glycoprotein is expressed by osteoblasts undergoing mineralization and is also found in tumors undergoing mineralization [104]. To date, the exact function of tetranectin in bone metabolism is not known; however, this protein might be involved in matrix mineralization. Overexpression of tetranectin by tumor cells causes an increase in ECM mineralization upon implantation into nude mice [104]. Tetranectin-deficient mice show delayed fracture healing, indicating that tetranectin may have a role during the early stage of the fracture healing process [105]. Additionally, tetranectin-knockout mouse present a phenotype with severe spinal deformities [106].

Although tetranectin has been thought to have a potential role in mineralization during osteogenesis in vivo and in vitro [104], this protein has not been applied to engineer bone. Although not within the scope of this review, tetranectins have been engineered to target proteins on tumor cells with greater specificity than on normal cells [161].

4.2.4. RGD-Containing Glycoproteins

Bone ECM also contains some glycoproteins with amino acid sequence RGD. These RGD sequences can be recognized by cell surface receptors and promote attachment between ECM and cells [162]. The receptors on the cell surface are integrins formed by one α subunit and one β subunit. Each subunit has a cytoplasmic extension that is associated with intracellular signaling pathways, a transmembrane domain and an extracellular domain [162]. The extracellular domains of the α and β subunits configure a binding pocket that recognizes the RGD sequences in the ECM proteins and mediates the cell-matrix interactions [162]. In bone matrix, some RGD-containing proteins include thrombospondin, FN, vitronectin (VN) and a family of small integrin-binding ligand, N-linked glycoproteins (SIBLINGs). The SIBLINGs have been identified by a cluster of genes including OPN, BSP, dentin matrix protein 1 (DMP-1) and dentin sialophosphoprotein (DSPP) [163].

RGD sequence has been reported to provide support for cell adhesion for bone tissue engineering; however, the RGD sequence may also enhance other cellular processes that promote mineralization [58]. Because RGD peptides interact with multiple cell types, there is a great need to identify peptide sequences that elicit more specific responses from particular cell types [58].

Thrombospondin(s)

Thrombospondin is a glycoprotein with a molecular weight of approximately 450 kDa [107]. Unlike other glycoproteins present in the bone ECM, thrombospondins are less abundant in mineralized bone matrix, but these proteins can be found in several connective tissues. Although their role in bone is still not known, thrombospondins have been suggested to be important in bone development and remodeling, especially in collagen fibrillogenesis and ECM organization [108–110]. Moreover, thrombospondins can bind to several matrix proteins and cell surface proteins [31]. Studies with mice that lack thrombospondin (TSP-2 null) present disordered collagen in their soft tissues [111] with increased cortical bone thickness and density [108].

Thrombospondin 1 (TSP-1) has been shown to influence neovascularization when incorporated in porous polyethylene implants. These findings represent a new promising approach for enhancing angiogenic properties of biomaterials for bone tissue engineering applications [164].

Fibronectin

Fibronectin (FN) is one of the most abundant ECM proteins in bone with a molecular weight of approximately 400 kDa. Since it is produced by all connective tissue cells, this protein can be found in all ECM within the body [112]. Evidence suggests that FN plays an important role during bone development. It accumulates in mineralized matrix at an early stage of bone formation [113] and is highly up-regulated by osteoblasts. In vivo studies showed that FN is an essential component for the development of the connective tissue, since the elimination of the FN gene in transgenic animals is lethal in utero, due to the lack of formation of connective tissues [114].

Interestingly, the attachment of osteogenic cells to FN, in vitro, is independent of RGD-mechanism [113]. Moreover, for cell-matrix interactions mediated by FN, $\alpha_4\beta_1$ binding may play a role in the maturation sequence of cells in the osteogenic lineage [32].

Recently, some studies have been focused on improving scaffolds' ability to target and recruit stem cells in order to accelerate cell adhesion—a promising tool for bone tissue engineering. To this end, FN has been used for surface modification by coating this protein on bioactive scaffolds to promote cell adhesion and proliferation [165]. Although ECM's affinity for cell adhesion has been well reported, the optimal ECM coating for osteogenic differentiation remains unclear. Therefore, studies have combined more than one ECM protein to coat the biomaterial's surface with a goal to enhance cell adhesion and differentiation for developing functional constructs for clinical skeletal regeneration. Moreover, by using a robotic-dispensing technique, it is possible to generate FN-immobilized nanobioactive glass (nBG)/polycaprolactone (PCL) (FN-nBG/PCL) scaffolds with an open pore architecture. With the addition of these cell-adhesive motifs onto the surface of the scaffold, cellular adhesion and differentiation processes can be accelerated [56]. Recent strategies involve the combination of biomaterials with ECM proteins or growth factors and hydroxyapatite as a promising application for bone reconstruction [166]. For example, Toupkanlou and colleagues incorporated nanohydroxyapatite in electrospun nanofibrous polycaprolactone (PCL) scaffolds coated with FN, demonstrating the synergistic effect of FN and hydroxyapatite on enhancing calcium deposition, collagen synthesis and early ALP activity and upregulation of osteogenic specific genes in vitro and in vivo [166]. Similarly, Lee and co-workers constructed a novel osteoinductive FN matrix fusion protein (oFN) containing FN_{III9} and FN_{III10} modules, the key cell-binding domain of FN, and an osteoinductive sequence from BMP-2 [165]. The engineered oFN matrix fusion protein resulted in more effective bone regeneration via promotion of cellular adhesion and differentiation. These studies show that the design of fusion proteins could represent a highly relevant approach for bone tissue engineering.

Recently, Trujillo and colleagues engineered 3D hydrogels with full-length fibronectin that sequesters growth factors, such as VEGF and BMP-2, to enhance angiogenesis and bone regeneration. The physical properties of this hydrogel can be tuned to mimic the properties of the native ECM, providing a novel 3D environment with the potential to recruit and retain growth factors promoting bone regeneration and vascularization in vivo [167].

Recent advances in bone tissue engineering have varied ECM properties by incorporating decellularized ECM from tissues or cells onto different scaffolds. For example, silk fibroin scaffolds have been coated with decellularized pulp/collagen/FN and the modified scaffold has demonstrated promising results for bone tissue engineering applications [168].

Vitronectin

Vitronectin (VN) has a molecular weight of approximately 70 kDa and it is generally found in matrices containing fibrillary collagens. In vitro, VN is produced by osteoblasts [113]. This protein plays an important role in cell adhesion. Osteogenic cells, including osteoclasts, attach very strongly to VN, mainly via the receptor integrin $\alpha_v\beta_3$ [113]. Interestingly, VN increases its concentration in the unmineralized osteoid prior to mineral deposition [115]. This evidence indicates that VN can have an important role in preparing the matrix for mineralization. In vivo studies demonstrated that mice that lack VN gene presented a thrombolytic phenotype; however, skeletal defects were not observed in these mice [116].

Similar to FN, VN has been used to coat bioactive scaffolds. Cacchioli and colleagues investigated the effect of VN peptide conjugated onto titanium surfaces and showed an increased cell apposition rate, higher ratio of mineralized surface to bone surface and more extended bone-to-implant contact in in vivo models [57].

Osteopontin

OPN is part of a family of five integrin-binding glycophosphoproteins (SIBLINGs). Besides OPN, this family comprises BSP, DMP-1, DSPP and matrix extracellular phosphoglycoprotein. The genes coding for members of the SIBLING protein family are similarly organized. SIBLINGs interact with cell surface receptors, such as integrins, mainly through a RGD sequence and function as modulators of cell adhesion as well as autocrine and paracrine soluble factors [169].

All of these proteins undergo similar post-translational modifications such as phosphorylation and glycosylation, the extent of which is crucial in determining their function [170].

OPN, also called secreted phosphoprotein (SPP), is an acidic glycoprotein that consists of about 300 amino acids, with a molecular weight of 34 kDa [117]. OPN was first identified in bone matrix acting as the bridge between cells and hydroxyapatite in bone ECM [118]. However OPN can also be detected in other tissues and plasma, such as dentin, cartilage, kidney and vascular tissues. In these tissues, OPN mediate communication between cells, suggesting that OPN could act both as a structural molecule and as a cytokine [119,120]. In bone, OPN is produced by osteoblasts during the pre-mineralization and at late stages of osteoblastic maturation [117]. OPN binds to $\alpha_v\beta_1$, $\alpha_v\beta_3$, $\alpha_v\beta_5$, $\alpha_4\beta_1$, $\alpha_5\beta_1$, and $\alpha_9\beta_1$ integrins [118,121], through their RGD domain. Additionally, OPN can also present a RGD-independent mechanism, in which OPN may engage CD44 [117,118]. However, details of the interaction of CD44 with OPN remain to be studied. The presence of additional cell receptors, the various isoforms of CD44 and variable post-translational modifications (phosphorylation and glycosylation) of OPN are all complicating factors. Moreover, $\alpha_v\beta_3$ integrin is considered to be responsible for major signals in response to the binding of OPN [122]. OPN has been proposed to regulate many physiological processes such as collagen organization, cell adhesion, cell viability, cell migration, angiogenesis, and calcification [117,123]. OPN interacts with several molecules present in the bone matrix. OPN is known to bind covalently to FN via transglutamination, and transglutamination of OPN increases its binding to collagen [124]. Ritter and colleagues demonstrated that OPN specifically associates with OC, forming stable complexes between OPN and OC [27]. In fact, the mechanisms responsible for bone formation and remodeling likely involve the association of bone matrix proteins into specific complexes that helps the organization of the matrix [27]. Moreover, the phosphorylation of OPN has shown significant effect on crystal growth [125], regulating bone crystal size. OPN has a high affinity to calcium, therefore it has been suggested to modulate the nucleation of calcium phosphate during mineralization [126]. Besides the RGD sequence (amino acid sequence Arg159-Asp159), OPN also contains an aspartate domain at its N-terminal composed of amino acid sequence Asp86-Asp89 and a calcium binding domain (amino acid sequence Asp216-Ser228) with high negative charge motifs that might be responsible for the binding of OPN to bone mineral [126,127]. However, initial studies on OPN-deficient mouse failed to indicate the presence of any major defect in mineralization [127]. Possibly, the role of OPN in bone mineralization is compensated for by other regulatory systems for mineralization or by other non-collagenous bone ECM proteins.

Further studies of genetically OPN-knockout mice showed that these mice presented larger crystal size and an increased mineral content [128], suggesting the inhibitory role of OPN in hydroxyapatite formation and growth [128,129]. Analysis of the OPN$^{-/-}$ mouse has also demonstrated that OPN is important for the function and activity of osteoclasts, specifically in osteoclast attachment and bone resorption [127,130]. Studies of ectopic bone implantation demonstrated that bone from OPN-deficient mice implanted into OPN-deficient mice exhibited significantly less resorption and lower number of osteoclasts attached to the surface of the bone compared to wild-type bone implanted intramuscularly in the back of the wild-type mice (5% vs. 25%) [130]. However, it is not known whether OPN promotes bone resorption by stimulating angiogenesis or by stimulating bone resorption via signaling through the bone matrix. Asou and colleagues demonstrated a relationship between OPN and bone resorption associated with vascularization [130], since the number of CD34$^+$ vessels near the bone implanted in OPN-deficient mice were reduced compared to wild-type, suggesting that OPN deficiency may lead to a reduction in neovascularization of ectopically implanted bones, and a consequent reduction in the number of osteoclasts and bone resorption efficiency. It is also possible that OPN may promote the survival of endothelial cells on bone matrix, facilitating the vascularization of bone tissue. However, further investigation is required to elucidate the molecular mechanisms of OPN action in mediating responses to inflammation, mechanical stress, angiogenesis, and accelerated bone resorption.

Regarding bone tissue engineering, OPN plays an important role in cell adhesion, remodeling and osseointegration at the biomaterial/tissue interface that is commonly found surrounding mineralized tissues [171]. More specifically, OPN is responsible for the recruitment of osteoblasts during the early stage of bone formation [128]. Somerman and co-workers showed that the attachment of osteoblasts to OPN was dose-dependent and was mediated by a conservative RGD peptide sequence [172]. Some peptides were already derived from OPN and have been shown to bind to collagen and promote biomineralization [117,171]. Work by Shin and colleagues showed that oligo(poly(ethylene glycol)) fumarate hydrogels modified with OPN-derived peptide influenced osteoblast proliferation and migration, demonstrating a high potential of these biomimetic materials [53,58]. OPN can also interact with multiple cell surface receptors and play an active role in many physiological processes like wound healing, bone turnover, inflammation and angiogenesis [117]. Moreover, we demonstrated that OPN dose-dependently increased the proliferation of MSCs, as well as improved the angiogenic properties of human umbilical vein endothelial cells by increasing the capillary-like tube formation in vitro [173]. For bone tissue engineering applications, a new biomimetic strategy to rapidly form mineralized bone tissue and secure a sustained bone formation response by MSCs was then developed by incorporating OPN and OC in type I collagen hydrogels [174]. He and colleagues described a hydrogel that combined OPN, BMP-2 and RGD sequence to provide a favorable microenvironment for osteogenic and vasculogenic differentiation of MSCs [175]. Lee and colleagues showed that a collagen-binding motif (GLRSKSKKFRRPDIQYPDATDEDITSHM) found in OPN could specifically bind to collagen without chemical conjugation and also demonstrated its capacity to promote hydroxyapatite formation in vitro and in vivo [176]. OPN has been reported to induce bone formation by inhibiting osteoclast resorption and, more importantly, by increasing angiogenesis [118]. In fact, it has been reported that the peptide SVVYGLR, corresponding to amino acids residues 162–168 of OPN, induces vasculogenic differentiation of MSCs [59,177,178]. This peptide induces tube formation by progenitor endothelial cells in 3D collagen gels with as much potency as VEGF [177]. Hamada and co-workers developed CO_3 apatite-collagen sponges containing the SVVYGLR motif and implemented the strategy to repair defect created in rat tibia. The scaffold promoted angiogenesis inside the graft, highlighting the importance of OPN incorporation into biomaterials for bone tissue engineering applications [59]. Zhu and colleagues developed 3D printed scaffolds composed by OPN sequenced peptide SVVYGLR grafted into mesoporous calcium silicate. Interestingly, the peptide motifs can be accessed on the surface of the scaffolds and can be released by the scaffold. In vitro and in vivo studies revealed enhanced angiogenic and osteogenic properties of these scaffolds [179].

Recently, plant-derived recombinant human OPN (p-rhOPN) has been conjugated to different scaffolds for bone tissue engineering applications [180,181]. Klinthoopthamrong and co-workers developed an active non-resorbable guided tissue regeneration membrane from bacterial cellulose combined with p-rhOPN [181]. These p-rhOPN membranes elicited biological functions leading to the enhancement of osteogenic differentiation of human periodontal ligament stem cells [181].

Bone Sialoprotein

In addition to OPN, BSP is another major SIBLING that accumulates in cement lines and in spaces between mineralized collagen fibrils [131]. BSP has an apparent molecular weight of approximately 75 kDa and its expression is exclusively located to the mineralized tissues, such as bone, dentin, cementum and certain regions of hypertrophic chondrocytes [131,132]. In bone, BSP is highly expressed by osteoblasts, osteoclasts, osteocytes and chondrocytes [132]. In fact, BSP expression indicates a late stage of osteogenic differentiation and an early stage of matrix mineralization. In addition, BSP has a high affinity for calcium, indicating that BSP is important for matrix mineralization [169]. In vitro, BSP might act as a hydroxyapatite nucleator [133]. It has been shown that a concentration as little as 9 nM BSP is sufficient to nucleate hydroxyapatite. The overexpression of BSP in osteoblasts has been shown to enhance mineralization [134]. Similarly, osteoblasts grown in the presence of an anti-BSP antibody exhibit reduced mineralization [135]. Interestingly, a cooperative relationship between BSP and collagen has been reported to increase nucleation potency when both proteins are linked [136]. In vitro assays also confirmed the role of BSP in mediating cell attachment, most likely through interaction with the $\alpha_v\beta_3$ receptor, facilitating the in vitro attachment of fibroblasts, osteoblasts and osteoclasts [169]. Moreover, BSP increases osteoclasts formation and bone resorption [137]. First in vivo studies demonstrated that BSP-deficient mouse does not exhibit an altered skeletal phenotype, possibly due to compensation of BSP function by other SIBLING proteins, revealing no differences in mineral crystal characteristics relative to controls [138]. However, more recent studies with BSP-deficient mouse show that it displays shorter, hypomineralized bones with associated higher trabecular bone mass and low bone turnover [137].

Many groups have explored the use of BSP to repair bone defects by enhancing their osteoinductive capacity [137,138]. In fact, BSP implants are attractive candidates for bone applications since BSP plays an important role in osteogenic differentiation by binding to type I collagen and to $\alpha_v\beta_3$ and $\alpha_v\beta_5$ integrins, and by mediating cell signaling and differentiation [134]. Moreover, BSP has been reported to enhance osteogenic differentiation of MSCs cultured on type I collagen [135]. Due to its high affinity for collagen, BSP combined with collagen facilitates cell migration, attachment, proliferation and differentiation through RGD and non-RGD binding of integrins [182]. Instead of using the whole protein, some investigators have been exploring the properties of the amino acids sequence 35–62 of rat BSP, corresponding to the collagen-binding peptide derived from BSP [182]. In vivo studies demonstrated that hydroxyapatite implants containing BSP-derived collagen-binding peptide implanted into rabbit calvarial defects promote new bone formation within two weeks after implantation as compared to untreated or hydroxyapatite scaffolds alone [183]. Several groups have also shown that the presence of BSP in some collagen implants, in vivo, stimulates osteogenic differentiation and bone repair [182–184] by up-regulating the expression of osteogenic genes associated with early differentiation. Interestingly, implanted BSP-collagen scaffolds, by day 7 after surgery, promote cell proliferation, matrix mineralization, and vascular invasion thereby extending bone formation into the central regions of the BSP-collagen implants. In contrast, when using only collagen scaffolds, the central regions of the implant are not affected [182]. Thus, defects where BSP-collagen scaffolds were implanted, presented new bone formation and remodeling in the whole areas of the defect, whereas defects that were implanted with collagen alone only demonstrated new bone formation in the areas near the host bone [182]. Moreover, another group has shown that functionalized silk-BSP scaffold enhanced osteogenesis, inducing mineralization and osteogenic differentiation of human MSCs, when compared to cells cultured in the presence of silk alone [185]. Furthermore, using synthetic polymers, Chan and colleagues have shown that some biomaterials that were initially thought to be used for cartilage tissue engineering applications can also be used in bone tissue engineering field [186]. Therefore, they modified polycaprolactone/poly(2-hydroxyethyl methacrylate) (PCL/pHEMA) surfaces with BSP, demonstrating an enhancement in cell adhesion, likely mediated through cell-surface receptors for RGD sequences. These enhanced cell-surface interactions found on BSP surfaces not only promote regeneration of bone, but also assist other important cellular events, such as proliferation, differentiation and matrix synthesis. Furthermore, other investigations have focused on the BSP-RGD peptide, instead of the entire protein. For example, Rezania and colleagues have covalently grafted BSP-RGD peptide from rat/mouse BSP onto quartz surfaces and shown that this peptide promotes ECM mineralization [187]. In a different study, Drevelle and co-workers used PCL films functionalized with BSP-RGD peptide and demonstrated an enhancement of cell spreading using MC3T3-E1 mouse pre-osteoblasts with an improvement of their responsiveness to recombinant human BMP-2 [188]. More recently, this BSP-RGD peptide has been reported to increase mineralization of human MSCs cultured on hydrogels, enhancing the expression of osteogenic gene markers [189]. Similarly, Rapuano used a fragment from human BSP corresponding to residues 278–293 and showed that BSP-coated plastics have better adhesion capacity, since more MC3T3-E1 cells were found attached to these surfaces [190]. In vivo studies evaluated BSP coating of 3D printed calcium phosphate scaffolds in a calvarial defect model in mice. Although histological analyses revealed that BSP-coated scaffolds had a better integration in the bone defect, no significant increase in bone formation was observed in BSP-coated scaffolds [60,61].

Dentin Matrix Proteins

Dentin matrix proteins (DMPs) are a group of non-collagenous proteins found in the ECM of dentin and bone, namely dentin matrix protein 1 (DMP-1), dentin phosphophoryn (DPP) or dentin matrix protein 2 (DMP-2), dentin sialoprotein (DSP), and dentin matrix protein 4 (DMP-4) [191]. Over the years, these DMPs have been shown to play multiple roles, such as in cell attachment, proliferation, differentiation, and matrix mineralization.

DMP-1 was first isolated from dentin, however, it can also be found in bone [139]. In bone, DMP-1 is expressed specifically in mineralized tissues by hypertrophic chondrocytes, osteoblasts, and osteocytes [140]. DMP-1 is a highly phosphorylated protein with a strong affinity for calcium. DMP-1 has been reported to influence mineralization, facilitating nucleation of hydroxyapatite crystals [141]. In fact, MC3T3 cells overexpressing DMP-1 promoted ECM mineralization [142]. Recombinant human DMP-1 has been shown to induce the osteogenic differentiation of human periodontal ligament cells [143].

Moreover, when phosphorylated, full-length DMP-1 inhibits the formation and growth of hydroxyapatite; however, when dephosphorylated, DMP-1 acts as a nucleator of hydroxyapatite formation [144]. Additionally, DMP-1 can bind specifically to the N-telopeptide region of type I collagen and, interestingly, nucleation of hydroxyapatite was exclusively found in regions where DMP-1 is bound to type I collagen [145]. DMP-1 has been reported to play a specific role in angiogenesis [146]. DMP-1-knockout mouse has further confirmed the potential role of DMP-1 in bone mineralization, since these mice have significant lower mineral content when compared to wild-type control mice [147,148].

Regarding tissue engineering, dentin has been explored to engineer the tooth structure. Compared to enamel, dentin is less mineralized and more elastic. It contains, approximately, 70% hydroxyapatite [192]. In fact, ECM proteins of human dentin are known to be necessary for dentinogenesis [193]. Several studies have attempted to create scaffolds that mimic the structure of natural dentin aiming to regenerate dentin [194]. Indeed, some reports have already demonstrated that treated dentin matrix could induce precursor cells to differentiate [192]. Dentin matrix has been fabricated to provide a natural, biocompatible scaffold, giving the appropriate microenvironment to induce complete human dentin tissue regeneration in vivo following implantation.

Current efforts are directed to study the role of the dentin matrix, specifically in tissue regeneration [195,196]. Since this is a new field in tissue engineering, there are a handful of studies incorporating specific dentin proteins/peptides into scaffolds for bone tissue engineering. However preliminary reports show that DMP-1 plays multifunctional roles and is potentially very attractive for tissue engineering applications [191]. To this end, Alsanea and colleagues demonstrated that dental pulp stem cells incorporated within a collagen scaffold in the presence of DMP-1 can differentiate into odontoblast-like cells, secreting a highly vascularized collagenous matrix [196]. These findings could be successfully applied to address clinical problems such as endodontic perforations, where the perforation site could be repaired using collagen scaffolds combined with dental pulp stem cells and DMP-1 signaling molecule.

Although DMP-1 was originally discovered from dentin matrix, it was also found in bone matrix and applied for bone tissue regeneration. Indeed, DMP-1-derived peptides have been shown to induce transformation of amorphous calcium phosphate to crystalline hydroxyapatite, demonstrating that this signaling molecule incorporated into a biomimetic scaffold could enhance nucleation of crystalline hydroxyapatite, generating high quality engineered tissues capable of withstanding the mechanical loading that bones are normally subjected to [197].

Using important domains of DMP-1, several peptides have been synthesized for incorporation into scaffolds designed for tissue engineering applications [191]. Besides the nucleating motifs of DMP-1, synthetic peptides that also contain type I collagen binding domain were generated for repair of carious dentin [191]. Thus, it may be possible to enhance the link between DMP-1 and collagen scaffolds and take advantage of the DMP-1 nucleating domain in order to facilitate calcium binding and mineralization. Recently, DMP-1 has been shown to stimulate osteogenic differentiation of MSCs [191]. Therefore, DMP-1 could be further explored in bone tissue engineering applications to enhance osteogenesis, by incorporating the whole protein or the nucleating domain of DMP-1. However, further studies are required to understand the potential role of DMP-1 as a signaling molecule for bone tissue engineering applications.

Dentin Sialophosphoprotein

DSPP is expressed in dentin, bone, cementum, and non-mineralized tissues including the lung and kidney [149]. As a single gene, an intact protein has not been isolated. However, two DSPP products, DSP and DPP, are co-expressed by odontoblasts and pre-ameloblasts at the same time as predentin is secreted [149]. Only DPP has been reported to regulate type I collagen fibrillogenesis [150] and acts as an effective nucleator for hydroxyapatite formation at lower concentrations and as an inhibitor at higher concentrations [151]. In contrast, DSP is not an effective modulator of in vitro mineralization. Some studies have shown that, when incubated with simulated body fluid solution, type I collagen scaffolds containing DPP are able to produce mineralized nodules similar to those found in bone [152], demonstrating the potential of DPP in bone tissue engineering applications for enhancing nucleation and growth of hydroxyapatite. Regarding DSP, this protein has not yet been used in bone tissue engineering applications, since its main function seems to be inhibition of mineralization [191]. In fact, negative regulators may be relevant for bone tissue engineering to allow for appropriate mineral deposition and for production of high quality mineral in cooperation with the positive regulators. Thus, DSP could be used with DPP to control the amount and quality of mineral produced for bone tissue engineering applications [191].

DSPP-knockout mice have shown decreased mineral content in dentin and bones [198,199], confirming importance of DSPP in mineralization. In humans, a mutation in the DSPP gene results in dentinogenesis imperfecta, characterized by dentin hypomineralization and significant tooth decay [200]. Some studies suggest that DSPP has roles not only in the initial mineralization of bone but also in the remodeling of the skeleton and therefore on bone turnover [201].

4.3. Gla-Containing Proteins

Bone contains several proteins that are post-translationally modified by vitamin K-dependent enzymes to form the amino acid, Gla [31] (Tables 1 and 4). Osteocalcin is the major Gla-containing protein, playing an important role in mineralization of bone, whereas matrix Gla protein is known to be more involved in regulating the calcification of cartilage.

Table 4. γ Carboxy glutamic acid-containing proteins in bone matrix: Protein functions and in vivo studies.

Gla-proteins.	In Vivo Studies	Functions	References
Osteocalcin	Osteocalcin deficient mice presented increased bone formation without impairing bone resorption.	Can influence bone mineralization. It has high affinity to calcium, accelerating nucleation of hydroxyapatite and playing an active role in the early stages of bone healing. Can regulate activity of osteoclasts and bone resorption. Acts as a hormone regulating insulin secretion and glucose homeostasis.	[202–213]
Matrix Gla Protein	Matrix Gla Protein deficient mice died prematurely due to massive calcification of their tracheal cartilage and blood vessels, indicating an important role in preventing mineralization.	Can function in cartilage metabolism inhibiting mineralization.	[214–227]

4.3.1. Osteocalcin

Osteocalcin (OC), also known as bone gama carboxyglutamic acid-containing protein (BGLAP), is an approximately 5.8 kDa protein consisting of a single chain of 49–50 amino acids, being the most abundant non-collagenous protein in bone, comprising about 20% of the non-collagenous matrix proteins [202]. OC is secreted by osteoblasts and is present in dentine and calcified matrix. This protein has three glutamic acid residues at positions 17, 21, and 24 that bind calcium and it is vitamin K-dependent [203]. Before being released into the bone ECM, OC is carboxylated on its three glutamine acid residues within the osteoblasts, however, both the carboxylated and uncarboxylated forms of OC can be found in the circulation [203,204]. Its concentration in serum is closely linked to bone metabolism, being used clinically as a marker of osteoblast activity for the assessment of bone diseases [205]. During bone development, OC production is very low and does not reach maximal levels until late stages of mineralization [206]. Although its precise mechanism of action is unclear, OC is presumed to influence bone mineralization, in part through its ability to bind with high affinity to the mineral component of bone and due to its acidic character [207]. By binding hydroxyapatite, OC accelerates the nucleation of hydroxyapatite and plays an active role in the early stages of bone healing [208]. In addition to binding to hydroxyapatite, OC functions in cell signaling and in the recruitment of osteoclasts [209] and osteoblasts [210], which have active roles in bone resorption and formation, respectively. Subcutaneous implantation of bone particles that were 99% deficient in OC leads to poor bone resorption, suggesting that OC might function as a matrix signal in recruitment and differentiation of osteoclasts [211]. In vivo data showed that OC deficient mice present increased bone formation without impairing bone resorption [198]. Although the exact mechanism is still unknown, new studies have shown that the uncarboxylated form of OC may also act as a hormone regulating insulin secretion and glucose homeostasis [212,213]. However, its physiological role in mineralization remains uncertain.

An OC-derived scaffold was described by Rammelt and colleagues in which they investigated the addition of OC enhanced bone healing around hydroxyapatite/collagen composites in a rat tibia model. They demonstrated that OC activates both osteoclasts and osteoblasts during early bone formation [208]. A different study used recombinant human $OC/FN_{III9-10}$ fusion protein to functionalize a collagen matrix for bone tissue engineering [214]. They demonstrated that $OC/FN_{III9-10}$–functionalized collagen matrices are more effective in the osteogenic differentiation than non-treated collagen matrices or even $FN_{III9-10}$–functionalized collagen matrices. These scaffolds could enhance cell adhesion, mostly by the FN domain but also were found to improve osteogenic differentiation. Regarding the effect of OC on angiogenesis, Cantatore and colleagues showed that OC exogenously applied to chick embryo chorioallantoic membrane stimulates angiogenesis and that the observed response was similar to that obtained with b-FGF [215]. Therefore, OC might be applied in bone regeneration to enhance angiogenesis in a defect site, improving the efficiency of bone healing. However, further assays are required to evaluate the angiogenic capacity of OC and its incorporation in scaffolds for bone healing applications.

4.3.2. Matrix Gla Protein

Besides OC, matrix Gla protein (MGP) is the other major Gla-containing protein in the skeleton which was first isolated from bone [216] but can also be expressed in other soft tissues [207]. It has a molecular weight of approximately 15 kDa. MGP is known to be more abundant in cartilage than in bone [216]. In the skeleton, MGP expression appears early and remains at the same level at all stages of development [217]. There is evidence that MGP is an in vivo inhibitor of mineralization of cartilage. In vivo data shows that MGP-knockout mice die prematurely because of massive calcification of tracheal cartilage and blood vessels [218].

Although the precise mechanism through which MGP regulates bone metabolism is unknown, a recent study suggested that MGP may promote osteoblast proliferation, differentiation and mineralization via the Wnt/β-catenin signaling pathway [219].

4.4. Serum Proteins

Besides non-collagenous bone proteins, there are other proteins that are not synthesized in bone but can also be found in bone matrix, such as immunoglobulins, cytokines, chemokines, and growth factors. These proteins are mostly synthesized in the liver and in the hematopoietic tissue. Through circulation, these proteins accumulate in bone, especially due to their adsorption into bone ECM by hydroxyapatite [220]. Although these serum proteins are not produced locally, they may play an important role in bone metabolism. One example of a serum protein that can be found in bone ECM is albumin that is synthesized by the liver. In vitro studies show that albumin inhibits hydroxyapatite growth and influences hydroxyapatite formation [220].

4.5. Synergistic Biomimetic Strategies: Combination of ECM Proteins/Peptides to Elicit Bone Tissue Regeneration Responses

As it was mentioned before, depending on the end use, several combinations of functional domains of non-collagenous bone ECM proteins can be incorporated into scaffolds to elicit responses for bone tissue regeneration. Indeed, an additive or sometimes synergistic effect has been reported when combining more than one ECM protein/peptide. Most of the peptides incorporated into the scaffolds have integrin-binding RGD sequences to enhance cell binding to scaffolds; however, the addition of other ECM proteins, like BSP, OPN, or OC, has been shown to enhance mineralization, to accelerate bone healing, and to induce angiogenesis [173,174]. In that sense, a combination of non-collagenous bone ECM proteins might be an impressive strategy to improve the properties of a scaffold, giving it "ideal" cues to accelerate the process of bone healing. In fact, our group demonstrated that the loss of OC and OPN reduces stem cells self-renewal potential, osteogenic differentiation, and angiogenic potential [221]. Moreover, loss of OC and OPN compromises the ECM integrity and maturation, observed by an unexpected enhancement of glycosaminoglycans content that are associated with a more primitive skeletal connective tissue, and by a delay on the maturation of mineral species produced [221]. Thus, our group has developed a synergistic biomimetic strategy to develop matrices for bone tissue engineering applications by enhancing collagen matrices with OC and OPN, increasing MSCs proliferation and accelerating osteogenic differentiation [174].

4.6. Native ECM as a Biomaterial Source

Although progress has been made, new strategies are being developed by applying the native ECM as a biomaterial source to achieve the molecular complexity and organization of native tissue [222]. Indeed, native ECM can be obtained from allogeneic tissues (living donor/cadaver) or xenogenic tissues (animals) after cleaning and decellularization. In a graft obtained from decellularization of whole mature organ, the structure and tissue architecture are preserved [223] making it suitable for tissue engineering application [224]. However, there are some limitations of using native ECM as a graft. In fact, to reduce the risk of disease transmission, harsh decellularization treatments [225] are often required, leading to the loss of bioactive components in the ECM. Another limitation is the uncontrolled tissue variability that occurs due to the age, health or gender of the tissue donor [226].

Regarding bone tissue, bone ECM can be processed by treatment with acid [227] to generate demineralized bone matrix (DBM). DBM has a gel-like consistency that can be processed as powder or granules and, therefore, can be used as bone filling material, since it does not offer a structural support. Nevertheless, this material contains collagenous proteins and growth factors, such as BMPs, FGFs, and transforming growth factors. As mentioned before as a limitation of decellularized native ECM, DBM also has uncontrolled variability depending not only on the donor [228] but also on the sterilization method [229]. Therefore, it is quite impossible to predict and guarantee the osteoinductive properties of this material.

Decellularized ECM derived from cultured cells (cell-derived ECM) appears as an alternative to decellularized native tissue-derived ECM. Cell-derived ECM is composed by a complex and organized mixture of macromolecules that mimic the native tissue microenvironment and can be obtained by decellularization of in vitro cell cultures. Cell-derived ECM acts as a reservoir of multiple growth factors, such as factors involved in inflammation (i.e., monocyte chemoattractant protein 1, macrophage colony-stimulating factor, interleukin 8), angiogenesis (i.e., VEGF), and remodeling (i.e., matrix metalloproteinase 13, osteoprotegerin) [230].

Cell-derived ECM has greater ability for customization, in contrast to tissue-derived ECM. Cell-derived ECM allows for selecting the desired cell types and the culture system (2D vs. 3D, static vs. perfusion) to optimize ECM production. Furthermore, decellularized cell-derived ECM can be fabricated with specific properties by genetically modifying the cell sources to enhance the expression or sub-express some specific molecules [226]. Cell-derived ECM has been used also as coating by depositing molecules on the scaffolds surface to enhance their bioactivity and osteoinductive properties [231,232]. However, cell-derived ECM has also some limitations with respect to tissue-derived matrices. In general, cell-derived ECM has poorer mechanical properties making them unsuitable for some applications [231].

In vitro, cell-derived ECM has been shown to sustain cell expansion and to enhance MSCs osteogenic differentiation [232–235] and has been used for bone tissue engineering applications via incorporation into scaffolds and electrospun fibers [232,236]. In vivo, cell-derived ECM presents good vascularization [237] and is able to undergo remodeling onto an immature osteoid tissue [238].

5. Concluding Remarks and Future Perspective

Interest in the bone and dental tissue engineering field has seen tremendous growth over the years. Studies have focused on strategies that can ideally eliminate the drawbacks of current clinical approaches. Understanding bone/tooth structure, mechanics and tissue regeneration is also essential to successfully regenerate functional tissue. Much progress on coupling engineering with biology has been made over the last years but many challenges remain before bone tissue engineering becomes a true clinical reality.

Currently, great efforts are being made to find new solutions for designing and developing novel biomaterials that mimic ECM and recreate the appropriate bone/tooth niche, accelerating the healing process in a defect site. The potential of using biomimetic ECM peptides for bone tissue engineering applications has been investigated and confirmed by the increasing number of published works that report the effects of various ECM peptides on cells and bone regeneration. Furthermore, addition of peptides with different activities on the same scaffold might help in directing bone formation and healing more efficiently. However, better characterization of the different peptides is necessary in order to understand their synergistic effect. With this in mind, optimization of the administration strategy and control of the release rate of those peptides are extremely important and should be further investigated.

So far, very few studies using different non-collagenous proteins from bone matrix have been performed, although the increasing popularity of ECM peptides might help advance this field. Moreover, non-collagenous bone ECM proteins can be exploited as versatile tools for functionalizing scaffolds with osteoinductive signals to enhance cell adhesion, osteogenic differentiation and angiogenesis. Further studies need to be done to understand the mechanism of action of non-collagenous bone ECM proteins and how to use them to engineer new bone tissue for tissue engineering applications.

Author Contributions: Conceptualization, M.S.C., D.V.; Writing—original draft preparation, M.S.C.; Writing—review and editing, M.S.C., J.M.S.C., C.L.d.S., D.V.; Supervision, C.L.d.S., D.V. All authors have read and agreed to the published version of the manuscript.

Funding: This research was funded by Fundação para a Ciência e Tecnologia (FCT), Portugal from individual fellowship SFRH/BD/52478/2014 awarded to Marta S. Carvalho and from the FCT project "DentalBioMatrix" (PTDC/BTM-BTM/3538/2020). Funding was provided by iBB-Institute for Bioengineering and Biosciences from FCT (UIDB/04065/2020), NIH grant NIAMS 49635, Sand Family for Orthopaedic Research and by the Center for Biotechnology and Interdisciplinary Studies, Rensselaer Polytechnic Institute.

Institutional Review Board Statement: Not applicable.

Informed Consent Statement: Not applicable.

Data Availability Statement: The data presented in this study are available on request from the corresponding author.

Conflicts of Interest: The authors declare no conflict of interest.

References

1. Yaszemski, M.J.; Payne, R.G.; Hayes, W.C.; Langer, R.; Mikos, A.G. Evolution of bone transplantation: Molecular, cellular and tissue strategies to engineer human bone. *Biomaterials* **1996**, *17*, 175–185. [CrossRef]
2. Koond, G.L.; Diba, M.; Mikos, A.G. Materials design for bone-tissue engineering. *Nat. Rev. Mater.* **2020**, *5*, 584–603. [CrossRef]

3. Stock, U.A.; Vacanti, J.P. Tissue engineering: Current state and prospects. *Annu. Rev. Med.* **2001**, *52*, 443–451. [CrossRef] [PubMed]
4. Giannoudis, P.V.; Einhorn, T.A.; Marsh, D. Fracture healing: The diamond concept. *Injury* **2007**, *38*, S3–S6. [CrossRef]
5. Fernandez-Yague, M.A.; Abbah, S.A.; McNamara, L.; Zeugolis, D.I.; Pandit, A.; Biggs, M.J. Biomimetic approaches in bone tissue engineering: Integrating biological and physicomechanical strategies. *Adv. Drug Deliv. Rev.* **2015**, *84*, 1–29. [CrossRef] [PubMed]
6. Lutolf, M.P.; Weber, F.E.; Schmoekel, H.G.; Schense, J.C.; Kohler, T.; Müller, R.; Hubbell, J.A. Repair of bone defects using synthetic mimetics of collagenous extracellular matrices. *Nat. Biotechnol.* **2003**, *21*, 513–518. [CrossRef]
7. Zhou, X.; Feng, W.; Qiu, K.; Chen, L.; Wang, W.; Nie, W.; Mo, X.; He, C. BMP-2-derived peptide and dexamethasone incorporated mesoporous silica nanoparticles for enhanced osteogenic differentiation of bone mesenchymal stem cells. *ACS Appl. Mater. Interfaces* **2015**, *7*, 15777–15789. [CrossRef] [PubMed]
8. Lin, Z.Y.; Duan, Z.X.; Guo, X.D.; Li, J.F.; Lu, H.W.; Zheng, Q.X.; Quan, D.P.; Yang, S.H. Bone induction by biomimetic PLGA-(PEG-ASP)n copolymer loaded with a novel synthetic BMP-2-related peptide in vitro and in vivo. *J. Control. Release* **2010**, *144*, 190–195. [CrossRef]
9. Wang, E.A.; Rosen, V.; D'Alessandro, J.S.; Bauduy, M.; Cordes, P.; Harada, T.; Israel, D.I.; Hewick, R.M.; Kerns, K.M.; LaPan, P.; et al. Recombinant human bone morphogenetic protein induces bone formation. *Proc. Natl. Acad. Sci. USA* **1990**, *87*, 2220–2224. [CrossRef]
10. Friedlaender, G.E.; Perry, C.R.; Cole, J.D.; Cook, S.D.; Cierny, G.; Muschler, G.F.; Zych, G.A.; Calhoun, J.H.; LaForte, A.J.; Yin, S. Osteogenic protein-1 (bone morphogenetic protein-7) in the treatment of tibial nonunions. *J. Bone Joint Surg. Am.* **2001**, *83*, S151–S158. [CrossRef]
11. Boden, S.D.; Kang, J.; Sandhu, H.; Heller, J.G. Use of recombinant human bone morphogenetic protein-2 to achieve posterolateral lumbar spine fusion in humans: A prospective, randomized clinical pilot trial: 2002 Volvo Award in clinical studies. *Spine* **2002**, *27*, 2662–2673. [CrossRef]
12. James, A.W.; LaChaud, G.; Shen, J.; Asatrian, G.; Nguyen, V.; Zhang, X.; Ting, K.; Soo, C. A review of the clinical side effects of bone morphogenetic protein-2. *Tissue Eng. Part. B Rev.* **2016**, *22*, 284–297. [CrossRef]
13. Woo, E.J. Adverse events after recombinant human BMP2 in nonspinal orthopaedic procedures. *Clin. Orthop. Relat. Res.* **2013**, *471*, 1707–1711. [CrossRef]
14. Mi, J.; Xu, J.; Yao, H.; Li, X.; Tong, W.; Li, Y.; Dai, B.; He, X.; Chow, D.H.K.; Li, G.; et al. Calcitonin gene-related peptide enhances distraction osteogenesis by increasing angiogenesis. *Tissue Eng. Part. A* **2021**, *27*, 87–102. [CrossRef]
15. Lai, M.; Yan, X.; Shen, K.; Tang, Q.; Fang, X.; Zhang, C.; Zhu, Z.; Hou, Y. The effect of calcitonin gene-related peptide functionalized TiO_2 nanotubes on osteoblast and osteoclast differentiation in vitro. *Colloids Surf. A* **2020**, *600*, 124899. [CrossRef]
16. Karadag, A.; Iqbai, H.; Yazici, H. Peptide-mediated bone tissue engineering. In *Racing of the Surface*; Li, B., Moriarty, T., Webster, T., Xing, M., Eds.; Springer: Cham, Switzerland, 2020. [CrossRef]
17. Sroga, G.E.; Karim, L.; Colon, W.; Vashishth, D. Biochemical Characterization of Major Bone-Matrix Proteins Using Nanoscale-Size Bone Samples and Proteomics Methodology. *Mol. Cell. Proteomics* **2011**, *10*, M110.006718. [CrossRef] [PubMed]
18. Vashishth, D. The role of the collagen matrix in skeletal fragility. *Curr. Osteoporos. Rep.* **2007**, *5*, 62–66. [CrossRef]
19. Herring, G.M.; Ashton, B.A. The isolation of soluble proteins, glycoproteins, and proteoglycans from bone. *Prep. Biochem.* **1974**, *4*, 179–200. [CrossRef]
20. Roach, H.I. Why does bone matrix contain non-collagenous proteins? The possible roles of osteocalcin, osteonectin, osteopontin and bone sialoprotein in bone mineralisation and resorption. *Cell Biol. Int.* **1994**, *18*, 617–628. [CrossRef]
21. Viguet-Carrin, S.; Garnero, P.; Delmas, P.D. The role of collagen in bone strength. *Osteoporos. Int.* **2006**, *17*, 319–336. [CrossRef] [PubMed]
22. Yerramshetty, J.S.; Akkus, O. The associations between mineral crystallinity and the mechanical properties of human cortical bone. *Bone* **2008**, *42*, 476–482. [CrossRef]
23. Fratzl, P.; Paris, O.; Klaushofer, K.; Landis, W.J. Bone mineralization in an osteogenesis imperfect mouse model studied by small-angle x-ray scattering. *J. Clin. Invest.* **2006**, *97*, 396–402. [CrossRef] [PubMed]
24. Dong, X.N.; Zoghi, M.; Ran, Q.; Wang, X. Collagen mutation causes changes of the microdamage morphology in bone of an OI mouse model. *Bone* **2002**, *47*, 1071–1075. [CrossRef]
25. Jämsä, T.; Rho, J.Y.; Fan, Z.; MacKay, C.A.; Marks, S.C., Jr.; Tuukkanen, J. Mechanical properties in long bones of rat osteopetrotic mutations. *J. Biomech.* **2002**, *35*, 161–165. [CrossRef]
26. Fantner, G.E.; Adams, J.; Turner, P.; Thurner, P.J.; Fisher, L.W.; Hansma, P.K. Nanoscale ion mediated networks in bone: Osteopontin can repeatedly dissipate large amounts of energy. *Nano Letters* **2007**, *7*, 2491–2498. [CrossRef] [PubMed]
27. Ritter, N.M.; Farach-Carson, M.C.; Butler, W.T. Evidence for the formation of a complex between osteopontin and osteocalcin. *J. Bone Miner. Res.* **1992**, *7*, 877–885. [CrossRef] [PubMed]
28. Poundarik, A.A.; Diab, T.; Sroga, G.E.; Ural, A.; Boskey, A.L.; Gundberg, C.M.; Vashishth, D. Dilatational band formation in bone. *Proc. Natl. Acad. Sci. USA* **2012**, *109*, 19178–19183. [CrossRef]
29. Boskey, A.L. Noncollagenous matrix proteins and their role in mineralization. *Bone Miner.* **1989**, *6*, 111–123. [CrossRef]
30. Ninomiya, J.T.; Tracy, R.P.; Calore, J.D.; Gendreau, M.A.; Kelm, R.J.; Mann, K.G. Heterogeneity of human bone. *J. Bone Min. Res.* **1990**, *5*, 933–938. [CrossRef]
31. Marcus, R.; Dempster, D.; Cauley, J.; Feldman, D. *Osteoporosis*, 4th ed.; Academic Press: Cambridge, MA, USA, 2013.
32. Bilezikian, J.P.; Raisz, L.G.; Martin, T.J. *Principles of Bone Biology*, 3rd ed.; Academic Press: Cambridge, MA, USA, 2008.

33. Kesireddy, V.; Kasper, F.K. Approaches for building bioactive elements into synthetic scaffolds for bone tissue engineering. *J. Mater. Chem. B.* **2016**, *4*, 6773–6786. [CrossRef]

34. Mano, J.F.; Silva, G.A.; Azevedo, H.S.; Malafaya, P.B.; Sousa, R.A.; Silva, S.S.; Boesel, L.F.; Oliveira, J.M.; Santos, T.C.; Marques, A.P.; et al. Natural origin biodegradable systems in tissue engineering and regenerative medicine: Present status and some moving trends. *J.R. Soc. Interface* **2007**, *4*, 999–1030. [CrossRef] [PubMed]

35. Maisani, M.; Pezzoli, D.; Chassande, O.; Mantovani, D. Cellularizing hydrogel-based scaffolds to repair bone tissue: How to create a physiologically relevant micro-environment? *J. Tissue Eng.* **2017**, *8*, 2041731417712073. [CrossRef] [PubMed]

36. Zhu, L.; Luo, D.; Liu, Y. Effect of the nano/microscale structure of biomaterial scaffolds on bone regeneration. *Int. J. Oral Sci.* **2020**, *12*. [CrossRef]

37. Harbers, G.M.; Healy, K.E. The effect of ligand type and density on osteoblast adhesion, proliferation, and matrix mineralization. *J. Biomed. Mater. Res. A* **2005**, *75*, 855–869. [CrossRef]

38. Barber, T.A.; Harbers, G.M.; Park, S.; Gilbert, M.; Healy, K.E. Ligand density characterization of peptide-modified biomaterials. *Biomaterials* **2005**, *26*, 6897–6905. [CrossRef]

39. Nicolas, J.; Magli, S.; Rabbachin, L.; Sampaolesi, S.; Nicotra, F.; Russo, L. 3D extracellular matrix mimics: Fundamental concepts and role of materials chemistry to influence stem cell fate. *Biomacromolecules* **2020**, *21*, 1968–1994. [CrossRef]

40. Lee, J.Y.; Choi, Y.S.; Lee, S.J.; Chung, C.P.; Park, Y.J. Bioactive peptide-modified biomaterials for bone regeneration. *Curr. Pharm. Des.* **2011**, *17*, 2663–2676. [CrossRef]

41. Koivunen, E.; Wang, B.; Dickinson, C.D.; Ruoslahti, E. Peptides in cell adhesion research. *Methods Enzymol.* **1994**, *245*, 346–369. [CrossRef]

42. Shin, H.; Jo, S.; Mikos, A.G. Modulation of marrow stromal osteoblast adhesion on biomimetic oligo[poly(ethylene glycol) fumarate] hydrogels modified with Arg-Gly-Asp peptides and a poly(ethyleneglycol) spacer. *J. Biomed. Mater. Res.* **2002**, *61*, 169–179. [CrossRef]

43. Schaffner, P.; Dard, M.M. Structure and function of RGD peptides involved in bone biology. *Cell Mol. Life Sci.* **2003**, *60*, 119–132. [CrossRef]

44. Wang, X.; Yan, C.; Ye, K.; He, Y.; Li, Z.; Ding, J. Effect of RGD nanospacing on differentiation of stem cells. *Biomaterials* **2013**, *34*, 2865–2874. [CrossRef] [PubMed]

45. Wang, X.; Ye, K.; Li, Z.; Yan, C.; Ding, J. Adhesion, proliferation, and differentiation of mesenchymal stromal cells on RGD nanopatterns of varied nanospacings. *Organogenesis* **2013**, *9*, 280–286. [CrossRef]

46. Vukicevic, S.; Luyten, F.P.; Kleinman, H.K.; Reddi, A.H. Differentiation of canalicular cell processes in bone cells by basement membrane matrix components: Regulation by discrete domains of laminin. *Cell* **1990**, *63*, 437–445. [CrossRef]

47. Aota, S.; Nagai, T.; Yamada, K.M. Characterization of regions of fibronectin besides the arginine-glycine-aspartic acid sequence required for adhesive function of the cell-binding domain using site-directed mutagenesis. *J. Biol. Chem.* **1991**, *266*, 15938–15943. [CrossRef]

48. Bhatnagar, R.S.; Qian, J.J.; Wedrychowska, A.; Sadeghi, M.; Wu, Y.M.; Smith, N. Design of biomimetic habitats for tissue engineering with P-15, a synthetic peptide analogue of collagen. *Tissue Eng.* **1999**, *5*, 53–65. [CrossRef]

49. Reyes, C.D.; García, A.J. Alpha2beta1 integrin-specific collagen-mimetic surfaces supporting osteoblastic differentiation. *J. Biomed. Mater. Res. A* **2004**, *69*, 591–600. [CrossRef]

50. Reyes, C.D.; Petrie, T.A.; Burns, K.L.; Schwartz, Z.; García, A.J. Biomolecular surface coating to enhance orthopaedic tissue healing and integration. *Biomaterials* **2007**, *28*, 3228–3235. [CrossRef]

51. Aziz, A.H.; Wilmoth, R.L.; Ferguson, V.L.; Bryant, S.J. IDG-SW3 osteocyte differentiation and bone extracellular matrix deposition are enhanced in a 3D matrix metalloproteinase-sensitive hydrogel. *ACS Appl. Biol. Mater.* **2020**, *16*, 1666–1680. [CrossRef] [PubMed]

52. Johnson, E.E.; Urist, M.R. Human bone morphogenetic protein allografting for reconstruction of femoral nonunion. *Clin. Orthop. Rel. Res.* **2000**, *371*, 61–74. [CrossRef] [PubMed]

53. Sun, Y.; Jiang, Y.; Liu, Q.; Gao, T.; Feng, J.Q.; Dechow, P.; D'Souza, R.N.; Qin, C.; Liu, X. Biomimetic engineering of nanofibrous gelatin scaffolds with noncollagenous proteins for enhanced bone regeneration. *Tissue Eng. Part. A* **2013**, *19*, 1754–1763. [CrossRef] [PubMed]

54. Osathanon, T.; Giachelli, C.M.; Somerman, M.J. Immobilization of alkaline phosphatase on microporous nanofibrous fibrin scaffolds for bone tissue engineering. *Biomaterials* **2009**, *30*, 4513–4521. [CrossRef] [PubMed]

55. Klontzas, M.E.; Reakasame, S.; Silva, R.; Morais, J.C.F.; Vernardis, S.; MacFarlane, R.J.; Heliotis, M.; Tsiridis, E.; Panoskaltsis, N.; Boccaccini, A.R.; et al. Oxidized alginate hydrogels with the GHK peptide enhance cord blood mesenchymal stem cell osteogenesis: A paradigm for metabolomics-based evaluation of biomaterial design. *Acta Biomater.* **2019**, *88*, 224–240. [CrossRef] [PubMed]

56. Won, J.E.; Mateos-Timoneda, M.A.; Castano, O.; Planell, J.A.; Seo, S.J.; Lee, E.J.; Han, C.M.; Kim, H.W. Fibronectin immobilization on to robotic-dispensed nanobioactive glass/polycaprolactone scaffolds for bone tissue engineering. *Biotechnol. Lett.* **2015**, *37*, 935–942. [CrossRef]

57. Cacchioli, A.; Ravanetti, F.; Bagno, A.; Dettin, M.; Gabbi, C. Human vitronectin-derived peptide covalently grafted onto titanium surface improves osteogenic activity: A pilot in vivo study on rabbits. *Tissue Eng. Part. A* **2009**, *15*, 2917–2926. [CrossRef] [PubMed]

58. Shin, H.; Zygourakis, K.; Farach-Carson, M.C.; Yaszemski, M.J.; Mikos, A.G. Attachment, proliferation, and migration of marrow stromal osteoblasts cultured on biomimetic hydrogels modified with an osteopontin-derived peptide. *Biomaterials* **2004**, *25*, 895–906. [CrossRef]

59. Hamada, Y.; Egusa, H.; Kaneda, Y.; Hirata, I.; Kawaguchi, N.; Hirao, T.; Matsumoto, T.; Yao, M.; Daito, T.; Suzuki, M.; et al. Synthetic Osteopontin-derived peptide SVVYGLR can induce neovascularization in artificial bone marrow scaffold biomaterials. *Dent. Mater. J.* **2007**, *26*, 487–492. [CrossRef] [PubMed]

60. Baranowski, A.; Klein, A.; Ritz, U.; Götz, H.; Mattyasovsky, S.G.; Rommens, P.M.; Hofmann, A. Evaluation of bone sialoprotein coating of three-dimensional printed calcium phosphate scaffolds in a calvarial defect model in mice. *Materials* **2018**, *11*, e2336. [CrossRef] [PubMed]

61. Klein, A.; Baranowski, A.; Ritz, U.; Götz, H.; Heinemann, S.; Mattyasovszky, S.; Rommens, P.M.; Hofmann, A. Effect of bone sialoprotein coated three-dimensional printed calcium phosphate scaffolds on primary human osteoblasts. *J. Biomed. Mater. Res. B Appl. Biomater.* **2018**, *106*, 2565–2575. [CrossRef] [PubMed]

62. Hunter, G.H. Role of proteoglycan in the provisional calcification of cartilage. *Clin. Orthop. Rel. Res.* **1991**, *262*, 256–263. [CrossRef]

63. Chen, C.C.; Boskey, A.L.; Rosenberg, L.C. The inhibitory effect of cartilage proteoglycans on hydroxyapatite growth. *Calcif. Tissue Int.* **1984**, *36*, 285–290. [CrossRef]

64. Fedarko, N.S. Isolation and purification of proteoglycans. *EXS* **1994**, *70*, 9–35. [CrossRef]

65. Weyers, A.; Linhardt, R.J. Neoproteoglycans in tissue engineering. *FEBS J.* **2013**, *280*, 2511–2522. [CrossRef] [PubMed]

66. Paderi, J.E.; Panitch, A. Design of a synthetic collagen-binding peptidoglycan that modulates collagen fibrillog.enesis. *Biomacromolecules* **2008**, *9*, 2562–2566. [CrossRef] [PubMed]

67. Sistiabudi, R.; Paderi, J.; Panitch, A.; Ivanisevic, A. Modification of native collagen with cell-adhesive peptide to promote RPE cell attachment on Bruch's membrane. *Biotechnol. Bioeng.* **2009**, *102*, 1723–1729. [CrossRef] [PubMed]

68. Stuart, K.; Paderi, J.; Snyder, P.W.; Freeman, L.; Panitch, A. Collagen-binding peptidoglycans inhibit MMP mediated collagen degradation and reduce dermal scarring. *PLoS ONE* **2011**, *6*, e22139. [CrossRef]

69. Pieper, J.S.; Oosterhof, A.; Dijkstra, P.J.; Veerkamp, J.H.; van Kuppervelt, T.H. Preparation and characterization of porous crosslinked collagenous matrices containing bioavailable chondroitin sulphate. *Biomaterials* **1999**, *20*, 847–858. [CrossRef]

70. Farrell, E.; O'Brien, F.J.; Doyle, P.; Fischer, J.; Yannas, I.; Harley, B.A.; O'Connell, B.; Prendergast, P.J.; Campbell, V.A. A collagen-glycosaminoglycan scaffold supports adult rat mesenchymal stem cell differentiation along osteogenic and chondrogenic routes. *Tissue Eng.* **2006**, *12*, 459–468. [CrossRef]

71. Caliari, S.R.; Harley, B.A.C. Collagen-GAG scaffold biophysical properties bias MSC lineage choice in the presence of mixed soluble signals. *Tissue Eng. Part. A* **2014**, *20*, 2463–2472. [CrossRef]

72. Tierney, C.M.; Jaasma, M.J.; O'Brien, F.J. Osteoblast activity on collagen-GAG scaffolds is affected by collagen and GAG concentration. *J. Biomed. Mater. Res. A* **2009**, *91*, 92–101. [CrossRef]

73. Caliari, S.R.; Weisgerber, D.W.; Ramirez, M.A.; Kelkhoff, D.O.; Harley, B.A. The influence of collagen-glycosaminoglycan scaffold relative density and microstructural anisotropy on tenocyte bioactivity and transcriptomic stability. *J. Mech. Behav. Biomed. Mater.* **2012**, *11*, 27–40. [CrossRef]

74. Pieper, J.S.; van Wachem, P.B.; Mja, V.L.; Brouwer, L.A.; Hafmans, T.; Veerkamp, J.H.; van Kuppevelt, T.H. Attachment of glycosaminoglycans to collagenous matrices modulates the tissue response in rats. *Biomaterials* **2000**, *21*, 1689–1699. [CrossRef]

75. Harley, B.A.; Lynn, A.K.; Wissner-Gross, Z.; Bonfield, W.; Yannas, I.V.; Gibson, L.J. Design of a multiphase osteochondral scaffold. II. Fabrication of a mineralized collagen-glycosaminoglycan scaffold. *J. Biomed. Mater. Res. A* **2010**, *92*, 1066–1077. [CrossRef] [PubMed]

76. Lee, J.C.; Pereira, C.T.; Ren, X.; Huang, W.; Bischoff, D.; Weisgerber, D.W.; Yamguchi, D.T.; Harley, B.A.; Miller, T.A. Optimizing collagen scaffolds for bone engineering: Effects of cross-linking and mineral content on structural contraction and osteogenesis. *J. Craniofac. Surg.* **2015**, *26*, 1992–1996. [CrossRef] [PubMed]

77. Yang, S.; Guo, Q.; Shores, L.S.; Aly, A.; Ramakrishnan, M.; Kim, G.H.; Lu, Q.; Su, L.; Elisseeff, J.H. Use of a chondroitin sulfate bioadhesive to enhance integration of bioglass particles for repairing critical-size bone defects. *J. Biomed. Mater. Res. A* **2015**, *103*, 235–242. [CrossRef] [PubMed]

78. Keskin, D.S.; Tezcaner, A.; Korkusuz, P.; Korkusuz, F.; Hasirci, V. Collagen-chondroitin sulfate-based PLLA-SAIB-coated rhBMP-2 delivery system for bone repair. *Biomaterials* **2005**, *26*, 4023–4034. [CrossRef]

79. Fujioka-Kobayashi, M.; Schaller, B.; Kobayashi, E.; Hernandez, M.; Zhang, Y.; Miron, R.J. Hyaluronic acid gel-based scaffolds as potential carrier for growth factors: An in vitro bioassay on its osteogenic potential. *J. Clin. Med.* **2016**, *5*, 112. [CrossRef]

80. Watanabe, H.; Yamada, Y. Chondrodysplasia of gene knockout mice for aggrecan and link protein. *Glycoconjugate J.* **2002**, *19*, 269–273. [CrossRef] [PubMed]

81. Nakamyra, M.; Sone, S.; Takahashi, I.; Mizoguchi, I.; Echigo, S.; Sasano, Y. Expression of versican and ADAMTS1, 4, and 5 during bone development in the rat mandible and hind limb. *J. Histochem. Cytochem.* **2005**, *53*, 1553–1562. [CrossRef]

82. Zhang, Y.; Cao, L.; Kiani, C.; Yang, B.L.; Hu, W.; Yang, B.B. Promotion of chondrocyte proliferation by versican mediated by G1 domain and EGF-like motifs. *J. Cell. Biochem.* **1999**, *73*, 445–457. [CrossRef]

83. Kjellän, L.; Lindahl, U. Proteoglycans: Structures and interactions. *Annu. Rev. Biochem.* **1991**, *60*, 443–475. [CrossRef]

84. Boskey, A.L.; Spevak, L.; Doty, S.B.; Rosenberg, L. Effects of bone CS-proteoglycans, DS-decorin, and DS-biglycan on hydroxyapatite formation in a gelatin gel. *Calcif. Tissue Int.* **1997**, *61*, 298–305. [CrossRef] [PubMed]

85. Fleischmajer, R.A.; Fisher, L.W.; MacDonald, E.D.; Jacobs, L., Jr.; Perlish, J.S.; Termine, J.D. Decorin interacts with fibrillary collagen of embryonic and adult human skin. *J. Struct. Biol.* **1991**, *106*, 82–90. [CrossRef]
86. Goldberg, M.; Septier, D.; Rapoport, O.; Iozzo, R.V.; Young, M.F.; Ameye, L.G. Targeted disruption of two small-leucine-rich proteoglycans, biglycan and decorin, excerpts divergent effects on enamel and dentin formation. *Calcif. Tissue Int.* **2005**, *77*, 297–310. [CrossRef]
87. Ameye, L.; Young, M.F. Mice deficient in small leucine-rich proteoglycans: Novel in vivo models for osteoporosis, osteoarthritis, Ehlers-Danlos syndrome, muscular dystrophy, and corneal diseases. *Glycobiology* **2002**, *12*, 107R–116R. [CrossRef] [PubMed]
88. Corsi, A.; Xu, T.; Chen, X.D.; Boyde, A.; Liang, J.; Mankani, M.; Sommer, B.; Iozzo, R.V.; Eichstetter, I.; Robey, P.G.; et al. Phenotypic effects of biglycan deficiency are linked to collagen fibril abnormalities, are synergized by decorin deficiency, and mimic Ehlers-Danlos-like changes in bone and other connective tissues. *J. Bone Miner. Res.* **2002**, *17*, 1180–1189. [CrossRef] [PubMed]
89. Xu, T.; Bianco, P.; Fisher, L.W.; Longenecker, G.; Smith, E.; Goldstein, S.; Bonadio, J.; Boskey, A.L.; Heegaard, A.M.; Sommer, B.; et al. Targeted disruption of the biglycan gene leads to an osteoporosis-like phenotype in mice. *Nat. Genet.* **1998**, *20*, 78–82. [CrossRef] [PubMed]
90. Fisher, L.W.; Hawkins, G.R.; Tuross, N.; Termine, J.D. Purification and partial characterization of small proteoglycans I and II, bone sialoproteins I and II, and osteonectin from the mineral compartment of developing human bone. *J. Biol. Chem.* **1987**, *262*, 9702–9708. [CrossRef]
91. Roach, H.I. Association of matrix acid and alkaline phosphatases with mineralization of cartilage and endochondral bone. *Histochem. J.* **1999**, *31*, 53–61. [CrossRef] [PubMed]
92. Kirsch, T.; Nah, H.D.; Shapiro, I.M.; Pacifici, M. Regulated production of mineralization-competent matrix vesicles in hypertrophic chondrocytes. *J. Cell. Biol.* **1997**, *137*, 1140–1160. [CrossRef]
93. Collin, P.; Nefussi, J.R.; Wetterwald, A.; Nikolas, V.; Boy-Lefever, M.L.; Fleisch, H.; Forest, N. Expression of collagen, osteocalcin and bone alkaline phosphatase in a mineralizing rat osteoblastic cell culture. *Calcif. Tissue Int.* **1992**, *50*, 175–183. [CrossRef]
94. Fedde, K.N.; Blair, L.; Silverstein, J.; Coburn, S.P.; Ryan, L.M.; Weinstein, R.S.; Waymire, K.; Narisawa, S.; Millan, J.L.; MacGregor, G.R.; et al. Alkaline phosphatase knockout mice recapitulate the metabolic and skeletal defects of infantile hypophosphatasia. *J. Bone Miner. Res.* **1999**, *14*, 2015–2026. [CrossRef]
95. Yoon, K.; Golub, E.; Rodan, G.A. Alkaline phosphatase transfected cells promote calcium and phosphate deposition. *Connect. Tissue Res.* **1989**, *22*, 17–25. [CrossRef]
96. Rosset, E.M.; Bradshaw, A.D. SPARC/Osteonectin in mineralized tissue. *Matrix Biol.* **2016**, *52*, 78–87. [CrossRef]
97. Stenner, D.D.; Tracy, R.G.; Riggs, B.L.; Mann, K.G. Human platelets contain and secrete Osteonectin. *Proc. Natl. Acad. Sci. USA* **1986**, *83*, 6892–6896. [CrossRef] [PubMed]
98. Kelm, R.J., Jr.; Mann, K.G. The collagen binding specificity of bone and platelet osteonectin is related to differences in glycosylation. *J. Biol. Chem.* **1991**, *266*, 9632–9639. [CrossRef]
99. Clezardin, P.; Malavel, L.; Ehrensperger, A.S.; Delmas, P.; Dechavanne, M.; McGregor, J.L. Complex formation of human thrombospondin with osteonectin. *Eur. J. Biochem.* **1988**, *175*, 275–284. [CrossRef]
100. Termine, J.D.; Kleinman, H.K.; Whitson, S.W.; Conn, K.M.; McGarvey, M.L.; Martin, G.R. Osteonectin, a bone-specific protein linking mineral to collagen. *Cell* **1981**, *26*, 99–105. [CrossRef]
101. Delany, A.M.; Amling, M.; Priemel, M.; Howe, C.; Baron, R.; Canalis, E. Osteopenia and decreased bone formation in osteonectin-deficient mice. *J. Clin. Invest.* **2000**, *105*, 915–923. [CrossRef]
102. Boskey, A.L.; Moore, D.J.; Amling, M.; Canalis, E.; Delany, A.M. Infrared analysis of the mineral and matrix in bones of osteonecin-null mice and their wild-type controls. *J. Bone Miner. Res.* **2003**, *18*, 1005–1011. [CrossRef]
103. Yan, Q.; Sage, E.H. SPARC, a matricellular glycoprotein with important biological functions. *J. Histochem. Cytochem.* **1999**, *47*, 1495–1506. [CrossRef]
104. Wewer, U.M.; Ibaraki, K.; Schjorring, P.; Durkin, M.E.; Young, M.F.; Albrechtsen, R. A potential role for tetranectin in mineralization during osteogenesis. *J. Cell. Biol.* **1994**, *127*, 1767–1775. [CrossRef] [PubMed]
105. Iba, K.; Abe, Y.; Chikenji, T.; Kanaya, K.; Chiba, H.; Sasaki, K.; Dohke, T.; Wada, T.; Yamashita, T. Delayed fracture healing in tetranectin-deficient mice. *J. Bone Miner. Metab.* **2013**, *31*, 399–408. [CrossRef] [PubMed]
106. Iba, K.; Durkin, M.E.; Johnsen, L.; Hunziker, E.; Damgaard-Pedersen, K.; Zhang, H.; Engvall, E.; Albrechtsen, R.; Wewer, U.M. Mice with a targeted deletion of the tetranectin gene exhibit a spinal deformity. *Mol. Cell. Biol.* **2001**, *21*, 7817–7825. [CrossRef]
107. Adams, J.C.; Lawler, J. The thrombospondins. *Int. J. Biochem. Cell. Biol.* **2004**, *36*, 961–968. [CrossRef]
108. Hankenson, K.D.; Bain, S.D.; Kyriakides, T.R.; Smith, E.A.; Goldstein, S.A.; Bornstein, P. Increased marrow-derived osteoprogenitor cells and endosteal bone formation in mice lacking thrombospondin 2. *J. Bone Miner. Res.* **2000**, *15*, 851–862. [CrossRef]
109. Carron, J.A.; Fraser, W.C.; Gallagher, J.A. Thrombospondin promotes resorption by osteoclasts in vitro. *Biochem. Biophys. Res. Commun.* **1995**, *213*, 1017–1025. [CrossRef]
110. Bornstein, P.; Kyriakides, T.R.; Yang, Z.; Armstrong, L.C.; Birk, D.E. Thrombospondin 2 modulates collagen fibrillogenesis and angiogenesis. *J. Invest. Dermatol. Symp. Proc.* **2000**, *5*, 61–66. [CrossRef]
111. Kyriakides, T.R.; Zhu, Y.H.; Smith, L.T.; Bain, S.D.; Yang, Z.; Lin, M.T.; Danielson, K.G.; Iozzo, R.V.; LaMarca, M.; McKinney, C.E.; et al. Mice that lack thrombospondin 2 display connective tissue abnormalities that are associated with disordered collagen fibrillogenesis, an increased vascular density, and a bleeding diathesis. *J. Cell. Biol.* **1998**, *140*, 419–430. [CrossRef]

112. Stenman, S.; Vaheri, A. Distribution of a major connective tissue protein, fibronectin, in normal human tissues. *J. Exp. Med.* **1978**, *147*, 1054–1064. [CrossRef]

113. Grzesik, W.J.; Robey, P.G. Bone matrix RGD glycoproteins: Immunolocalization and interaction with human primary osteoblastic bone cells in vitro. *J. Bone Miner. Res.* **1994**, *9*, 487–496. [CrossRef]

114. George, E.L.; Georges-Labouesse, E.N.; Patel-King, R.S.; Rayburn, H.; Hynes, R.O. Defects in mesoderm, neural tube and vascular development in mouse embryos lacking fibronectin. *Development* **1993**, *119*, 1079–1091.

115. Kumagai, T.; Lee, I.; Ono, Y.; Maeno, M.; Takagi, M. Ultrastuctural localization and biochemical characterization of vitronectin in developing rat bone. *Histochem. J.* **1998**, *30*, 111–119. [CrossRef]

116. Koschnick, S.; Konstantinides, S.; Schafer, K.; Crain, K.; Loskutoff, D.J. Thrombotic phenotype of mice with a combined deficiency in plasminogen activator inhibitor 1 and vitronectin. *J. Thromb. Haemost.* **2005**, *3*, 2290–2295. [CrossRef]

117. Denhardt, D.T.; Guo, X. OPN, a protein with diverse functions. *FASEB J.* **1993**, *7*, 1475–1482. [CrossRef]

118. Denhardt, D.T.; Noda, M. OPN expression and function: Role in bone remodeling. *J. Cell. Biochem.* **1998**, *72*, 92–102. [CrossRef]

119. Rittling, S.R.; Denhardt, D.T. OPN function in pathology: Lessons from OPN-deficient mice. *Exp. Nephrol.* **1999**, *7*, 103–113. [CrossRef]

120. Denhardt, D.T.; Noda, M.; O'Regan, A.W.; Pavlliln, D.; Berman, J.S. Osteopontin as a means to cope with environmental insults: Regulation of inflammation, tissue remodeling, and cell survival. *J. Clin. Invest.* **2001**, *107*, 1055–1061. [CrossRef]

121. Barry, S.T.; Ludbrook, S.B.; Murrison, E.; Horgan, C.M. Analysis of the alpha4beta1 integrin-OPN interaction. *Exp. Cell Res.* **2000**, *258*, 342–351. [CrossRef]

122. Miyauchi, A.; Alvarez, J.; Greenfield, E.M.; Teti, A.; Grano, M.; Colucci, S.; Zambonin-Zallone, A.; Ross, F.P.; Teitelbaum, S.L.; Cheresh, D. Recognition of OPN and related peptides by an v3 integrin stimulates immediate cell signals in osteoclasts. *J. Biol. Chem.* **1991**, *266*, 20369–20374. [CrossRef]

123. Rodriguez, D.E.; Thula-Mata, T.; Toro, E.J.; Gower, L.B. Multifunctional role of osteopontin in directing intrafibrillar mineralization of collagen and activation of osteoclasts. *Acta Biomater.* **2014**, *10*, 494–507. [CrossRef]

124. Kaartinen, M.T.; Pirhonen, A.; Linnala-Kankkunen, A.; Maenpaa, P.H. Cross-linking of OPN by tissue transglutaminase increases its collagen binding properties. *J. Biol. Chem.* **1999**, *274*, 1729–1735. [CrossRef]

125. Wang, L.; Guan, X.; Tang, R.; Hoyer, J.R.; Wierzbicki, A.; de Yoreo, J.J.; Nancollas, G.H. Phosphorylation of osteopontin is required for inhibition of calcium oxalate crystallization. *J. Phys. Chem. B* **2008**, *112*, 9151–9157. [CrossRef]

126. Boskey, A.L. Osteopontin and related phosphorylated sialoproteins, effects on mineralization. *Ann. N. Y. Acad. Sci.* **1995**, *760*, 249–256. [CrossRef]

127. Rittling, S.R.; Matsumoto, H.N.; McKee, M.D.; Nanci, A.; An, X.R.; Novick, K.E.; Kowalski, A.J.; Noda, M.; Denhardt, D.T. Mice lacking OPN show normal development and bone structure but display altered osteoclast formation in vitro. *J. Bone Miner. Res.* **1998**, *13*, 1101–1111. [CrossRef]

128. Boskey, A.L.; Spevak, L.; Paschalis, E.; Doty, S.B.; McKee, M.D. Osteopontin deficiency increases mineral content and mineral crystallinity in mouse bone. *Calcif. Tissue Inter.* **2002**, *71*, 145–154. [CrossRef] [PubMed]

129. Boskey, A.L.; Christensen, B.; Taleb, H.; Sørensen, E.S. Post-translational modification of osteopontin: Effects on in vitro hydroxyapatite formation and growth. *Biochem. Biophys. Res. Commun.* **2012**, *419*, 333–338. [CrossRef]

130. Asou, Y.; Rittling, S.R.; Yoshitake, H.; Tsuji, K.; Shinomiya, K.; Nifuji, A.; Denhardt, D.T.; Noda, M. Osteopontin facilitates angiogenesis, accumulation of osteoclasts and resorption in ectopic bone. *Endocrinology* **2001**, *142*, 1325–1332. [CrossRef]

131. Franzen, A.; Heinegard, D. Isolation and characterization of two sialoproteins present only in bone calcified matrix. *Biochem. J.* **1985**, *232*, 715–724. [CrossRef]

132. Bianco, P.; Fisher, L.W.; Young, M.F.; Termine, J.D.; Robey, P.G. Expression of bone sialoprotein (BSP) in developing human tissues. *Calcif. Tissue Int.* **1991**, *49*, 421–426. [CrossRef]

133. Gordon, J.A.; Tye, C.E.; Sampaio, A.V.; Underhill, T.M.; Hunter, G.K.; Goldberg, H.A. Bone sialoprotein expression enhances osteoblast differentiation and matrix mineralization in vitro. *Bone* **2007**, *41*, 462–473. [CrossRef]

134. Hunter, G.K.; Goldberg, H.A. Nucleation of hydroxyapatite by bone sialoproteins. *Proc. Natl. Acad. Sci. USA* **1993**, *90*, 8562–8565. [CrossRef] [PubMed]

135. Mizuno, M.; Imai, T.; Fujisawa, R.; Tani, H.; Kuboki, Y. Bone sialoprotein (BSP) is a crucial factor for the expression of osteoblastic phenotypes of bone marrow cells cultured on type I collagen matrix. *Calcif. Tissue Int.* **2000**, *66*, 388–396. [CrossRef]

136. Baht, G.S.; Hunter, G.K.; Goldberg, H.A. Bone sialoprotein–collagen interaction promotes hydroxyapatite nucleation. *Matrix Biol.* **2008**, *27*, 600–608. [CrossRef]

137. Malaval, L.; Wade-Gueye, N.M.; Boudiffa, M.; Fei, J.; Zirngibl, R.; Chen, F.; Laroche, N.; Roux, J.P.; Burt-Pichat, B.; Duboeuf, F.; et al. Bone sialoprotein plays a functional role in bone formation and osteoclastogenesis. *J. Exp. Med.* **2008**, *205*, 1145–1153. [CrossRef]

138. Bouleftour, W.; Boudiffa, M.; Wade-Gueye, N.M.; Bouët, G.; Cardelli, M.; Laroche, N.; vanden-Bossche, A.; Thomas, M.; Bonnelye, E.; Aubin, J.E.; et al. Skeletal development of mice lacking bone sialoprotein (BSP)-impairment of long bone growth and progressive establishment of high trabecular bone mass. *PLoS ONE* **2014**, *9*, e95144. [CrossRef]

139. Sun, Y.; Chen, L.; Ma, S.; Zhou, J.; Zhang, H.; Feng, J.Q.; Qin, C. Roles of DMP1 processing in osteogenesis, dentinogenesis and chondrogenesis. *Cells Tissues Organs* **2011**, *194*, 199–204. [CrossRef] [PubMed]

140. Feng, J.Q.; Huang, H.; Lu, Y.; Ye, L.; Xie, Y.; Tsutsui, T.W.; Kunieda, T.; Castranio, T.; Scott, G.; Bonewald, L.B.; et al. The dentin matrix protein 1 (Dmp1) is specifically expressed in mineralized, but not soft, tissues during development. *J. Dent. Res.* **2003**, *82*, 776–780. [CrossRef]

141. Bhatia, A.; Albazzaz, M.; Espinoza Orías, A.A.; Inoue, N.; Miller, L.M.; Acerbo, A.; George, A.; Sumner, D.R. Overexpression of DMP1 accelerates mineralization and alters cortical bone biomechanical properties in vivo. *J. Mech. Behav. Biomed. Mater.* **2012**, *5*, 1–8. [CrossRef] [PubMed]

142. Narayanan, K.; Srinivas, R.; Ramachandran, A.; Hao, J.; Quinn, B.; George, A. Differentiation of embryonic mesenchymal cells to odontoblast-like cells by overexpression of dentin matrix protein 1. *Proc. Natl. Acad. Sci. USA* **2001**, *98*, 4516–4521. [CrossRef] [PubMed]

143. Ahmad, A.R.; Kaewpungsup, P.; Khorattanakulchai, N.; Rattanapisit, K.; Pavasant, P.; Phoolcharoen, W. Recombinant human dentin matrix protein 1 (DMP1) induces the osteogenic differentiation of human periodontal ligament cells. *Biotechnol. Rep. (Amst).* **2019**, *23*, e00348. [CrossRef] [PubMed]

144. Gericke, A.; Qin, C.; Sun, Y.; Redfern, R.; Redfern, D.; Fujimoto, Y.; Taleb, H.; Butler, W.T.; Boskey, A.L. Different forms of DMP1 play distinct roles in mineralization. *J. Dent. Res.* **2010**, *89*, 355–359. [CrossRef]

145. He, G.; George, A. Dentin matrix protein 1 immobilized on type I collagen fibrils facilitates apatite deposition in vitro. *J. Biol. Chem.* **2004**, *279*, 11649–11656. [CrossRef]

146. Pirotte, S.; Lamour, V.; Lambert, V.; Alvarez Gonzalez, M.L.; Ormenese, S.; Noel, A.; Mottet, D.; Castronovo, V.; Bellahcene, A. Dentin matrix protein 1induces membrane expression of VE-cadherin on endothelial cells and inhibits VEGF-induced angiogenesis by blocking VEGFR-2 phosphorylation. *Blood* **2011**, *117*, 2515–2526. [CrossRef]

147. Ling, Y.; Rios, H.F.; Myers, E.R.; Lu, Y.; Feng, J.Q.; Boskey, A.L. DMP1 depletion decreases bone mineralization in vivo: An FTIR imaging analysis. *J. Bone Miner. Res.* **2005**, *20*, 2169–2177. [CrossRef] [PubMed]

148. Feng, J.Q.; Ward, L.M.; Liu, S.; Lu, Y.; Xie, Y.; Yuan, B.; Yu, X.; Rauch, F.; Davis, S.I.; Zhang, S.; et al. Loss of DMP1 causes rickets and osteomalacia and identifies a role for osteocytes in mineral metabolism. *Nat. Genet.* **2006**, *38*, 1310–1315. [CrossRef]

149. Qin, C.; Brunn, J.C.; Cadena, E.; Ridall, A.; Tsujigiwa, H.; Nagatsuka, H.; Nagai, N.; Butler, W.T. The expression of dentin sialophosphoprotein gene in bone. *J. Dent. Res.* **2002**, *81*, 392–394. [CrossRef] [PubMed]

150. Traub, W.; Jodaikin, A.; Arad, T.; Veis, A.; Sabsay, B. Dentin phosphophoryn binding to collagen fibrils. *Matrix* **1992**, *12*, 197–201. [CrossRef]

151. Boskey, A.L.; Maresca, M.; Doty, S.; Sabsay, B.; Veis, A. Concentration-dependent effects of dentin phosphophoryn in the regulation of in vitro hydroxyapatite formation and growth. *Bone Miner.* **1990**, *11*, 55–65. [CrossRef]

152. Zurick, K.M.; Qin, C.; Bernards, M.T. Mineralization induction effects of osteopontin, bone sialoprotein, and dentin phosphoprotein on a biomimetic collagen substrate. *J. Biomed. Mater. Res. A* **2013**, *101*, 1571–1581. [CrossRef] [PubMed]

153. Sengottuvelan, A.; Balasubramanian, P.; Will, J.; Boccaccini, A.R. Bioactivation of titanium dioxide scaffolds by ALP-functionalization. *Bioact. Mater.* **2017**, *2*, 108–115. [CrossRef]

154. Jafaary, F.; Hanachi, P.; Gorjipour, K. Osteoblast differentiation on collagen scaffold with immobilized alkaline phosphatase. *Int. J. Organ. Transplant. Med.* **2017**, *8*, 195–202.

155. Detsch, R.; Sarker, B.; Zehnder, T.; Boccaccini, A.R.; Douglas, T.E. Additive manufacturing of cell-loaded alginate enriched with alkaline phosphatase for bone tissue engineering application. *BioNanoMaterials* **2014**, *15*, 79–87. [CrossRef]

156. Pietraszek, A.; Karewicz, A.; Widnic, M.; Lachowicz, D.; Gajewska, M.; Bernasik, A.; Nowakowska, M. Halloysite-alkaline phosphatase system—A potential bioactive component of scaffold for bone tissue engineering. *Colloids Surf. B Biointerfaces* **2019**, *173*, 1–8. [CrossRef]

157. Pataquiva-Mateus, A.Y.; Wu, H.C.; Lucchesi, C.; Ferraz, M.P.; Monteiro, F.J.; Spector, M. Supplementation of collagen scaffolds with SPARC to facilitate mineralization. *J. Biomed. Mater. Res. B Appl. Biomater.* **2012**, *100*, 862–870. [CrossRef]

158. Liao, S.; Ngiam, M.; Chan, C.K.; Ramakrishna, S. Fabrication of nano-hydroxyapatite/collagen/osteonectin composites for bone graft applications. *Biomed. Mater.* **2009**, *4*, 025019. [CrossRef]

159. Sarvestani, A.S.; He, X.; Jabbari, E. Osteonectin-derived peptide increases the modulus of a bone-mimetic nanocomposite. *Eur. Biophys. J.* **2008**, *37*, 229–234. [CrossRef] [PubMed]

160. Jose, S.; Hughbanks, M.L.; Binder, B.Y.; Ingavle, G.C.; Leach, J.K. Enhanced trophic factor secretion by mesenchymal stem/stromal cells with Glycine-Histidine-Lysine (GHK)-modified alginate hydrogels. *Acta Biomater.* **2014**, *10*, 1955–1964. [CrossRef]

161. Mogues, T.; Etzerodt, M.; Hall, C.; Engelich, G.; Graversen, J.H.; Hartshorn, K.L. Tetranectin binds to the kringle 1-4 form of angiostatin and modifies its functional activity. *J. Biomed. Biotechnol.* **2004**, *2004*, 73–78. [CrossRef]

162. Hynes, R.O. The emergence of integrins: A personal and historical perspective. *Matrix Biol.* **2004**, *23*, 333–340. [CrossRef]

163. Fisher, L.W.; Fedarko, N.S. Six genes expressed in bones and teeth encode the current members of the SIBLING family of proteins. *Connect. Tissue Res.* **2003**, *44*, 33–40. [CrossRef]

164. Eckrich, J.; Maas, A.; Jurk, K.; Strieth, S.; Kumm, E.; Brieger, J. The role of thrombospondin-1 in biomaterial integration of porous polyethylene implants in vivo. *Laryngorhinootologie* **2019**, *98*, S199–S200.

165. Lee, S.; Lee, D.S.; Choi, I.; Pham, I.B.H.; Jang, J.H. Design of an osteoinductive extracellular fibronectin matrix protein for bone tissue engineering. *Int. J. Mol. Sci.* **2015**, *16*, 7672–7681. [CrossRef]

166. Mohamadyar-Toupkanlou, F.; Vasheghani-Farahani, E.; Hanaee-Ahvaz, H.; Soheimani, M.; Dodel, M.; Havasi, P.; Ardeshirylakimi, A.; Taherzadeh, E.S. Osteogenic differentiation of MSCs on fibronectin-coated and nHA-modified scaffolds. *ASAIO J.* **2017**, *63*, 684–691. [CrossRef]

167. Trujillo, S.; Gonzalez-Garcia, C.; Rico, P.; Reid, A.; Windmill, J.; Dalby, M.J.; Salmeron-Sanchez, M. Engineered 3D hydrogels with full-length fibronectin that sequester and present growth factors. *Biomaterials* **2020**, *252*, 120104. [CrossRef]

168. Sangkert, S.; Kamonmattayakul, S.; Chai, W.L.; Jirut, M. A biofunctional-modified silk fibroin scaffold with mimic reconstructed extracellular matrix of decellularized pulp/collagen/fibronectin for bone tissue engineering in alveolar bone resorption. *Matt. Lett.* **2016**, *166*, 30–34. [CrossRef]

169. Staines, K.A.; MacRae, V.E.; Farquharson, C. The importance of the SIBLING family of proteins on skeletal mineralisation and bone remodelling. *J. Endocrinol.* **2012**, *214*, 241–255. [CrossRef]

170. Boskey, A.L.; Chiang, P.; Fermanis, A.; Brown, J.; Taleb, H.; David, V.; Rowe, P.S. MEPE's diverse effects on mineralization. *Calcif. Tissue Int.* **2009**, *86*, 42–46. [CrossRef]

171. McKee, M.D.; Nanci, A. Osteopontin at mineralized tissue interfaces in bone, teeth, and osseointegrated implants: Ultra-structural distribution and implications for mineralized tissue formation, turnover, and repair. *Microsc. Res. Tech.* **1996**, *33*, 141–164. [CrossRef]

172. Somerman, M.J.; Prince, C.W.; Butler, W.T.; Foster, R.A.; Moehring, J.M.; Sauk, J.J. Cell attachment activity of the 44 kilodalton bone phosphoprotein is not restricted to bone cells. *Matrix* **1989**, *9*, 49–54. [CrossRef]

173. Carvalho, M.S.; Cabral, J.M.S.; da Silva, C.L.; Vashishth, D. Synergistic effect of extracellularly supplemented osteopontin and osteocalcin on stem cell proliferation, osteogenic differentiation, and angiogenic properties. *J. Cell. Biochem.* **2019**, *120*, 6555–6569. [CrossRef] [PubMed]

174. Carvalho, M.S.; Poundarik, A.A.; Cabral, J.M.S.; da Silva, C.L.; Vashishth, D. Biomimetic matrices for rapidly forming mineralized bone tissue based on stem cell-mediated osteogenesis. *Sci. Rep.* **2018**, *8*, 14388. [CrossRef]

175. He, X.; Yang, X.; Jabbar, E. Combined effect of osteopontin and BMP-2 derived peptides grafted to an adhesive hydrogel on osteogenic and vasculogenic differentiation of marrow stromal cells. *Langmuir* **2012**, *28*, 5387–5397. [CrossRef]

176. Lee, S.H.; Shin, H. Matrices and scaffold for delivery of bioactive molecules in bone and cartilage tissue engineering. *Adv. Drug. Deliv. Rev.* **2007**, *59*, 339–359. [CrossRef] [PubMed]

177. Egusa, H.; Kaneda, Y.; Akashi, Y.; Hamada, Y.; Matsumoto, T.; Saeki, M.; Thakor, D.K.; Tabata, Y.; Matsuura, N.; Yatani, H. Enhanced bone regeneration via multimodal actions of synthetic peptide SVVYGLR on osteoprogenitors and osteoclasts. *Biomaterials* **2009**, *30*, 4676–4686. [CrossRef] [PubMed]

178. Hamada, Y.; Nokihara, K.; Okazaki, M.; Fujitani, W.; Matsumoto, T.; Matsuo, M.; Umakoshi, Y.; Takahashi, J.; Matsuura, N. Angio-genic activity of osteopontin-derived peptide SVVYGLR. *Biochem. Biophy. Res. Commun.* **2003**, *310*, 153–157. [CrossRef] [PubMed]

179. Zhu, M.; He, H.; Meng, Q.; Zhu, Y.; Ye, X.; Xu, N.; Yu, J. Osteopontin sequence modified mesoporous calcium silicate scaffolds to promote angiogenesis in bone tissue regeneration. *J. Mater. Chem. B* **2020**, *15*, 5849–5861. [CrossRef] [PubMed]

180. Damsongsang, P.; Chaikiawkeaw, D.; Phoolcharoen, W.; Rattanapisit, K.; Kaewpungsup, P.; Pavasant, P.; Hoven, V.P. Surface-immobilized plant-derived osteopontin as an effective platform to promote osteoblast adhesion and differentiation. *Colloids Surf. B Biointerfaces* **2019**, *173*, 816–824. [CrossRef]

181. Klinthoopthamrong, N.; Chaikiawkeaw, D.; Phoolcharoen, W.; Rattanapisit, K.; Kaewpungsup, P.; Pavasant, P.; Hoven, V.P. Bacte-rial cellulose membrane conjugated with plant-derived osteopontin: Preparation and its potential for bone tissue regeneration. *Int. J. Biol. Macromol.* **2020**, *149*, 50–59. [CrossRef]

182. Kruger, T.E.; Miller, A.H.; Wang, J. Collagen scaffolds in bone sialoprotein-mediated bone regeneration. *Sci. World J.* **2013**, *2013*, 812718. [CrossRef] [PubMed]

183. Choi, Y.J.; Lee, J.Y.; Chung, C.P.; Park, Y.J. Enhanced osteogenesis by collagen-binding peptide from bone sialoprotein in vitro and in vivo. *J. Biomed. Mater. Res. A* **2013**, *101*, 547–554. [CrossRef]

184. Xu, L.; Anderson, A.L.; Lu, Q.; Wang, J. Role of fibrillary structure of collagenous carrier in bone sialoprotein-mediated matrix mineralization and osteoblast differentiation. *Biomaterials* **2007**, *28*, 750–761. [CrossRef]

185. Gomes, S.; Leonor, I.B.; Mano, J.F.; Reis, R.L.; Kaplan, D.L. Spider silk-bone sialoprotein fusion proteins for bone tissue engineering. *Soft Matter.* **2011**, *7*, 4964–4973. [CrossRef]

186. Chan, W.D.; Goldberg, H.A.; Hunter, G.K.; Dixon, S.J.; Rizkalla, A.S. Modification of polymer networks with bone sialoprotein promotes cell attachment and spreading. *J. Biomed. Mater. Res. A* **2010**, *94*, 945–952. [CrossRef]

187. Rezania, A.; Healy, K.E. The effect of peptide surface density on mineralization of a matrix deposited by osteogenic cells. *J. Biomed. Mater. Res.* **2000**, *52*, 595–600. [CrossRef]

188. Drevelle, O.; Bergeron, E.; Senta, H.; Lauzon, M.A.; Roux, S.; Grenier, G.; Faucheux, N. Effect of functionalized polycaprolactone on the behaviour of murine preosteoblasts. *Biomaterials* **2010**, *31*, 6468–6476. [CrossRef]

189. Jha, A.K.; Jackson, W.M.; Healy, K.E. Controlling osteogenic stem cell differentiation via soft bioinspired hydrogels. *PLoS ONE* **2014**, *9*, e98640. [CrossRef]

190. Rapuano, B.E.; Wu, C.; MacDonald, D.E. Osteoblast-like cell adhesion to bone sialoprotein peptides. *J. Orthop. Res.* **2004**, *22*, 353–361. [CrossRef]

191. Ravindran, S.; Georg, A. Dentin matrix proteins in bone tissue engineering. *Adv. Exp. Med. Biol.* **2015**, *881*, 129–142. [CrossRef]

192. Yang, B.; Chen, G.; Li, J.; Zou, Q.; Xie, D.; Chen, Y.; Wang, H.; Zheng, X.; Long, J.; Tang, W.; et al. Tooth root regeneration using dental follicle cell sheets in combination with a dentin matrix-based scaffold. *Biomaterials* **2012**, *33*, 2449–2461. [CrossRef]
193. Li, R.; Guo, W.; Yang, B.; Guo, L.; Sheng, L.; Chen, G.; Li, Y.; Zou, Q.; Xie, D.; An, X.; et al. Human treated dentin matrix as a natural scaffold for complete human dentin tissue regeneration. *Biomaterials* **2011**, *32*, 4525–4538. [CrossRef]
194. Chun, S.Y.; Lee, H.J.; Choi, Y.A.; Kim, K.M.; Baek, S.H.; Park, H.S.; Kim, J.Y.; Ahn, J.M.; Cho, J.Y.; Cho, D.W.; et al. Analysis of the soluble human tooth proteome and its ability to induce dentin/tooth regeneration. *Tissue Eng. Part. A* **2011**, *17*, 181–191. [CrossRef]
195. Guo, W.; Gong, K.; Shi, H.; Zhu, G.; He, Y.; Ding, B.; Wen, L.; Jin, Y. Dental follicle cells and treated dentin matrix scaffold for tissue engineering the tooth root. *Biomaterials* **2012**, *33*, 1291–1302. [CrossRef]
196. Alsanea, R.; Ravindran, S.; Fayad, M.I.; Johnson, B.R.; Wenckus, C.S.; Hao, J.; George, A. Biomimetic approach to perforation repair using dental pulp stem cells and dentin matrix protein 1. *J. Endod.* **2011**, *37*, 1092–1097. [CrossRef] [PubMed]
197. Xu, H.; Cao, B.; George, A.; Mao, C. Self-assembly and mineralization of genetically modifiable biological nanofibers driven by beta- structure formation. *Biomacromolecules* **2011**, *12*, 2193–2199. [CrossRef]
198. Sreenath, T.; Thyagarajan, T.; Hall, B.; Longenecker, G.; D'Souza, R.; Hong, S.; Wright, J.T.; MacDougall, M.; Sauk, J.; Kulkarni, A.B. Dentin sialophosphoprotein knockout mouse teeth display widened predentin zone and develop defective dentin mineralization similar to human dentinogenesis imperfecta type III. *J. Biol. Chem.* **2003**, *278*, 24874–24880. [CrossRef]
199. Verdelis, K.; Ling, Y.; Sreenath, T.; Haruyama, N.; MacDougall, M.; van der Meulen, M.C.; Lukashova, L.; Spevak, L.; Kulkarni, A.B.; Boskey, A.L. DSPP effects on in vivo bone mineralization. *Bone* **2008**, *43*, 983–990. [CrossRef]
200. Kim, J.W.; Hu, J.C.; Lee, J.I.; Moon, S.K.; Kim, Y.J.; Jang, K.T.; Lee, S.H.; Kim, C.C.; Hahn, S.H.; Simmer, J.P. Mutational hot spot in the DSPP gene causing dentinogenesis imperfecta type II. *Hum. Genet.* **2005**, *116*, 186–191. [CrossRef]
201. Boskey, A.L.; Spevak, L.; Tan, M.; Doty, S.B.; Butler, W.T. Dentin sialoprotein (DSP) has limited effects on in vitro apatite formation and growth. *Calcif. Tissue Int.* **2000**, *67*, 472–478. [CrossRef]
202. Hauschka, P.V.; Lian, J.B.; Cole, D.E.C.; Gundberg, C.M. Osteocalcin and matrix Gla protein: Vitamin K-dependent proteins in bone. *Physiol. Rev.* **1989**, *69*, 990–1047. [CrossRef]
203. Poser, J.W.; Esch, F.S.; Ling, N.C.; Price, P.A. Isolation and sequence of the vitamin K-dependent protein from human bone. Undercarboxylation of the first glutamic acid residue. *J. Biol. Chem.* **1980**, *255*, 8685–8691. [CrossRef]
204. Calvo, M.S.; Eyre, D.R.; Caren, M.G. Molecular basis and clinical application of biological markers of bone turnover. *Endocr. Rev.* **1996**, *17*, 333–368. [CrossRef] [PubMed]
205. Gundberg, C.M. Biochemical markers of bone formation. *Clin. Lab. Med.* **2000**, *20*, 489–501. [CrossRef]
206. Hauschka, P.V.; Reid, M.L. Timed appearance of a calcium-binding protein containing g-carboxyglutamic acid in developing chick bone. *Dev. Biol.* **1978**, *65*, 431–436. [CrossRef]
207. Ducy, P.; Desbois, C.; Boyce, B.; Pinero, G.; Story, B.; Dunstan, C.; Smith, E.; Bonadio, J.; Goldstein, S.; Gundberg, C.; et al. Increased bone formation in osteocalcin-deficient mice. *Nature* **1996**, *382*, 448–452. [CrossRef]
208. Rammelt, S.; Neumann, M.; Hanisch, U.; Reinstorf, A.; Pompe, W.; Zwipp, H.; Biewener, A. Osteocalcin enhances bone remodeling around hydroxyapatite/collagen composites. *J. Biomed. Mater. Res. A* **2005**, *73*, 284–294. [CrossRef]
209. Chenu, C.; Colucci, S.; Grano, M.; Zigrino, P.; Barattolo, R.; Zambonin, G.; Baldini, N.; Vergnaud, P.; Delmas, P.D.; Zallone, A.Z. Osteocalcin induces chemotaxis, secretion of matrix proteins, and calcium-mediated intracellular signaling in human osteoclast-like cells. *J. Biol. Chem.* **1994**, *127*, 1149–1158. [CrossRef]
210. Bodine, P.V.; Komm, B.S. Evidence that conditionally immortalized human osteoblasts express an osteocalcin receptor. *Bone* **1999**, *25*, 535–543. [CrossRef]
211. Glowacki, J.; Lian, J.B. Impaired recruitment and differentiation of osteoclast progenitors by osteocalcin-deplete bone implants. *Cell. Differ.* **1987**, *21*, 247–254. [CrossRef]
212. Ferron, M.; Hinoi, E.; Karsenty, G.; Ducy, P. Osteocalcin differentially regulates beta cell and adipocyte gene expression and affects the development of metabolic diseases in wild-type mice. *Proc. Natl. Acad. Sci. USA* **2008**, *105*, 5266–5270. [CrossRef]
213. Wei, J.; Karsenty, G. An overview of the metabolic functions of osteocalcin. *Rev. Endocr. Metab. Disord.* **2015**, *16*, 93–98. [CrossRef] [PubMed]
214. Kim, S.G.; Lee, D.S.; Lee, S.; Jang, J.H. Osteocalcin/fibronectin-functionalized collagen matrices for bone tissue engineering. *J. Biomed. Mater. Res. A* **2015**, *103*, 2133–2140. [CrossRef]
215. Cantatore, F.P.; Crivellato, E.; Nico, B.; Ribatti, D. Osteocalcin is angiogenic in vivo. *Cell Biol. Inter.* **2005**, *29*, 583–585. [CrossRef]
216. Price, P.A.; Williamson, M.K. Primary structure of bovine matrix Gla protein, a new vitamin K-dependent bone protein. *J. Biol. Chem.* **1985**, *260*, 14791–14975. [CrossRef]
217. Murshed, M.; Schinke, T.; McKee, M.D.; Karsenty, G. Extracellular matrix mineralization is regulated locally: Different roles of two Gla-containing proteins. *J. Cell. Biol.* **2004**, *165*, 625–630. [CrossRef] [PubMed]
218. Luo, G.; Ducy, P.; McKee, M.D.; Pinero, G.J.; Loyer, E.; Behringer, R.R.; Karsenty, G. Spontaneous calcification of arteries and cartilage in mice lacking matrix GLA protein. *Nature* **1997**, *386*, 78–81. [CrossRef] [PubMed]
219. Zhang, J.; Ma, Z.; Yan, K.; Wang, Y.; Yang, Y.; Wu, X. Matrix gla protein promotes the bone formation by up-regulating Wnt/β-catenin signaling pathway. *Front. Endocrinol. (Lausanne)* **2019**, *10*, 891. [CrossRef]
220. Garnett, J.; Dieppe, P. The effects of serum and human albumin on calcium hydroxyapatite crystal growth. *Biochem. J.* **1990**, *266*, 863–868.

221. Carvalho, M.S.; Silva, J.C.; Hoff, C.M.; Cabral, J.M.S.; Linhardt, R.J.; da Silva, C.L.; Vashishth, D. Loss and rescue of osteocalcin and osteopontin modulate osteogenic and angiogenic features of mesenchymal stem/stromal cells. *J. Cell. Physiol.* **2020**, *235*, 7496–7515. [CrossRef]
222. Ott, H.C.; Matthiesen, T.S.; Goh, S.K.; Black, L.D.; Kren, S.M.; Netoff, T.I.; Taylor, D.A. Perfusion-decellularized matrix:usingnature's platform to engineer a bioartificial heart. *Nat. Med.* **2008**, *14*, 213–221. [CrossRef]
223. Petersen, T.H.; Calle, E.A.; Zhao, L.; Lee, E.J.; Gui, L.; Raredon, M.B.; Gavrilov, K.; Yi, T.; Zhuang, Z.W.; Breuer, C.; et al. Tissue-engineered lungs for in vivo implantation. *Science* **2010**, *329*, 538–541. [CrossRef]
224. Papadimitropoulos, A.; Scotti, C.; Bourgine, P.; Scherberich, A.; Martin, I. Engineered decellularized matrices to instruct bone regeneration processes. *Bone* **2015**, *70*, 66–72. [CrossRef]
225. Laurencin, C.T.; Khan, Y. Regenerative engineering. *Sci. Transl. Med.* **2012**, *4*, 160ed9. [CrossRef] [PubMed]
226. Fitzpatrick, L.E.; McDevitt, T.C. Cell-derived matrices for tissue engineering and regenerative medicine application. *Biomater. Sci.* **2015**, *3*, 12–24. [CrossRef] [PubMed]
227. Urist, M.R.; Silverman, B.F.; Buring, K.; Dubuc, F.L.; Rosenberg, J.M. The bone induction principle. *Clin. Orthop. Relat. Res.* **1967**, *53*, 243–283. [CrossRef] [PubMed]
228. Schwartz, Z.; Somers, A.; Mellonig, J.T.; Carnes, D.L., Jr.; Dean, D.D.; Cochran, D.L.; Boyan, B.D. Ability of commercial demineralized freeze-dried bone allograft to induce new bone formation is dependent on donor age but not gender. *J. Periodontol.* **1998**, *69*, 470–478. [CrossRef]
229. Munting, E.; Wilmart, J.F.; Wijne, A.; Hennebert, P.; Delloye, C. Effect of sterilization on osteoinduction. Comparison of five methods in demineralized rat bone. *Acta Orthop. Scand.* **1988**, *59*, 34–38. [CrossRef]
230. Bourgine, P.E.; Scotti, C.; Pigeot, S.; Tchang, L.A.; Todorov, A.; Martin, I. Osteoinductivity of engineered cartilaginous templates devitalized by inducible apoptosis. *Proc. Natl. Acad. Sci. USA* **2014**, *111*, 17426–17431. [CrossRef]
231. Carvalho, M.S.; Silva, J.C.; Cabral, J.M.S.; da Silva, C.L.; Vashishth, D. Cultured cell-derived extracellular matrices to enhance the osteogenic differentiation and angiogenic properties of human mesenchymal stem/stromal cells. *J. Tissue Eng. Regen. Med.* **2019**, *13*, 1544–1558. [CrossRef] [PubMed]
232. Silva, J.C.; Carvalho, M.S.; Udangawa, R.N.; Moura, C.S.; Cabral, J.M.S.; da Silva, C.L.; Ferreira, F.C.; Vashishth, D.; Linhardt, R.J. Extracellular matrix decorated polycaprolactone scaffolds for improved mesenchymal stem/stromal cell osteogenesis towards a patient-tailored bone tissue engineering approach. *J. Biomed. Mater. Res. B Appl. Biomater.* **2020**, *108*, 2153–2166. [CrossRef]
233. Lin, H.; Yang, G.; Tan, J.; Tuan, R.S. Influence of decellularized matrix derived from human mesenchymal stem cells on their proliferation, migration and multi-lineage differentiation potential. *Biomaterials* **2012**, *33*, 4480–4489. [CrossRef]
234. Datta, N.; Holtorf, H.L.; Sikavitsas, V.I.; Jansen, J.A.; Mikos, A.G. Effect of bone extracellular matrix synthesized in vitro on the osteoblastic differentiation of marrow stromal cells. *Biomaterials* **2005**, *26*, 971–977. [CrossRef]
235. Datta, N.; Pham, Q.P.; Sharma, U.; Sikavitsas, V.I.; Jansen, J.A.; Mikos, A.G. In vitro generated extracellular matrix and fluid shear stress synergistically enhance 3D osteoblastic differentiation. *Proc. Natl. Acad. Sci. USA* **2006**, *103*, 2488–2493. [CrossRef]
236. Carvalho, M.S.; Silva, J.C.; Udangawa, R.N.; Cabral, J.M.S.; Ferreira, F.C.; da Silva, C.L.; Linhardt, R.J.; Vashishth, D. Co-culture cell-derived extracellular matrix loaded electrospun microfibrous scaffolds for bone tissue engineering. *Mater. Sci. Eng. C* **2019**, *99*, 479–490. [CrossRef] [PubMed]
237. Pham, Q.P.; Kasper, F.K.; Mistry, A.S.; Sharma, U.; Yasko, A.W.; Jansen, J.A. Analysis of the osteoinductive capacity and angiogenicity of an in vitro generated extracellular matrix. *J. Biomed. Mater. Res. A* **2008**, *88*, 295–303. [CrossRef]
238. Sadr, N.; Pippenger, B.E.; Scherberich, A.; Wendt, D.; Mantero, S.; Martin, I. Enhancing the biological performance of synthetic polymeric materials by decoration with engineered, decellularized extracellular matrix. *Biomaterials* **2012**, *33*, 5085–5093. [CrossRef] [PubMed]

polymers

MDPI

Review

A Polymer-Biologic Hybrid Hernia Construct: Review of Data and Early Experiences

Michael Sawyer [1,*], Stephen Ferzoco [2] and George DeNoto III [3]

1   Department of Surgery, Oklahoma State University, Comanche County Memorial Hospital, Lawton, OK 73505, USA
2   Department of Surgery, Atrius Health, Dedham, MA 02026, USA; steven_ferzoco@atriushealth.org
3   General Surgery Department, St. Francis Hospital, Roslyn, NY 11576, USA; George.Denoto@chsli.org
*   Correspondence: michael.sawyer@ccmhhealth.com

Abstract: Surgical mesh reinforcement of the human abdominal wall has been found to reduce the chance of recurrence in hernia repairs. While traditionally polymer meshes have been used in hernia repair, alternative mesh options have been engineered to prevent the inflammatory foreign body response invoked by polymers. A reinforced tissue matrix (RTM) mesh has been developed by embedding a polymer within a decellularized extracellular matrix. This combination has been attributed to the recruitment of host cells, a pro-healing response, and attenuation of the foreign body response. This has been observed to lead to the regeneration of functional tissue within the repair site that is reinforced by the polymer to offload abdominal pressures over time. This manuscript presents the review of OviTex, an RTM, in several types of hernia repair. The authors have found that the use of RTM in hernia repair is effective in preventing foreign body response, promoting wound healing, and providing reinforcement to lower the risk of hernia recurrence.

Keywords: reinforced tissue matrix; decellularized extracellular matrix; synthetic mesh; ventral hernia repair; inguinal hernia repair; hiatal hernia repair

Citation: Sawyer, M.; Ferzoco, S.; DeNoto, G., III. A Polymer-Biologic Hybrid Hernia Construct: Review of Data and Early Experiences. *Polymers* **2021**, *13*, 1928. https://doi.org/10.3390/polym13121928

Academic Editor: João Carlos Silva

Received: 1 May 2021
Accepted: 7 June 2021
Published: 10 June 2021

Publisher's Note: MDPI stays neutral with regard to jurisdictional claims in published maps and institutional affiliations.

1. Introduction

1.1. Abdominal Wall Hernia and Treatment Options

Abdominal wall hernias are common clinical entities and represent a major source of disability and morbidity in patients. Until the mid-20th century, primary suture repair, where the defect in the abdomen is closed with polymeric sutures, was the only available option for their surgical treatment. Reported recurrence rates with this approach were relatively high. The advent and clinical application of a synthetic polymer-based mesh represented a significant advance in the surgical treatment of hernias. The mesh is sutured into the defect area, placed within planes of muscle and fascia in the abdominal wall, acting as a scaffold to provide mechanical support and a structure for tissue to "scar" into to reduce the chance of recurrence. A study comparing suture to synthetic mesh-reinforced repair showed a reduction from 63 to 32% for synthetic mesh-reinforced repairs, and, for this reason, almost 90% of hernia procedures today involve some sort of mesh [1,2]. While this reduced recurrence, the repaired "scar" does not have the same dynamic qualities as human fascia [3]. It also became recognized that the application of synthetic mesh was also associated with its own set of serious complications, including chronic wound infections, adhesions to and erosion into intra-abdominal viscera, enterocutaneous fistulae, and others. Most of these complications can be attributed to the prolonged foreign body exposure inherent to the amount, density, and type of permanent synthetic polymer in the mesh.

1.2. Wound Healing

The implantation of any foreign material in a human body elicits a foreign body response, although the extent and severity are dependent on the amount of foreign material and, in the case of a mesh, the density of the structure [4–7]. The response is characterized by an inflammatory reaction followed by the recruitment of fibroblasts and myofibroblasts, whereby the first mainly lay down new collagen and the latter generate contractile forces [8,9]. In a typical wound healing sequence, the fibroblasts and myofibroblasts in the end undergo apoptosis, and the newly laid collagen remains. However, in the presence of a continued stimulus caused by a foreign body, such as a permanent synthetic mesh, they may remain active. In the event of prolonged foreign body exposure, host cells wall off the material with fibrotic tissue characterized by disorganized collagen and inflammatory cells [10–12]. The ongoing struggle has been to design a mesh that balances defect area support while reducing the inflammatory response.

1.3. Mesh Landscape

Over the years, the mesh landscape has evolved to address this issue with the creation of resorbable polymer and biologic mesh materials. Resorbable polymer materials have been developed in combinations of polyglycolic acid, polylactic acid, trimethylene carbonate, poly-4-hydroxybutyrate (P4HB), and other natural and synthetic polymers. Biologic materials are derived from mammalian (e.g., human, ovine, bovine, or porcine) source tissues taken from organs, such as the dermis, intestines, stomach, or bladder, and decellularized to retain the extracellular matrix as a scaffold for remodeling [13].

Polymeric meshes continue to be chosen for most hernia repairs due to their low cost and ability to withstand intrabdominal wall pressures, ensuring the defect does not re-herniate. These meshes are knit from polymer fibers forming porous structures classified based on their surface area density and pore size. Lightweight meshes are classified as less than 35 g/m^2, midweight between 50 and 80 g/m^2, and heavy weight greater than 80 g/m^2 [14], with lightweight meshes yielding the lowest dose-related inflammatory response [4,15]. A balance of pore size is also important to create holes large enough to promote cell infiltration, limit adhesions, and reduce the dose of polymer, while also remaining small enough to provide strength and prevent bulging [15,16].

The most common construct is a knit mesh made of monofilament fibers and, less often, multifilament fibers. Multifilament fibers are used less due to the spaces between the fibers being large enough for bacteria, yet too small to allow white blood cells, preventing an immune response to fight the formation of a biofilm and slime, which can lead to surgical site infections (SSI) [17–19]. Incisional ventral hernia repair has a four times larger incidence of infection than inguinal/femoral or umbilical hernias [20]. For this reason, when multifilament fibers are used, they are typically constructed of a polymer with a relatively short resorption time, such as polyglycolic acid.

Synthetic meshes have also been combined with barriers to prevent some of the most common complications associated with the prolonged foreign body response, such as viscera adhesions. Seprafilm (Genzyme, Cambridge, MA, USA), made of hyaluronic acid and carboxymethylcellulose, is an anti-adhesive bioresorbable barrier coating first used for abdominal and pelvic laparotomy and later as a coating for Phasix and Ventralight. While the barrier reduced the chance of adhesions [21], it also prevented tissue ingrowth, limiting the ability of the mesh to create a fascia-like layer of tissue in the defect [22].

P4HB (Phasix) is one of the more promising fully resorbable meshes, undergoing hydrolysis and degrading within 12–18 months. Roth et al. recently published on its clinical performance from a prospective, multi-center single arm study showing a three-year recurrence rate of 18% [23]. While this recurrence rate is significantly less than early permanent repairs, it has demonstrated localization of histiocytic cells walling off the polymer fibers of the mesh resulting in comparable inflammatory levels to two other permanent synthetics at the late chronic time point in a full thickness abdominal wall repair model in non-human primates [24].

Biologic materials exist as an alternative to synthetic materials due to their inherently lower inflammatory response and resilience, particularly in the repair of infected or contaminated fields and other high-risk patients undergoing hernia repair [25]. These materials are harvested from xenogeneic or allogeneic organs to isolate the extracellular matrix (ECM), a three-dimensional structure possessing proteins and signaling molecules [13,26]. The biologic materials are processed to remove cells and DNA of the source species to create an immunologically inert scaffold for remodeling, though the quality and extent of this is dependent upon the processing methods [27,28]. While these materials show promise, unfortunately, they are expensive and due to the retention of the protein elastin, may stretch over time [29–31].

1.4. Ideal Mesh

The ideal mesh would possess the strength and handling of a polymer mesh and possess the proteins and signaling molecules of a biologic ECM to attenuate the foreign body reaction and promote wound healing. The creation of a polymer mesh loaded with proteins and signaling molecules from an ECM is a challenge not only in creation but also in cost, scalability, and clinical utility. Studies have been conducted to include these proteins within polymeric materials but have largely been limited to lab settings with only in vitro results. Furthermore, the clinical utility of these have not been proven beyond in vitro studies.

1.5. Hybrid Meshes

A potential alternative solution to this is the embedment of a polymer mesh in a biologic material forming a hybrid composite mesh. In a study by Liang et al., researchers showed that a polypropylene mesh can behave like a biologic material by embedding it in a biologic extracellular matrix, reducing the inflammatory response. The result showed a mature elastin content, sulfated glycosaminoglycan content, and collagen subtype III/I ratio. These findings were attributed to the recruitment of pro-healing M2 macrophages by the ECM rather than proinflammatory M1 macrophages [32–35]. Similar designs have also been made by embedding a polypropylene mesh within an acellular dermal matrix showing a reduction in adhesions, a common complication associated with synthetic meshes [29].

1.6. Reinforced Tissue Matrices

Two relatively new hybrid meshes, consisting of a polypropylene or polyglycolic acid mesh, interwoven within layers of ovine-derived ECM, have shown promise as a suitable alternative both preclinically and clinically. The hybrid device creates a three-dimensional scaffold, interweaving polymer throughout layers of ovine (sheep) forestomach (rumen)-derived ECM. The polymer provides strength and compliance comparable to that of the human abdominal wall [24,36–38], critical to offloading the wound during the acute healing phase. The layers of the ECM and the channels and pores both inherent to the structure of the ECM and created by the interwoven polymer promote fluid exchange and have been attributed to the rapid host cellular response and infiltration of the network [24]. This rapid recruitment is critical in the attenuation of the foreign body reaction to the polymer, surrounding the fibers with host cells, resulting in a functional tissue repair both directly around the polymer fibers and within the ECM, with no encapsulation of the fibers nor "scar"-like collagen as is seen in synthetic devices [24]. It is believed that these properties unique to reinforced tissue matrices provide an advantage over synthetics in the reduction of SSOs, infection, recurrence, and inflammatory response in comparison to purely polymeric materials.

Reinforced tissue matrices (OviTex Core, 1S and 2S Resorbable and Permanent) are produced by interweaving polypropylene (PP) and polyglycolic acid (PGA) filament through layers of ovine (sheep) rumen, specifically the forestomach, as depicted in Figures 1 and 2. The ovine forestomach is processed to isolate the propria submucosa from the tunica muscularis and lamina epithelia. The extracellular matrix (ECM) of the propria submucosa is gently decellularized via a proprietary processing method (Aroa Biosurgery, New Zealand) to give ovine forestomach matrix (OFM) [39]. OFM is composed of structural and functional proteins, including collagen I and III, fibronectin, glycosaminoglycans, elastin, fibroblast growth factor basic, laminin, and collagen IV [39]. A recent proteomic analysis has identified more than 150 different ECM proteins that exist in OFM [40], and the matrix structure is essentially identical to tissue ECM [41,42]. Reinforced tissue matrices are made in several configurations, primarily differing in the number of OFM layers (four to eight) as well as the amount and density of polymer. The four-layer device is engineered with either PP or PGA, interwoven as a 6 mm grid pattern, while the six- and eight-layer devices add an additional two layers to one or both sides with a less dense 25 mm PP or PGA interwoven grid to reduce the risk of adhesions. All devices are terminally sterilized with ethylene oxide prior to use. OviTex devices are cleared for hernia repair and soft tissue reinforcement and have been used since July 2016.

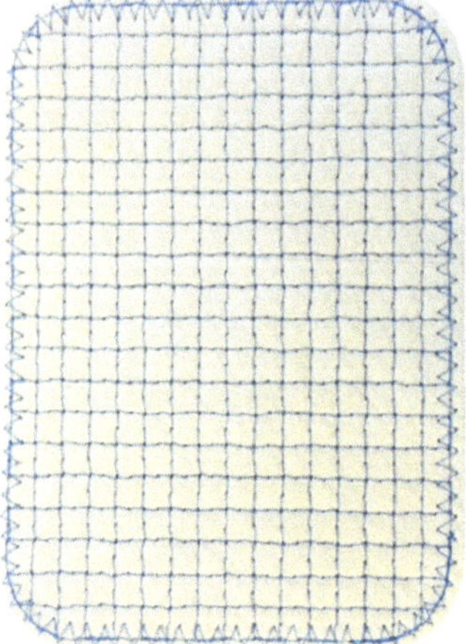

Figure 1. OviTex RTM mesh.

Figure 2. OviTex RTM cross section of four-layer device with polypropylene.

Review data for published works on RTM in ventral, hiatal, and inguinal hernias are presented to evaluate the effectiveness of RTMs in preventing foreign body response, promoting wound healing, and providing reinforcement to lower the risk of hernia recurrence.

2. Review of Reinforced Tissue Matrices in the Repair of Various Hernia Types

2.1. Ventral Hernia Repair

Ventral hernias are defects of the fascia abdominal wall, which allow for projection of abdominal organs and tissues through this wall [43]. Ventral hernia repair (VHR) commonly requires surgical intervention to place these protruding tissues and organs back into the abdominal cavity and close the defect in the abdominal wall [44]. While in some cases ventral hernia defects can be closed through suture alone, use of mesh supports has been shown to decrease incidence of recurrence [44]. As described above, several types of mesh are available in order to repair ventral hernias. The type of mesh used for repair is selected based upon durability, sterility, cost, foreign body reaction, host tissue ingrowth, patient comorbidity, surgeon preference, etc. [45]. The reinforced tissue matrix, OviTex, has shown promise in preventing recurrence in ventral hernia repair in preclinical and early clinical studies [24]. Recent studies have corroborated these early indications.

A single institution retrospective review by Parker et al. published in September of 2020 showed OviTex to be effective in high-risk patients who underwent VHR (Table 1) [46]. This study compared the surgical outcomes when OviTex was used to repair ventral hernias in high-risk patients compared to lower risk patients in which synthetic polypropylene mesh with a barrier was utilized [46]. Patient risk was determined based on hernia severity, classified on two different scales [46]. The modified Ventral Hernia Working Group (VHWG) grading classification system uses three grade levels to indicate hernia severity, grade 1 being the lowest risk, least severe hernias and grade 3 being contaminated, high-risk hernias [47]. The Center for Disease Control (CDC) wound classification scale ranks hernia severity from class I–IV, with class I being clean, less severe hernias and class IV being dirty/infected, high-risk hernias [48]. In this study, OviTex was used to repair ventral hernias in 50 consecutive patients with mostly VHWG grade 3 (61%) and CDC wound class III (61%) hernias [46]. Synthetic mesh was used to repair ventral hernias in 45 patients with mostly VHWG grade 2 (91%) and CDC wound class I (91%) hernias [46]. After 12 months, the hernia recurrence rate was lower in the OviTex group despite having a more challenging population than the synthetic group, 6 vs. 12% for OviTex and synthetic groups, respectively [46]. This was also true when recurrence rates were evaluated in a subset of patients who had developed a complication or infection at the surgical site, known as a surgical site occurrence (SSO). The rate of recurrence in this OviTex group was lower at 17% compared to the synthetic mesh group at 55%, a result that is counter-intuitive when considering SSOs typically increase the risk of recurrence [46]. This study indicates that recurrence rates in challenging patients, including those with SSOs, are lowered when OviTex is used compared to synthetic mesh. [46].

Table 1. Follow-up data and results of various surgical hernia repair studies where RTMs were used for repair.

Study	Parker et al., 2020	Sawyer 2018	Ferzoco 2018
Number of Patients	50	25	31
Hernia type	Ventral	Hiatal	Inguinal
Hernia severity	68% modified VHWG grade 3, 70% CDC wound class > I	56% CDC wound class > I	N/A
Months to follow-up	12	14.2	12.6
Surgical site occurrence (SSO)	36%	N/A	0% at 30 days
Recurrence rate	6%	0%	0%
Postoperative pain	N/A	6/7 symptoms resolved between 85.7 and 100%	0%

2.2. Hiatal Hernia Repair

Hiatal hernias occur when a portion or all of the stomach protrudes through a defect in the esophageal hiatus of the diaphragm, the portion of the diaphragm in which the esophagus passes from the thoracic to the abdominal cavity [49]. As with ventral hernias, surgical intervention is often needed to correct this defect and use of mesh decreases recurrence rate. Case studies have demonstrated that permanent synthetic meshes, used to repair these defects, can erode into the esophagus [50–52]. Reinforced tissue matrices could play a role in the repair of these hernias for surgeons.

Previously, Sawyer conducted a retrospective study to document their surgical experience in using resorbable OviTex to repair symptomatic hiatal hernias (Table 1) [53]. Surgical repair was performed on 25 patients who had a high incidence of comorbidities and 52% who had type III and IV hiatal hernias [53]. These hernias were considered high-risk as the anatomic classification scale for hiatal hernias classifies type III and IV hiatal hernias as most severe [49]. Type III hiatal hernias occur when both the fundus and the gastroesophageal junction herniate through the hiatus [49]. Type IV hernias are characterized by an organ other than the stomach protruding through the thoracic cavity [49]. After a mean follow-up time of 14.2 months, there were no recurrences [53]. Complications, such as contraction of repaired defect site leading to recurrence of symptoms, were low with only 8% of patients needing intervention to induce esophageal dilation [53]. Good-to-excellent relief of symptoms, specifically heartburn (20 of 21 patients; 95%) and dysphagia (18 of 19 patients; 94.7%), was achieved in all but one patient, indicating improved quality of life [53].

2.3. Inguinal Hernia Repair

Inguinal hernias are defects at the inner groin through which fat or part of the small intestine partially protrude through the lower abdominal wall [54]. As with the aforementioned hernia types, surgical repair is often necessary and the addition of mesh support reduces recurrence rate. These hernias are notable for a high incidence of chronic pain after surgical repair, known as chronic postoperative inguinal pain (CPIP) [55]. While the etiology of CPIP is unknown, it is hypothesized that repair with synthetic mesh causes high inflammation, which irritates surrounding nerves leading to increased pain sensation [56]. Use of a reinforced tissue matrix, such as OviTex, may therefore help to reduce CPIP after inguinal hernia repair. In a single surgeon study, Ferzoco investigated whether use of OviTex in inguinal hernia repair was effective in repairing the defect, reducing recurrence, and preventing chronic postoperative pain (Table 1) [56]. Thirty-one (31) patients, treated on an outpatient basis, underwent inguinal hernia repair with OviTex. There were no reported surgical site infections (SSI) during the initial 30 days postoperatively. At an average 12.6 month follow-up there were no reported recurrences [56]. Additionally, no patients suffered postoperative complications, including onset of CPIP, and there were no requests for postoperative narcotic refills [56]. OviTex seems especially effective in repairing inguinal hernias, as it is not associated with recurrence and its low inflammatory response may prevent occurrence of chronic postoperative pain [56].

3. Discussion

Hernia surgery has evolved over the last decade with new techniques that allow for primary closure of fascia under minimal tension. These procedures have shifted from using the mesh as a plug or patch to using the mesh to offload or splint the primary fascial repairs. This provides support of the defect area offloading abdominal pressures and allowing the repair to heal without re-herniation. While the technique and many mesh materials have been developed, the incorporation of polymeric materials still results in inflammation, foreign body response, and complications or recurrences [30,57,58]. Biologic materials have been shown to have lower inflammatory responses, though over time they may stretch and bulge or herniate [30].

The ideal mesh would have the ability to remodel or regenerate into tissue similar to native fascia, without generating a chronic foreign body response resulting in stiffening, contraction, and fibrotic scar tissue. This mesh would also need to possess the proper strength and compliance to maintain the integrity of the repair. A polymeric scaffold that possesses signaling molecules to orchestrate host immune cells and fibroblasts to regenerate human fascia would be the ideal solution. However, it needs to be capable of offloading significant tension while undergoing the remodeling process. Knitted and other polymeric scaffolds either lack the compliance properties necessary for proper offloading in the short-term healing phase [38] or the signaling molecules to guide fascial tissue regeneration in the long term [24]. The incorporation of a polymer within the ECM of reinforced tissue matrices has offered a promising alternative.

The polymer, interwoven through layers of the biologic component of RTMs, provides a reinforcement structure for the repair. This reinforcement is engaged at pressures defined by Junge et al. to mimic the compliance of the abdominal wall, between 11 and 32% [37,38]. The compliance of RTMs has been characterized between 11 and 15% to match the compliance of the native tissue, which in turn transfers abdominal pressures to the mesh rather than the repair site [38]. This reinforcement and offloading of the repair is especially important in patients who have poor collagen formation. The increase in collagen III in the ratio of collagen I and III has been shown to reduce the mechanical strength of connective tissues and may contribute to hernia recurrence [59,60]. If a patient suffering from this is treated with a biologic that is expected to remodel into host collagen over time, the repair may be replaced with poor host collagen, thus decreasing the strength over time from surgery. The polymer structure in RTMs provides a reinforcement network to avoid this and offload the wound, which has yielded low recurrence rates at both the short term and, per early reports, the long term, indicating a promising repair [46,53,56]. This is further evidenced by the results of the Parker ventral hernia repair series evaluating complex patients with lower recurrence rates than other similar studies [23,61,62].

The ECM acts as a signaling device to promote host cellular infiltration within the scaffold and promote remodeling of the matrix to attenuate the foreign body response to the polymer and generate functional collagen [24]. The result has shown a low inflammatory response similar to a biologic, lower than that observed in permanent and resorbable synthetic devices [24]. This recruitment of host cells and remodeling is attributed to the porous structure of the ECM and its retention of 153 unique matrisome proteins in the ECM of RTMs [24,40]. The remodeling properties of RTMs have also been observed in patient clinical biopsies, as depicted in Figure 3a–f. The RTM at 6 months, 7 months, and 23 months exhibited low inflammation and end-stage remodeling with the repair exhibiting mature, lamellar, and aponeurosis-like connective tissue. Furthermore, the polypropylene fibers of the 23 month biopsy were immediately surrounded by host cells and tissue with minimal inflammation. It is believed that the embedment of the polymer in a neovascularized biologic tissue allows for direct contact of the polymer with host immune cells, offering the mesh protection from contamination and infection. The ECM combined with the polymeric reinforcement offers a hybrid device providing short- and long-term strength to resist recurrence and a minimal foreign body footprint to limit complications and infections. The results continue to be confirmed in an ongoing 24 month clinical study.

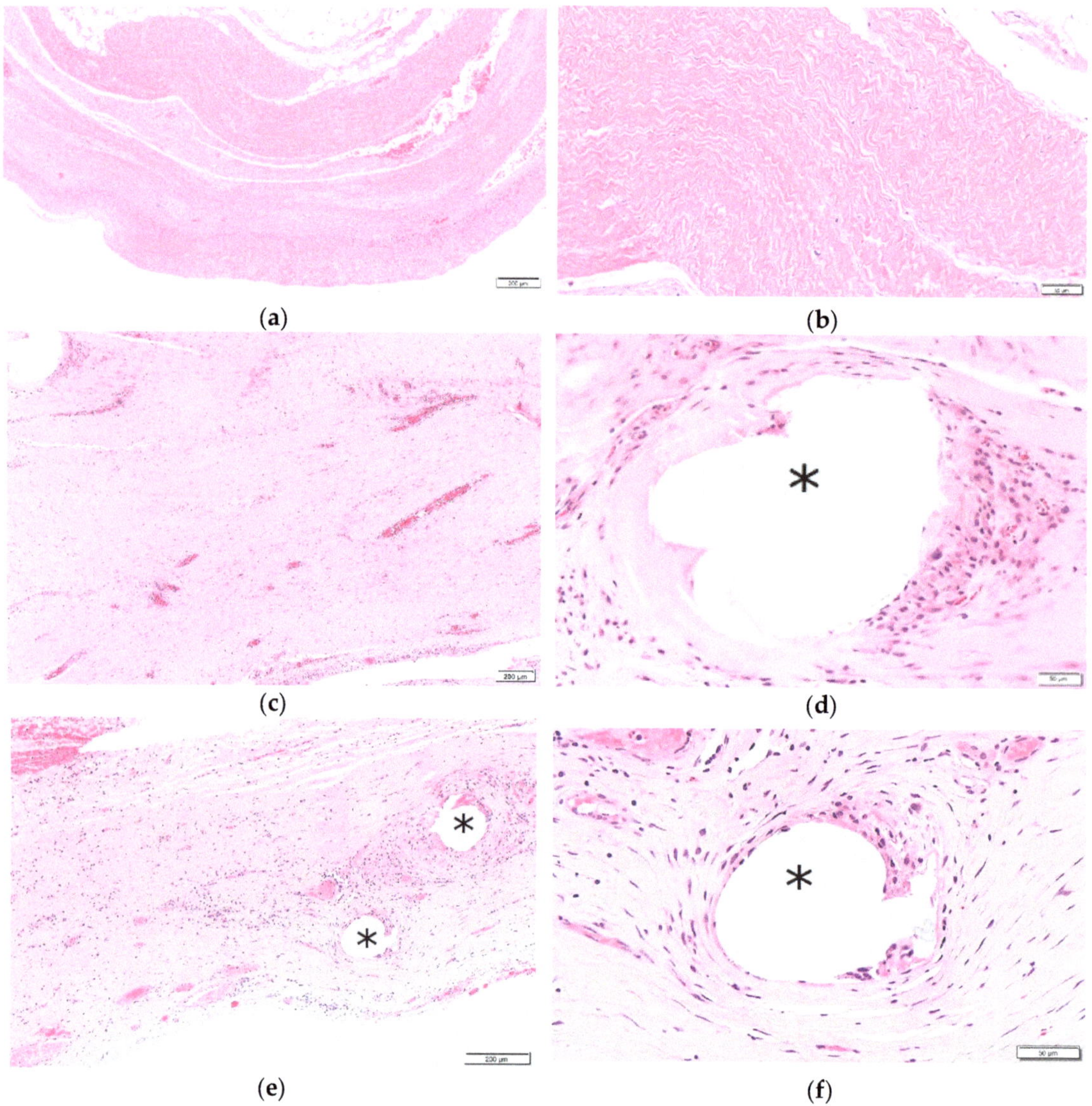

Figure 3. RTM clinical biopsies. * Denotes polypropylene fiber; (**a**) 6 month RTM clinical biopsy, scale bar is 200 μm; (**b**) 6 month RTM clinical biopsy, scale bar is 50 μm; (**c**) 7 month RTM clinical biopsy, scale bar is 200 μm; (**d**) 7 month clinical biopsy, scale bar is 50 μm; (**e**) 23 month RTM clinical biopsy, scale bar is 200 μm; (**f**) 23 month RTM clinical biopsy, scale bar is 50 μm.

4. Conclusions

This review provides an early assessment of the clinical utility of RTMs in an assortment of hernia repairs, where synthetic mesh has historically been used, including high-risk complex ventral hernias and minimally invasive procedures. Use of RTMs results in a decrease in chronic postoperative inguinal pain, improvement of preoperative hiatal hernia symptoms, and a lower incidence of recurrence compared to use of synthetic mesh in a more challenging ventral hernia repair population. These early, yet positive, results are attributed to the embeddment of polymer within a decellularized extracellular matrix, which allows for the construct to be remodeled and form new tissue around the filaments as part of a pro-healing response. The results of the use of OviTex in treatment of hernias continues to be monitored in clinical trials and preclinical studies.

Author Contributions: Conceptualization, M.S.; methodology, M.S., S.F. and G.D.III; formal analysis, M.S., S.F. and G.D.III; writing—original draft preparation, M.S.; writing—review and editing, M.S., S.F. and G.D.III; All authors have read and agreed to the published version of the manuscript.

Funding: This research received no external funding.

Institutional Review Board Statement: Ethical review and approval were waived for this study due to this being a review article. Details on the approval of the reviewed protocols by Institutional Review Board is contained within reviewed publications.

Informed Consent Statement: Patient consent was waived due to this being a review article. Details on the informed consent of the reviewed studies is contained within the reviewed publications.

Data Availability Statement: No new data were created or analyzed in this study. Data sharing is not applicable to this article.

Conflicts of Interest: M.S., S.F., and G.D.III are all consultants for TELA Bio, Inc.

References

1. Burger, J.W.; Luijendijk, R.W.; Hop, W.C.; Halm, J.A.; Verdaasdonk, E.G.; Jeekel, J. Long-term Follow-up of a Randomized Controlled Trial of Suture Versus Mesh Repair of Incisional Hernia. *Ann. Surg.* **2004**, *240*, 578–585. [CrossRef]
2. Köckerling, F.; Hoffmann, H.; Mayer, F.; Zarras, K.; Reinpold, W.; Fortelny, R.; Weyhe, D.; Lammers, B.; Adolf, D.; Schug-Pass, C. What are the trends in incisional hernia repair? Real-world data over 10 years from the Herniamed registry. *Hernia* **2021**, *25*, 255–265. [CrossRef] [PubMed]
3. Corr, D.T.; Hart, D.A. Biomechanics of Scar Tissue and Uninjured Skin. *Adv. Wound Care* **2013**, *2*, 37–43. [CrossRef] [PubMed]
4. Orenstein, S.B.; Saberski, E.R.; Kreutzer, N.L.; Novitsky, Y.W. Comparative Analysis of Histopathologic Effects of Synthetic Meshes Based on Material, Weight, and Pore Size in Mice. *J. Surg. Res.* **2012**, *176*, 423–429. [CrossRef] [PubMed]
5. Cobb, W.S.; Kercher, K.W.; Heniford, B.T. The Argument for Lightweight Polypropylene Mesh in Hernia Repair. *Surg. Innov.* **2005**, *12*, 63–69. [CrossRef] [PubMed]
6. Klinge, U.; Klosterhalfen, B.; Birkenhauer, V.; Junge, K.; Conze, J.; Schumpelick, V. Impact of Polymer Pore Size on the Interface Scar Formation in a Rat Model. *J. Surg. Res.* **2002**, *103*, 208–214. [CrossRef]
7. Christensen, M.B.; Tresco, P.A. The foreign body response and morphometric changes associated with mesh-style peripheral nerve cuffs. *Acta Biomater.* **2018**, *67*, 79–86. [CrossRef]
8. Hwang, K.; Sim, H.B.; Huan, F.; Kim, D.J. Myofibroblasts and Capsular Tissue Tension in Breast Capsular Contracture. *Aesthetic Plast. Surg.* **2010**, *34*, 716–721. [CrossRef]
9. Schuster, R.; Rockel, J.S.; Kapoor, M.; Hinz, B. The inflammatory speech of fibroblasts. *Immunol. Rev.* **2021**. [CrossRef]
10. Anderson, J.M.; Rodriguez, A.; Chang, D.T. Foreign body reaction to biomaterials. *Semin. Immunol.* **2008**, *20*, 86–100. [CrossRef]
11. Major, M.R.; Wong, V.W.; Nelson, E.R.; Longaker, M.T.; Gurtner, G.C. The Foreign Body Response. *Plast. Reconstr. Surg.* **2015**, *135*, 1489–1498. [CrossRef] [PubMed]
12. Molina, C.P.; Giglio, R.; Gandhi, R.M.; Sicari, B.M.; Londono, R.; Hussey, G.S.; Bartolacci, J.G.; Luque, L.M.Q.; Cramer, M.C.; Dziki, J.L.; et al. Comparison of the host macrophage response to synthetic and biologic surgical meshes used for ventral hernia repair. *J. Immunol. Regen. Med.* **2019**, *3*, 13–25. [CrossRef]
13. Brown, B.N.; Badylak, S.F. Extracellular matrix as an inductive scaffold for functional tissue reconstruction. *Transl. Res.* **2014**, *163*, 268–285. [CrossRef] [PubMed]
14. Deeken, C.R.; Abdo, M.S.; Frisella, M.M.; Matthews, B.D. Physicomechanical Evaluation of Polypropylene, Polyester, and Polytetrafluoroethylene Meshes for Inguinal Hernia Repair. *J. Am. Coll. Surg.* **2011**, *212*, 68–79. [CrossRef] [PubMed]
15. See, C.W.; Kim, T.; Zhu, D. Hernia Mesh and Hernia Repair: A Review. *Eng. Regen.* **2020**, *1*, 19–33. [CrossRef]
16. White, A.R.; Hirose, F.M.; Sproat, R.W.; Lawrence, R.S.; Nelson, R.J. Histopathologic observations after short-term implantation of two porous elastomers in dogs. *Biomaterials* **1981**, *2*, 171–176. [CrossRef]
17. Engelsman, A.F.; van der Mei, H.C.; Busscher, H.J.; Ploeg, R.J. Morphological aspects of surgical meshes as a risk factor for bacterial colonization. *Br. J. Surg.* **2008**, *95*, 1051–1059. [CrossRef]
18. Brown, C.N.; Finch, J.G. Which mesh for hernia repair? *Ann. R. Coll. Surg. Engl.* **2010**, *92*, 272–278. [CrossRef]
19. Verhorstert, K.W.J.; Guler, Z.; De Boer, L.; Riool, M.; Roovers, J.-P.W.R.; Zaat, S.A.J. In Vitro Bacterial Adhesion and Biofilm Formation on Fully Absorbable Poly-4-hydroxybutyrate and Nonabsorbable Polypropylene Pelvic Floor Implants. *ACS Appl. Mater. Interfaces* **2020**, *12*, 53646–53653. [CrossRef]
20. Olsen, M.A.; Nickel, K.B.; Wallace, A.E.; Mines, D.; Fraser, V.J.; Warren, D.K. Stratification of surgical site infection by operative factors and comparison of infection rates after hernia repair. *Infect. Control Hosp. Epidemiol.* **2014**, *36*, 329–335. [CrossRef]

21. Diamond, M.P.; Burns, E.L.; Accomando, B.; Mian, S.; Holmdahl, L. Seprafilm® adhesion barrier: (1) a review of preclinical, animal, and human investigational studies. *Gynecol. Surg.* **2012**, *9*, 237–245. [CrossRef] [PubMed]

22. Park, H.; Baek, S.; Kang, H.; Lee, D. Biomaterials to Prevent Post-Operative Adhesion. *Materials* **2020**, *13*, 3056. [CrossRef]

23. Roth, J.S.; Anthone, G.J.; Selzer, D.J.; Poulose, B.K.; Pierce, R.A.; Bittner, J.G.; Hope, W.W.; Dunn, R.M.; Martindale, R.G.; Goldblatt, M.I.; et al. Prospective, multicenter study of P4HB (Phasix™) mesh for hernia repair in cohort at risk for complications: 3-Year follow-up. *Ann. Med. Surg.* **2021**, *61*, 1–7. [CrossRef] [PubMed]

24. Overbeck, N.; Nagvajara, G.M.; Ferzoco, S.; May, B.C.H.; Beierschmitt, A.; Qi, S. In-vivo evaluation of a reinforced ovine biologic: A comparative study to available hernia mesh repair materials. *Hernia* **2020**, *24*, 1293–1306. [CrossRef] [PubMed]

25. Garvey, P.B.; Giordano, S.A.; Baumann, D.P.; Liu, J.; Butler, C.E. Long-Term Outcomes after Abdominal Wall Reconstruction with Acellular Dermal Matrix. *J. Am. Coll. Surg.* **2017**, *224*, 341–350. [CrossRef]

26. Hynes, R.O.; Naba, A. Overview of the Matrisome–An Inventory of Extracellular Matrix Constituents and Functions. *Cold Spring Harb. Perspect. Biol.* **2011**, *4*, a004903. [CrossRef]

27. Annor, A.H.; Tang, M.E.; Pui, C.L.; Ebersole, G.C.; Frisella, M.M.; Matthews, B.D.; Deeken, C.R. Effect of enzymatic degradation on the mechanical properties of biological scaffold materials. *Surg. Endosc.* **2012**, *26*, 2767–2778. [CrossRef]

28. Gómez-Gil, V.; Pascual, G.; Bellón, J.M. Biomaterial Implants in Abdominal Wall Hernia Repair: A Review on the Importance of the Peritoneal Interface. *Processes* **2019**, *7*, 105. [CrossRef]

29. Butler, C.E.; Prieto, V.G. Reduction of Adhesions with Composite AlloDerm/Polypropylene Mesh Implants for Abdominal Wall Reconstruction. *Plast. Reconstr. Surg.* **2004**, *114*, 464–473. [CrossRef]

30. Kumar, A.S.; Fitzgerald, J.F. Biologic versus Synthetic Mesh Reinforcement: What are the Pros and Cons? *Clin. Colon Rectal Surg.* **2014**, *27*, 140–148. [CrossRef]

31. Tognetti, L.; Pianigiani, E.; Ierardi, F.; Lorenzini, G.; Casella, D.; Liso, F.G.; De Pascalis, A.; Cinotti, E.; Rubegni, P. The use of human acellular dermal matrices in advanced wound healing and surgical procedures: State of the art. *Derm Ther.* **2021**, *34*, e14987. [CrossRef]

32. Liang, R.; Knight, K.M.; Barone, W.; Powers, R.W.; Nolfi, A.; Palcsey, S.; Abramowitch, S.; Moalli, P.A. Extracellular matrix regenerative graft attenuates the negative impact of polypropylene prolapse mesh on vagina in rhesus macaque. *Am. J. Obs. Gynecol.* **2017**, *216*, 153.e1–153.e9. [CrossRef]

33. Brown, B.N.; Ratner, B.D.; Goodman, S.B.; Amar, S.; Badylak, S.F. Macrophage polarization: An opportunity for improved outcomes in biomaterials and regenerative medicine. *Biomaterials* **2012**, *33*, 3792–3802. [CrossRef] [PubMed]

34. Londono, R.; Badylak, S.F. Biologic Scaffolds for Regenerative Medicine: Mechanisms of In vivo Remodeling. *Ann. Biomed. Eng.* **2014**, *43*, 577–592. [CrossRef]

35. Kim, H.; Wang, S.Y.; Kwak, G.; Yang, Y.; Kwon, I.C.; Kim, S.H. Exosome-Guided Phenotypic Switch of M1 to M2 Macrophages for Cutaneous Wound Healing. *Adv. Sci.* **2019**, *6*, 1900513. [CrossRef] [PubMed]

36. Deeken, C.R.; Lake, S.P. Mechanical properties of the abdominal wall and biomaterials utilized for hernia repair. *J. Mech. Behav. Biomed. Mater.* **2017**, *74*, 411–427. [CrossRef]

37. Junge, K.; Klinge, U.; Prescher, A.; Giboni, P.; Niewiera, M.; Schumpelick, V. Elasticity of the anterior abdominal wall and impact for reparation of incisional hernias using mesh implants. *Hernia* **2001**, *5*, 113–118. [CrossRef] [PubMed]

38. 2019 AHS Annual Meeting. *Hernia* **2019**, *23*, 1–109. [CrossRef]

39. Lun, S.; Irvine, S.M.; Johnson, K.D.; Fisher, N.J.; Floden, E.W.; Negron, L.; Dempsey, S.; McLaughlin, R.J.; Vasudevamurthy, M.; Ward, B.R.; et al. A functional extracellular matrix biomaterial derived from ovine forestomach. *Biomaterials* **2010**, *31*, 4517–4529. [CrossRef]

40. Dempsey, S.; Miller, C.H.; Hill, R.C.; Hansen, K.C.; May, B.C.H. Functional Insights from the Proteomic Inventory of Ovine Forestomach Matrix. *J. Proteome Res.* **2019**, *18*, 1657–1668. [CrossRef]

41. Sizeland, K.H.; Wells, H.C.; Kelly, S.J.; Nesdale, K.E.; May, B.C.H.; Dempsey, S.; Miller, C.H.; Kirby, N.; Hawley, A.; Mudie, S.; et al. Collagen Fibril Response to Strain in Scaffolds from Ovine Forestomach for Tissue Engineering. *Acs. Biomater. Sci. Eng.* **2017**, *3*, 2550–2558. [CrossRef]

42. Dempsey, S.G.; Miller, C.H.; Schueler, J.; Veale, R.W.F.; Day, D.J.; May, B.C.H. A novel chemotactic factor derived from the extracellular matrix protein decorin recruits mesenchymal stromal cells in vitro and in vivo. *PLoS ONE* **2020**, *15*, e0235784. [CrossRef]

43. Smith, J.; Parmely, J.D. *Ventral Hernia*; StatPearls Publishing: Treasure Island, FL, USA, 2021.

44. Faylona, J.M. Evolution of ventral hernia repair. *Asian J. Endosc. Surg.* **2017**, *10*, 252–258. [CrossRef] [PubMed]

45. Bringman, S.; Conze, J.; Cuccurullo, D.; Deprest, J.; Junge, K.; Klosterhalfen, B.; Parra-Davila, E.; Ramshaw, B.; Schumpelick, V. Hernia repair: The search for ideal meshes. *Hernia* **2009**, *14*, 81–87. [CrossRef] [PubMed]

46. Parker, M.J.; Barrio, M.; House, M.G.; Socas, J.; Reed, R.L.; Nakeeb, A.; Ceppa, E.P. Reinforced BioScaffold Mesh Lowers Recurrent Hernia Rate in High-Risk Ventral Hernia Repair with Surgical Site Occurrences. In Proceedings of the Americas Hernia Society (AHS) Annual Meeting, Las Vegas, NV, USA, 11–14 March 2019.

47. Kanters, A.E.; Krpata, D.M.; Blatnik, J.A.; Novitsky, Y.M.; Rosen, M.J. Modified Hernia Grading Scale to Stratify Surgical Site Occurrence after Open Ventral Hernia Repairs. *J. Am. Coll. Surg.* **2012**, *215*, 787–793. [CrossRef]

48. Onyekwelu, I.; Yakkanti, R.; Protzer, L.; Pinkston, C.M.; Tucker, C.; Seligson, D. Surgical Wound Classification and Surgical Site Infections in the Orthopaedic Patient. *Glob. Res. Rev.* **2017**, *1*, e022. [CrossRef] [PubMed]

49. Sfara, A.; Dumitrașcu, D.L. The management of hiatal hernia: An update on diagnosis and treatment. *Med. Pharm. Rep.* **2019**, *92*, 321–325. [CrossRef]
50. Schneider, R.; Herrington, J.L.; Granda, A.M. Marlex mesh in repair of a diaphragmatic defect later eroding into the distal esophagus and stomach. *Am. Surg.* **1979**, *45*, 337–339.
51. Hazebroek, E.J.; Leibman, S.; Smith, G.S. Erosion of a Composite PTFE/ePTFE Mesh After Hiatal Hernia Repair. *Surg. Laparosc. Endosc. Percutaneous Tech.* **2009**, *19*, 175–177. [CrossRef]
52. Stadlhuber, R.J.; El Sherif, A.; Mittal, S.K.; Jr, R.J.F.; Brunt, L.M.; Hunter, J.G.; Demeester, T.R.; Swanstrom, L.L.; Smith, C.D.; Filipi, C.J. Mesh complications after prosthetic reinforcement of hiatal closure: A 28-case series. *Surg. Endosc.* **2008**, *23*, 1219–1226. [CrossRef]
53. Sawyer, M.A.J. New Ovine Polymer-Reinforced Bioscaffold in Hiatal Hernia Repair. *JSLS J. Soc. Laparoendosc. Surg.* **2018**, *22*. [CrossRef] [PubMed]
54. Ramanan, B.; Maloley, B.J.; Fitzgibbons, R.J. Inguinal Hernia. *Adv. Surg.* **2014**, *48*, 1–11. [CrossRef] [PubMed]
55. Ferzli, G.S.; Edwards, E.D.; Khoury, G.E. Chronic Pain after Inguinal Herniorrhaphy. *J. Am. Coll. Surg.* **2007**, *205*, 333–341. [CrossRef]
56. Ferzoco, S.J. Early experience outcome of a reinforced Bioscaffold in inguinal hernia repair: A case series. *Int. J. Surg. Open* **2018**, *12*, 9–11. [CrossRef]
57. Kokotovic, D.; Bisgaard, T.; Helgstrand, F. Long-term Recurrence and Complications Associated with Elective Incisional Hernia Repair. *JAMA* **2016**, *316*, 1575–1582. [CrossRef]
58. Hawn, M.T.; Gray, S.; Snyder, C.W.; Graham, L.A.; Finan, K.R.; Vick, C.C. Predictors of mesh explantation after incisional hernia repair. *Am. J. Surg.* **2011**, *202*, 28–33. [CrossRef]
59. Klosterhalfen, B.; Lynen, P.; Schumpelick, V.; Junge, K.; Klinge, U.; Rosch, R.; Mertens, P.R.; Kirch, J. Decreased collagen typeI/III ratio in patients with recurring hernia after implantation of alloplastic prostheses. *Langenbeck's Arch. Surg.* **2003**, *389*, 17–22. [CrossRef]
60. Klinge, U.; Si, Z.; Zheng, H.; Schumpelick, V.; Bhardwaj, R.; Klosterhalfen, B. Abnormal collagen I to III distribution in the skin of patients with incisional hernia. *Eur. Surg. Res.* **2000**, *32*, 43–48. [CrossRef]
61. Itani, K.M.; Rosen, M.; Vargo, D.; Awad, S.S.; DeNoto, G.; Butler, C.E. Prospective study of single-stage repair of contaminated hernias using a biologic porcine tissue matrix: The RICH Study. *Surgery* **2012**, *152*, 498–505. [CrossRef]
62. Rosen, M.J.; Bauer, J.J.; Harmaty, M.; Carbonell, A.M.; Cobb, W.S.; Matthews, B.; Goldblatt, M.I.; Selzer, D.J.; Poulose, B.K.; Hansson, B.M.E.; et al. Multicenter, Prospective, Longitudinal Study of the Recurrence, Surgical Site Infection, and Quality of Life After Contaminated Ventral Hernia Repair Using Biosynthetic Absorbable Mesh. *Ann. Surg.* **2017**, *265*, 205–211. [CrossRef]

Review

A Short Review on Nanostructured Carbon Containing Biopolymer Derived Composites for Tissue Engineering Applications

Mattia Bartoli [1,2], **Erik Piatti** [3] **and Alberto Tagliaferro** [2,3,4,*]

[1] Center for Sustainable Future Technologies (CSFT), Istituto Italiano di Tecnologia (IIT), Via Livorno 60, 10144 Turin, Italy; mattia.bartoli@polito.it
[2] Consorzio Interuniversitario Nazionale per la Scienza e Tecnologia dei Materiali (INSTM), Via G. Giusti 9, 50121 Florence, Italy
[3] Department of Applied Science and Technology, Politecnico di Torino, Corso Duca degli Abruzzi 24, 10129 Turin, Italy; erik.piatti@polito.it
[4] Faculty of Science, Ontario Tech University, 2000 Simcoe Street North, Oshawa, ON L1G 0C5, Canada
[*] Correspondence: alberto.tagliaferro@polito.it; Tel.: +39-011-0907347

Abstract: The development of new scaffolds and materials for tissue engineering is a wide and open realm of material science. Among solutions, the use of biopolymers represents a particularly interesting area of study due to their great chemical complexity that enables creation of specific molecular architectures. However, biopolymers do not exhibit the properties required for direct application in tissue repair—such as mechanical and electrical properties—but they do show very attractive chemical functionalities which are difficult to produce through in vitro synthesis. The combination of biopolymers with nanostructured carbon fillers could represent a robust solution to enhance composite properties, producing composites with new and unique features, particularly relating to electronic conduction. In this paper, we provide a review of the field of carbonaceous nanostructure-containing biopolymer composites, limiting our investigation to tissue-engineering applications, and providing a complete overview of the recent and most outstanding achievements.

Keywords: graphene; carbon nanotubes; biopolymers; scaffolds; tissue engineering

check for **updates**

Citation: Bartoli, M.; Piatti, E.; Tagliaferro, A. A Short Review on Nanostructured Carbon Containing Biopolymer Derived Composites for Tissue Engineering Applications. *Polymers* **2023**, *15*, 1567. https://doi.org/10.3390/polym15061567

Academic Editor: João Carlos Silva

Received: 17 February 2023
Revised: 16 March 2023
Accepted: 19 March 2023
Published: 21 March 2023

1. Introduction

Complex chemical architectures represent a challenge for material science and synthetic chemistry. Several approaches can be used to achieve the complexity that naturally occurs in living organisms during anabolic metabolism. The level of complexity shown by biopolymers is astonishing and represents an opportunity to apply ready-to- use complex building blocks for the production of new materials [1]. This is particularly interesting for all those applications that require replicable polymeric matrixes with morphology and chemical functionalities that are hard to synthesize in vitro. Among possible applications, tissue repair and engineering is one of the most challenging [2], requiring high biocompatibility and an ability to mimic the original tissue texture. Biopolymers could be successfully used for this application to achieve remarkable results [3–5]. Nonetheless, native biopolymers do not exhibit suitable mechanical properties and electrical conductivity, excluding their use for both structural and neuronal/nerve tissue treatments. An elegant solution is represented by mixing neat biopolymers with fillers that are able to improve their mechanical and electrical features [6]. Good candidates for this task are nanostructured carbonaceous materials [7]. Nanostructured carbon is a wide family that comprises several species, from carbon black to the most cutting-edge materials, such as carbon nanotubes (CNTs) and graphene-related materials. These last two are the most remarkable with regard to both electrical and mechanical properties, as described in the recent literature [8,9], though their production is expensive and their dispersion into polymers is still rather challenging and costly. Nevertheless, the great challenge represented by the use of biopolymers in tissue engineering requires solutions beyond the state of the art—coupling with nanostructured carbon is the most promising route to fulfill this task [10–12]. In this work, we review the use of both CNTs and graphene-related materials as fillers for biopolymers used in tissue engineering. We lay out a clear picture of the latest achievements in the field, providing an introduction to the main topic before presenting more in-depth discussion. We consider biopolymer composites in two main categories: protein-derived and polysaccharide-based.

2. A Brief Overview on the Use of Polymers in Tissue Engineering

Tissue engineering is one of the most cutting-edge applications of polymer science, aiming to repair and replace damaged biological tissues using polymers and polymer composites by means of several production routes, as illustrated in Figure 1.

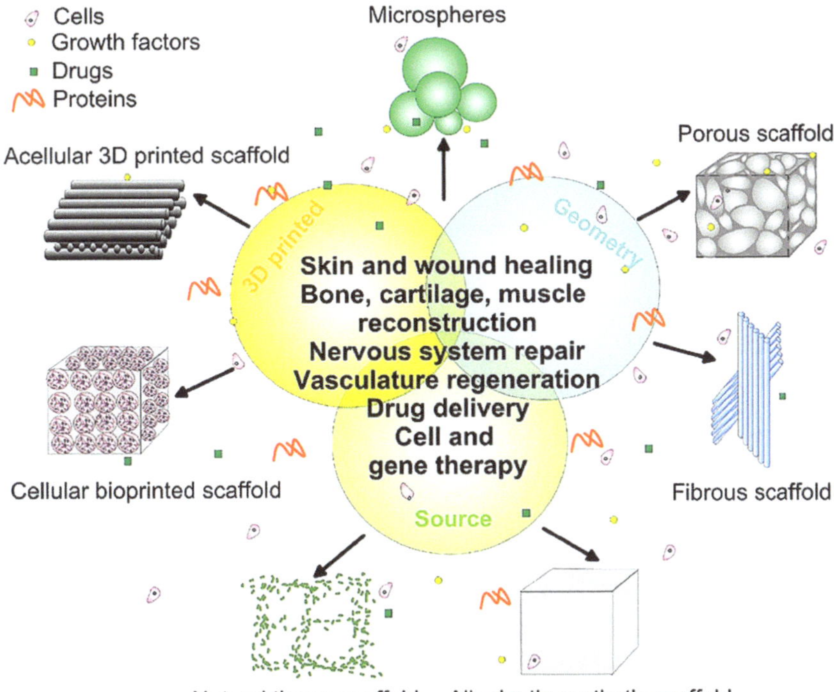

Figure 1. Multiple polymer- and polymer-composite-based routes to tissue engineering, highlighting the main morphologies used. Reprint with permission from Nikolova et al. [13] under CC BY license.

Over the last two decades, the realm of tissue engineering underwent rapid developments and several experimental routes for the reparation and regeneration of organs, tissues and cells were developed [2]. Furthermore, the use of the tissue-engineering approach has proved a promising alternative response to pathologies when traditional chemotherapy has failed [14]. Such a promising tool has inspired researchers to invest effort in quickly reaching the highest technological readiness level achievable. Nonetheless, this frontier is still far away due to the complexity of biological systems and their interaction with chemicals. Accordingly, the first experimental efforts were focused on cell implantation or replacement, creating artificial replacements for damaged tissues [15]. Cell-based tissue engineering remains quite complex, whereas the use of biopolymer-formulated scaffolds is cheaper and more tunable. Generally, a polymeric scaffold is defined as an intrinsic porous three-dimensional solid that interacts with cell lines and can be used to repair damage by merging with native tissue. Biopolymers play a crucial role due to their biocompatibility with biological matrixes, promoting the transport of solutions and gases and the proliferation of cell lines [5]. Biopolymer scaffolds are generally biodegradable, are able to match the same rate of regeneration of the native tissue, and enable its replacement with new undamaged tissues without releasing harmful species [16]. The first report of such materials was a 1975 study by Yannas et al. [17] based on the use of native collagen functionalized with glycosaminoglucans, followed by a systematic investigation of its use as a skin repair agent [18]. Biopolymers are also used as suitable replacements for damaged tissues, such as in the use of ocular, bone, vascular and dental implants [19].

The great advantage of biopolymers compared with synthetic polymers is associated with the complexity of their molecular architectures and their reproducibility resulting from biological pathways. The great potential for the application of biologically synthesized polymers, together with the role of purification procedures, is evidenced by considering cellulose as a case study. The structure of cellulose was determined by Staudinger in the early years of the 20th century [20], but only in 2001 was it fully synthesized in vitro [21]. Even now, more complex biopolymers, such as chitin or lignin, cannot be feasibly produced using chemical synthesis. Different considerations apply to protein synthesis, where a solid-state automatized synthetic approach provides new means for the in vitro production of peptides and proteins [22]. Biopolymer use has extended to several fields of the biomedical sciences, from drug delivery {Catania, 2021 #10529} to wound dressing [23].

Nevertheless, several biological applications—ranging from scaffolding to prosthetics—require superior conductive and mechanical properties. Neat biopolymers do not perform well enough to be used without mixing them with fillers. Nanostructured fillers, such as graphene derivatives [24] and CNTs [25], have found several applications as fillers for the production of scaffolds and prosthetics. Nevertheless, the interaction between the cellular structure and both CNTs and graphene-like materials is a very complex topic, as shown in Figure 2 [26]. All the interaction mechanisms shown are of major importance for the design of carbon-based biocompatible/bioactive composites [12] and are discussed in the next sections.

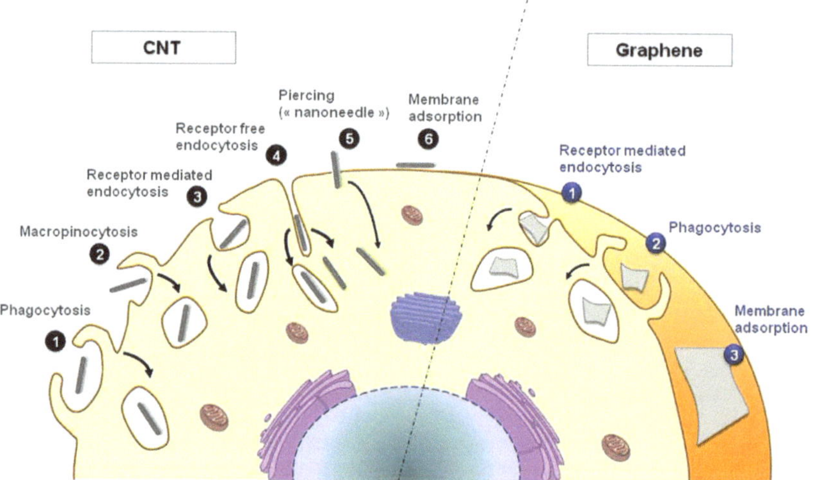

Figure 2. Schematic route of the cellular uptake of both neat CNTs and graphene-related materials. Reprinted with permission from Bussy et al. [27] (Copyright 2012 American Chemical Society).

3. Nanostructured Carbon Materials

3.1. Graphene and Graphene-Like Materials

Graphene, graphene oxide (GO) and reduced GO (rGO) are briefly illustrated in Figure 3.

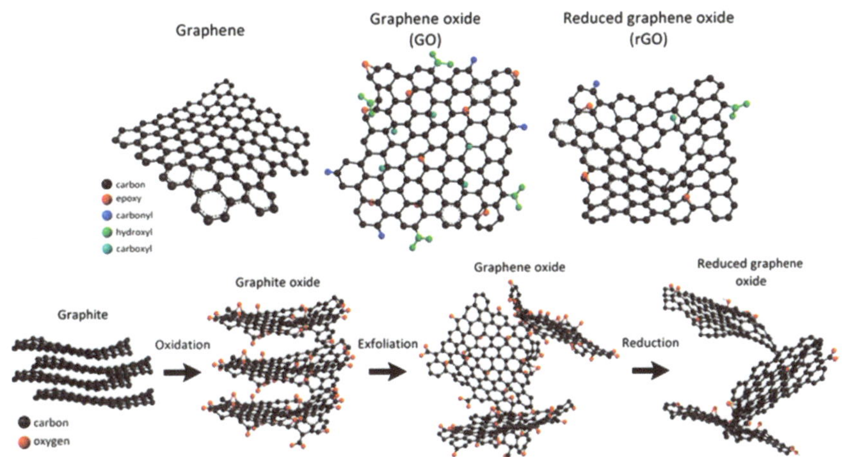

Figure 3. Methodologies for the production of graphene and related material structures through the oxidation of graphite. Reprinted from Jimenez-Cervantes et al. [28] under CC BY license.

In its ideal form, graphene is a two-dimensional crystal of carbon atoms bound in a honeycomb lattice and can be thought as a single plane of graphite. In each elementary cell of graphene, each carbon atom is linked by three in-plane σ bonds to its neighboring atoms with sp^2 hybridization, while the p orbitals perpendicular to the sp^2 plane do not participate in the bonding, hosting the conduction electrons. As a consequence, the electrons in the π bond are delocalized over the entire lattice structure and are able to move freely in the graphene plane [29–31], enabling high electrical conductivity despite the monolayer nature of the material [4]. The peculiar structure of graphene gives rise to unique electronic properties, including low-energy electrons behaving as massless Dirac fermions, ultra-high carrier mobilities up to $\sim 10^6$ cm^2 V^{-1} s^{-1}, room-temperature ballistic transport and half-integer quantum Hall effects [32]. These features have made graphene extremely appealing for high-frequency electronics and optoelectronics, while its ultrahigh tensile strength and flexibility make it very well suited for being incorporated in flexible devices [33].

The astonishing properties exhibited by ideal graphene are counterbalanced by its poor compatibility with established technological pipelines, as well as its scarce availability [34]. Indeed, the term graphene is often used in the literature to refer to other related carbon allotropes, including few-layer graphene and nanographite [33,35–37]. Several different materials have been proposed as alternatives that are able to overcome the issues associated with the use of "real" graphene. A first alternative is GO (graphene oxide), which is an oxidized graphene compound rich in oxygen functionalities [38]. Among these, epoxide and hydroxyl residual groups are predominant on the GO basal plane, whereas carbonyl and carboxylic groups are mostly localized along the edges. However, while the structure of pure graphene is relatively robust against different production methods, the structure of GO is deeply affected by its production process.

Due to this sensitivity to the production path, several structures have been proposed for GO in different studies, such as the Hofmann, Ruess, Scholz–Boehm, Nakajima–Matsuo, Lerf–Klinowski, and Szabo models [39,40].

Among these, the Lerf–Klinowski model is usually considered to represent the most realistic description of actual GO [16], attributing the presence of defects in the final material—such as holes, wrinkles, and cracks—mostly to the oxidation process.

GO is not the only graphene-derived material currently under extensive scientific investigation. Another very promising graphene-based material is reduced graphene oxide (rGO), which is obtained through a reduction process of GO by means of several reducing agents [41]. This process is necessary to eliminate the oxygenated functional groups of GO, which improve its handleability, but also result in an almost complete suppression of the high conductivity of neat graphene. The different production paths of GO result in different profiles of oxygen-containing functionalities and deeply alter the final properties of rGO [42]. High electrical conductivities are usually achieved in rGO, which is usually formed with oxygen amounts of between 1.5 and 0.4 wt.% [43].

Despite their astonishing properties, graphene-related materials can damage cellular structures, promoting oxidative degradation, apoptosis and cell-membrane disruption [27], mainly mediated by oxidation induced by peroxidase [44]. Modelling and tuning the interactions between graphene-like materials and cells is a challenging task and requires very good knowledge of the graphene-like flake sizes and thicknesses [45] as well as their functionalities [46].

The use of graphene and related materials for biological applications can exploit several interesting features [47,48]. The addition of GO or rGO to several polymeric matrixes has been reported to be effective for the replacement of damaged tissues through surface interactions [49]. The immobilization of graphene-related materials inside a polymeric host avoids the immunogenic [50] and inflammatory [51] effects they promote after cellular uptake.

As reported by Bahrami et al. [52], graphene flakes dispersed into polyurethane promote the growth of fibroblasts and endothelial cells on membranes containing graphene flakes. The findings were particularly interesting regarding endothelial cells. These could be grown on the inner surface of tubular scaffolds by mimicking the native blood vessel structure. The authors of the study demonstrated the ability of graphene-based composites to support the attachment, spreading and proliferation of cell lines. A similar approach was used by Pant et al. [53] for the production of coated stents using graphene oxide mixed with polyurethane.

3.2. Carbon Nanotubes

CNTs are another carbon allotrope based on the honeycomb lattice of sp^2-hybridized carbon atoms and can be thought of as cylindrical graphite sheets rolled up in a tubular structure. Different types of folding are possible for CNTs according to the so-called chiral vector $\vec{C} = n\,\vec{a}_1 + m\,\vec{a}_2$, where n and m are the positive integer chiral indexes, and $\vec{a}_1$, $\vec{a}_2$ are the graphene in-plane lattice vectors. As shown in Figure 4, the chiral indexes enable identification of three different orientations of the longitudinal axis of the nanotube with respect to the hexagonal lattice of graphene, resulting in three different CNT types termed zig-zag ($m = 0$), armchair ($m = n$), and chiral ($n \neq m \neq 0$). Chiral indexes have also been shown to enable prediction of whether the electronic properties of CNTs are metallic or semiconducting [54]; specifically, if $|n - m| = 3q$ with $n \neq m$ (where q is a positive integer) the CNT is metallic, otherwise it is semiconducting. Armchair CNTs always show metallic behavior, whereas chiral CNTs exhibit the same geometric properties as enantiomeric molecules.

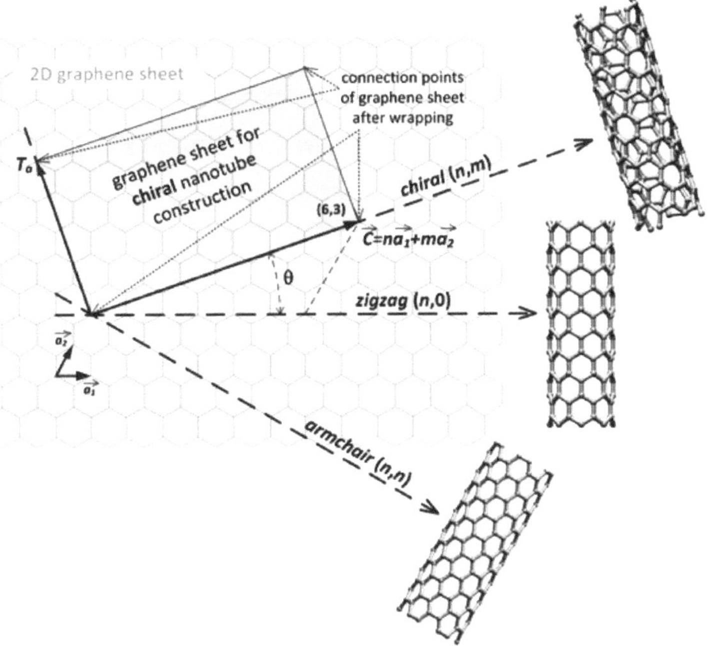

Figure 4. Theoretical folding of a perfect graphene sheet according to different linear combinations of the chiral vectors, leading to the three main CNT structures, named armchair, zig-zag and chiral. Reprinted with permission from Sanginario et al. [55] under CC BY license.

Furthermore, CNT models can be obtained by folding both single-layer and multi-layer graphene sheets, resulting in single-walled CNTs (SWCNTs) and multi-walled CNTs (MWCNTs), respectively, even if real CNT structure is more complex [56]. This choice affects the structural properties of CNTs, with SWCNTs exhibiting diameters ranging between 3 Å and 1 nm [57], whereas the diameter of MWCNTs can exceed 100 nm [58,59]. The CNT length, on the other hand, is highly dependent on the production process for both types and can range from a few nanometers up to a few tens of cm [60–62]. The extremities of CNTs can either be open or closed with fullerene-type caps [30] and variously functionalized [63]. Similar to graphene, the sp^2-hybridized honeycomb carbon lattice endows CNTs with excellent electrical [64,65], mechanical [66] and thermal [67,68] properties.

Similarly to graphene-like materials, CNTs are harmful for tissues and cells due to intrinsic defects [69,70] and metallic catalyst impurities [70,71], slowing down their use for neuronal tissue repair where their electric conductivity could otherwise enable game-changing approaches. The cytotoxicity and harmful effects of CNTs may be explained by the *"pathogenic fiber paradigm"* model which states that thin, long and persistent fibrous structures promote both inflammation and carcinogenesis [72]. Nevertheless, the real effects of CNT radical activity is still controversial, as shown by the scavenging effect reported by Fenoglio et al. [73]. After incorporation into a polymeric matrix, CNT toxicity can be minimized, as already reported for graphene-related materials [74,75]. This has enabled their use for the production of membranes with low filler loading (up to 0.5 wt.%) for neuronal tissue repair [76]. Furthermore, CNT-filled scaffolds can be used effectively for electrical stimulation to promote neurite overgrowth [77].

4. Carbon-Containing Biopolymer Composites for Biomedical Applications: An Overview

4.1. Nanostructured Carbon-Containing Proteinaceous Composites

4.1.1. Collagen

Collagen is a fibrous protein exhibiting strong hydrophilic behavior, which is insoluble in the majority of organic solvents, and represents the most abundant component of connective tissue. Remarkably, collagen accounts for more than 30 wt.% of total animal proteins, representing the main component of the extracellular matrix in animals. Collagen shows very interesting biological activities, such as the inhibition of tissue damage induced by stretching stress [78], preserving the structure of cartilage, bones and blood vessels.

The medical use of collagen started between the end of the 19th and the beginning of the 20th century with the development of biodegradable sutures for intestine surgery [79]. Neat collagen is widely used as a replacement for repairing corneal tissue due to its high bioadhesivity, biocompatibility with reduced immunogenicity, and acceptable mechanical properties [80]. Additionally, collagen fibrils can pass through epithelial tissue of the cornea stacking themselves onto native collagen [81]. In addition, collagen promotes the cellular adhesion, proliferation and differentiation of multiple cellular lines, promoting enhanced regeneration of several tissues [82].

The complex mechanisms of collagen degradation are central issues in the use of neat collagen associated with its high hydrophilicity that induces denaturation pathways and swelling in in vivo experiments [83]. Agarwal et al. [84] surmounted this issue by entrapping graphene sheets into a highly crosslinked collagen cryogel. The graphene-related composite was studied using X-ray tomography, showing improvement in both porosity and the porous connections of collagen cryogels induced by graphene flakes, together with thermal stability [85]. Nonetheless, the most used filler among the graphene-like materials for the production of composites remains GO, which is able to drive the self-assembly of hydrogels using a loading of up to 4 wt.% [86].

A cryogel based on mixed rGO flakes and collagen has found a promising application in bone restoration, as reported by Bahrami et al. [87]. The authors demonstrated that the mechanical strength of rGO-coated collagen cryogel was enhanced by three times compared with the uncoated cryogel. Furthermore, the immobilized rGO flakes promoted total biocompatibility of the collagen-based scaffold, enhancing mesenchymal stem cell proliferation. The authors also showed the ability of this scaffold to promote bone regeneration in a rabbit animal model.

A simple sol-gel production route is also effective for GO-containing collagen composites, as demonstrated by Liu et al. [88]. The authors produced a hydrophilic scaffold with good bone mineralization by adding 0.1 wt/v% GO to type I collagen during in vitro experiments. The osteogenesis promoted by the GO-based collagen scaffold could be boosted by adding inorganics, such as strontium, to the mixture [89], activating a protein complex responsible for bone-cell proliferation and differentiation. GO-based collagen composites can also be coupled with hydroxyapatite, as reported by Zhou et al. [90]. The authors crafted a multi-layer mineralized composite able to upregulate genes involved in osteogenesis, providing a comfortable microenvironment for hosting the cells of bone tissues. Similar results were obtained using bone apatite [91] due to the ability of the carboxylic residues on GO edges to provide a gripping point for inducing mineralization. These materials may also be used for cartilage repair, as reported by Lyu et al. [92]. The authors demonstrated the ability of GO-based collagen composites to induce the growth of chondrocytes, forming a cartilage matrix through the upregulation of metalloproteinases or their inhibitors. This double regulation enhanced the formation of homogenous cartilage tissue with regular distribution and porosity. A very advanced application of graphene-derivative-based collagen materials is related to the regeneration of neuronal tissue and damaged nerves. As reported by Guo and co-workers [93], rGO-based collagen scaffolds were able to enhance neuronal differentiation due to high conductivity. Agarwal et al. [94] followed similar protocols using amine-tailored graphene flakes dispersed into collagen to match both the electrical conductivity of 3 S/m and the Young's modulus of around

350 kPa of spinal cord tissue. The authors successfully suppressed the inflammatory response in rats, showing a remarkable ability of the scaffold to avoid macrophage proliferation. A more traditional use of GO-containing collagen composites is represented by using them as a coating for inorganic biocompatible alloys [95], inducing useful antifibrotic [96] and antibacterial [97] effects.

Interestingly, CNTs containing collagen scaffolds are more effective as substrates for the proliferation and differentiation of stem cells [98–100] and could be internalized into the cells without inducing disruption of the cellular membrane [101]. CNT-based collagen scaffolds have been used for applications that require very high electrical conductivity together with enhanced stiffness [102,103].

Chi et al. [104] reported the possibility of using CNT-based collagen composites for the stimulation of fibroblasts by simply preparing an electrospun mate. Considering its high customizability, the authors proposed the use of such produced materials for the treatment of dysfunctional fibroblasts associated with cartilaginous disorders. The conductivity of CNT-based collagen composites has been exploited in the production of advanced procedures for the treatment of neural degeneration [105], efficiently promoting neuronal growth and achieving remarkable results in nerve regeneration. The high conductivity of CNT-containing collagen also represents a tool for the production of electrically responsive cardiac patches, though the mechanism of action is still debated [106].

4.1.2. Keratin

Keratin is a protein in which cysteine reaches up to 13 wt.%; it is the major constituent of structures such as feathers and horns. The use of keratin as a biological scaffold is quite complex and requires multistep chemical modification of the native keratin structure.

The first step of this process is the extraction of keratin performed in strong chemical conditions using thermochemical or microbial approaches [107]. When the extracted keratin is ready to be processed, it forms stable transparent films that show high biocompatibility and biostability, and that are able to induce cell adhesion in living rabbits [108]. Nevertheless, keratin cytocompatibility is debated as the role of keratin in hemostasis processes is not totally clear [109]. The most accepted hemostatic effect of keratin hydrogel is the formation of a wound seal by reticulated coagulated keratin that induces the formation of granulose tissue upon it.

Nonetheless, keratin hydrogels have been used as effective hemostatic solutions in rabbits affected by lethal liver injury [110]. In addition, several commercial keratin-based hemostatic products are available, including HemCon® (Portland, OR, USA) and QuickClot® (Morrisville, NC, USA).

Keratin and keratin hydrogels have also been used for the regeneration of peripheral nerves, as reported by Sierpinski et al. [111] using mice models. The authors used human keratin extracted from hair, improving Schwann cell proliferation, together with upregulation of their gene expression, improving neuronal functionality. Furthermore, the authors described enhanced axon regeneration in damaged tibial nerves. Similarly, Apel et al. [112] induced the acceleration of nerve regeneration using keratin hydrogels, improving axon density. Neat keratin and keratin hydrogels have been shown to induce neovascularization and inflammation with reduction in soft tissue adhesion on polymeric scaffolds [113]. The addition of nanostructured carbon improved the interfacial properties and magnified the mechanical and electrical performance [114].

Song et al. [115] used an electrospinning technique to produce a composite based on rGO, keratin and poly(caprolactone)(PCL) for the treatment of damaged wounds that intrinsically show poor self-healing capacity. The material produced was composed of fibers with an average diameter of 240 nm, which was able to enhance cell adhesion and proliferation, accelerating wound re-epithelialization. Interestingly, the formulation increased the potential of mitochondrial membranes, improving cell viability. Additionally, the authors observed an increase in moisture absorption and the permeability of the rGO scaffold, supporting the use of graphene-based keratin scaffolds for the treatment of skin wounds.

Considering the number of studies reported in the literature, graphene-containing keratin materials represent a new and not-entirely-mature technology, whereas the combination of fibrous keratin with CNT has been more extensively studied. Mahmoodi et al. [116] followed a protocol close to the one previous described for the production of a PCL-keratin matrix through electrospinning using carboxylated MWCNTs. The composites produced showed very thin fibers with an average diameter of up to 50 nm, with both the tensile strength and tensile modulus increased by up to 260% compared to neat PCL-keratin material. The CNT-based composite induced relevant enhancement of osteogenic differentiation in mesenchymal stem cells with significant mineralization on the scaffold surface.

Similarly, Asl et al. [117] replaced PCL with poly(hydroxybutyrate), reporting overexpression of alkaline phosphatase with a related increase in mineralization rate. The combination of keratin with nanostructured carbon appears to be a reliable route for the reduction of an inflammatory response to CNTs, neat keratin and graphene-like materials, together with improving both interfacial properties and bone tissue healing.

4.1.3. Silk and Fibroin

Silk is a biopolymer composed mainly of two proteins, fibroin and sericin. Fibroin is a structural protein responsible for the strength and stiffness of silk, while sericin is a jelly protein that promotes adhesion of fibroin fibers together [118].

Fibroin is structured in long parallel β-sheet fibers interacting through hydrogen bonds. Fibroin is a hydrophobic amino-acid-rich material; its ordered folding process underlies the impressive mechanical properties of silk residues, which enable the protein to self-assemble into a stable, ordered structure [119]. As discussed by Salehi et al. [120], silk is highly biocompatible and its use in tissue repair is related to its morphology. Accordingly, thin silk fibers are currently used as surgical sutures, while porous silk represents a promising template for the production of scaffolds able to tune the conformation of silk [121].

Izyan et al. [122] evaluated the properties of graphene flakes dispersed into silk films, reporting an increase in β-sheet conformation, increasing both the crystallinity and the elongation at failure compared with neat silk materials. The modulation of the properties of silk can also be performed by directly acting on the feed of silk worms, as reported by Qu et al. [123]. The authors fed silkworms with mulberry leaves coated with 2 wt.% of GO, showing an increase in the elongation of recovered silk of up to 60% due to increase in α-helical regions. Furthermore, the authors reported that a reduction in the GO amount down to 1 wt.% promoted a selective increase in the Young's modulus. Similar results were achieved by adding up to 1 wt.% of rGO [124]. The graphene-based silk composite can also be used as a template for the production of biocompatible, size-controlled pore hydrogels with good cell viability [125]. Wang et al. [126] also demonstrated that the direction of channels could be organized in the presence of GO-based silk materials.

There are three main production routes for graphene-like materials containing silk composites: the sol-gel approach [127], reverse phase casting [128] and electrospinning [129]. Reverse phase casting is used for the production of homogeneous hydrogel layers, while electrospinning enables creation of a proper fibrous mate. The sol-gel approach is used for the production of injectable and conformable silk scaffolds. Together with the flexibility in production routes, graphene-like-containing silk composites show remarkable biocompatibility and cell viability [124]. The use of such composites has found many applications in the production of biomaterials [130–132]. Pathmanapan and coworkers [133] dispersed GO into fibroin hydrogel as scaffolds for bone repair, showing enhanced levels of alkaline phosphatase activity. This approach can be boosted by adding hydroxyapatite to the mixture, promoting mesenchymal stem cell adhesion and proliferation with more effective bone repair [134]. Similar results were obtained by mixing GO-containing silk composites with poly(l-lactic-co-glycolic acid) up to 10 wt.%.

Furthermore, Eivazzadeh-Keihan et al. [135] tested GO-based silk composites mixed with alginate, demonstrating their compatibility with blood, reporting a hemolytic effect below 6% for a concentration of up to 1 mg/mL. Li et al. [136] moved a step forward by modifying GO-containing silk composites with tannic acid and strontium cations, eliminating any inflammatory response mediated by both cytokines and reactive oxygen species, for the treatment of osteoarthritis. Furthermore, the GO-based scaffolds promoted the formation of cartilage tissue in injured rat knees. However, the exploitation of the electrical conductivity of graphene and rGO has found a main application in the treatment of neuronal and nerve tissues. Magaz et al. [137] were able to regulate the neuronal cell response by using GO mixed with an rGO-containing silk microcomposite. The prepared material had a conductivity of up to 3 μS/m after hydration and was able to mimic the arrangement of the extracellular matrix. The authors used a neuronal cell line, reporting the extension of neurites by up to 250 μm after 5 days. Similar results were achieved using graphene flakes dispersed into spun silk [129,138], mat [139] or gel [140], though the dispersibility of graphene flakes remained challenging.

Colloidal composites based on GO and silk were used to repair myocardial tissue after infarction by mixing with growth factors [141].

CNTs can also be mixed with silk using electrospinning for the production of biocompatible composites with high electronic conductivity and improved mechanical properties [142], showing that, even when using a high loading of MWCNTs, the effect on silk fiber morphology remained negligible. Shrestha et al. [143] dispersed functionalized MWCNTs into silk mixed with poly(urethane) producing a highly conductive scaffold in which CNTs were aligned. The authors reported the proliferation of Schwann cells, together with the spontaneous outgrowth of neurites along the fiber direction. Additionally, injectable shape memory CNT-based silk composites were used to treat ischemic stroke, promoting the customized growth of new neuronal tissue [144].

4.2. Nanostructured Carbon-Containing Polysaccharide Composites

4.2.1. Chitosan

Chitin is a functionalized polysaccharide containing units of β-(1,4)-N-acetyl-*d*-glucosamine produced as crystalline ordered microfibrils and is a crucial component in the formation of the cell walls of fungi, yeast and arthropod exoskeletons. Chitin has been widely used as an electrospun fiber mat by improving its mechanical properties by functionalization and coupling with polymers [145]. The as-produced chitin-derived fibers were shown to be biocompatible and able to promote both cell adhesion and proliferation without appreciable cytotoxicity.

Among chitin derivatives, chitosan is the most used and is produced by the simple partial deacetylation of neat chitin through alkaline or enzymatic hydrolysis [146]. Chitosan can be combined with GO, producing films with a low inflammatory response, advanced cicatrization, and good resorption in subdermal rat tissue [147]. As reported by Yilmaz et al. [148], chitosan combined with GO and hydroxyapatite showed a sponge-like morphology, together with both high cell viability and mechanical compressive strength, suggesting possible use as a bone replacement. This is supported also by research conducted by Depan et al. [149] in which GO-containing chitosan composites stimulated osteoblast growth. The authors suggested that osteoblast proliferation was due to both high water resorption and porous connection reducing their enzymatic degradation. Furthermore, the osteoblasts established intercellular connection mimicking real bone tissue. Liu et al. [150] demonstrated that GO-based chitosan composites with a lamellar conformation were able to promote the ordered growth of bone cell line MC3T3-E1 along the longitudinal direction. GO-based chitosan composites can simultaneously host toughening agents, such as hydroxyapatite [151] or another macrobiomolecule [152,153], to enhance its performance as a bone scaffold, without producing collateral reactions. Similar materials have been used for both cartilage [154,155] and dentary pulp [156] replacements. As reported by Fenge et al. [157], the unique weak and hydrogen-bond-type interactions occurring in GO-based chitosan composites presented a perfect case study of the realization of innovative wound dressing. The authors demonstrated that the reversible breakage–reformation of non-covalent interactions inside the composite allowed their adhesiveness to changing tissues, such as in the case of healing wounds. Additionally, GO-containing chitosan was shown to be able to stimulate nerve growth up to 20% more than neat chitosan scaffolds.

Similarly, Gupta et al. [158] dispersed both graphene nanoflakes and MWCNTs inside a chitosan matrix, promoting neurite outgrowth due to improved neural cell adhesion on the scaffold. CNT addition to a chitosan matrix modified the properties of chitosan hydrogels, improving the adhesion of several cell lines, as reported by Garnica-Palafox et al. [159]. CNT-based chitosan materials have also been used in bone tissue engineering due to improved cellular adhesion, together with high modulus and compressive strength, as reported by many authors [160–162], even without hydroxyapatite addition [163,164]. The superior mechanical properties of CNT-based chitosan represents a sound choice for cartilage replacement due to the high stress that this tissue can be exposed to. As reported by Mirmusavi et al. [165], PCL blended with chitosan and filled with 0.5 wt.% MWCNTs achieved a tensile strength of 34 MPa with high porosity and good compatibility with chondrocytes, enabling use in in vitro tests with good results.

4.2.2. Hyaluronic Acid

Hyaluronic acid is a biopolymer composed of disaccharide units based on N-acetyl-D-glucosamine and glucuronic acid dimers with a molecular weight up to MDa; it can be found in the majority of biological tissues.

Hyaluronic acid use has been explored in viscosurgery for tissue protection during ophthalmological medical procedures, such as lens implantations or cataract surgery, and as a humor vitreous replacement [166].

Outside of surgical applications, hyaluronic acid can be used as a protective agent for wounds due to the moisturizing effect that accelerates tissue healing. Edmonds and coworkers [167] described the positive role of hyaluronic acid in the treatment of both wound infections and ulcers, preventing the formation of large scars [168].

Additionally, neat hyaluronic acid was used as a synovial fluid replacement and as an additive in cosmetic surgical treatments, inducing increase in the inter-tissue space with a filling effect, as reported by several studies [169,170].

Neat hyaluronic acid can promote the proliferation of chondrocytes, slowing down cartilage degradation and related ageing problems [171].

Furthermore, neat hyaluronic acid is able to reduce the inflammatory response by modulating cytokine metabolism, with a massive reduction in reactive oxygen species [172]. Neat hyaluronic acid carboxylic residues have been used for the production of quite stable hydrogels for the delivery of several types of active molecule, such as drugs, nucleic acids and antibodies [173]. Several authors [174,175] have described GO-based hyaluronic composites for osteogenic differentiation and drug release without causing degradative processes in relevant tissues [176].

Accordingly, Umar et al. [177] combined GO-based hyaluronic acid composite with bacterial cellulose for the production of a tough bone scaffold. The authors successfully produced a highly interconnected porous material able to inhibit bacterial growth and to promote osteocyte proliferation.

The addition of CNTs has created the possibility of producing hyaluronic composites enriched with pluripotent stem cells able to induce neurogenesis together with upregulation of calcium channels. CNT-containing hyaluronic fibers have been used for neural engineering [178]. The authors of this study investigated the stimulation of cultured neurons, showing that CNT-based composite stimulation prolonged growth, without suggesting a mechanism for this phenomenon.

4.2.3. Cellulose

Cellulose is the most abundant polysaccharide and is composed of linear chains of glucose units connected to each other through a β 1-4 glycosidic chemical bond. Cellulose is a highly tunable template and can be shaped from nanocrystals to microfibrils based on an isolation methodology involving rearrangement of its hydrogen bond network. However, neat cellulose is not a good biological scaffold due to issues related to cellular interaction. Human cells are unable to adhere and to proliferate on neat cellulose films due to the hydrophilicity of the cellulose surface. This issue has been tackled by tailoring the cellulose hydroxyl residues by inserting species through simple chemical tailoring [179]. Cellulose tailoring is an easy and useful process able to produce a wide range of functionalized cellulose derivatives [180] used as biological scaffolds [181]. The most used tailoring processes are focused on the production of alkyl (i.e., methyl-, ethyl cellulose) or acetate derivatives [182].

Alternatively, neat cellulose can be blended with other polymers, as reported by Mao et al. [183]. The authors mixed neat cellulose with (poly)lactic acid filled with hydroxyapatite, demonstrating its effectiveness as a bone-repairing solution.

The main feature of nanostructured carbon-based cellulose composites is represented by the strong interaction between cellulose hydroxyl groups and the residual functionalities of carbon materials that tune cell adhesion, growth and proliferation [184].

Graphene and GO flakes dispersed into cellulose matrixes resulted in interesting properties, such as stimulation of osteogenesis [185,186], neural stem cell proliferation [187] and wound healing [188,189].

Similar results have been achieved using functionalized MCNTs, as reported by Khalid and co-workers [190].

5. Nanostructured Carbonaceous Materials and Related Biopolymer Composites for Tissue Engineering Applications: Weakness and Strengths

Nanostructured carbonaceous materials and related biopolymers represent an advanced means of addressing the many issues related to their use as neat materials. CNTs and graphene-related materials are the best candidates to improve both the mechanical and electrical properties of composites. However, several other solutions exist based on nanostructured carbon particles, which are particularly attractive for biomedical applications, such as carbon nano-onions [191]. This particular class of compounds can be used to produce drug delivery systems [192], and reinforce bulk [193] or film [194] materials, but the magnitude of improvement is still far from that achieved by adding CNTs or graphene-related materials. Opposite considerations apply when considering the use of the same nanostructured fillers combined with synthetic polymers, that show superior performance, such as ultra-high density poly(ethylene) (UHMWPE), as reported by Diabb Zavala et al. [195]. Here, the authors reported an increase in yield strength of up 65% of CNT-based UHMWPE composites, reaching a value beyond any biopolymer-based material. Nevertheless, UHMWPE and other olefins have issues related to their degradative routes due to friction and biological degradation [196,197], while biopolymers show higher biocompatibility with the production of less harmful degradation species [198,199].

A further important point to be considered is biopolymer isolation, that can represent a cost-determining step in biopolymer production. Similar considerations apply to the production of both CNTs and graphene materials that are still far from being large-scale and sufficiently standardized to meet biomedical industry requirements [200–202].

Nevertheless, the great economical value of biomedical solutions represents a driving force for rapidly achieving significant advancements in the large-scale industrial production of biopolymers, nanostructured carbon species and their composites.

This can lead to a new generation of multiresponsive materials useful for applications at the neural interface due to both their chemical and electrical properties [203,204]. Furthermore, new hybrid materials can open the way to the development of theranostic platforms based on electrical signal monitoring [205].

6. Conclusions

The use of biopolymers represents one of the most interesting frontiers in tissue engineering due to the great potential of biopolymer architectures naturally produced by the molecular engines of biological systems. Their combination with nanostructured carbon represents an additional step towards the production of complex and resilient species with enhanced electrical conductivity and mechanical properties, which will enable the creation of multifunctional composites for many applications in biological fields, such as regenerative medicine, scaffolds, and prosthesis production. Nonetheless, several issues remain unsolved, such as the high cost of biopolymers, CNTs and graphene and related materials, together with their difficult purification, processing, and standardization. Additionally, the degradative pathways of biopolymers are a key point to be considered for in vivo applications, particularly for prosthetic implants where they could represent a considerable advancement compared with the degradative pathways occurring with synthetic-polymer-based composites. Nevertheless, the full exploitation of nanostructured carbon-based biopolymer composites will surely lead to discoveries in many fields, such as regenerative medicine and neuronal implantation.

Author Contributions: Conceptualization, M.B. and A.T.; writing—original draft preparation, M.B., E.P. and A.T.; writing—review and editing, M.B., E.P. and A.T.; visualization, M.B. and E.P.; supervision, A.T. All authors have read and agreed to the published version of the manuscript.

Funding: This research received no external funding.

Conflicts of Interest: The authors declare no conflict of interest.

References

1. Guo, L.; Liang, Z.; Yang, L.; Du, W.; Yu, T.; Tang, H.; Li, C.; Qiu, H. The role of natural polymers in bone tissue engineering. *J. Control. Release* **2021**, *338*, 571–582. [CrossRef] [PubMed]
2. Ikada, Y. Challenges in tissue engineering. *J. R. Soc. Interface* **2006**, *3*, 589–601. [CrossRef] [PubMed]
3. Biswal, T. Biopolymers for tissue engineering applications: A review. *Mater. Today Proc.* **2021**, *41*, 397–402. [CrossRef]
4. Marras, E.; Bartoli, M.; Tagliaferro, A. 16—Drug delivery. In *Biopolymeric Nanomaterials*; Kanwar, S., Kumar, A., Nguyen, T.A., Sharma, S., Slimani, Y., Eds.; Elsevier: Amsterdam, The Netherlands, 2021; pp. 347–362.
5. Catania, F.; Bartoli, M.; Tagliaferro, A. Biopolymer-nanoparticles hybrids. In *Biopolymeric Nanomaterials*; Elsevier: Amsterdam, The Netherlands, 2021; pp. 293–309.
6. Bendrea, A.-D.; Cianga, L.; Cianga, I. Progress in the field of conducting polymers for tissue engineering applications. *J. Biomater. Appl.* **2011**, *26*, 3–84. [CrossRef] [PubMed]
7. Greil, P. Perspectives of nano-carbon based engineering materials. *Adv. Eng. Mater.* **2015**, *17*, 124–137. [CrossRef]
8. Soni, S.K.; Thomas, B.; Kar, V.R. A comprehensive review on CNTs and CNT-reinforced composites: Syntheses, characteristics and applications. *Mater. Today Commun.* **2020**, *25*, 101546. [CrossRef]
9. Razaq, A.; Bibi, F.; Zheng, X.; Papadakis, R.; Jafri, S.H.M.; Li, H. Review on graphene-, graphene oxide-, reduced graphene oxide-based flexible composites: From fabrication to applications. *Materials* **2022**, *15*, 1012. [CrossRef]
10. Amiryaghoubi, N.; Fathi, M.; Barzegari, A.; Barar, J.; Omidian, H.; Omidi, Y. Recent advances in polymeric scaffolds containing carbon nanotube and graphene oxide for cartilage and bone regeneration. *Mater. Today Commun.* **2021**, *26*, 102097. [CrossRef]
11. Ioniță, M.; Vlăsceanu, G.M.; Watzlawek, A.A.; Voicu, S.I.; Burns, J.S.; Iovu, H. Graphene and functionalized graphene: Extraordinary prospects for nanobiocomposite materials. *Compos. Part B Eng.* **2017**, *121*, 34–57. [CrossRef]
12. Mamidi, N.; Velasco Delgadillo, R.M.; Barrera, E.V.; Ramakrishna, S.; Annabi, N. Carbonaceous nanomaterials incorporated biomaterials: The present and future of the flourishing field. *Compos. Part B Eng.* **2022**, *243*, 110150. [CrossRef]
13. Nikolova, M.P.; Chavali, M.S. Recent advances in biomaterials for 3D scaffolds: A review. *Bioact. Mater.* **2019**, *4*, 271–292. [CrossRef]
14. Chapekar, M.S. Tissue engineering: Challenges and opportunities. *J. Biomed. Mater. Res. Off. J. Soc. Biomater. Jpn. Soc. Biomater. Aust. Soc. Biomater. Korean Soc. Biomater.* **2000**, *53*, 617–620. [CrossRef]
15. Bian, W.; Bursac, N. Cellular/tissue engineering. *IEEE Eng. Med. Biol. Mag.* **2008**, *27*, 109–113. [CrossRef]
16. Reddy, M.S.B.; Ponnamma, D.; Choudhary, R.; Sadasivuni, K.K. A comparative review of natural and synthetic biopolymer composite scaffolds. *Polymers* **2021**, *13*, 1105. [CrossRef]
17. Yannas, I. Suppression of in vivo degradability and of immunogenicity of collagen by reaction with glycosaminoglycans. *Polym. Prepr. Am. Chem. Soc.* **1975**, *16*, 209–214.
18. Yannas, I.; Burke, J.F. Design of an artificial skin. I. Basic design principles. *J. Biomed. Mater. Res.* **1980**, *14*, 65–81. [CrossRef]

19. Patel, N.R.; Gohil, P.P. A review on biomaterials: Scope, applications & human anatomy significance. *Int. J. Emerg. Technol. Adv. Eng.* **2012**, *2*, 91–101.

20. Staudinger, H. On Polymerization. In *A Source Book in Chemistry, 1900–1950*; Harvard University Press: Cambridge, MA, USA, 2013; pp. 259–264.

21. Kobayashi, S.; Sakamoto, J.; Kimura, S. In vitro synthesis of cellulose and related polysaccharides. *Prog. Polym. Sci.* **2001**, *26*, 1525–1560. [CrossRef]

22. Raibaut, L.; El Mahdi, O.; Melnyk, O. Solid phase protein chemical synthesis. In *Protein Ligation and Total Synthesis II*; Springer: Cham, Switzerland, 2015; pp. 103–154.

23. Bartoli, M.; Tagliaferro, A. 25 Recent Polymers Advancements for Biomedical in Applications. In *Specialty Polymers: Fundamentals, Properties, Applications and Advances*; CRC Press: Boca Raton, FL, USA, 2023; p. 383.

24. Catania, F.; Marras, E.; Giorcelli, M.; Jagdale, P.; Lavagna, L.; Tagliaferro, A.; Bartoli, M. A Review on Recent Advancements of Graphene and Graphene-Related Materials in Biological Applications. *Appl. Sci.* **2021**, *11*, 614. [CrossRef]

25. Yang, W.; Thordarson, P.; Gooding, J.J.; Ringer, S.P.; Braet, F. Carbon nanotubes for biological and biomedical applications. *Nanotechnology* **2007**, *18*, 412001. [CrossRef]

26. Narsimha, M. Cytotoxicity Evaluation of Carbon Nanotubes for Biomedical and Tissue Engineering Applications. In *Perspective of Carbon Nanotubes*; Hosam El-Din, S., Said Moawad Mohamed, E.-S., Eds.; IntechOpen: Rijeka, Croatia, 2019; p. 12.

27. Bussy, C.; Ali-Boucetta, H.; Kostarelos, K. Safety Considerations for Graphene: Lessons Learnt from Carbon Nanotubes. *Acc. Chem. Res.* **2013**, *46*, 692–701. [CrossRef] [PubMed]

28. Edgar Jimenez-Cervantes, A.; Juventino, L.B.; Ana Laura, M.H.; Carlos, V.S. Graphene-Based Materials Functionalization with Natural Polymeric Biomolecules. In *Recent Advances in Graphene Research*; Pramoda Kumar, N., Ed.; IntechOpen: Rijeka, Croatia, 2016; p. 12.

29. Mintmire, J.W.; Dunlap, B.I.; White, C.T. Are fullerene tubules metallic? *Phys. Rev. Lett.* **1992**, *68*, 631. [CrossRef] [PubMed]

30. Yan, J.-A.; Ruan, W.; Chou, M. Electron-phonon interactions for optical-phonon modes in few-layer graphene: First-principles calculations. *Phys. Rev. B* **2009**, *79*, 115443. [CrossRef]

31. Dresselhaus, M.; Jorio, A.; Saito, R. Characterizing graphene, graphite, and carbon nanotubes by Raman spectroscopy. *Annu. Rev. Condens. Matter Phys.* **2010**, *1*, 89–108. [CrossRef]

32. Castro Neto, A.H.; Guinea, F.; Peres, N.M.R.; Novoselov, K.S.; Geim, A.K. The electronic properties of graphene. *Rev. Mod. Phys.* **2009**, *81*, 109–162. [CrossRef]

33. Piatti, E.; Arbab, A.; Galanti, F.; Carey, T.; Anzi, L.; Spurling, D.; Roy, A.; Zhussupbekova, A.; Patel, K.A.; Kim, J.M. Charge transport mechanisms in inkjet-printed thin-film transistors based on two-dimensional materials. *Nat. Electron.* **2021**, *4*, 893–905. [CrossRef]

34. Lee, H.C.; Liu, W.-W.; Chai, S.-P.; Mohamed, A.R.; Lai, C.W.; Khe, C.-S.; Voon, C.; Hashim, U.; Hidayah, N. Synthesis of single-layer graphene: A review of recent development. *Procedia Chem.* **2016**, *19*, 916–921. [CrossRef]

35. Narita, A.; Wang, X.-Y.; Feng, X.; Müllen, K. New advances in nanographene chemistry. *Chem. Soc. Rev.* **2015**, *44*, 6616–6643. [CrossRef]

36. Sun, Z.; Fang, S.; Hu, Y.H. 3D Graphene Materials: From Understanding to Design and Synthesis Control. *Chem. Rev.* **2020**, *120*, 10336–10453. [CrossRef]

37. Piatti, E.; Galasso, S.; Tortello, M.; Nair, J.; Gerbaldi, C.; Bruna, M.; Borini, S.; Daghero, D.; Gonnelli, R. Carrier mobility and scattering lifetime in electric double-layer gated few-layer graphene. *Appl. Surf. Sci.* **2017**, *395*, 37–41. [CrossRef]

38. Brisebois, P.; Siaj, M. Harvesting graphene oxide—years 1859 to 2019: A review of its structure, synthesis, properties and exfoliation. *J. Mater. Chem. C* **2020**, *8*, 1517–1547. [CrossRef]

39. Szabó, T.; Berkesi, O.; Forgó, P.; Josepovits, K.; Sanakis, Y.; Petridis, D.; Dékány, I. Evolution of surface functional groups in a series of progressively oxidized graphite oxides. *Chem. Mater.* **2006**, *18*, 2740–2749. [CrossRef]

40. Lavagna, L.; Meligrana, G.; Gerbaldi, C.; Tagliaferro, A.; Bartoli, M. Graphene and Lithium-Based Battery Electrodes: A Review of Recent Literature. *Energies* **2020**, *13*, 4867. [CrossRef]

41. Ahmed, A.; Singh, A.; Young, S.-J.; Gupta, V.; Singh, M.; Arya, S. Synthesis Techniques and Advances in Sensing Applications of Reduced Graphene Oxide (rGO) Composites: A Review. *Compos. Part A Appl. Sci. Manuf.* **2022**, *165*, 107373–107394. [CrossRef]

42. Poh, H.L.; Šaněk, F.; Ambrosi, A.; Zhao, G.; Sofer, Z.; Pumera, M. Graphenes prepared by Staudenmaier, Hofmann and Hummers methods with consequent thermal exfoliation exhibit very different electrochemical properties. *Nanoscale* **2012**, *4*, 3515–3522. [CrossRef]

43. Lee, X.J.; Hiew, B.Y.Z.; Lai, K.C.; Lee, L.Y.; Gan, S.; Thangalazhy-Gopakumar, S.; Rigby, S. Review on graphene and its derivatives: Synthesis methods and potential industrial implementation. *J. Taiwan Inst. Chem. Eng.* **2019**, *98*, 163–180. [CrossRef]

44. Kotchey, G.P.; Hasan, S.A.; Kapralov, A.A.; Ha, S.H.; Kim, K.; Shvedova, A.A.; Kagan, V.E.; Star, A. A Natural Vanishing Act: The Enzyme-Catalyzed Degradation of Carbon Nanomaterials. *Acc. Chem. Res.* **2012**, *45*, 1770–1781. [CrossRef]

45. Yue, H.; Wei, W.; Yue, Z.; Wang, B.; Luo, N.; Gao, Y.; Ma, D.; Ma, G.; Su, Z. The role of the lateral dimension of graphene oxide in the regulation of cellular responses. *Biomaterials* **2012**, *33*, 4013–4021. [CrossRef]

46. Mu, Q.; Su, G.; Li, L.; Gilbertson, B.O.; Yu, L.H.; Zhang, Q.; Sun, Y.-P.; Yan, B. Size-dependent cell uptake of protein-coated graphene oxide nanosheets. *ACS Appl. Mater. Interfaces* **2012**, *4*, 2259–2266. [CrossRef]

47. Munir, K.S.; Wen, C.; Li, Y. Carbon nanotubes and graphene as nanoreinforcements in metallic biomaterials: A review. *Adv. Biosyst.* **2019**, *3*, 1800212. [CrossRef]

48. Thompson, B.C.; Murray, E.; Wallace, G.G. Graphite oxide to graphene. Biomaterials to bionics. *Adv. Mater.* **2015**, *27*, 7563–7582. [CrossRef] [PubMed]

49. Wang, Q.; Wang, M.; Wang, K.; Sun, Y.; Zhang, H.; Lu, X.; Duan, K. Molecular mechanisms of interactions between BMP-2 and graphene: Effects of functional groups and microscopic morphology. *Appl. Surf. Sci.* **2020**, *525*, 146636. [CrossRef]

50. Ni, G.; Wang, Y.; Wu, X.; Wang, X.; Chen, S.; Liu, X. Graphene oxide absorbed anti-IL10R antibodies enhance LPS induced immune responses in vitro and in vivo. *Immunol. Lett.* **2012**, *148*, 126–132. [CrossRef] [PubMed]

51. Di Ianni, E.; Møller, P.; Vogel, U.B.; Jacobsen, N.R. Pro-inflammatory response and genotoxicity caused by clay and graphene nanomaterials in A549 and THP-1 cells. *Mutat. Res. Genet. Toxicol. Environ. Mutagen.* **2021**, *872*, 503405. [CrossRef]

52. Bahrami, S.; Solouk, A.; Mirzadeh, H.; Seifalian, A.M. Electroconductive polyurethane/graphene nanocomposite for biomedical applications. *Compos. Part B Eng.* **2019**, *168*, 421–431. [CrossRef]

53. Pant, H.R.; Pokharel, P.; Joshi, M.K.; Adhikari, S.; Kim, H.J.; Park, C.H.; Kim, C.S. Processing and characterization of electrospun graphene oxide/polyurethane composite nanofibers for stent coating. *Chem. Eng. J.* **2015**, *270*, 336–342. [CrossRef]

54. Belin, T.; Epron, F. Characterization methods of carbon nanotubes: A review. *Mater. Sci. Eng. B* **2005**, *119*, 105–118. [CrossRef]

55. Sanginario, A.; Miccoli, B.; Demarchi, D. Carbon nanotubes as an effective opportunity for cancer diagnosis and treatment. *Biosensors* **2017**, *7*, 9. [CrossRef]

56. Dresselhaus, M.; Dresselhaus, G.; Jorio, A. Unusual properties and structure of carbon nanotubes. *Annu. Rev. Mater. Res.* **2004**, *34*, 247–278. [CrossRef]

57. Fagan, J.A.; Hároz, E.H.; Ihly, R.; Gui, H.; Blackburn, J.L.; Simpson, J.R.; Lam, S.; Hight Walker, A.R.; Doorn, S.K.; Zheng, M. Isolation of >1 nm diameter single-wall carbon nanotube species using aqueous two-phase extraction. *ACS Nano* **2015**, *9*, 5377–5390. [CrossRef]

58. Liu, W.-W.; Chai, S.-P.; Mohamed, A.R.; Hashim, U. Synthesis and characterization of graphene and carbon nanotubes: A review on the past and recent developments. *J. Ind. Eng. Chem.* **2014**, *20*, 1171–1185. [CrossRef]

59. Navas, H.; Picher, M.; Andrieux-Ledier, A.; Fossard, F.; Michel, T.; Kozawa, A.; Maruyama, T.; Anglaret, E.; Loiseau, A.; Jourdain, V. Unveiling the Evolutions of Nanotube Diameter Distribution during the Growth of Single-Walled Carbon Nanotubes. *ACS Nano* **2017**, *11*, 3081–3088. [CrossRef]

60. Inam, F.; Reece, M.J.; Peijs, T. Shortened carbon nanotubes and their influence on the electrical properties of polymer nanocomposites. *J. Compos. Mater.* **2011**, *46*, 1313–1322. [CrossRef]

61. Zhang, R.; Zhang, Y.; Zhang, Q.; Xie, H.; Qian, W.; Wei, F. Growth of Half-Meter Long Carbon Nanotubes Based on Schulz–Flory Distribution. *ACS Nano* **2013**, *7*, 6156–6161. [CrossRef]

62. Zhu, Z.; Wei, N.; Cheng, W.; Shen, B.; Sun, S.; Gao, J.; Wen, Q.; Zhang, R.; Xu, J.; Wang, Y.; et al. Rate-selected growth of ultrapure semiconducting carbon nanotube arrays. *Nat. Commun.* **2019**, *10*, 4467. [CrossRef]

63. Tasis, D.; Tagmatarchis, N.; Bianco, A.; Prato, M. Chemistry of Carbon Nanotubes. *Chem. Rev.* **2006**, *106*, 1105–1136. [CrossRef]

64. Thess, A.; Lee, R.; Nikolaev, P.; Dai, H.; Petit, P.; Robert, J.; Xu, C.; Lee, Y.H.; Kim, S.G.; Rinzler, A.G. Crystalline ropes of metallic carbon nanotubes. *Science* **1996**, *273*, 483–487. [CrossRef]

65. Mansfield, E.; Feldman, A.; Chiaramonti, A.N.; Lehman, J.; Curtin, A.E. Morphological and Electrical Characterization of MWCNT Papers and Pellets. *J. Res. Natl. Inst. Stand. Technol.* **2015**, *120*, 304–315. [CrossRef]

66. Han, J. Structures and properties of carbon nanotubes. In *Carbon Nanotubes*; CRC Press: Boca Raton, FL, USA, 2004; pp. 16–45.

67. Choi, T.Y.; Poulikakos, D.; Tharian, J.; Sennhauser, U. Measurement of thermal conductivity of individual multiwalled carbon nanotubes by the 3-ω method. *Appl. Phys. Lett.* **2005**, *87*, 013108. [CrossRef]

68. Cao, J.; Yan, X.; Xiao, Y.; Ding, J. Thermal conductivity of zigzag single-walled carbon nanotubes: Role of the umklapp process. *Phys. Rev. B* **2004**, *69*, 073407. [CrossRef]

69. Fenoglio, I.; Greco, G.; Tomatis, M.; Muller, J.; Raymundo-Piñero, E.; Béguin, F.; Fonseca, A.; Nagy, J.B.; Lison, D.; Fubini, B. Structural Defects Play a Major Role in the Acute Lung Toxicity of Multiwall Carbon Nanotubes: Physicochemical Aspects. *Chem. Res. Toxicol.* **2008**, *21*, 1690–1697. [CrossRef] [PubMed]

70. Muller, J.; Huaux, F.; Fonseca, A.; Nagy, J.B.; Moreau, N.; Delos, M.; Raymundo-Pinñero, E.; Béguin, F.; Kirsch-Volders, M.; Fenoglio, I. Structural defects play a major role in the acute lung toxicity of multiwall carbon nanotubes: Toxicological aspects. *Chem. Res. Toxicol.* **2008**, *21*, 1698–1705. [CrossRef] [PubMed]

71. Visalli, G.; Facciolà, A.; Iannazzo, D.; Piperno, A.; Pistone, A.; Di Pietro, A. The role of the iron catalyst in the toxicity of multi-walled carbon nanotubes (MWCNTs). *J. Trace Elem. Med. Biol.* **2017**, *43*, 153–160. [CrossRef]

72. Kostarelos, K. The long and short of carbon nanotube toxicity. *Nat. Biotechnol.* **2008**, *26*, 774–776. [CrossRef] [PubMed]

73. Fenoglio, I.; Tomatis, M.; Lison, D.; Muller, J.; Fonseca, A.; Nagy, J.B.; Fubini, B. Reactivity of carbon nanotubes: Free radical generation or scavenging activity? *Free Radic. Biol. Med.* **2006**, *40*, 1227–1233. [CrossRef]

74. Huang, B. Carbon nanotubes and their polymeric composites: The applications in tissue engineering. *Biomanuf. Rev.* **2020**, *5*, 3. [CrossRef]

75. Mamidi, N.; Leija, H.M.; Diabb, J.M.; Lopez Romo, I.; Hernandez, D.; Castrejón, J.V.; Martinez Romero, O.; Barrera, E.V.; Elias Zúñiga, A. Cytotoxicity evaluation of unfunctionalized multiwall carbon nanotubes-ultrahigh molecular weight polyethylene nanocomposites. *J. Biomed. Mater. Res. Part A* **2017**, *105*, 3042–3049. [CrossRef]

76. Vicentini, N.; Gatti, T.; Salerno, M.; Gomez, Y.S.H.; Bellon, M.; Gallio, S.; Marega, C.; Filippini, F.; Menna, E. Effect of different functionalized carbon nanostructures as fillers on the physical properties of biocompatible poly (l-lactic acid) composites. *Mater. Chem. Phys.* **2018**, *214*, 265–276. [CrossRef]

77. Koppes, A.; Keating, K.; McGregor, A.; Koppes, R.; Kearns, K.; Ziemba, A.; McKay, C.; Zuidema, J.; Rivet, C.; Gilbert, R. Robust neurite extension following exogenous electrical stimulation within single walled carbon nanotube-composite hydrogels. *Acta Biomater.* **2016**, *39*, 34–43. [CrossRef]

78. Sionkowska, A.; Skrzyński, S.; Śmiechowski, K.; Kołodziejczak, A. The review of versatile application of collagen. *Polym. Adv. Technol.* **2017**, *28*, 4–9. [CrossRef]

79. Chattopadhyay, S.; Raines, R.T. Collagen-based biomaterials for wound healing. *Biopolymers* **2014**, *101*, 821–833. [CrossRef]

80. Shoulders, M.D.; Raines, R.T. Collagen structure and stability. *Annu. Rev. Biochem.* **2009**, *78*, 929–958. [CrossRef]

81. Meek, K.M.; Boote, C. The organization of collagen in the corneal stroma. *Exp. Eye Res.* **2004**, *78*, 503–512. [CrossRef]

82. Ferreira, A.M.; Gentile, P.; Chiono, V.; Ciardelli, G. Collagen for bone tissue regeneration. *Acta Biomater.* **2012**, *8*, 3191–3200. [CrossRef]

83. Laurent, G. Dynamic state of collagen: Pathways of collagen degradation in vivo and their possible role in regulation of collagen mass. *Am. J. Physiol. Cell Physiol.* **1987**, *252*, C1–C9. [CrossRef]

84. Agarwal, G.; Agrawal, A.K.; Fatima, A.; Srivastava, A. X-ray tomography analysis reveals the influence of graphene on porous morphology of collagen cryogels. *Micron* **2021**, *150*, 103127. [CrossRef]

85. Ilya Syafiqa Zulkifli, N.; Ibrahim, N.; Jaafar, M.; Teramoto, N. The properties of the modified fish collagen peptide hydrogel. *Mater. Today Proc.* **2022**, *66*, 2738–2741. [CrossRef]

86. Girão, A.F.; Gonçalves, G.; Bhangra, K.S.; Phillips, J.B.; Knowles, J.; Irurueta, G.; Singh, M.K.; Bdkin, I.; Completo, A.; Marques, P.A.A.P. Electrostatic self-assembled graphene oxide-collagen scaffolds towards a three-dimensional microenvironment for biomimetic applications. *RSC Adv.* **2016**, *6*, 49039–49051. [CrossRef]

87. Bahrami, S.; Baheiraei, N.; Shahrezaee, M. Biomimetic reduced graphene oxide coated collagen scaffold for in situ bone regeneration. *Sci. Rep.* **2021**, *11*, 16783. [CrossRef]

88. Liu, S.; Zhou, C.; Mou, S.; Li, J.; Zhou, M.; Zeng, Y.; Luo, C.; Sun, J.; Wang, Z.; Xu, W. Biocompatible graphene oxide–collagen composite aerogel for enhanced stiffness and in situ bone regeneration. *Mater. Sci. Eng. C* **2019**, *105*, 110137. [CrossRef]

89. Chen, Y.; Zheng, Z.; Zhou, R.; Zhang, H.; Chen, C.; Xiong, Z.; Liu, K.; Wang, X. Developing a Strontium-Releasing Graphene Oxide-/Collagen-Based Organic–Inorganic Nanobiocomposite for Large Bone Defect Regeneration via MAPK Signaling Pathway. *ACS Appl. Mater. Interfaces* **2019**, *11*, 15986–15997. [CrossRef]

90. Zhou, C.; Luo, C.; Liu, S.; Jiang, S.; Liu, X.; Li, J.; Zhang, X.; Wu, X.; Sun, J.; Wang, Z. Pearl-inspired graphene oxide-collagen microgel with multi-layer mineralization through microarray chips for bone defect repair. *Mater. Today Bio* **2022**, *15*, 100307. [CrossRef] [PubMed]

91. Zhou, C.; Liu, S.; Li, J.; Guo, K.; Yuan, Q.; Zhong, A.; Yang, J.; Wang, J.; Sun, J.; Wang, Z. Collagen Functionalized With Graphene Oxide Enhanced Biomimetic Mineralization and in Situ Bone Defect Repair. *ACS Appl. Mater. Interfaces* **2018**, *10*, 44080–44091. [CrossRef] [PubMed]

92. Lyu, C.; Cheng, C.; He, Y.; Qiu, L.; He, Z.; Zou, D.; Li, D.; Lu, J. Graphene Hydrogel as a Porous Scaffold for Cartilage Regeneration. *ACS Appl. Mater. Interfaces* **2022**, *14*, 54431–54438. [CrossRef] [PubMed]

93. Guo, W.; Wang, S.; Yu, X.; Qiu, J.; Li, J.; Tang, W.; Li, Z.; Mou, X.; Liu, H.; Wang, Z. Construction of a 3D rGO–collagen hybrid scaffold for enhancement of the neural differentiation of mesenchymal stem cells. *Nanoscale* **2016**, *8*, 1897–1904. [CrossRef]

94. Agarwal, G.; Kumar, N.; Srivastava, A. Highly elastic, electroconductive, immunomodulatory graphene crosslinked collagen cryogel for spinal cord regeneration. *Mater. Sci. Eng. C* **2021**, *118*, 111518. [CrossRef]

95. Yılmaz, E.; Çakıroğlu, B.; Gökçe, A.; Findik, F.; Gulsoy, H.O.; Gulsoy, N.; Mutlu, Ö.; Özacar, M. Novel hydroxyapatite/graphene oxide/collagen bioactive composite coating on Ti16Nb alloys by electrodeposition. *Mater. Sci. Eng. C* **2019**, *101*, 292–305. [CrossRef]

96. Chen, C.-Y.; Tsai, P.-H.; Lin, Y.-H.; Huang, C.-Y.; Chung, J.H.Y.; Chen, G.-Y. Controllable graphene oxide-based biocompatible hybrid interface as an anti-fibrotic coating for metallic implants. *Mater. Today Bio* **2022**, *15*, 100326. [CrossRef]

97. Liu, J.; Wang, X.; Saberi, A.; Heydari, Z. The effect of Co-encapsulated GNPs-CNTs nanofillers on mechanical properties, degradation and antibacterial behavior of Mg-based composite. *J. Mech. Behav. Biomed. Mater.* **2023**, *138*, 105601. [CrossRef]

98. Kim, T.; Sridharan, I.; Zhu, B.; Orgel, J.; Wang, R. Effect of CNT on collagen fiber structure, stiffness assembly kinetics and stem cell differentiation. *Mater. Sci. Eng. C* **2015**, *49*, 281–289. [CrossRef]

99. Lee, J.H.; Lee, J.-Y.; Yang, S.H.; Lee, E.-J.; Kim, H.-W. Carbon nanotube–collagen three-dimensional culture of mesenchymal stem cells promotes expression of neural phenotypes and secretion of neurotrophic factors. *Acta Biomater.* **2014**, *10*, 4425–4436. [CrossRef]

100. Madhusoodan, A.P.; Das, K.; Mili, B.; Kumar, K.; Kumar, A.; Saxena, A.C.; Singh, P.; Dutt, T.; Bag, S. In vitro proliferation and differentiation of canine bone marrow derived mesenchymal stem cells over hydroxyl functionalized CNT substrates. *Biotechnol. Rep.* **2019**, *24*, e00387. [CrossRef] [PubMed]

101. Mao, H.; Kawazoe, N.; Chen, G. Uptake and intracellular distribution of collagen-functionalized single-walled carbon nanotubes. *Biomaterials* **2013**, *34*, 2472–2479. [CrossRef]

102. MacDonald, R.A.; Laurenzi, B.F.; Viswanathan, G.; Ajayan, P.M.; Stegemann, J.P. Collagen–carbon nanotube composite materials as scaffolds in tissue engineering. *J. Biomed. Mater. Res. Part A Off. J. Soc. Biomater. Jpn. Soc. Biomater. Aust. Soc. Biomater. Korean Soc. Biomater.* **2005**, *74*, 489–496. [CrossRef]

103. Dong, C.; Lv, Y. Application of collagen scaffold in tissue engineering: Recent advances and new perspectives. *Polymers* **2016**, *8*, 42. [CrossRef]

104. Chi, N.; Wang, R. Electrospun protein-CNT composite fibers and the application in fibroblast stimulation. *Biochem. Biophys. Res. Commun.* **2018**, *504*, 211–217. [CrossRef]

105. Ghosh, S.; Roy, P.; Lahiri, D. Enhanced neurogenic differentiation on anisotropically conductive carbon nanotube reinforced polycaprolactone-collagen scaffold by applying direct coupling electrical stimulation. *Int. J. Biol. Macromol.* **2022**, *218*, 269–284. [CrossRef]

106. Yu, H.; Zhao, H.; Huang, C.; Du, Y. Mechanically and Electrically Enhanced CNT–Collagen Hydrogels As Potential Scaffolds for Engineered Cardiac Constructs. *ACS Biomater. Sci. Eng.* **2017**, *3*, 3017–3021. [CrossRef]

107. Shavandi, A.; Silva, T.H.; Bekhit, A.A.; Bekhit, A.E.-D.A. Keratin: Dissolution, extraction and biomedical application. *Biomater. Sci.* **2017**, *5*, 1699–1735. [CrossRef]

108. Borrelli, M.; Joepen, N.; Reichl, S.; Finis, D.; Schoppe, M.; Geerling, G.; Schrader, S. Keratin films for ocular surface reconstruction: Evaluation of biocompatibility in an in-vivo model. *Biomaterials* **2015**, *42*, 112–120. [CrossRef]

109. Rahmany, M.B.; Hantgan, R.R.; Van Dyke, M. A mechanistic investigation of the effect of keratin-based hemostatic agents on coagulation. *Biomaterials* **2013**, *34*, 2492–2500. [CrossRef] [PubMed]

110. Aboushwareb, T.; Eberli, D.; Ward, C.; Broda, C.; Holcomb, J.; Atala, A.; Van Dyke, M. A keratin biomaterial gel hemostat derived from human hair: Evaluation in a rabbit model of lethal liver injury. *J. Biomed. Mater. Res. Part B Appl. Biomater.* **2009**, *90*, 45–54. [CrossRef] [PubMed]

111. Sierpinski, P.; Garrett, J.; Ma, J.; Apel, P.; Klorig, D.; Smith, T.; Koman, L.A.; Atala, A.; Van Dyke, M. The use of keratin biomaterials derived from human hair for the promotion of rapid regeneration of peripheral nerves. *Biomaterials* **2008**, *29*, 118–128. [CrossRef] [PubMed]

112. Apel, P.J.; Garrett, J.P.; Sierpinski, P.; Ma, J.; Atala, A.; Smith, T.L.; Koman, L.A.; Van Dyke, M.E. Peripheral nerve regeneration using a keratin-based scaffold: Long-term functional and histological outcomes in a mouse model. *J. Hand Surg.* **2008**, *33*, 1541–1547. [CrossRef]

113. Loschke, F.; Seltmann, K.; Bouameur, J.-E.; Magin, T.M. Regulation of keratin network organization. *Curr. Opin. Cell Biol.* **2015**, *32*, 56–64. [CrossRef]

114. Li, Y.B.; Liu, H.H.; Wang, X.C.; Zhang, X.X. Fabrication and performance of wool keratin–functionalized graphene oxide composite fibers. *Mater. Today Sustain.* **2019**, *3*, 100006. [CrossRef]

115. Song, Z.; Wang, J.; Tan, S.; Gao, J.; Wang, L. Conductive biomimetic bilayer fibrous scaffold for skin regeneration. *Colloids Surf. A Physicochem. Eng. Asp.* **2023**, *656*, 130211. [CrossRef]

116. Mahmoodi, M.; Haghighi, V.; Mirhaj, M.; Tavafoghi, M.; Shams, F.; Darabi, A. Highly osteogenic and mechanically strong nanofibrous scaffolds based on functionalized multi-walled carbon nanotubes-reinforced electrospun keratin/poly(ε-caprolactone). *Mater. Today Commun.* **2021**, *27*, 102401. [CrossRef]

117. Asl, M.A.; Karbasi, S.; Beigi-Boroujeni, S.; Benisi, S.Z.; Saeed, M. Polyhydroxybutyrate-starch/carbon nanotube electrospun nanocomposite: A highly potential scaffold for bone tissue engineering applications. *Int. J. Biol. Macromol.* **2022**, *223*, 524–542. [CrossRef]

118. Liu, B.; Song, Y.-W.; Jin, L.; Wang, Z.-J.; Pu, D.-Y.; Lin, S.-Q.; Zhou, C.; You, H.-J.; Ma, Y.; Li, J.-M. Silk structure and degradation. *Colloids Surf. B Biointerfaces* **2015**, *131*, 122–128. [CrossRef]

119. Römer, L.; Scheibel, T. The elaborate structure of spider silk: Structure and function of a natural high performance fiber. *Prion* **2008**, *2*, 154–161. [CrossRef]

120. Salehi, S.; Koeck, K.; Scheibel, T. Spider silk for tissue engineering applications. *Molecules* **2020**, *25*, 737. [CrossRef]

121. Altman, G.H.; Diaz, F.; Jakuba, C.; Calabro, T.; Horan, R.L.; Chen, J.; Lu, H.; Richmond, J.; Kaplan, D.L. Silk-based biomaterials. *Biomaterials* **2003**, *24*, 401–416. [CrossRef]

122. Izyan Syazana Mohd Yusoff, N.; Uzir Wahit, M.; Jaafar, J.; Wong, T.-W. Characterization of Graphene-Silk Fibroin Composites Film. *Mater. Today: Proc.* **2018**, *5*, 21853–21860. [CrossRef]

123. Qu, J.; Dai, M.; Ye, W.; Fang, Y.; Bian, D.; Su, W.; Li, F.; Sun, H.; Wei, J.; Li, B. Study on the effect of graphene oxide (GO) feeding on silk properties based on segmented precise measurement. *J. Mech. Behav. Biomed. Mater.* **2021**, *113*, 104147. [CrossRef]

124. Nalvuran, H.; Elçin, A.E.; Elçin, Y.M. Nanofibrous silk fibroin/reduced graphene oxide scaffolds for tissue engineering and cell culture applications. *Int. J. Biol. Macromol.* **2018**, *114*, 77–84. [CrossRef]

125. Dorishetty, P.; Balu, R.; Gelmi, A.; Mata, J.P.; Quigley, A.; Dutta, N.K.; Choudhury, N.R. Microporosity engineered printable silk/graphene hydrogels and their cytocompatibility evaluations. *Mater. Today Adv.* **2022**, *14*, 100233. [CrossRef]

126. Wang, L.; Lian, J.; Xia, Y.; Guo, Y.; Xu, C.; Zhang, Y.; Xu, J.; Zhang, X.; Li, B.; Zhao, B. A study on in vitro and in vivo bioactivity of silk fibroin/nano-hydroxyapatite/graphene oxide composite scaffolds with directional channels. *Colloids Surf. A Physicochem. Eng. Asp.* **2022**, *652*, 129886. [CrossRef]

127. Wang, L.; Lu, R.; Hou, J.; Nan, X.; Xia, Y.; Guo, Y.; Meng, K.; Xu, C.; Wang, X.; Zhao, B. Application of injectable silk fibroin/graphene oxide hydrogel combined with bone marrow mesenchymal stem cells in bone tissue engineering. *Colloids Surf. A Physicochem. Eng. Asp.* **2020**, *604*, 125318. [CrossRef]

128. Wang, S.-D.; Ma, Q.; Wang, K.; Ma, P.-B. Strong and biocompatible three-dimensional porous silk fibroin/graphene oxide scaffold prepared by phase separation. *Int. J. Biol. Macromol.* **2018**, *111*, 237–246. [CrossRef]

129. Aznar-Cervantes, S.; Pagán, A.; Martínez, J.G.; Bernabeu-Esclapez, A.; Otero, T.F.; Meseguer-Olmo, L.; Paredes, J.I.; Cenis, J.L. Electrospun silk fibroin scaffolds coated with reduced graphene promote neurite outgrowth of PC-12 cells under electrical stimulation. *Mater. Sci. Eng. C* **2017**, *79*, 315–325. [CrossRef] [PubMed]

130. Xu, Z.; Ma, Y.; Dai, H.; Tan, S.; Han, B. Advancements and Applications in the Composites of Silk Fibroin and Graphene-Based Materials. *Polymers* **2022**, *14*, 3110. [CrossRef]

131. López Barreiro, D.; Yeo, J.; Tarakanova, A.; Martin-Martinez, F.J.; Buehler, M.J. Multiscale Modeling of Silk and Silk-Based Biomaterials—A Review. *Macromol. Biosci.* **2019**, *19*, 1800253. [CrossRef] [PubMed]

132. Sun, J.; Shakya, S.; Gong, M.; Liu, G.; Wu, S.; Xiang, Z. Combined application of graphene-family materials and silk fibroin in biomedicine. *ChemistrySelect* **2019**, *4*, 5745–5754. [CrossRef]

133. Pathmanapan, S.; Periyathambi, P.; Anandasadagopan, S.K. Fibrin hydrogel incorporated with graphene oxide functionalized nanocomposite scaffolds for bone repair—In vitro and in vivo study. *Nanomed. Nanotechnol. Biol. Med.* **2020**, *29*, 102251. [CrossRef]

134. Wang, Q.; Chu, Y.; He, J.; Shao, W.; Zhou, Y.; Qi, K.; Wang, L.; Cui, S. A graded graphene oxide-hydroxyapatite/silk fibroin biomimetic scaffold for bone tissue engineering. *Mater. Sci. Eng. C* **2017**, *80*, 232–242. [CrossRef]

135. Eivazzadeh-Keihan, R.; Radinekiyan, F.; Madanchi, H.; Aliabadi, H.A.M.; Maleki, A. Graphene oxide/alginate/silk fibroin composite as a novel bionanostructure with improved blood compatibility, less toxicity and enhanced mechanical properties. *Carbohydr. Polym.* **2020**, *248*, 116802. [CrossRef]

136. Li, Y.; Chen, M.; Yan, J.; Zhou, W.; Gao, S.; Liu, S.; Li, Q.; Zheng, Y.; Cheng, Y.; Guo, Q. Tannic acid/Sr^{2+}-coated silk/graphene oxide-based meniscus scaffold with anti-inflammatory and anti-ROS functions for cartilage protection and delaying osteoarthritis. *Acta Biomater.* **2021**, *126*, 119–131. [CrossRef]

137. Magaz, A.; Li, X.; Gough, J.E.; Blaker, J.J. Graphene oxide and electroactive reduced graphene oxide-based composite fibrous scaffolds for engineering excitable nerve tissue. *Mater. Sci. Eng. C* **2021**, *119*, 111632. [CrossRef]

138. Zhang, C.; Fan, S.; Shao, H.; Hu, X.; Zhu, B.; Zhang, Y. Graphene trapped silk scaffolds integrate high conductivity and stability. *Carbon* **2019**, *148*, 16–27. [CrossRef]

139. Zhang, C.; Wang, X.; Fan, S.; Lan, P.; Cao, C.; Zhang, Y. Silk fibroin/reduced graphene oxide composite mats with enhanced mechanical properties and conductivity for tissue engineering. *Colloids Surf. B Biointerfaces* **2021**, *197*, 111444. [CrossRef]

140. Wang, L.; Song, D.; Zhang, X.; Ding, Z.; Kong, X.; Lu, Q.; Kaplan, D.L. Silk–Graphene Hybrid Hydrogels with Multiple Cues to Induce Nerve Cell Behavior. *ACS Biomater. Sci. Eng.* **2019**, *5*, 613–622. [CrossRef]

141. Yuan, Z.; Qin, Q.; Yuan, M.; Wang, H.; Li, R. Development and novel design of clustery graphene oxide formed Conductive Silk hydrogel cell vesicle to repair and routine care of myocardial infarction: Investigation of its biological activity for cell delivery applications. *J. Drug Deliv. Sci. Technol.* **2020**, *60*, 102001. [CrossRef]

142. Gandhi, M.; Yang, H.; Shor, L.; Ko, F. Post-spinning modification of electrospun nanofiber nanocomposite from Bombyx mori silk and carbon nanotubes. *Polymer* **2009**, *50*, 1918–1924. [CrossRef]

143. Shrestha, S.; Shrestha, B.K.; Lee, J.; Joong, O.K.; Kim, B.-S.; Park, C.H.; Kim, C.S. A conducting neural interface of polyurethane/silk-functionalized multiwall carbon nanotubes with enhanced mechanical strength for neuroregeneration. *Mater. Sci. Eng. C* **2019**, *102*, 511–523. [CrossRef]

144. Wang, J.; Li, X.; Song, Y.; Su, Q.; Xiaohalati, X.; Yang, W.; Xu, L.; Cai, B.; Wang, G.; Wang, Z.; et al. Injectable silk sericin scaffolds with programmable shape-memory property and neuro-differentiation-promoting activity for individualized brain repair of severe ischemic stroke. *Bioact. Mater.* **2021**, *6*, 1988–1999. [CrossRef]

145. Shalumon, K.; Binulal, N.; Selvamurugan, N.; Nair, S.; Menon, D.; Furuike, T.; Tamura, H.; Jayakumar, R. Electrospinning of carboxymethyl chitin/poly (vinyl alcohol) nanofibrous scaffolds for tissue engineering applications. *Carbohydr. Polym.* **2009**, *77*, 863–869. [CrossRef]

146. Elieh-Ali-Komi, D.; Hamblin, M.R. Chitin and chitosan: Production and application of versatile biomedical nanomaterials. *Int. J. Adv. Res.* **2016**, *4*, 411.

147. Valencia, A.M.; Valencia, C.H.; Zuluaga, F.; Grande-Tovar, C.D. Synthesis and fabrication of films including graphene oxide functionalized with chitosan for regenerative medicine applications. *Heliyon* **2021**, *7*, e07058. [CrossRef]

148. Yılmaz, P.; Öztürk Er, E.; Bakırdere, S.; Ülgen, K.; Özbek, B. Application of supercritical gel drying method on fabrication of mechanically improved and biologically safe three-component scaffold composed of graphene oxide/chitosan/hydroxyapatite and characterization studies. *J. Mater. Res. Technol.* **2019**, *8*, 5201–5216. [CrossRef]

149. Depan, D.; Girase, B.; Shah, J.S.; Misra, R.D.K. Structure–process–property relationship of the polar graphene oxide-mediated cellular response and stimulated growth of osteoblasts on hybrid chitosan network structure nanocomposite scaffolds. *Acta Biomater.* **2011**, *7*, 3432–3445. [CrossRef] [PubMed]

150. Liu, Y.; Fang, N.; Liu, B.; Song, L.; Wen, B.; Yang, D. Aligned porous chitosan/graphene oxide scaffold for bone tissue engineering. *Mater. Lett.* **2018**, *233*, 78–81. [CrossRef]

151. Prakash, J.; Prema, D.; Venkataprasanna, K.S.; Balagangadharan, K.; Selvamurugan, N.; Venkatasubbu, G.D. Nanocomposite chitosan film containing graphene oxide/hydroxyapatite/gold for bone tissue engineering. *Int. J. Biol. Macromol.* **2020**, *154*, 62–71. [CrossRef]

152. Sivashankari, P.R.; Prabaharan, M. Three-dimensional porous scaffolds based on agarose/chitosan/graphene oxide composite for tissue engineering. *Int. J. Biol. Macromol.* **2020**, *146*, 222–231. [CrossRef] [PubMed]

153. Souza, A.P.C.; Neves, J.G.; Navarro da Rocha, D.; Lopes, C.C.; Moraes, Â.M.; Correr-Sobrinho, L.; Correr, A.B. Chitosan/Xanthan membrane containing hydroxyapatite/Graphene oxide nanocomposite for guided bone regeneration. *J. Mech. Behav. Biomed. Mater.* **2022**, *136*, 105464. [CrossRef]

154. Shamekhi, M.A.; Mirzadeh, H.; Mahdavi, H.; Rabiee, A.; Mohebbi-Kalhori, D.; Baghaban Eslaminejad, M. Graphene oxide containing chitosan scaffolds for cartilage tissue engineering. *Int. J. Biol. Macromol.* **2019**, *127*, 396–405. [CrossRef]

155. Cao, L.; Zhang, F.; Wang, Q.; Wu, X. Fabrication of chitosan/graphene oxide polymer nanofiber and its biocompatibility for cartilage tissue engineering. *Mater. Sci. Eng. C* **2017**, *79*, 697–701. [CrossRef]

156. Amiryaghoubi, N.; Noroozi Pesyan, N.; Fathi, M.; Omidi, Y. Injectable thermosensitive hybrid hydrogel containing graphene oxide and chitosan as dental pulp stem cells scaffold for bone tissue engineering. *Int. J. Biol. Macromol.* **2020**, *162*, 1338–1357. [CrossRef]

157. Feng, W.; Wang, Z. Shear-thinning and self-healing chitosan-graphene oxide hydrogel for hemostasis and wound healing. *Carbohydr. Polym.* **2022**, *294*, 119824. [CrossRef]

158. Gupta, P.; Agrawal, A.; Murali, K.; Varshney, R.; Beniwal, S.; Manhas, S.; Roy, P.; Lahiri, D. Differential neural cell adhesion and neurite outgrowth on carbon nanotube and graphene reinforced polymeric scaffolds. *Mater. Sci. Eng. C* **2019**, *97*, 539–551. [CrossRef]

159. Garnica-Palafox, I.M.; Estrella-Monroy, H.O.; Vázquez-Torres, N.A.; Álvarez-Camacho, M.; Castell-Rodríguez, A.E.; Sánchez-Arévalo, F.M. Influence of multi-walled carbon nanotubes on the physico-chemical and biological responses of chitosan-based hybrid hydrogels. *Carbohydr. Polym.* **2020**, *236*, 115971. [CrossRef]

160. Ali, A.; Bano, S.; Priyadarshi, R.; Negi, Y.S. Effect of carbon based fillers on properties of Chitosan/PVA/βTCP based composite scaffold for bone tissue engineering. *Mater. Today Proc.* **2019**, *15*, 173–182. [CrossRef]

161. Venkatesan, J.; Qian, Z.-J.; Ryu, B.; Ashok Kumar, N.; Kim, S.-K. Preparation and characterization of carbon nanotube-grafted-chitosan—Natural hydroxyapatite composite for bone tissue engineering. *Carbohydr. Polym.* **2011**, *83*, 569–577. [CrossRef]

162. Türk, S.; Altınsoy, I.; Çelebi Efe, G.; Ipek, M.; Özacar, M.; Bindal, C. 3D porous collagen/functionalized multiwalled carbon nanotube/chitosan/hydroxyapatite composite scaffolds for bone tissue engineering. *Mater. Sci. Eng. C* **2018**, *92*, 757–768. [CrossRef]

163. Venkatesan, J.; Ryu, B.; Sudha, P.N.; Kim, S.-K. Preparation and characterization of chitosan–carbon nanotube scaffolds for bone tissue engineering. *Int. J. Biol. Macromol.* **2012**, *50*, 393–402. [CrossRef]

164. Gholizadeh, S.; Moztarzadeh, F.; Haghighipour, N.; Ghazizadeh, L.; Baghbani, F.; Shokrgozar, M.A.; Allahyari, Z. Preparation and characterization of novel functionalized multiwalled carbon nanotubes/chitosan/β-Glycerophosphate scaffolds for bone tissue engineering. *Int. J. Biol. Macromol.* **2017**, *97*, 365–372. [CrossRef]

165. Mirmusavi, M.H.; Ahmadian, M.; Karbasi, S. Polycaprolactone-chitosan/multi-walled carbon nanotube: A highly strengthened electrospun nanocomposite scaffold for cartilage tissue engineering. *Int. J. Biol. Macromol.* **2022**, *209*, 1801–1814. [CrossRef]

166. Kretz, F.T.; Limberger, I.-J.; Auffarth, G.U. Corneal endothelial cell coating during phacoemulsification using a new dispersive hyaluronic acid ophthalmic viscosurgical device. *J. Cataract Refract. Surg.* **2014**, *40*, 1879–1884. [CrossRef]

167. Edmonds, M.; Bates, M.; Doxford, M.; Gough, A.; Foster, A. New treatments in ulcer healing and wound infection. *Diabetes Metab. Res. Rev.* **2000**, *16*, S51–S54. [CrossRef]

168. Namazi, M.R.; Fallahzadeh, M.K.; Schwartz, R.A. Strategies for prevention of scars: What can we learn from fetal skin? *Int. J. Dermatol.* **2011**, *50*, 85–93. [CrossRef]

169. Monheit, G.D.; Coleman, K.M. Hyaluronic acid fillers. *Dermatol. Ther.* **2006**, *19*, 141–150. [CrossRef] [PubMed]

170. Chhetri, D.K.; Mendelsohn, A.H. Hyaluronic acid for the treatment of vocal fold scars. *Curr. Opin. Otolaryngol. Head Neck Surg.* **2010**, *18*, 498–502. [CrossRef] [PubMed]

171. Ishida, O.; Tanaka, Y.; Morimoto, I.; Takigawa, M.; Eto, S. Chondrocytes are regulated by cellular adhesion through CD44 and hyaluronic acid pathway. *J. Bone Miner. Res.* **1997**, *12*, 1657–1663. [CrossRef] [PubMed]

172. Gotoh, S.; Onaya, J.; Abe, M.; Miyazaki, K.; Hamai, A.; Horie, K.; Tokuyasu, K. Effects of the molecular weight of hyaluronic acid and its action mechanisms on experimental joint pain in rats. *Ann. Rheum. Dis.* **1993**, *52*, 817–822. [CrossRef]

173. Huang, G.; Huang, H. Application of hyaluronic acid as carriers in drug delivery. *Drug Deliv.* **2018**, *25*, 766–772. [CrossRef]

174. Lee, S.J.; Nah, H.; Heo, D.N.; Kim, K.-H.; Seok, J.M.; Heo, M.; Moon, H.-J.; Lee, D.; Lee, J.S.; An, S.Y.; et al. Induction of osteogenic differentiation in a rat calvarial bone defect model using an In situ forming graphene oxide incorporated glycol chitosan/oxidized hyaluronic acid injectable hydrogel. *Carbon* **2020**, *168*, 264–277. [CrossRef]

175. Rajan Unnithan, A.; Ramachandra Kurup Sasikala, A.; Park, C.H.; Kim, C.S. A unique scaffold for bone tissue engineering: An osteogenic combination of graphene oxide–hyaluronic acid–chitosan with simvastatin. *J. Ind. Eng. Chem.* **2017**, *46*, 182–191. [CrossRef]

176. Patil, R.; Kansara, V.; Ray, D.; Aswal, V.K.; Jha, P.K.; Bahadur, P.; Tiwari, S. Slow degrading hyaluronic acid hydrogel reinforced with cationized graphene nanosheets. *Int. J. Biol. Macromol.* **2019**, *141*, 232–239. [CrossRef]

177. Umar Aslam Khan, M.; Haider, S.; Haider, A.; Izwan Abd Razak, S.; Rafiq Abdul Kadir, M.; Shah, S.A.; Javed, A.; Shakir, I.; Al-Zahrani, A.A. Development of porous, antibacterial and biocompatible GO/n-HAp/bacterial cellulose/β-glucan biocomposite scaffold for bone tissue engineering. *Arab. J. Chem.* **2021**, *14*, 102924. [CrossRef]

178. Steel, E.M.; Azar, J.-Y.; Sundararaghavan, H.G. Electrospun hyaluronic acid-carbon nanotube nanofibers for neural engineering. *Materialia* **2020**, *9*, 100581. [CrossRef]

179. Kalia, S.; Boufi, S.; Celli, A.; Kango, S. Nanofibrillated cellulose: Surface modification and potential applications. *Colloid Polym. Sci.* **2014**, *292*, 5–31. [CrossRef]

180. Jedvert, K.; Heinze, T. Cellulose modification and shaping—A review. *J. Polym. Eng.* **2017**, *37*, 845–860. [CrossRef]

181. Seddiqi, H.; Oliaei, E.; Honarkar, H.; Jin, J.; Geonzon, L.C.; Bacabac, R.G.; Klein-Nulend, J. Cellulose and its derivatives: Towards biomedical applications. *Cellulose* **2021**, *28*, 1893–1931. [CrossRef]

182. Tavakolian, M.; Jafari, S.M.; van de Ven, T.G. A review on surface-functionalized cellulosic nanostructures as biocompatible antibacterial materials. *Nano Micro Lett.* **2020**, *12*, 1–23. [CrossRef]

183. Mao, D.; Li, Q.; Bai, N.; Dong, H.; Li, D. Porous stable poly(lactic acid)/ethyl cellulose/hydroxyapatite composite scaffolds prepared by a combined method for bone regeneration. *Carbohydr. Polym.* **2018**, *180*, 104–111. [CrossRef]

184. Luo, H.; Ao, H.; Peng, M.; Yao, F.; Yang, Z.; Wan, Y. Effect of highly dispersed graphene and graphene oxide in 3D nanofibrous bacterial cellulose scaffold on cell responses: A comparative study. *Mater. Chem. Phys.* **2019**, *235*, 121774. [CrossRef]

185. Li, J.; Liu, X.; Crook, J.M.; Wallace, G.G. Electrical stimulation-induced osteogenesis of human adipose derived stem cells using a conductive graphene-cellulose scaffold. *Mater. Sci. Eng. C* **2020**, *107*, 110312. [CrossRef]

186. Liu, X.; Shen, H.; Song, S.; Chen, W.; Zhang, Z. Accelerated biomineralization of graphene oxide-incorporated cellulose acetate nanofibrous scaffolds for mesenchymal stem cell osteogenesis. *Colloids Surf. B Biointerfaces* **2017**, *159*, 251–258. [CrossRef]

187. Guo, R.; Li, J.; Chen, C.; Xiao, M.; Liao, M.; Hu, Y.; Liu, Y.; Li, D.; Zou, J.; Sun, D.; et al. Biomimetic 3D bacterial cellulose-graphene foam hybrid scaffold regulates neural stem cell proliferation and differentiation. *Colloids Surf. B Biointerfaces* **2021**, *200*, 111590. [CrossRef]

188. Aly, A.A.; Ahmed, M.K. Nanofibers of cellulose acetate containing ZnO nanoparticles/graphene oxide for wound healing applications. *Int. J. Pharm.* **2021**, *598*, 120325. [CrossRef]

189. Soliman, M.; Sadek, A.A.; Abdelhamid, H.N.; Hussein, K. Graphene oxide-cellulose nanocomposite accelerates skin wound healing. *Res. Vet. Sci.* **2021**, *137*, 262–273. [CrossRef] [PubMed]

190. Khalid, A.; Madni, A.; Raza, B.; Islam, M.U.; Hassan, A.; Ahmad, F.; Ali, H.; Khan, T.; Wahid, F. Multiwalled carbon nanotubes functionalized bacterial cellulose as an efficient healing material for diabetic wounds. *Int. J. Biol. Macromol.* **2022**, *203*, 256–267. [CrossRef] [PubMed]

191. Mamidi, N.; Delgadillo, R.M.V.; González-Ortiz, A. Engineering of carbon nano-onion bioconjugates for biomedical applications. *Mater. Sci. Eng. C* **2021**, *120*, 111698. [CrossRef] [PubMed]

192. Mamidi, N.; González-Ortiz, A.; Lopez Romo, I.; Barrera, E.V. Development of Functionalized Carbon Nano-Onions Reinforced Zein Protein Hydrogel Interfaces for Controlled Drug Release. *Pharmaceutics* **2019**, *11*, 621. [CrossRef]

193. Mamidi, N.; Villela Castrejón, J.; González-Ortiz, A. Rational design and engineering of carbon nano-onions reinforced natural protein nanocomposite hydrogels for biomedical applications. *J. Mech. Behav. Biomed. Mater.* **2020**, *104*, 103696. [CrossRef]

194. Mamidi, N.; Velasco Delgadillo, R.M.; Gonzáles Ortiz, A.; Barrera, E.V. Carbon Nano-Onions Reinforced Multilayered Thin Film System for Stimuli-Responsive Drug Release. *Pharmaceutics* **2020**, *12*, 1208. [CrossRef]

195. Diabb Zavala, J.M.; Leija Gutiérrez, H.M.; Segura-Cárdenas, E.; Mamidi, N.; Morales-Avalos, R.; Villela-Castrejón, J.; Elías-Zúñiga, A. Manufacture and mechanical properties of knee implants using SWCNTs/UHMWPE composites. *J. Mech. Behav. Biomed. Mater.* **2021**, *120*, 104554. [CrossRef]

196. Choudhury, D.; Ranuša, M.; Fleming, R.A.; Vrbka, M.; Křupka, I.; Teeter, M.G.; Goss, J.; Zou, M. Mechanical wear and oxidative degradation analysis of retrieved ultra high molecular weight polyethylene acetabular cups. *J. Mech. Behav. Biomed. Mater.* **2018**, *79*, 314–323. [CrossRef]

197. Diabb, J.; Juarez-Hernandez, A.; Reyes, A.; González-Rivera, C.; Hernandez-Rodriguez, M. Failure analysis for degradation of a polyethylene knee prosthesis component. *Eng. Fail. Anal.* **2009**, *5*, 1770–1773. [CrossRef]

198. Sell, S.A.; McClure, M.J.; Garg, K.; Wolfe, P.S.; Bowlin, G.L. Electrospinning of collagen/biopolymers for regenerative medicine and cardiovascular tissue engineering. *Adv. Drug Deliv. Rev.* **2009**, *61*, 1007–1019. [CrossRef]

199. Rebelo, R.; Fernandes, M.; Fangueiro, R. Biopolymers in medical implants: A brief review. *Procedia Eng.* **2017**, *200*, 236–243. [CrossRef]

200. Paradise, M.; Goswami, T. Carbon nanotubes—production and industrial applications. *Mater. Des.* **2007**, *28*, 1477–1489. [CrossRef]

201. Phiri, J.; Gane, P.; Maloney, T.C. General overview of graphene: Production, properties and application in polymer composites. *Mater. Sci. Eng. B* **2017**, *215*, 9–28. [CrossRef]

202. Kauling, A.P.; Seefeldt, A.T.; Pisoni, D.P.; Pradeep, R.C.; Bentini, R.; Oliveira, R.V.; Novoselov, K.S.; Castro Neto, A.H. The worldwide graphene flake production. *Adv. Mater.* **2018**, *30*, 1803784. [CrossRef]

203. Guo, Y.; Jiang, S.; Grena, B.J.; Kimbrough, I.F.; Thompson, E.G.; Fink, Y.; Sontheimer, H.; Yoshinobu, T.; Jia, X. Polymer composite with carbon nanofibers aligned during thermal drawing as a microelectrode for chronic neural interfaces. *Acs Nano* **2017**, *11*, 6574–6585. [CrossRef]

204. Fattahi, P.; Yang, G.; Kim, G.; Abidian, M.R. A review of organic and inorganic biomaterials for neural interfaces. *Adv. Mater.* **2014**, *26*, 1846–1885. [CrossRef]

205. Kim, M.; Jang, J.; Cha, C. Carbon nanomaterials as versatile platforms for theranostic applications. *Drug Discov. Today* **2017**, *22*, 1430–1437. [CrossRef]

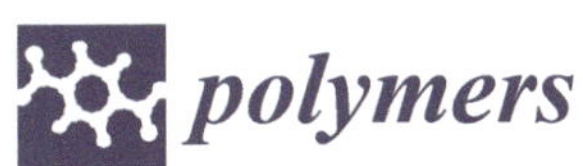

polymers

Article

Development of Conductive Gelatine-Methacrylate Inks for Two-Photon Polymerisation

Paola Sanjuan-Alberte [1,2,*,†] , Jayasheelan Vaithilingam [3,†], Jonathan C. Moore [4] , Ricky D. Wildman [3], Christopher J. Tuck [3], Morgan R. Alexander [5], Richard J. M. Hague [3] and Frankie J. Rawson [1,*]

1 Regenerative Medicine and Cellular Therapies, Biodiscovery Institute, School of Pharmacy, University of Nottingham, University Park, Nottingham NG7 2RD, UK
2 Department of Bioengineering and Institute for Bioengineering and Biosciences, Instituto Superior Técnico, Universidade de Lisboa, Av. Rovisco Pais, 1049-001 Lisboa, Portugal
3 Centre for Additive Manufacturing, Faculty of Engineering, University of Nottingham, University Park, Nottingham NG7 2RD, UK; ezajv4@exmail.nottingham.ac.uk (J.V.); ezzrdw@exmail.nottingham.ac.uk (R.D.W.); ezzcjt@exmail.nottingham.ac.uk (C.J.T.); ezzrjh@exmail.nottingham.ac.uk (R.J.M.H.)
4 School of Chemistry, University of Nottingham, University Park, Nottingham NG7 2RD, UK; j.moore@nottingham.ac.uk
5 Advanced Materials and Healthcare Technologies, School of Pharmacy, University of Nottingham, University Park, Nottingham NG7 2RD, UK; pazma1@exmail.nottingham.ac.uk
* Correspondence: Paola.Alberte@tecnico.ulisboa.pt (P.S.-A.); Frankie.Rawson@nottingham.ac.uk (F.J.R.)
† These authors have contributed equally.

Citation: Sanjuan-Alberte, P.; Vaithilingam, J.; Moore, J.C.; Wildman, R.D.; Tuck, C.J.; Alexander, M.R.; Hague, R.J.M.; Rawson, F.J. Development of Conductive Gelatine-Methacrylate Inks for Two-Photon Polymerisation. *Polymers* **2021**, *13*, 1038. https://doi.org/10.3390/polym13071038

Academic Editor: Ilaria Armentano

Received: 24 February 2021
Accepted: 23 March 2021
Published: 26 March 2021

Publisher's Note: MDPI stays neutral with regard to jurisdictional claims in published maps and institutional affiliations.

Abstract: Conductive hydrogel-based materials are attracting considerable interest for bioelectronic applications due to their ability to act as more compatible soft interfaces between biological and electrical systems. Despite significant advances that are being achieved in the manufacture of hydrogels, precise control over the topographies and architectures remains challenging. In this work, we present for the first time a strategy to manufacture structures with resolutions in the micro-/nanoscale based on hydrogels with enhanced electrical properties. Gelatine methacrylate (GelMa)-based inks were formulated for two-photon polymerisation (2PP). The electrical properties of this material were improved, compared to pristine GelMa, by dispersion of multi-walled carbon nanotubes (MWCNTs) acting as conductive nanofillers, which was confirmed by electrochemical impedance spectroscopy and cyclic voltammetry. This material was also confirmed to support human induced pluripotent stem cell-derived cardiomyocyte (hPSC-CMs) viability and growth. Ultra-thin film structures of 10 μm thickness and scaffolds were manufactured by 2PP, demonstrating the potential of this method in areas spanning tissue engineering and bioelectronics. Though further developments in the instrumentation are required to manufacture more complex structures, this work presents an innovative approach to the manufacture of conductive hydrogels in extremely low resolution.

Keywords: conductive hydrogels; GelMa; carbon nanotubes; two-photon polymerization; nanofabrication

1. Introduction

Conductive hydrogels are gaining significant interest from the bioelectronics community as they can act as soft interfaces between biological and electronic systems, offering an alternative to traditional inorganic materials [1].

Hydrogels are highly hydrated polymeric networks with water contents similar to that of soft tissues [2]. To mimic in vivo conditions more accurately, it is common to develop three-dimensional (3D) cellular scaffolds based on hydrogels for regenerative medicine and tissue engineering applications. This is due to their biocompatibility and because they can create porous, soft and elastic interfaces [3]. Furthermore, the mechanical properties of hydrogels are comparable to biological tissues. Some examples of biomimetic hydrogel scaffolds with therapeutic applications include alginate and collagen-based hydrogels for applications in wound healing, cartilage repair, bone regeneration or drug delivery [4,5].

Strategies to enhance the electrical properties of hydrogels include the integration of conductive polymers and conductive nano-fillers into the hydrogel matrix [6,7]. Some examples include the incorporation of poly(3,4-ethylenedioxythiophene) (PEDOT): polystyrene sulfonate (PSS) to polyethylene glycol diacrylate (PEGDA) for neural tissue engineering applications [8] and the dispersion of gold nanorods into gelatine for the engineering of cardiac constructs [9]. The overall conductivity of these materials combines the electronic functionality of the conductive elements and the ionic contribution from inter- and intra-chain charge transfer throughout the matrix/networks [10].

Cells can respond to different types of external stimuli, including the chemical composition of their surroundings [11], stiffness [12], topography [13] or electrical stimulation [14]. Therefore, controlling and combining these different aspects is important to create biomimetic architectures with the capacity to promote tissue organisation and response after injury. Additive manufacturing (AM) techniques, notably inkjet-based 3D printing, extrusion printing and vat polymerisation AM techniques offer the possibility of creating structures that can be easily tuned and fitted for a specific application [15–17]. In particular, two-photon polymerisation (2PP) holds great potential in the engineering of the local cellular microenvironments due to the high resolution (<100 nm) of printed structures that can be achieved with this technique [18,19].

Photo-polymerisation of hydrogels leads to the formation of a polymeric network with high water content [20]. Photopolymerisable formulations often consist of a photoinitiator, cross-linkers and a solvent [20]. Most of the hydrogel inks developed for 2PP are based on poly(ethylene glycol) diacrylate (PEGDA) [21–23], bovine serum albumin (BSA) [24] or gelatine-methacrylate (GelMa) [25–27]. Despite the biochemical and biomechanical control that can be achieved using these materials, there are no known reports of conductive gels developed for 2PP being investigated for increased electrical properties.

In previous work, we have demonstrated the ability of multi-walled carbon nanotubes (MWNCTs) to act as conductive nano-fillers of 2PP-processed polymeric structures based on pentaerythritol triacrylate (PETrA) [18]. Others have also reported the use of MWCNTs to fabricate 3D conductive structures by 2PP based on thiol-acrylate composites [28] and the acrylate monomer R712 [29,30]. However, this has not been attempted with hydrogels.

For the first time, this work presents a strategy to manufacture 3D microstructures by 2PP with enhanced electrical properties for potential applications in electronics and bioelectronics based on conductive hydrogels. To achieve this, a conductive photocurable ink was formulated by dispersing MWCNTs into a GelMa solution incorporating a water-soluble photoinitiator. Electrical properties of the hydrogels were assessed by impedance spectroscopy and cyclic voltammetry, demonstrating an enhanced electrical conductivity and electron transfer of the MWCNTs-modified GelMa (MWCNTs-GelMa). This material was also compatible with human pluripotent stem cell-derived cardiomyocytes (hPSC-CMs), demonstrating its potential in tissue engineering and regenerative medicine. The last part of the work presented herein was to develop micron-sized biomimetic scaffolds by 2PP using the conductive hydrogel inks.

2. Materials and Methods

2.1. GelMa Preparation

Gelatine from porcine skin (Sigma Aldrich, St. Louis, MO, USA) was added to warm Dulbecco's phosphate buffer saline (DPBS) (Thermo Fischer, Waltham, MA, USA) at a concentration of 100 mg mL^{-1} and stirred until fully dissolved at 50 °C. When the gelatine solution was clear, methacrylic anhydride (MAA, Sigma Aldrich) at a concentration of 8% (v/v) was added dropwise. After this, the solution was left stirring for 3 more h at 50 °C. The solution was then diluted 1:5 in DBPS to reach a final gelatine concentration of 20 mg mL^{-1}. Gelatine methacrylate (GelMa) solution was placed in 2000 MWCO dialysis tubes

(Sigma Aldrich) and dialysed in ultrapure water at 45–50 °C. Water was changed three times a day for 1 week. Once GelMa solution was dialysed, it was placed at −80 °C for 24 h, lyophilised and stored until further used.

2.2. Ink Development

Sodium 3,3′-((((1E,1′E)-(2-oxocyclopentane-1,3-diylidene) bis (methanylylidene))bis(4,1-phenylene))bis(methylazanediyl)) dipropanoate (P2CK) was used as photoinitiator. Preparation of P2CK was performed following previous protocols [31]. Lyophilised GelMa was reconstituted in DPBS at a concentration of 5% (*w/v*) at 50 °C for 10 min. In solutions containing carboxyl-functionalised multi-walled carbon nanotubes (MWCNTs, OD 15 ± 5 nm, length 1–5 μm, Nano-Lab), a concentration of 5 mg mL^{-1} MWCNTs was used and sonicated for 1 hour, centrifuged at 40 °C for 30 min at 6000 rpm and used as a stock solution. Selected amounts of this solution were added to 5% GelMa solutions containing 0.5% w/v P2CK to make the different inks.

For SEM imaging, materials were lyophilised using the previous procedure, placed in a titanium pin stub with a carbon tab and coated with a thin platinum layer before SEM imaging in a field-emission scanning electron microscope (FEG-SEM JOEL 7100F) in secondary electron mode with an acceleration voltage of 5 kV and a working distance of 10 mm.

For cell tests, GelMa and MWCNTs-GelMa solutions were added to a 96-well plate and UV-cross-linked using a LED lamp (25 mm × 10 mm FireFly UV LED lamp from Phoseon Technology, Hillsboro, OR, USA) with a wavelength of 385 nm for 3 min. This lamp consists of an array of LEDs with a total power density of 4 W cm^{-2}. The distance between the LED lamp and the 96-Well plate was 1 cm.

Gels used in subsequent experiments with cells were washed in a 1% antibiotic solution in DBPS (penicillin/streptomycin solution 10,000 units penicillin 10 mg streptomycin/mL; Sigma Aldrich) for 1 h and thoroughly washed in PBS. Samples were sealed with parafilm and stored hydrated in the fridge when not immediately used.

2.3. NMR Characterisation

The degree of conversion was quantified by characterising the samples using ^{1}H Nuclear Magnetic Resonance at 400 MHz (NMR, DPX UltraShield 400 MHz, Bruker UK ltd, Coventry, UK). A sample of lyophilised gelatine and 2 batches of GelMa (GB1 and GB2) prepared separately were used for this characterisation. Characterisation samples were prepared using deuterium oxide as a solvent. The area under the peak of lysine methylene proton of GelMa (X) and the lysine methylene proton of gelatine (Y) was used to compute the degree of conversion as follows:

$$Degree\ of\ conversion\ (DoC) = \left[1 - \left(\frac{X}{Y}\right)\right] \times 100. \qquad (1)$$

2.4. Electrochemical Characterisation of Hydrogels

Electrochemical impedance spectroscopy was performed on a PGSTAT potentiostat including a FRA32M module (Metrohm Autolab, Utrecht, The Netherlands) and interfaced with a personal computer including the NovaLab software. Briefly, 1 cm diameter gels were photopolymerised in situ and sandwiched between two gold electrodes (Georg-Albert PVD, Silz, Germany) in a two-electrode configuration at room temperature. Frequencies ranging from 10 to 1,000,000 Hz were applied. For this, 1 mg mL^{-1} MWCNTs-GelMa hydrogels were used. Each measurement was performed in triplicate.

Cyclic voltammetry was performed in the aforementioned instrument. Gels were formed in situ on a gold electrode connected to an electrochemical cell and used as a working electrode. A silver/silver chloride electrode was used as a reference electrode and a platinum wire was used to close the circuit. Potentials were applied from 0.8 to −0.2 V at a scan rate of 100 mV s^{-1}. Phosphate buffered saline (PBS) and 1 mM potassium hexacyanoferrate (II) (Sigma-Aldrich) in PBS solutions were used as supporting electrolyte and redox-active species, respectively. For this, 1 mg mL^{-1} MWCNTs-GelMa hydrogels were used. Each measurement was performed in triplicate.

2.5. hPSC-CMs Differentiation

A REBL-PAT hPSC line was derived from a skin punch biopsy from a male subject. Procedures of isolation, culture, differentiation and dissociation are described elsewhere [32]. Early hPSC-CMs were seeded on the photo-cured gels. Dissociation of cells took place 6–8 weeks after the differentiation process and cells were seeded on hydrogels at an approximate concentration of 2 million cells mL^{-1} in basal RPMI medium (Life Technologies #11875093) supplemented with B27 (Life Technologies #17504044, Carlsbad, CA, USA), Y-27632 ROCK inhibitor (20 µM; Tocris #1254) and 10% foetal bovine serum (Sigma-Aldrich). The medium was changed after 24 h to RPMI/B27 medium.

2.6. hPSC-CMs Viability and Immunostaining

Viability studies were performed in gels with different GelMa concentrations (5 and 10 wt. %) and different MWCNTs concentrations of 0.5 and 1 mg mL^{-1}. hPSC-CM were washed in complete DPBS and incubated for 30 min in 1 µM acetoxymethyl (AM) calcein solution (Sigma Aldrich #C1359) in DPBS to stain viable cells and dead cells were stained with 5 µM ethidium homodimer I (Sigma Aldrich #E1903) in PBS. Fluorescence images were taken on an inverted EVOS FL Fluorescence microscope. Statistical significance was calculated using a *t*-test (N = 3, *n* = 2; ±SD)

Immunostaining of hPSC-CMs was performed by fixing cells in 2% formaldehyde (VWR) and permeabilised with 0.1% Triton-X 100 (Sigma-Aldrich). Non-specific binding was blocked with 4% foetal bovine serum (FBS) (Sigma-Aldrich) in DPBS for 1 h. Samples were immunostained with a monoclonal primary antibody against sarcomeric alpha-actinin produced in mouse (Abcam #ab9465, Cambridge, UK) at a concentration of 1:400 in DPBS and incubated overnight at 4 °C. A solution of 0.05% of Tween 20 was used to wash the samples and a solution of goat anti-mouse secondary antibody IgG was added (1:1000; Abcam # ab6785) and incubated for 2 h. Samples were washed with a washing solution and exposed to Hoechst 33258 (5 µg mL^{-1}; Sigma Aldrich) and phalloidin (1:150, Thermo Fischer Scientific) for 2 h. Fluorescence images were taken on an inverted EVOS FL Fluorescence microscope.

2.7. Two-Photon Polymerisation of Hydrogels

A Nanoscribe Photonic Professional GT fitted with a fibre laser was used for the fabrication following similar protocols developed previously [27]. The wavelength of the laser is 780 nm with a pulse frequency of 80 MHz and a pulse duration of 120 fs. An oil immersion objective was used for the printing (63×, NA = 1.4, WD = 190 µm). The laser power used for the work varied from 30 to 100 mW and the scan speed varied from 500 to 50,000 µm/s for printing condition optimisation. The whole printing and developing process was carried out in a UV-free environment.

Briefly, 0.5 wt% of the photoinitiator P2CK was added to MWCNTs-GelMa and dissolved by magnetic stirring for 15 min before printing in a water bath (40 °C). Droplets of GelMa and MWCNTs-GelMa inks were placed in glass coverslips previously cleaned with ethanol and dried. These were then placed in the equipment printing stage. The optimised process condition used for printing pristine GelMa was 90 mW laser power and 10,000 µm/s print speed. The hatch spacing was 0.2 and 0.4 µm layer thickness. For the MWCNTs-GelMa, the print parameters were 40 mW laser power, 10,000 µm/s print speed and 0.2 and 0.3 µm slicing. The laser power for MWCNTs containing GelMa had to be significantly reduced to prevent boiling of the material due to high laser energy absorption and conduction within the material.

After fabrication of the different structures, excess material was gently removed by immersing the coverslip in a beaker containing DPBS maintained at 40 °C. Upon developing, the samples were stored in DPBS to prevent dehydration. Printed samples were wet-imaged using an optical microscope with the sample immersed in DPBS. For SEM imaging, a Hitachi TableTop SEM was used. Before imaging, the sample was dehydrated by freeze drying.

3. Results and Discussion

Gelatine was chemically modified using MAA in order to introduce methacrylic groups to the gelatine backbone with the potential to be UV-cross-linked. MAA binds to amine groups present in the amino acid lysine (Lys) (Figure 1a). Once this reaction was completed, solutions were dialysed to eliminate unreacted MAA. To induce the photo-cross-linking of the GelMa, P2CK is added to the solution, which then gelates the material after exposure with the 2PP laser (Figure 1b).

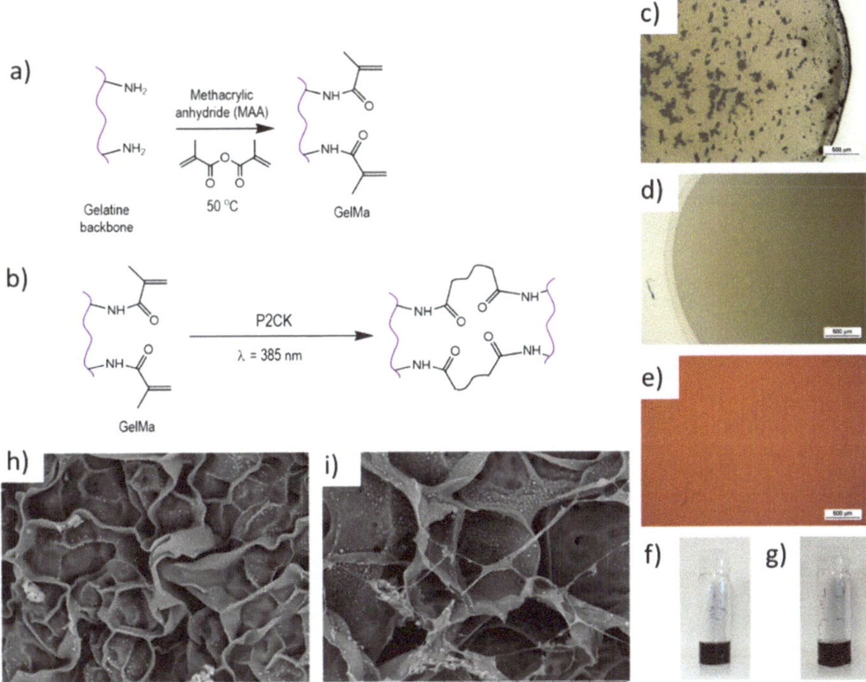

Figure 1. Mechanisms of (**a**) modification of gelatine with methacrylate groups to obtain gelatine methacrylate (GelMa) and (**b**) photo-polymerisation of GelMa groups to obtain a cross-linked gel. GelMa pre-polymer hydrogel (5 wt. %) with 1 mg mL^{-1} multi-walled carbon nanotubes (MWCNTs-GelMa) (**c**) before and (**d**) after centrifugation and (**e**) ink formulation after addition of photoinitiator. Stability of MWCNTs-GelMa solutions (**f**) 3 and (**g**) 45 h after MWCNTs dispersion. SEM imaging of lyophilised (**h**) GelMa and (**i**) MWCNTs-GelMa structures.

The degree of conversion/methacrylation was then calculated from the lysine methylene peak of gelatine and the GelMa to confirm the presence of methacrylate groups in the gelatine. It is expected that the methylene lysine proton peak intensity would decrease with an increase in methacrylation and an increase in methacrylamide, with respect to conversion. The results (Figure S1) show a clear indication of the decrease in lysine groups and an increase in acrylic proton. This was evident from two individual batches of GelMa prepared using this protocol. Upon calculating the degree of conversion, GelMa batch 1 (GB1) showed 86.9% and GelMa batch 2 (GB2) showed 88.4% conversion.

To increase the conductivity of GelMa, 1 mg mL^{-1} MWCNTs were used as conductive nano-fillers. Initial experiments were performed to compare the dispersion of MWCNT functionalised with carboxylic groups (-COOH) to those not functionalised. It became apparent that without functionalisation, the MWCNTs aggregated and were difficult to disperse in the GelMa (Figure S2) even with several hours of ultrasonication. These larger aggregates in the order of hundreds of microns can prevent the nano-structures to be fabricated in 2PP as the energy absorption is non-uniform, which can affect the polymerisation. A similar effect has also been reported in single-walled carbon nanotubes (SWCNTs), where aggregation and sedimentation were observed 3 h after ink preparation [30]. This study also indicated a direct correlation between SWCNTs concentration used and the aggregation effects.

Although aggregates were also observed with -COOH functionalised MWCNTs (Figure 1c), a uniform dispersion was achieved after ultrasonication for 15 min in an ice bath (to prevent heating-up of the material) and centrifugation at 6000 rpm for 30 min at 40 °C (to prevent gelation). Centrifuging resulted in a clear solution, ideal for subsequent fabrication of three-dimensional structures using the two-photon polymerisation technique due to its improved dispersion and homogenous distribution of the material, which affects its performance. -COOH functionalised MWCNTs were then selected for subsequent dispersion in GelMa and ink formulation (MWCNTs-GelMa). To complete the ink formulation for photopolymerisation, P2CK was added to the MWCNTs-GelMa solution. No further aggregation was noted upon adding the photoinitiator (Figure 1e).

The dispersion stability of MWCNTs-GelMa solutions was investigated to analyse whether this was adequate for 2PP fabrication by leaving the dispersion stationary after sonication and centrifugation. No sedimentation was observed in MWCNTs-GelMa solutions after 3 and 45 h of the dispersion of MWCNTs (Figure 1f,g). This allows sufficient time for 2PP fabrication of the structures and indicates that sedimentation of MWCNTs should not occur during the manufacturing of the 2PP structures. Conversely, the composition containing non-functionalised MWCNTs started to settle within an hour for the 1 mg.mL^{-1} composition (Figure S3).

Lyophilised samples of GelMa and MWCNTs-GelMa (Figure 1h,i) were obtained, where MWCNTs can be seen forming an interconnecting network inside GelMa pores. We hypothesise that this demonstrates the ability of MWCNTs to form an interconnected conductive mesh that contributes to increased conductivity.

To confirm this, the electrochemical activity of the hydrogels was studied using electrochemical impedance spectroscopy and cyclic voltammetry (CV) to determine the conductivity and ability to exchange electrons of the MWCNT-GelMa. Electrochemical characterisation was performed on non-2PP printed structures, as the micron range dimensions (100 μm × 100 μm × 10 μm) of the actual printed structures is not characterizable with the systems used. Significant differences in the impedance values of the different materials can be seen in Figure 2a. GelMa was the most resistive material as expected, as collagen molecules typically present resistance values in the order of MΩ [33]. Suspension of MWCNTs into hydrogels decreased the impedance of the materials, which depended on the concentration of MWCNTs. This effect was more evident at lower frequencies (10–100 Hz) as materials behave as capacitors, where impedance values are inversely proportional to frequency. To compare our soft conductive hydrogels to a more traditional conductive material, the impedance of bare gold was also measured, with the lowest impedance of 4.8 Ω. From this, it can be seen that MWCNT-modified hydrogels have intermediate values of impedance when compared to highly conductive materials (gold) and resistive materials (GelMa). For instance, at 100 Hz, the impedance of MWCNTs-GelMa corresponds to 2745.06 Ω. This contrasts the values of pristine GelMa (6209.67 Ω) and bare gold (4.8 Ω).

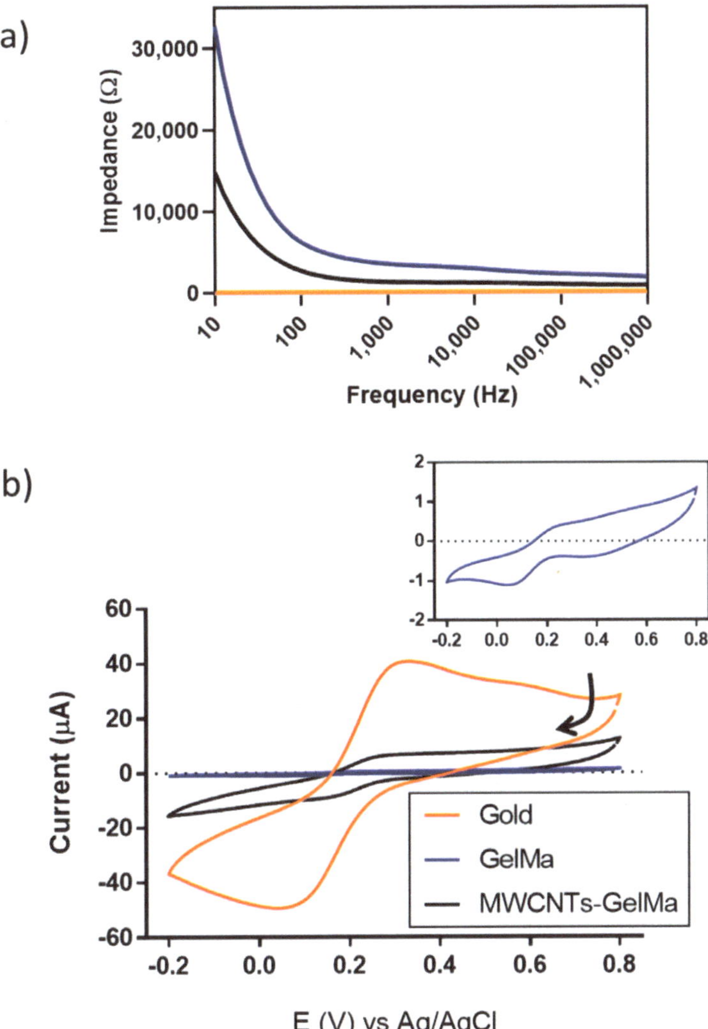

Figure 2. Electrochemical characterisation of gels. (**a**) Impedance spectroscopy measurements of GelMa and MWCNTs-GelMa (1 mg mL^{-1}) compared to gold surfaces. (**b**) Typical cyclic voltammograms (CVs) of gels developed in situ on gold surfaces. Potassium hexacyanoferrate (II) (1 mM) in phosphate buffer saline (PBS) was used as an electrolyte. CVs were measured from 0.8 to −0.2 at 100 mV s^{-1} vs. an Ag/AgCl reference electrode. Arrow indicates the direction of the CV. All measurements were performed in triplicate. Inset CV corresponds to GelMa, shown for clarity.

Cyclic voltammograms (CVs) of GelMa, MWCNTs-GelMa and gold in the presence of the well-known redox-active molecule potassium hexacyanoferrate (II) can be seen in Figure 2b. On bare gold, a reduction peak corresponding to gold was detected at 0.436 V. This peak was more evident in the absence of the redox-active molecule (Figure S4). Peaks corresponding to reduction and oxidation of potassium hexacyanoferrate (II), respectively, were found at 0.043 and 0.322 V. These peaks were also detected in the presence of GelMa at 0.044 and 0.286 V and MWCNTs-GelMa at 0.153 and 0.271 V. It is important to note that the lowest peak separation corresponded to MWCNTs-GelMa (0.118 vs. 0.279 V in gold and 0.242 V in GelMa). Peak separation can be used to determine the electrochemical reversibility of a system. Lower values of peak separation can be interpreted as an increase in the electrochemical reversibility, where electrochemical reversible systems possess a value of peak separation of 0.059 V [34]. This could be due to the ability of the material to exchange electrons more easily. A gold peak can also be seen in the CVs of GelMa and MWCNTs-GelMa at 0.400 V as gels were formed in situ on gold electrodes before performing the measurements (Figure 2b and Figure S4). The peak height of CVs can be related to the number of electrons exchanged, being higher at the gold surfaces followed by MWCNTs-GelMa and GelMa. It can be concluded that the addition of MWCNTs to GelMa increases the ability to exchange electrons within the hydrogel.

The next part of the investigation was aimed at determining whether GelMa and MWCNTs-GelMa could be used as a potential material in 2PP in the manufacturing of biomimetic scaffolds. To do this, materials were incubated with human pluripotent stem cell-derived cardiomyocytes (hPSC-CMs), after which the viability and structural morphology of the cells were studied. Details of the hPSC-CMs and differentiation protocols used can be found elsewhere [32].

A live/dead fluorescence staining (Figure S5) was performed on hPSC-CMs at different GelMa concentrations (5 and 10 wt.%), to identify the optimal concentrations that support cell viability. hPSC-CMs were incubated on GelMa structures for 5 days, where no significant differences were found between concentrations of GelMa with viabilities above 90% (Figure 3a). The lowest concentration of GelMa (5%) was selected in the following experiments, as higher concentrations increase the viscosity of the solutions, thus affecting the dispersion of MWCNTs. Effects of MWCNTs on cell viability were also studied at different MWCNTs concentrations of 0.5 and 1 mg mL^{-1} since certain cytotoxic effects can be associated with MWCNTs [35]. However, as MWCNTs were confined in the hydrogels, viability values were above 86% in all the cases (Figure 3b) and there were no significant differences with the viability of hPSC-CMs in GelMa without MWCNTs. This confirms previous reported results where 90% cardiomyocyte viability was achieved at 3 and 4 mg mL^{-1} MWCNTs concentration in GelMa [36].

Morphology of hPSC-CMs was assessed by fluorescence microscopy of intracellular structures on GelMa (Figure 3c) and MWCNTs-GelMa (Figure 3d) materials. As can be seen from the images, hPSCM-CMs tend to aggregate in individual clusters of cells. On the edges of the cellular agglomerations, it is possible to see elongated hPSCM-CMs with visible sarcomeres. No differences in cell morphology can be observed when using different materials.

Once it was demonstrated that materials support cell viability and development, optimisation of printing parameters was carried out. Initial experiments were performed with the pristine GelMa to find the optimal process conditions. Figure S6 shows the screenshots taken during the fabrication process of an ultra-thin film of GelMa (100 × 100 × 10 μm). The bright line in the image indicates the laser scanning the sample. At the start of the optimisation process, some material was being polymerised but structures did not adhere to the glass coverslips. Some burning and boiling effects were also observed, and black bubbles indicated an excess of heat/energy. Upon optimisation, it can be seen that a uniform structure can be obtained. Some of the values used in the optimisation process can be seen in Figure S7. It should be noted that printing hydrogels using 2PP is more complex than printing non-hydrogels. One of the challenges of 2PP printing of hydrogels is the drying of the ink droplet with time as shown in Figure S8. Materials need to be processed before water evaporates to avoid dehydration of the material and destruction of the structures, therefore, manufacturing of bigger constructs is limited with the current technology. The optimised process condition in this study to print pristine GelMa with 0.5% photoinitiator concentration was 90 mW laser power and 10,000 μm/s print speed. The hatch spacing was 0.2 and 0.4 μm layer thickness.

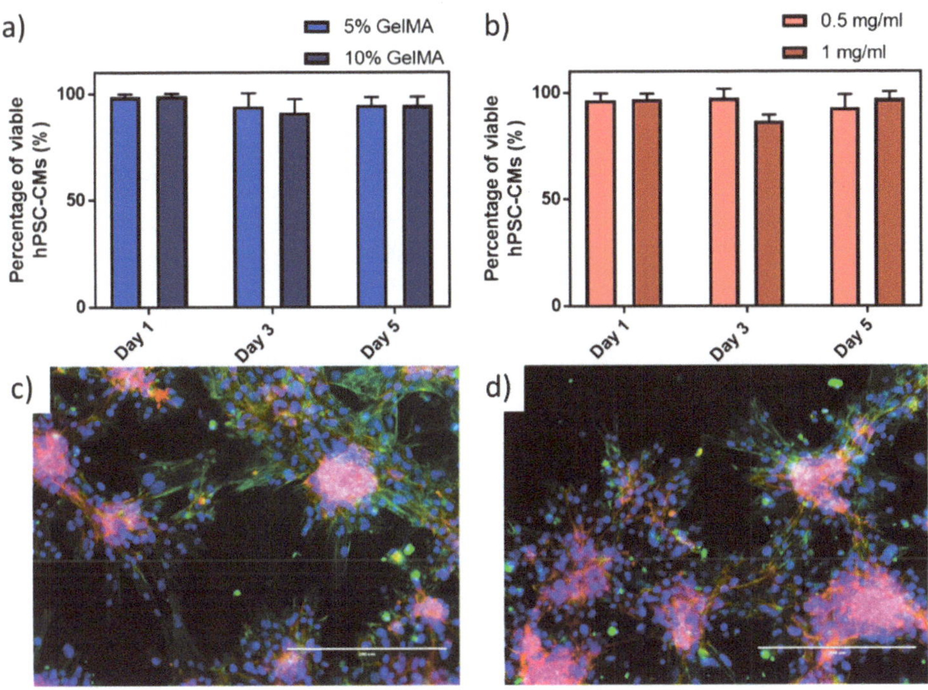

Figure 3. Viability studies performed on human pluripotent stem cell-derived cardiomyocytes (hPSC-CMs) on (**a**) 5% and 10% GelMa and (**b**) 0.5 mm mL^{-1} and 1 mg mL^{-1} MWCNTs in 5% GelMa. Fluorescence images of immunostained hPSC-CMs after 7 days of culture in (**c**) 5% GelMa and (**d**) 1 mg mL^{-1} MWCNTs in 5% GelMa. Nuclei were stained with a Hoechst probe (blue), actin fibres were stained with a phalloidin probe (red), and sarcomeres were immunostained with anti-alpha actinin (green). Scale bar 200 μm.

These optimal conditions were used to print a sample scaffold structure as shown in Figure 4a. The optical microscopic image was taken by soaking the sample in DPBS to prevent dehydration of the material. Hydrogels swelling can be seen as the printed material was left soaking in DPBS for a few hours before characterisation. In addition, slight damages might have occurred during the developing state where un-reacted material is removed from the structures. Since the concentration of GelMa is only 5% in this study, structures are highly prone to damage. However, the lower concentration of GelMa prevents the drying of the ink droplet during the 2PP process and demonstrated good MWCNTs dispersion. The equipment used in this study does not have any additional modification to prevent the drying of the material. This is important to consider while performing further developments. An SEM image of the dehydrated structure has also been shown to showcase the success of printing GelMa in 2PP (Figure 4b).

The 2PP capabilities of printing resolution in the micro and nanoscale allow us to introduce further complexity to our structures, which is key in developing scaffolds for tissue engineering and regenerative medicine applications. To test this, 120 μm × 100 μm scaffolds were manufactured including three 10 μm channels (Figure 4c). Despite some deformation that took place during material development and lyophilisation, the different elements of the structure could also be identified by SEM imaging (Figure 4d).

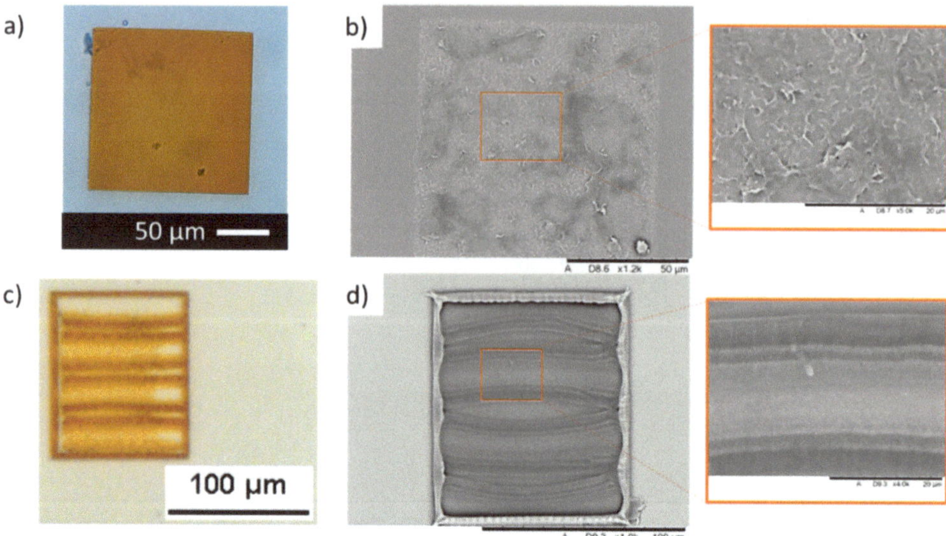

Figure 4. Hydrogel GelMa structures developed by 2PP. (**a**) Optical microscopy and (**b**) SEM imaging of hydrated and dehydrated GelMa ultra-thin films, respectively. (**c**) Optical microscopy and (**d**) SEM imaging of 120 μm × 100 μm complex GelMa scaffolds with 10 μm channels developed by 2PP.

During the MWCNTs-GelMa printing optimisation, a higher concentration of MWC-NTs was used to test the printing capabilities of the material. A concentration of 3 mg mL^{-1} MWCNTs was initially used. This ink proved challenging to print for two main reasons. First, the intense black colour of the material made it difficult to observe the printing area. Second, the MWCNTs absorbed significant laser energy resulting in localised boiling during the printing process that disrupted the manufacturing. Hence this MWCNTs concentration was no longer considered.

The 1 mg mL^{-1} concentration of MWCNTs-GelMa enabled significantly more straightforward printing as fewer boiling events were observed during the printing process, possibly due to the lower concentration of MWCNTs. However, it was difficult to visualise the printing process and the laser scanning over the material. Hence the printing conditions were pre-programmed with varied layer thickness, hatch distance, laser power and scan speed. This is shown in Figure 5a, where thin films (layer thickness 0.3 μm) were imaged by optical microscopy after printing and material development.

At lower scan rates, the material causes burns as indicated by the black appearance of the structures and this effect is particularly evident with increased laser power. This is probably due to the increased exposure of the material to the laser energy. Similar effects were also described on graphene oxide composites, attributed to localised excessive heat generated during a high rate of photothermal conversion under the laser radiation [37]. Conversely, higher scan rates lead to the structures not being formed. These effects were also observed at different layer thicknesses (0.2 and 0.4 μm) (Figure S9). Printing parameters of 40 mW laser power, 5000–10,000 μm/s print speed, 0.3 μm layer thickness and 0.2 μm hatch spacing rendered more stable structures.

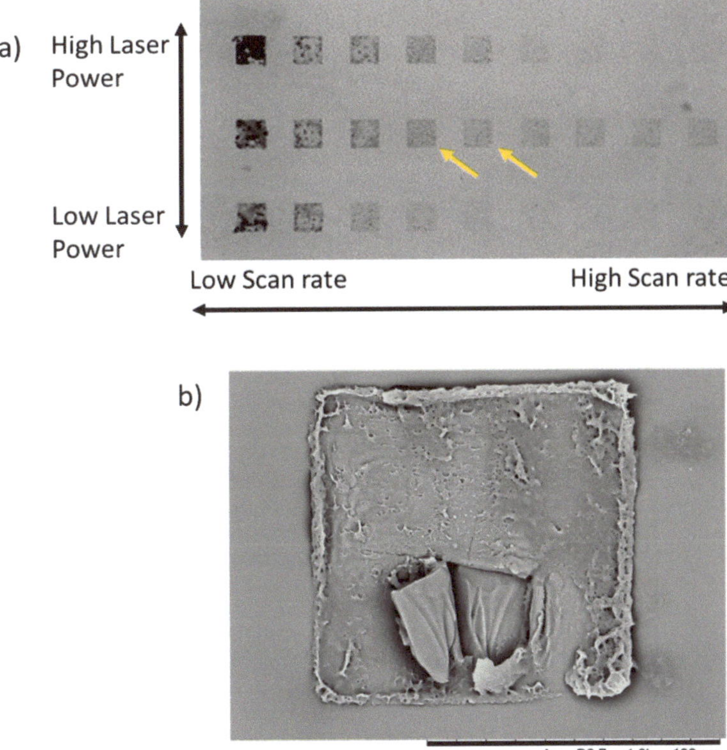

Figure 5. (**a**) Optimisation conditions of MWCNTs-GelMa inks to produced thin-films by 2PP. Laser power was varied between 50, 40 and 30 mW and the scan rate was varied between 500, 1000, 2000, 5000, 10,000, 20,000, 30,000, 40,000 and 50,000 µm/s. Layer distance corresponds to 0.3 µm. Hatch spacing corresponds to 0.2 µm. Yellow arrow indicates our selected optimal conditions. (**b**) SEM image of dehydrated MWCNTs-GelMa thin-film manufactured by 2PP.

It can be noted that the energy required for printing the MWCNTs-GelMa structures (40 mW) was significantly lower than the energy required to print the pristine GelMa structures (90 mW). This requirement for a lower energy density is mainly due to the material's properties, as increased energy absorption and conduction within the material can lead to overheating and boiling. Spatial resolution of MWCNTs composites have also been compared to pristine materials, where higher resolutions were achieved in the composites structures [30]. This could be due to the lower printing energy used as this study reported that lower laser intensities led to higher printing resolution.

SEM imaging of the dehydrated MWCNTs-GelMa structures (Figure 5b) shows that microstructures of this material can be manufactured. Some structural damage can be observed due to the low concentration of GelMa that makes structures extremely fragile and prone to damage during the cleaning, dehydration and sputter coating processes. This can be prevented in the future by increasing the adhesion of the structures to the glass coverslips through a silanisation process, creating covalent bonds between the structures and the substrate. Moving forward, modifications to the 2PP equipment that could allow the printing of larger structures through reduced water evaporation rates from inks could offer a significant advancement in the development of more complex conductive structures for bioelectronics and electronics applications.

4. Conclusions

In this study, we have presented a strategy of the manufacture of conductive hydrogel nano-/microstructures by 2PP for potential applications in bioelectronics. Conductive hydrogel inks (MWCNTs-GelMa) consisted of three components: (i) GelMa, as the main hydrogel component, (ii) a dispersion of -COOH functionalised MWCNTs, acting as conductive nanofillers; (iii) P2CK, as a water-soluble biocompatible photoinitiator. Electrical properties of the hydrogels were enhanced after the addition of the MWCNTs, showing lower impedance values and improved ability to exchange electrons compared to pristine GelMa. This material was also shown to support hPSC-CMs viability. Films with a thickness of 10 μm and scaffolds were fabricated using 2PP, where the optimal printing parameters corresponded to 40 mW laser power, 5000–10,000 μm/s print speed, 0.2 μm hatch spacing and 0.3 μm layer thickness. While further technological developments are necessary for the current embodiment to prevent ink dehydration during the printing process to fabricate more complex structures, this work opens up the opportunity to fabricate conductive biomimetic scaffolds in extremely low resolution using hydrogels.

Supplementary Materials: The following are available online at https://www.mdpi.com/article/10.3390/polym13071038/s1, Figure S1: ^{1}H NMR spectra of (a) gelatine and two different batches of gelatine methacrylate (GelMa), (b) GB1 and (c) GB2; Figure S2: (a) 5× and (b) 50× optical microscopy images of the GelMa after addition of multi-walled carbon nanotubes (MWCNTs) without cross-functionalisation; Figure S3: GelMa dispersions including 1 mg mL^{-1} of unfunctionalized MWCNTs after 0 (left) and 60 (right) min. Yellow arrow indicates the location of the deposits that lead to an uneven distribution of MWCNTs in the dispersion; Figure S4: Cyclic voltammograms (CVs) of gels developed in situ on gold surfaces using PBS as electrolyte. CVs were measured from 0.8 to −0.2 at 100 mV s^{-1} vs. an Ag/AgCl reference electrode (n = 3). Arrow indicates direction of the CV; Figure S5: Fluorescence images of hPSC-CMs after live/dead staining with ethidium homodimer 1 (dead cells) and calcein-am (viable cells) on (a) 5% GelMa, (b) 10% GelMa, (c) 5% GelMa modified with 0.5 mg mL^{-1} MWCNTs and (d) 5% GelMa modified with 1 mg mL^{-1} MWCNTs on days (i) one, (ii) three and (iii) five of incubation. Scale bar 400 μm; Figure S6: Screenshots taken during 2PP printing of GelMa ultra-thin films. Bright line shows laser photopolymerising GelMa structures and black dots indicate bubbles from burnt areas due to an excess of heat. Optimal conditions were used to print GelMa film in the lower right; Figure S7: Optical microscopy images of GelMa films and some of the parameters used during the optimisation; Figure S8: Printing stage of the 2PP showing the effect of water evaporation 60 min of addition of the ink droplet. Red line indicates the original area occupied by the ink; Figure S9: Optimisation conditions of MWCNTs-GelMa inks to produced thin-films by 2PP. Laser power was varied between 50, 40 and 30 mW and scan rate was varied between 500, 1000, 2000, 5000, 10,000, 20,000, 30,000, 40,000 and 50,000 μm/s. Layer thickness corresponds to (a) 0.4 and (b) 0.2 μm. Hatch spacing was kept constant at 0.2 μm.

Author Contributions: P.S.-A. performed the cardiomyocyte differentiation, cell studies and contributed to the electrochemical characterisation and analysis. J.V. performed the ink formulation and characterisation, design and 3D printing of structures. J.C.M. synthesised P2CK. P.S.-A., J.V. and F.J.R. wrote the initial manuscript. P.S.-A., J.V., J.C.M., R.D.W., C.J.T., M.R.A., R.J.M.H. and F.J.R. contributed to experiment design and interpretation of results. F.J.R. and P.S.-A. designed the experiments. All authors have read and agreed to the published version of the manuscript.

Funding: This work was supported by the Engineering and Physical Sciences Research Council (Grant numbers EP/R004072/1, EP/P027261/1 and EP/N006615/1).

Institutional Review Board Statement: Not Applicable.

Informed Consent Statement: Not Applicable.

Data Availability Statement: Data supporting results can be found at the University of Nottingham repository (https://nottingham-repository.worktribe.com/).

Polymers **2021**, 13, 1038

Acknowledgments: Laurence Burroughs is acknowledged for his assistance during NMR characterisation and analysis. Chris Denning is thanked for providing laboratory access. Part of this work was supported by an EPSRC Doctoral Prize award hosted by the University of Nottingham granted to Sanjuan-Alberte. The authors thank the Nanoscale and Microscale Research Centre (nmRC) for providing access to instrumentation.

Conflicts of Interest: The authors declare no conflict of interest.

References

1. Fu, F.; Wang, J.; Zeng, H.; Yu, J. Functional Conductive Hydrogels for Bioelectronics. *ACS Mater. Lett.* **2020**, *2*, 1287–1301. [CrossRef]
2. Saldin, L.T.; Cramer, M.C.; Velankar, S.S.; White, L.J.; Badylak, S.F. Extracellular Matrix Hydrogels from Decellularized Tissues: Structure and Function. *Acta Biomater.* **2017**, *49*, 1–15. [CrossRef]
3. Slaughter, B.V.; Khurshid, S.S.; Fisher, O.Z.; Khademhosseini, A.; Peppas, N.A. Hydrogels in Regenerative Medicine. *Adv. Mater.* **2009**, *21*, 3307–3329. [CrossRef]
4. Sun, J.; Tan, H. Alginate-Based Biomaterials for Regenerative Medicine Applications. *Materials* **2013**, *6*, 1285–1309. [CrossRef] [PubMed]
5. Wahl, D.A.; Czernuszka, J.T. Collagen-hydroxyapatite Composites for Hard Tissue Repair. *Eur. Cell Mater.* **2006**, *11*, 43–56. [CrossRef] [PubMed]
6. Yuk, H.; Lu, B.; Zhao, X. Hydrogel bioelectronics. *Chem. Soc. Rev.* **2019**, *48*, 1642–1667. [CrossRef]
7. Han, L.; Lu, X.; Wang, M.; Gan, D.; Deng, W.; Wang, K.; Fang, L.; Liu, K.; Chan, C.W.; Tang, Y. A mussel-inspired conductive, self-adhesive, and self-healable tough hydrogel as cell stimulators and implantable bioelectronics. *Small* **2017**, *13*, 1601916. [CrossRef]
8. Heo, D.N.; Lee, S.-J.; Timsina, R.; Qiu, X.; Castro, N.J.; Zhang, L.G. Development of 3D printable conductive hydrogel with crystallized PEDOT: PSS for neural tissue engineering. *Mater. Sci. Eng. C* **2019**, *99*, 582–590. [CrossRef]
9. Navaei, A.; Saini, H.; Christenson, W.; Sullivan, R.T.; Ros, R.; Nikkhah, M. Gold nanorod-incorporated gelatin-based conductive hydrogels for engineering cardiac tissue constructs. *Acta Biomater.* **2016**, *41*, 133–146. [CrossRef]
10. Athukorala, S.S.; Tran, T.S.; Balu, R.; Truong, V.K.; Chapman, J.; Dutta, N.K.; Roy Choudhury, N. 3D Printable Electrically Conductive Hydrogel Scaffolds for Biomedical Applications: A Review. *Polymers* **2021**, *13*, 474. [CrossRef]
11. Patel, A.K.; Celiz, A.D.; Rajamohan, D.; Anderson, D.G.; Langer, R.; Davies, M.C.; Alexander, M.R.; Denning, C. A defined synthetic substrate for serum-free culture of human stem cell derived cardiomyocytes with improved functional maturity identified using combinatorial materials microarrays. *Biomaterials* **2015**, *61*, 257–265. [CrossRef]
12. Mitrossilis, D.; Fouchard, J.; Pereira, D.; Postic, F.; Richert, A.; Saint-Jean, M.; Asnacios, A. Real-time single-cell response to stiffness. *Proc. Natl. Acad. Sci. USA* **2010**, *107*, 16518–16523. [CrossRef] [PubMed]
13. Nguyen, A.T.; Sathe, S.R.; Yim, E.K. From Nano to Micro: Topographical Scale and its Impact on Cell Adhesion, Morphology and Contact Guidance. *J. Phys. Condens. Matter.* **2016**, *28*, 183001. [CrossRef] [PubMed]
14. Llucia-Valldeperas, A.; Sanchez, B.; Soler-Botija, C.; Galvez-Monton, C.; Prat-Vidal, C.; Roura, S.; Rosell-Ferrer, J.; Bragos, R.; Bayes-Genis, A. Electrical Stimulation of Cardiac Adipose Tissue-Derived Progenitor Cells Modulates Cell Phenotype and Genetic Machinery. *J. Tissue Eng. Regen. Med.* **2015**, *9*, E76–E83. [CrossRef] [PubMed]
15. Malda, J.; Visser, J.; Melchels, F.P.; Jungst, T.; Hennink, W.E.; Dhert, W.J.; Groll, J.; Hutmacher, D.W. 25th Anniversary Article: Engineering Hydrogels for Biofabrication. *Adv. Mater.* **2013**, *25*, 5011–5028. [CrossRef] [PubMed]
16. Joung, D.; Truong, V.; Neitzke, C.C.; Guo, S.Z.; Walsh, P.J.; Monat, J.R.; Meng, F.; Park, S.H.; Dutton, J.R.; Parr, A.M. 3D Printed Stem-Cell Derived Neural Progenitors Generate Spinal Cord Scaffolds. *Adv. Funct. Mater.* **2018**, *28*, 1801850. [CrossRef] [PubMed]
17. Mannoor, M.S.; Jiang, Z.; James, T.; Kong, Y.L.; Malatesta, K.A.; Soboyejo, W.O.; Verma, N.; Gracias, D.H.; McAlpine, M.C. 3D Printed Bionic Ears. *Nano Lett.* **2013**, *13*, 2634–2639. [CrossRef]
18. Vaithilingam, J.; Sanjuan-Alberte, P.; Campora, S.; Rance, G.A.; Jiang, L.; Thorpe, J.; Burroughs, L.; Tuck, C.J.; Denning, C.; Wildman, R.D.; et al. Multifunctional Bioinstructive 3D Architectures to Modulate Cellular Behavior. *Adv. Funct. Mater.* **2019**, *29*, 1902016. [CrossRef]
19. Hu, Q.; Sun, X.Z.; Parmenter, C.D.J.; Fay, M.W.; Smith, E.F.; Rance, G.A.; He, Y.; Zhang, F.; Liu, Y.; Irvine, D.; et al. Additive manufacture of complex 3D Au-containing nanocomposites by simultaneous two-photon polymerisation and photoreduction. *Sci. Rep.* **2017**, *7*, 17150. [CrossRef]
20. Torgersen, J.; Qin, X.H.; Li, Z.; Ovsianikov, A.; Liska, R.; Stampfl, J. Hydrogels for Two-Photon Polymerization: A Toolbox for Mimicking the Extracellular Matrix. *Adv. Funct. Mater.* **2013**, *23*, 4542–4554. [CrossRef]
21. Hahn, M.S.; Miller, J.S.; West, J.L. Three-Dimensional Biochemical and Biomechanical Patterning of Hydrogels for Guiding Cell Behavior. *Adv. Mater.* **2006**, *18*, 2679–2684. [CrossRef]
22. Lee, S.H.; Moon, J.J.; West, J.L. Three-Dimensional Micropatterning of Bioactive Hydrogels via Two-Photon Laser Scanning Photolithography for Guided 3D Cell Migration. *Biomaterials* **2008**, *29*, 2962–2968. [CrossRef] [PubMed]
23. Hoffmann, J.C.; West, J.L. Three-Dimensional Photolithographic Patterning of Multiple Bioactive Ligands in Poly(Ethylene Glycol) Hydrogels. *Soft Matter.* **2010**, *6*, 5056–5063. [CrossRef]

24. Basu, S.; Campagnola, P.J. Properties of Crosslinked Protein Matrices for Tissue Engineering Applications Synthesized by Multiphoton Excitation. *J. Biomed. Mater. Res. A* **2004**, *71*, 359–368. [CrossRef] [PubMed]
25. Ovsianikov, A.; Deiwick, A.; van Vlierberghe, S.; Pflaum, M.; Wilhelmi, M.; Dubruel, P.; Chichkov, B. Laser Fabrication of 3D Gelatin Scaffolds for the Generation of Bioartificial Tissues. *Materials* **2011**, *4*, 288–299. [CrossRef]
26. Ovsianikov, A.; Deiwick, A.; van Vlierberghe, S.; Dubruel, P.; Moller, L.; Drager, G.; Chichkov, B. Laser Fabrication of Three-Dimensional CAD Scaffolds from Photosensitive Gelatin for Applications in Tissue Engineering. *Biomacromolecules* **2011**, *12*, 851–858. [CrossRef] [PubMed]
27. Prina, E.; Amer, M.H.; Sidney, L.; Tromayer, M.; Moore, J.; Liska, R.; Bertolin, M.; Ferrari, S.; Hopkinson, A.; Dua, H. Bioinspired Precision Engineering of Three-Dimensional Epithelial Stem Cell Microniches. *Adv. Biosyst.* **2020**, *4*, 2000016. [CrossRef] [PubMed]
28. Xiong, W.; Liu, Y.; Jiang, L.J.; Zhou, Y.S.; Li, D.W.; Jiang, L.; Silvain, J.F.; Lu, Y.F. Laser-Directed Assembly of Aligned Carbon Nanotubes in Three Dimensions for Multifunctional Device Fabrication. *Adv. Mater.* **2016**, *28*, 2002–2009. [CrossRef]
29. Ushiba, S.; Shoji, S.; Masui, K.; Kono, J.; Kawata, S. Direct laser writing of 3D architectures of aligned carbon nanotubes. *Adv. Mater.* **2014**, *26*, 5653–5657. [CrossRef]
30. Ushiba, S.; Shoji, S.; Masui, K.; Kuray, P.; Kono, J.; Kawata, S. 3D microfabrication of single-wall carbon nanotube/polymer composites by two-photon polymerization lithography. *Carbon* **2013**, *59*, 283–288. [CrossRef]
31. Li, Z.; Torgersen, J.; Ajami, A.; Mühleder, S.; Qin, X.; Husinsky, W.; Holnthoner, W.; Ovsianikov, A.; Stampfl, J.; Liska, R. Initiation efficiency and cytotoxicity of novel water-soluble two-photon photoinitiators for direct 3D microfabrication of hydrogels. *RSC Adv.* **2013**, *3*, 15939–15946. [CrossRef]
32. Mosqueira, D.; Mannhardt, I.; Bhagwan, J.R.; Lis-Slimak, K.; Katili, P.; Scott, E.; Hassan, M.; Prondzynski, M.; Harmer, S.C.; Tinker, A.; et al. CRISPR/Cas9 Editing in Human Pluripotent Stem Cell-Cardiomyocytes Highlights Arrhythmias, Hypocontractility, and Energy Depletion as Potential Therapeutic Targets for Hypertrophic Cardiomyopathy. *Eur. Heart J.* **2018**, *39*, 3879–3892. [CrossRef] [PubMed]
33. Tzukert, A.; Leviner, E.; Mahler, Y.; Shoshan, S. Electroconductivity of Collagen In Vitro and In Vivo. *Biochim. Biophys. Acta* **1980**, *627*, 276–280. [CrossRef]
34. Elgrishi, N.; Rountree, K.J.; McCarthy, B.D.; Rountree, E.S.; Eisenhart, T.T.; Dempsey, J.L. A Practical Beginner's Guide to Cyclic Voltammetry. *J. Chem. Educ.* **2017**, *95*, 197–206. [CrossRef]
35. Guo, Y.Y.; Zhang, J.; Zheng, Y.F.; Yang, J.; Zhu, X.Q. Cytotoxic and Genotoxic Effects of Multi-Wall Carbon Nanotubes on Human Umbilical Vein Endothelial Cells in Vitro. *Mutat. Res.* **2011**, *721*, 184–191. [CrossRef] [PubMed]
36. Shin, S.R.; Farzad, R.; Tamayol, A.; Manoharan, V.; Mostafalu, P.; Zhang, Y.S.; Akbari, M.; Jung, S.M.; Kim, D.; Comotto, M. A bioactive carbon nanotube-based ink for printing 2D and 3D flexible electronics. *Adv. Mater.* **2016**, *28*, 3280–3289. [CrossRef]
37. Yeung, K.-W.; Dong, Y.; Chen, L.; Tang, C.-Y.; Law, W.-C.; Tsui, G.C.-P.; Engstrøm, D.S. Printability of photo-sensitive nanocomposites using two-photon polymerization. *Nanotechnol. Rev.* **2020**, *9*, 418–425. [CrossRef]

Review

Designing Electrical Stimulation Platforms for Neural Cell Cultivation Using Poly(aniline): Camphorsulfonic Acid

Fábio F. F. Garrudo [1,2,3,4,*], Robert J. Linhardt [5], Frederico Castelo Ferreira [2,3,4] and Jorge Morgado [1,2,*]

1. Instituto de Telecomunicações, Instituto Superior Técnico, Universidade de Lisboa, Avenida Rovisco Pais, 1049-001 Lisbon, Portugal
2. Department of Bioengineering, Instituto Superior Técnico, Universidade de Lisboa, Avenida Rovisco Pais, 1049-001 Lisbon, Portugal; frederico.ferreira@tecnico.ulisboa.pt
3. iBB—Institute for Bioengineering and Biosciences, Instituto Superior Técnico, Universidade de Lisboa, Avenida Rovisco Pais, 1049-001 Lisbon, Portugal
4. Associate Laboratory i4HB, Institute for Health and Bioeconomy, Avenida Rovisco Pais, 1049-001 Lisbon, Portugal
5. Department of Chemical and Biological Engineering, Biology and Chemistry and Chemical Biology, Center for Biotechnology and Interdisciplinary Studies, Rensselaer Polytechnic Institute, 110 8th Street, Troy, NY 12180, USA; linhar@rpi.edu
* Correspondence: fabio.garrudo@tecnico.ulisboa.pt (F.F.F.G.); jmfmorgado@tecnico.ulisboa.pt (J.M.)

Abstract: Electrical stimulation is a powerful strategy to improve the differentiation of neural stem cells into neurons. Such an approach can be implemented, in association with biomaterials and nanotechnology, for the development of new therapies for neurological diseases, including direct cell transplantation and the development of platforms for drug screening and disease progression evaluation. Poly(aniline):camphorsulfonic acid (PANI:CSA) is one of the most well-studied electroconductive polymers, capable of directing an externally applied electrical field to neural cells in culture. There are several examples in the literature on the development of PANI:CSA-based scaffolds and platforms for electrical stimulation, but no review has examined the fundamentals and physico-chemical determinants of PANI:CSA for the design of platforms for electrical stimulation. This review evaluates the current literature regarding the application of electrical stimulation to neural cells, specifically reviewing: (1) the fundamentals of bioelectricity and electrical stimulation; (2) the use of PANI:CSA-based systems for electrical stimulation of cell cultures; and (3) the development of scaffolds and setups to support the electrical stimulation of cells. Throughout this work, we critically evaluate the revised literature and provide a steppingstone for the clinical application of the electrical stimulation of cells using electroconductive PANI:CSA platforms/scaffolds.

Keywords: bioelectricity; biophysical cues; additive manufacturing; electroconductive polymers; neurological diseases; nerve/neuron regeneration; voltage-dependent calcium channels

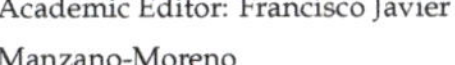

Citation: Garrudo, F.F.F.; Linhardt, R.J.; Ferreira, F.C.; Morgado, J. Designing Electrical Stimulation Platforms for Neural Cell Cultivation Using Poly(aniline): Camphorsulfonic Acid. *Polymers* **2023**, *15*, 2674. https://doi.org/10.3390/polym15122674

Academic Editor: Francisco Javier Manzano-Moreno

Received: 19 April 2023
Revised: 1 June 2023
Accepted: 9 June 2023
Published: 14 June 2023

1. Introduction

Neurological diseases, in particular neurodegenerative ones, are chronic, cause incapacity and their progression cannot be stopped or reversed. The onset of these diseases takes place in the brain tissue, in which many factors associated with cell senescence/death homeostasis are thrown out of balance. In these diseases, the basic neurological functions are compromised by massive or localized progressive cell death. Still, most symptoms only start to appear later. Due to the incapacity of the brain tissue to regenerate, patients afflicted with these diseases have no hope for a cure. Palliative treatments dealing with behavioral symptoms, motricity and pain are available. However, as these diseases progress, medication loses its effectiveness.

Electrical stimulation has emerged as a powerful therapeutic tool for the treatment of neurological diseases. Through deep-brain stimulation (DBS), it is possible to apply direct current into the brain tissue. This technique has numerous beneficial effects on the brain tissue, including on: (1) neurogenesis; (2) neurite ramification; (3) axonal remodeling; (4) synaptogenesis; (5) gliogenesis; (6) increase in neuronal size; and (7) increase in vascularization.

Some trials have shown evidence of positive effects on patients after treatment. In Alzheimer's disease patients, Sankar and colleagues [1] succeeded in retarding brain shrinkage. This was attributed to a reduction in Amyloid beta (Aβ) deposition and improved memory, as Mann and colleagues [2] observed for mice when DBS was applied for 25 days. Other trials were performed in Parkinson's disease (PD) patients, where direct stimulation of the subthalamic nucleus and *globus pallidus* was found to reduce rigidity, bradykinesia and iatrogenic dyskinesias [3,4]. The application of this technique is limited to early-to-mild cases when the nigrostrial pathway has some activity. However, this technique has limitations since the use of DBS (six weeks on/off) was not associated with improvements in cognition for cases of PD-derived dementia [5].

Cell therapy can provide a reliable methodology for substituting degenerated neurons with healthy ones, aiming to restore the normal function of the brain. Stem cells have been widely studied for this purpose and several clinical studies using embryonic stem cells (ESCs), induced pluripotent stem cells (iPSCs) [6] mesenchymal/stromal stem cells (MSCs) [7] and neural stem cells (NSCs), are underway [8]. The application of ESCs or autologous stem cells for cell therapy is compromised for both ethical and practical reasons. However, the discovery of a method to obtain, from adult somatic cells, patient-specific iPSCs from which NSC can then be obtained and maintained [9] opens the possibility for personalized NSC or trophic-factor therapies [10]. The use of differentiated/somatic cells is also possible, as reported by Ma and colleagues in their clinical trial [8]. The results obtained after a four-year follow-up of directly transplanted fetal dopaminergic cells revealed improved motor function in the patients. Transplant was also deemed successful based on the increased concentration of fluciclovine F 18 (18F-DOPA) in the engrafted area.

The next generation of devices for treating neurological diseases should harness the advantages of both stem cell therapy and DBS for a potential cure. For this, a multidisciplinary approach is necessary and knowledge from diverse areas, such as biomaterials, nanotechnology, materials science, neuroscience, and stem cell biology, must be wisely used. In this review, we propose to evaluate the physical and biological fundamentals necessary for developing therapeutic strategies that rely on electrical stimulation. Although several reviews on electrical stimulation, materials science, and neurological diseases exist, none of them approaches this topic in a wholistic and integrated manner. Here, we propose to evaluate: (1) the biological and physical fundamentals behind the generation of electrical fields by living cells; (2) the consequences of manipulating cell bioelectricity using externally applied electrical stimulation; (3) the development of platforms/scaffolds capable of directing electrical stimulation to the cells; and (4) the benefits of using electroconductive polymers in their design. Poly(aniline) doped with camphorsulfonic acid (PANI:CSA) is used as our model of an electroconductive polymer. Therefore, we also discuss (Section 3.1) the physico-chemical determinants for the correct PANI:CSA manipulation in the efficient manufacture of platforms/scaffolds, and provide examples of PANI:CSA-based platforms used for the electrical stimulation of neural cells (Section 3.2). With this broad and multidisciplinary review, we hope to provide other researchers a steppingstone for the development of future therapies and more reliable in vitro models for studying neurological disease progression and drug screening. Although neural cells are the focus of this review, including neural cell bioelectricity and basic neural cell biology/biochemistry, we also consider other off-topic case studies, when deemed necessary, to better understand the manipulation of bioelectricity and the design of electrical stimulation platforms (e.g., eye/retinal epithelial cell bioelectricity).

2. Bioelectricity and Electrical Stimulation

2.1. Bioelectricity in Living Organisms

Natural electric fields exist in all living cells. In fact, their existence is evidence of the functional capacity of cells to regulate their membrane potential through the transport of ions and other charged molecules across it. For this, cells make use of ion channels, mainly for fast adjustments, and transporter proteins, for slow and prolonged maintenance of membrane potential.

Bioelectricity is a physiological phenomenon that exists in all living cells and influences their physiology. Bacteria are also sensitive to electric fields. They can propagate potassium waves to communicate with each other when encapsulated in biofilms [11]. Moreover, certain anaerobic bacteria can control electron currents for nutrition and interaction with their environment, such as *Geobacter sulfurreducens* [12] and ones from the *Desulfobulbaceae* family [13].

Bioelectricity and regeneration in multicellular organisms are deeply connected. Cell-generated electricity can regulate morphogenesis in all living tissues. The regeneration of an organ involves careful regulation of cell division, movement, and positioning. It has been even proposed that, during embryogenesis, electrical cues serve to direct organ and limb formation. Moreover, bioelectrical fields can encode information regarding morphology and the patterning of tissue. For example, morphogenesis can be influenced by manipulating the membrane potential. The use of chemical cocktails that either depolarize or hyperpolarize the cell membranes of severed areas in planaria can induce either the formation of a new head or tail, respectively [11,14]. Bioelectricity is also important for the regeneration of limbs and lost organs in certain animals, such as frogs, lizards, and axolotls. Ion channels have been identified as important for tail regeneration in lizards [15].

Bioelectricity and genetic pathways are also interconnected in neural cells. The expression of ion channels (e.g., chloride channels) or transporters regulates the activity membrane's electrochemical potential and can be signaled by the cell [14,16]. Conversely, changes in the membrane's electrochemical potential can be induced by these proteins, triggering genetic responses for cell behavior changes and adaptation. In neurons, for example, the presence or lack of stimulation can determine cell survival and synapse formation or cell death [17]. Also, bioelectrical signals are proposed to encode the positioning and polarity of cells and are described as being critical in the regulation of many physiological phenomena, such as wound healing, neural-circuit shaping and body polarity in planarias, axolotls and frogs [11,18]. Therefore, we expect that the manipulation of bioelectric fields can have positive consequences for regenerative medicine. This is due to the importance of bioelectricity for different cells and organisms, as it is involved not only in homeostasis but also in morphogenesis. However, what are the general cell mechanisms involved in bioelectricity generation?

2.2. The Role of Membrane Potential on Bioelectricity

Bioelectricity is the result of the generation of electrical current by living cells and/or organisms. Bioelectricity depends solely on living cells, as it arises from gradients of charged species established across the intracellular and extracellular faces of their membrane (transmembrane electrochemical/voltage potential), a process intimately dependent on adenosine tri-phosphate (ATP) use. It starts with the necessity of cells to keep diverse ionic concentrations in check for the maintenance of cellular metabolism. First, cells express membrane transporters that either actively (ATP-dependent) or passively (ATP-independent) keep sodium, potassium, chloride, hydrogenocarbonate and hydronium ion gradients across the cellular membrane. This, in turn, allows cells to transport important nutrients intracellularly, such as glycose and aminoacids, and allows for the release of acidic metabolism byproducts, such as lactate, to the extracellular milieu [19,20]. The rapid manipulation of ionic concentrations and transmembrane electrochemical potential in cells can also be done actively by cells using ion channels. These are transmembrane proteins that control the rapid flow of ions between cells and/or between cells and the extracellular matrix (ECM), being sensitive to ligands (sodium channels) or membrane voltage changes (e.g., calcium channels). Cells that use these quick-response elements include nerve and muscle cells [21].

The eye's lens is an example of a well-studied organ with homeostasis and regeneration dependent on bioelectricity. The lens enables the passage of light into the retina, and, with the help of the ciliary muscles, is also responsible for focusing this light. Epithelial cells on the lens undergo migration and differentiation to generate the transparent structures in the eyes known as the lens. The differentiation process involves, in the first stage, the synthesis of lens-specific proteins (e.g., aquaporins, α- and β-crystalin and beaded filament structural protein 2 (BFSP2)), the elongation and orientation of epithelial cells, followed by the destruction of every cellular organelle present (including the nucleus), and finally the packing and water exclusion from the formed core. Cao and colleagues [22] hypothesized that this process is orchestrated naturally by a cell's own membrane potential. In fact, lens cells at a terminal-differentiation stage present a hyperpolarized membrane, while less mature epithelial and fibrous cells have a more depolarized membrane (difference of 32.5 ± 1.8 mV). This process is dependent of Na^+/K^+ ATPase activity, with higher expression in mature denucleated cells. The authors hypothesize that this naturally generated electric field is the main trigger for the differentiation of cells. Bioelectricity and genetic pathways are interconnected in a dynamic system. The expression of ion channels or transporters regulates the membrane's electrochemical potential and can be signaled by the cell. This occurs in neurons, which can increase the expression of chloride channels to increase their stimulation threshold in response to constant stimulation [15,16]. Conversely, changes in the membrane's electrochemical potential can be induced by these proteins, triggering genetic responses for cell behavior changes and adaptation. In addition to determining cell death/survival [17], bioelectrical signals are proposed to encode the positioning and polarity of neural cells and are described as being critical to the regulation of many physiological phenomena, such as neural development, wound healing, and neural-circuit shaping [11,18].

When multiple cells intimately interact to form specific domains or tissues, they establish more direct forms of communication. This is done by gap junctions, composed of channel proteins (connexins) that directly connect cell membranes. Using these gap junctions, cells can distribute and stabilize membrane electrochemical potentials between them, rapidly transmit bioelectrical changes (ionic currents), and coordinate cell responses and cell patterning. These are also known as electrical synapses, in opposition to conventional neural–neural synapses, due to the direct flow of ions/charged species between cells. In neural tissue, their main role is to synchronize cell activity [23,24]. For example, a recent study by Liu and colleagues illustrates that, in *C. elegans*, gap junctions can enhance synaptic transmission between interneurons and motor neurons [25].

Membrane potential is widely used by cells to exchange signals, especially through gap junctions and synaptic connections. Epithelial cells, for example, are known to establish two different membrane potentials across their apical and basal membranes (trans-epithelial potential—TEP), that are otherwise kept stable throughout a cell's life. When the cell layer is damaged, through wounding or the mechanical removal of some of the cells, there is a short circuit and a steady and long-lasting electric field of low-intensity (140 mV mm^{-1}) results. However, this field is strong enough to induce cell proliferation and migration to the lesion area. Once repairs are done and the cells re-establish the continuous cell layer, the electrical field is interrupted, making the cells return to their basal state. The same process occurs in damaged nerves. Overall, in the human body, the magnitudes of electric fields relevant for regeneration are between 1 and 100 mV mm^{-1}, 10-times less than what is necessary for the stimulation of neurons (1–2 V mm^{-1}) and 1000-times less than what is necessary for electroporation (100–500 V mm^{-1}) [26].

The transmembrane electrochemical potential in cells is dynamic and changes with metabolism, cell cycles, and differentiation states. This is true even between different types of stem cells, such as, iPSCs, human embryonic stem cells (hESCs) and MSCs. For example, although both iPSCs and hESCs are electrophysiologically stable, their potassium-channel transcriptomes are slightly different. Regarding overall cell-population membrane potential and electrophysiological activity, MSCs present great heterogeneity whereas iPSCs (obtained from fibroblasts) and hESCs are more homogeneous [27–29]. These differences also indicate differences in the metabolism and even differentiation potential of MSCs within the same pool, while for iPSCs and ESCs, this does not occur. The transmembrane potential of fully differentiated cells varies according to cell type. Differentiated cells, such as neurons, have more hyperpolarized (more negative) resting membrane potentials (-70 mV). In contrast, cells with higher proliferative potential (cancer cells, MSCs, hESCs) have higher (more positive) resting potential values [14].

2.3. Direct Effect of Electrical Stimulation on the Cells

The manipulation of the bioelectricity of cells and tissues is made possible by the use of externally applied electric fields. According to McCaig and colleagues [26], in the human body, the electric fields relevant for regeneration are in the range 1–100 mV mm^{-1}, which are 10-fold lower than the fields needed for stimulating neurons (1–2 V mm^{-1}) and 1000-fold lower than the field necessary for electroporation (100–500 V mm^{-1}). Most cell experiments are carried out with such field values, and the best value is usually chosen when visible changes in morphology (stretching and alignment) are observed.

Electrical stimulation protocols are usually optimized for each cell type and some of the optimized parameters include: (a) the stimulation regimen (e.g., direct current (DC), alternate current (AC), pulsatile direct current (pDC)) [30] (b) the respective signal frequency (e.g., 100 Hz) [31,32] (c) the electric field intensity (e.g., 1 V cm^{-1}) or applied current (e.g., 100 µA) [33,34] (d) the duration of the stimulation (e.g., 12, 24 or 48 h) [30,35] and (e) the phase of cell development at which the electrical stimulation is applied (e.g., proliferation and/or differentiation stage) [36–38]. Depending on the parameters used, the manipulation of bioelectrical fields can greatly affect cell homeostasis. This is true for both matured neural cells and stem cells. Current studies in the literature show that externally applied electrical stimulation can impact neural stem cell morphology, migration, proliferation, apoptosis, and differentiation.

Neural cells and neurites align when stimulated by an electric field, a common hallmark. However, there is no consensus concerning the optimal orientation and the importance of this orientation, as they can both align parallelly or perpendicularly to the direction of the electric field. The adoption of the perpendicular orientation has been attributed to a minimization of the induced voltage drop across the cell body. This was observed by Chen and colleagues [39] for PC12 cells grown on bacterial cellulose (BC)-poly(3,4-ethylenedioxythiophene) (PEDOT) microfibers. However, these authors did not consider the influence of topography on cell alignment. In fact, Koppes and colleagues [40] found that substrate topography motivation orientation, provided by electrospun microfibers (2 µm in diameter), is the main determinant factor for cell and neurite alignment of dorsal root ganglia (DRG) neurons. They found that electrical stimulation (DC, 1 mA, 0.5 V cm^{-1}, 8 h) only enhanced the length of the obtained neurites, while the direction was dependent on fiber orientation. Similar observations for the effect of combined topography and electrical stimulation were reported by other groups [38,41,42]. In another work, Koppes and colleagues [43] also observed specific changes on neurites. They observed that the electrical stimulation of dorsal root ganglion (DRG) neurons (DC, 0.1 mA, 0.1–1 V cm^{-1}, 8 h) enhanced neurite extension and alignment with the electrical field. Although similar neurite extensions were observed when DRG neurons were co-cultured with Schwann cells, neurite alignment did not occur. Overall, we can conclude, based on the gathered results, that cell and neurite alignment is determined by substrate topography, even when electrical stimulation is applied. However, when the influence of topography is minimized, cell and neurite alignment will depend on the direction of the electric field [37,42].

Sun and colleagues [44] explored the impact of electrical stimulation on co-cultures of DRG neurons and Schwann cells on graphene-oxide (GO):PEDOT films. Electrical stimulation was found to significantly increase neurite length when voltages of 15 and 30 mV were applied but decreased when 60 mV was applied (with a distance between electrodes of approximately 1 cm). Moreover, at 30 mV, they found a higher number of Schwann cells interacting with the neuronal axons, even at large distances (such as 1506 μm from the central aggregate), and the formation of higher number of synapses. This suggests that an increased myelination of the cultured neurons occurs with electrical stimulation. The controlled release of the TrkB agonist 7,8-dihydroxyflavone, a neuroprotective drug [45] led to the potentiation of the above-described effects of the electrical stimulation.

Electrical stimulation can promote cell proliferation. Cell-cycle progression appears to be intimately dependent on membrane potential. In frog embryos, depolarized cells can signal distant cells in the neurocrest to proliferate [14]. Similarly, only depolarized astrocytes can proliferate and migrate in response to injuries [46]. Ghasemi-Mobarakeh and colleagues [47] conducted a proliferation study with NSCs where a short electrical stimulation (1 V cm^{-1}) applied for 60 min was associated with an increase in cell number and higher neurite extension (30 μm for stimulated vs. 22 μm for control). Shorter stimulation periods (15 and 30 min) did not produce the same effect. Similar observations were made by Xu and colleagues [48]. When human neural stem cells (hNSCs) were electrically stimulated for seven days, they proliferated faster than their non-stimulated counterparts. Similar to the results of Ghasemi-Mobarakeh and colleagues, Xu and colleagues increasing voltages (15, 35 and 75 mV) induced an overall decrease in cell proliferation and an increase in neurite length.

Electrical stimulation can also enhance the expression and release of bioactive molecules by proliferating NSCs. Song and colleagues [49] observed an increase in the mRNA expression of brain-derived neurotrophic factor (BDNF), glial cell-line-derived neurotrophic factor (GDNF) and neurotrophin 3 (NT3) after one hour of electrical stimulation and 24 h of resting using a 2D experiment, and this effect was enhanced when cells were electrically stimulated in a 3D experiment. Other markers, such as heparin-binding epidermal growth factor-like growth factor (HBEGF) (important for cell growth and glia-derived progenitor formation) and heat shock protein family member 1 (HSPB1) (important as a protein chaperon, neurofilament homeostasis and axonal transport), were also upregulated after electrical stimulation, whereas vascular endothelium growth factor A (VEGF-A) (pro-angiogenic factor) expression did not change, and enolase 2 (ENO2) (associated with neural lineage in NSCs) expression decreased.

Electricity can improve the differentiation yield of neural stem cells. Yang and colleagues [42] showed that the highest expression of microtubule-associated protein 2 (*Map2*) and class III beta-tubulin (transcript) (*Tubb3*) occurred with electrical stimulation regardless of the topography used. Interestingly, the combination of electrical stimulation with topography allowed for the differentiation of neural cells capable of generating sodium current, with the highest expression of voltage-gated sodium channel subunit 1 alpha (*Scn1α*) and calcium voltage-gated channel subunit 1C (*Cacna1c*). The authors noted that electrical stimulation did not interfere with the expression of glial fibrillary acidic protein (*Gfap*), suggesting that astrocytic differentiation was not promoted. Finally, since the expression of octamer-binding transcription factor 4 (*Oct4*), homebox protein nanog (*Nanog*) and Nestin (*Nes*) was not impacted by electrical stimulation, the protocol used was considered efficient for cell differentiation. In fact, altering the cell membrane potential may not only interfere but also direct the fate of the differentiation of stem cells into specific cell types [15]. The positive effects of electricity on specific neural gene expressions in differentiating cells are also described in another work [48].

Other studies have also explored the impact of electrical stimulation on neural cell differentiation. Pires and colleagues [32] previously showed that the application of an electrical field (1 V cm^{-1}) to neural stem cells on cross-linked PEDOT doped with poly(styrene sulfonic acid) (PEDOT:PSS) films promoted elongation of the cells and increased neurite length and neuron-specific class III beta-tubulin (protein) (TUJ1) expression. Zhu and colleagues [33] also demonstrated increased TUJ1 and MAP2 expression and increased neurite length when cells were grown on cross-linked PAN electrospun fibers. Borah and colleagues [50] observed increasing neurite numbers and length when an electric field was applied to PC 12 pheochromocytoma cells on coaxial poly [2-methoxy-5-(2′-ethylhexyloxy)-1,4-phenylene vinylene] (MEH-PPV): poly(ε-caprolactone) (PCL) electrospun fibers. Finally, Koppes and colleagues [40] tested different electric field strengths (10, 50 and 100 mV mm^{-1}) for rat dorsal root ganglion neurons (DRG) cultured alone or in co-culture with Schwann cells. They found that, in the case of DRG neurons cultured alone, neurite outgrowth increased from 700 μm to 1260 μm when they applied an electric field of 50 mV mm^{-1}. Neurite length further increased to 2229 μm when DRG neurons were co-cultured with Schwann cells and subjected to an electric field of 50 mV mm^{-1}. The authors argued that the nerve/neurotrophic growth factor (NGF) secreted by the Schwann cells was responsible for this boost.

Electrical stimulation can also influence the neural lineage of differentiated NSCs. Yang and colleagues [42] indicated that the combination of nanotopography and an electroconductive surface can guide hNSC differentiation into inhibitory neurons (gamma-aminobutyric acid (*Gaba*)+) instead of excitatory neurons (vesicular glutamate transporter 1 (*Vglut1*)+). In the work of our group, Garrudo and colleagues [38] reported a similar trend for neural differentiating hiNPC. However, when electrical stimulation was applied (pDC, 1 V cm^{-1}, 100 Hz, 12 h per day, 30 days) excitatory neuron (VGLUT1) and maturation (neurofilament heavy polypeptide (*Nef-H*), synaptophysin (*Syp*), CACNA1C, SCN1α) markers are overexpressed, and early maturation (neuron-specific class III beta-tubulin (transcript) (*Tubb3*), doublecortin (*Dcx*), tubulin-associated unit (*Tau*)) and inhibitory neuron (glutamate decarboxylase 67 (*Gad67*), vesicular GABA transporter (*Vgat*)) markers were downregulated. Interestingly, scanning electron microscopy (SEM) images suggest that the axon diameter dramatically increases with electrical stimulation, which coincides with the upregulation of NEF-H. VGLUT1 fluorescence intensity, an indirect measurement of protein content, was also increased, which coincides with the upregulation of VGLUT1 (transcript).

The bioelectricity of cells, and their response to electrical stimulation, can also be modulated pharmacologically by using substances that interfere with: (1) connexins; (2) ionic channels; or (3) the release of neurotransmitters, which causes changes in the cells' transmembrane electric field and can lead to similar effects in gene expression. However, due to the ubiquity of these proteins and the number of different functions they regulate in different cells in the human body, these effects might be more unspecific. As such, there is a need to further study the biomolecular mechanisms behind the signal transduction of electrical stimulation and its effects.

Biochemical Pathways Involved in Signal Transduction of Electrical Stimulation

The effects of electrical stimulation on neural cells are well documented, and the literature will surely expand on this subject. However, the molecular mechanisms involved in the signal transduction from an electric field to the activation of specific/multiple molecular cascades inside the cells are yet to be conclusively established.

The current literature provides evidence that voltage-dependent calcium channels are the starting point for neural cell response to externally applied electric fields. This is evidenced by numerous reports in the literature where calcium levels increased in cells subjected to electrical stimulation when cultured on electroconductive substrates [39]. Calcium channels play an important role in the electric-chemical signal transduction in neurons and are important for both cell communication and adaptative responses. Voltage-dependent calcium channels enable extracellular calcium influx into the intracellular environment and lead to the release of synaptic vesicles containing neurotransmitters into the synaptic cleft. Intracellular calcium can also act as a secondary neurotransmitter, and, through calmodulin/calcineurin binding, can lead to the activation of proteins involved in the mitogen-activated protein kinase (MAPK)/extracellular signal-regulated kinase (ERK) pathway (Ras, MAPK, ERK1/2) and Ca^{2+}/calmodulin-dependent protein kinase (CAMK)-IV, leading to the consequent activation of the transcription factor cAMP-response element binding protein (CREB) [51,52]. This ultimately leads to the increased expression of several neural specific genes. It is also possible that voltage-dependent calcium channel activation occurs due to conformational changes, which can lead to an increased influx of calcium to the neurons.

Another potential mechanism that can explain the transduction of electrical stimulation into changes in neural cells is direct membrane depolarization. Cao and colleagues [22] observed such a phenomenon in differentiating human lens epithelial cells (HLECs). In a normal lens, the bioelectric field is established by the selective Na^+/K^+ transporter activity and leads to different intensities depending on the area and degree of differentiation of the cells. Although not directly quantified, the difference between the electric fields (V_{mem}) of the differentiating fiber zone and the fully differentiated fibers was estimated to be 32.5 mV, which was reduced to around 11 mV when ouabain (Na^+/K^+ transporter inhibitor) was used. The stimulation of cultured HLECs through the application of 1 V cm^{-1} on the culture medium significantly increased their differentiation. This was demonstrated by increasing intracellular amounts of phosphorylated forms (activated) of protein kinase B (Akt) (survival pathway) and cyclin-dependent kinase 2 (CDC2) (nuclear assembly during differentiation), and a rapid increase in β-crystallin and aquaporin expression. The authors attributed this effect to direct membrane depolarization of immature fiber HLECs, possibly caused by ion movement in the culture medium. Nevertheless, a lack of a pharmacological control and the presence of regular and voltage-dependent ion channels in HLECs limits the generalization of these conclusions [53,54].

Calcium channel downstream cascade activation and membrane depolarization appear to be opposing mechanisms to explain the improvement of differentiation in NSCs. However, these mechanisms might be intimately associated with, and may influence or lead to, the activation of each other. Wang and colleagues [55] describe one of the first pharmacological studies on the influence of selective voltage-dependent channel inhibition during NSC differentiation on a 3D scaffold. Three important conclusions were made: (1) the application of electrical stimulation enhanced NSC differentiation; (2) the selective inhibition of sodium (amiloride hydrochloride), potassium (4-aminopyridine), calcium (nifedipine) and chloride (4,4′-diisothiocyanostillbene-2,2′-disulfonic acid) voltage-dependent channels affected NSC normal differentiation; and (3) the NSC normal differentiation was rescued by the simultaneous application of electrical stimulation in all four treatments. The phenotype rescued with electrical stimulation was more pronounced on amiloride- and nifedipine-treated cells, but no statistically significant differences between the two conditions were reported.

Other possible mechanisms involve changes in cells biomechanics, including: (a) the reorganization of cell receptors (e.g., clustering of BDNF receptor or integrins) and consequently; (b) of the cytoskeleton; (c) modulation of the activity of other ion channels; (d) interference with mitochondrial metabolism; and (e) epigenetic changes [56]. However, these can also be a consequence of the mechanisms described before and may vary depending on the cell type and maturation stage. As such, there is still no direct evidence of a common transduction mechanism for the effect of electrical stimulation on neural cells, and more research needs to be conducted.

2.4. Methods for Electrical Stimulation—The Importance of Support Materials

Bioreactors need to be designed to perform the electrical stimulation of cells, as regular cell-culture materials are not designed for this. Both the culture medium and the stimulation process must coexist without either compromising each other's quality or negatively impacting the scaffold or the cultured cells. Different types of bioreactors are possible depending on the element that is directly subjected to electrical stimulation: (a) the culture medium; or (b) the supporting material (Figure 1) [57,58]. Scaffolds for neural tissue engineering can be stimulated with both types of designs.

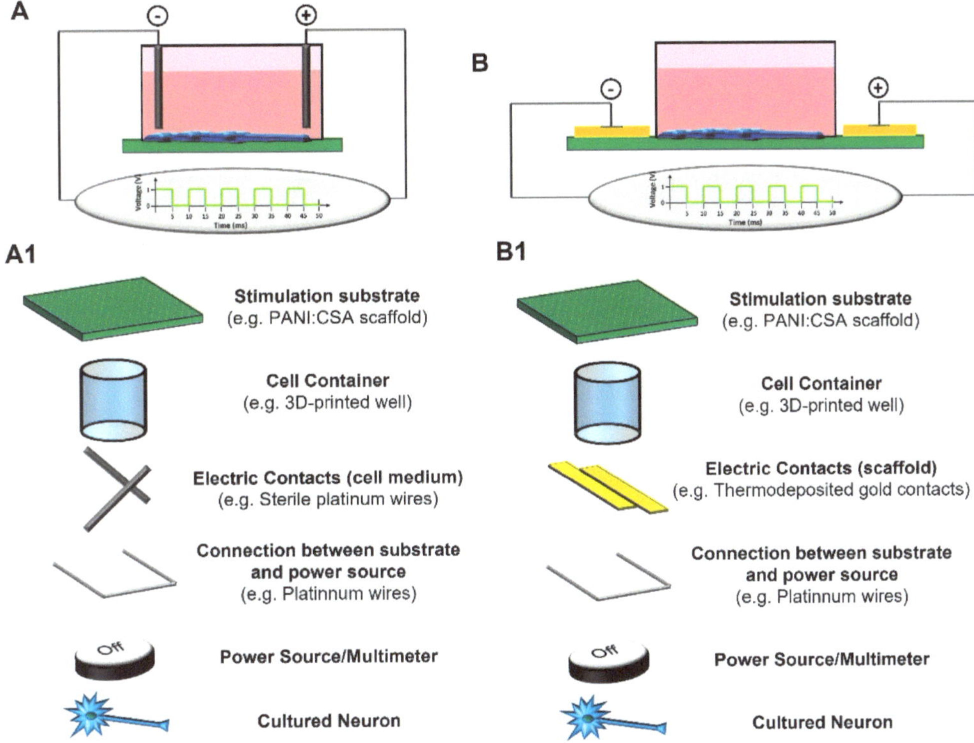

Figure 1. Schematics of the basic bioreactor for the electrical stimulation of neural cells: (**A**) culture medium bioreactor, and (**A1**) respective composition; and (**B**) supporting material bioreactor, and (**B1**) respective composition.

Culture media bioreactors (Figure 1A,(A1)) rely on the application of an electrical field to the culture media in which the cells are maintained. Ions and charged molecules are forced to migrate to either of the electrodes used depending on their charge. Once the stimulus is removed, the homogeneous distribution of charged molecules resumes. The only limitation of this method is the potential hydrolysis of water when the applied potential rises above 1.23 V, promoting the formation of charged chemical species such as OH^- and radicals $OH^{\bullet}$ (faradic products) [59,60].

We consider supporting material bioreactors (Figure 1B,(B1)) as those which have an electroconductive substrate that short-circuits electrons between the used electrodes. Typically, cells are cultured on top of the electroconductive material and are directly influenced by it. Since water hydrolysis is minimized, higher voltages can be applied to cells cultured directly over them. The electroconductivity of the support material is important in order to maximize current passing through the material or to minimize the applied voltage. Other key properties for the support material used include: (1) electrochemical properties; (2) structure; (3) stability in aqueous solutions; and (4) biocompatibility. Some examples of reported bioreactors are summarized in Table 1.

The applied electric field distribution on an electroconductive material should be homogeneous throughout the sample length, independently of its structure. This thematic was explored by Song and colleagues [49], using finite element simulation to evaluate the distribution of the applied electric field on a 2D poly(pyrrole) (PPY) film versus a 3D PPY-coated hydrogel. They observed that the effective electric field was evenly distributed through the PPY film/coating (0.536 $\pm$ 0.095 V cm^{-1} and 0.511 $\pm$ 0.025 V cm^{-1} for 2D and 3D experiments, respectively). The only exception was the area of the contact points with the electrodes where the electric field was twice the value (0.9 V cm^{-1}). For the 3D construct, the authors did not consider the influence of the ionic conductivity of the hydrogel used for cell encapsulation on the penetration of the electrical field. As such, no conclusions could be made about the distribution of the electric field inside the hydrogel and whether seeded cells could be influenced by it. Meneses and colleagues [61] modeled the spatial pattern of electric fields established in the culture medium surrounding a 3D scaffold with different electroconductivities, following electrical stimulation. Scaffolds with lower electroconductivities, including poly(lactide acid) (PLA) (1×10^{-12} S cm^{-1}, not electroconductive) and chitosan–graphene (2.5×10^{-3} S cm^{-1}, mild electroconductive) were associated with high culture-medium polarization (5–6 V m^{-1}), whereas titanium scaffolds with higher electroconductivity (7.5×10^3 S cm^{-1}, highly electroconductive) were not. Such observations can be attributed to a dominant electron conduction with increasing electroconductivity of the material used.

The production of pure electroconductive scaffolds is mostly confined to 2D films due to the conductive polymer's limited processability. Even so, research has shown positive results in the use of electroconductive scaffolds on cell growth and differentiation, especially those studies focused on the initial biocompatibility assessment of new materials or when studying the isolated effects of electroconductivity and/or electrical stimulation on cells [42,49]. Alternatively, the processability of electroconductive polymers into 3D structures can be enhanced by: (1) direct coating/electrodeposition on pre-made 3D structures [62,63]; or (2) blending with carrier polymers, allowing the production of many diverse structures, such as electrospun fibers [64], extruded 3D structures [65,66] and hydrogels [67,68]. Interestingly, the choice between a 2D and a 3D structure is not trivial and carries different consequences on the cell phenotype obtained. For example, Song and colleagues [49] studied the effect of a 2D vs. a 3D structure of PPY films on the phenotype of human-induced neural progenitor cells (hiNPCs) (encapsulated in alginate hydrogels) under electrical stimulation. Interestingly, after just one hour of electrical stimulation (0.4 V cm^{-1}), the authors observed an increased expression of HBEGF, correlated with an increased formation of glial precursor cells, HSPB1, associated with the stress response from cells after electrical stimulation, and NT3, a neurotrophin responsible for neuron maturation, for both the 2D and 3D structures. However, cells increased the expression of BDNF and GDNF (neurotrophic factors responsible for neurite elongation and neuron maturation) only when growing in the 3D structure. This effect might be the consequence of a more evenly distributed electrical field through the 3D structure, as predicted by the finite element analysis. Scaffold limitations, such as low levels of oxygen/glucose inside the hydrogel, should also be considered as possible explanations for the observed secretion of neurotrophic factors. However, electrical stimulation dramatically improved neurotrophic-factor secretion on both 2D and 3D samples. This also justifies the use of more complex and intricated tissue-mimicking structures as reliable platforms for electrical stimulation.

Table 1. Electrical stimulation devices used in the literature and relevant for neural cells: components, type of stimulation performed on neural cells and the observed outcomes.

Type	Cells Used	Stimulated Substrate	Types of Electrodes	Electrolyte Solution Used	Stimulation	Power Source Used	Signal Frequency (Hz)	Duration	Outcomes	References
Direct substrate stimulation	mNSCs	Cross-linked poly(acrylonitrile) (PAN) electrospun fibers	Platinum + printed circuit board	(not used)	100 µA (asymmetric biphasic)	asymmetric biphasic programmable electrical device and a printed circuit board	100	24 h, 1 day after seeding. Cells were allowed to differentiate for more 7 days.	When electrical stimulation was performed, it was observed at the end of the experiment: - Increased cell number; - Increased neurite length; - Increased *Tubb3* (qPCR) and *Map2*/MAP2 (qPCR and IF) and decreased expression of GFAP (IF).	Zhu and colleagues, 2017 [33]
Direct substrate stimulation	hNSC	Nanopatterned Titanium coated PUA	Copper wire + PDMS	(not used)	DC (pulsed), maximum 3 µA and 25 V,	programmable digital power supply MK3003P	1	30 min, 2 times a day, 5 days	Enhancement of hNSC differentiation, independently of the type of substrate used; Increased expression of TUJ1 and MAP2 on differentiated neurons	Yang and colleagues, 2017 [42]
Direct substrate stimulation	NSCs	PCL- PANI_Gelatin electrospun fibers	1 platinum and 1 silver electrodes	(not used)	DC (1 V cm^{-1})	(not referred)	(not referred)	15, 30 and 60 min.	Increased cell number 1, 3 and 5 days after stimulation. Longer neurite extent (30 µm vs. 22 µm)	Ghasemi-Mobarakeh and colleagues, 2009 [47]
Direct substrate stimulation	hNSC (ReN-VM), p3-5	PANI coated PVV hydrogel	PANI-coated Indium Tin Oxide	(not used)	AC—Charged-Balanced biphasic (15, 35 and 75 mV)	Agilent B2912A precision source/measure unit	200	Every 6 h for 1,3,5 and 7 days	Enhanced cell proliferation, that decreased for higher voltage values. Enhanced neurite extension with increasing voltage values. Enhanced neural gene expression with electrical stimulation	Xu and colleagues, 2016 [48]
Dual system of conductive stimulating layer and inert cell support	iNPCs	Alginate hydrogel with cells encapsulated on top of (2D) or surrounded by a PPY film (3D).	Silver wire + silver paste	(not used)	AC, 0.4 V cm^{-1}	(See Oh 2018)	100 Hz	1 h of stimulation + 24 h of resting period	Electrical stimulation enhanced the mRNA expression of HBEGF, HSPB1 and the neurotrophins BDNF, GDNF and NT3. 3D structure + electrical stimulation boosted mRNA expression of BDNF and GDNF	Song and colleagues, 2019 [49]
Direct substrate stimulation	mNSCs	PPY-coated PAN (no cross-linking) electrospun fibers	Stainless Steel	(not used)	100 mV cm^{-1},	AFG3022C, Tektronix, USA	100	4 h of stimulation, total of 7 days.	- Enhanced cell maturation through increased Tau protein expression; - Prevention of neurons growing and differentiating into clusters - Enhanced proliferation of glial cells	Xu and colleagues, 2018 [69]

Table 1. *Cont.*

Type	Cells Used	Stimulated Substrate	Types of Electrodes	Electrolyte Solution Used	Stimulation	Power Source Used	Signal Frequency (Hz)	Duration	Outcomes	References
Direct substrate stimulation	PC12	PLCA-SF-PANI electrospun fibers, monoaxial and hollow co-axial	Similar to Ghasemi-Mobarakeh L et al. 2009	(not used)	(100 mV cm^{-1})	(not referred)	(not referred)	1 h per day, 5 days in total.	Increased neurite-positive cells and respective length.	Zhang and colleagues, 2014 [70]
Direct substrate stimulation	PC12	indium doped tin oxide (ITO) needle coated with PANI	"Wires"	(not used)	AC ($100 \text{ }\mu\text{A}$)	TBSI Neural Stimulator ()V1.0.8, Triangle BioSystems, Durham, NC, USA	1Hz (1 s repeat interval)	1, 2 and 4 h + 24 h of resting	Cell density higher in the following order: $4 \text{ h} = 2 \text{ h} > 1 \text{ h} > 0 \text{ h}$ Increased neurite length with increased duration; Increased protein adsorption with electrical stimulation	Wang and colleagues, 2015 [71]
Stimulation of the culture media	Dissociated neurons from *Xenopus laevis*	Culture media (20% Liebowitz L15 culture medium, 2% penicillin/streptomycin, 1% fetal bovine serum) made in Steinberg's solution	Ag/AgCl electrodes, indirectly connected to culture through agar bridges.	Steinberg solution for the electrode solution and to prepare the culture media.	DC—50–133 mV mm^{-1} (low field strength) and 143–200 mV mm^{-1} (high field strength)	(not referred)	(not referred)	2–4 h after seeding + 5 h of stimulation	Electrical field induced neurite orientation to the cathode; Addition of CS-6S rich GAGs to culture media enhanced neurite turning, whereas CS-4S rich ones inhibited;	Erskine and colleagues, 1997 [72]

The Importance of Electroconductive Materials

Scaffold development for tissue-engineering applications relies on the optimization of biomaterials and their nanostructuring to support NSC growth, differentiation, and maturation. Biomaterials can be viewed as engineered materials (natural, synthetic or hybrid) with bioactive properties, and their therapeutic and toxicologic profiles must be compatible in order to directly contact living systems. In fact, their acceptance both by cells and tissues depends on how their physical (2D/3D structure, electroconductivity, hydrophilicity, mechanical properties, porosity, zeta potential), chemical (chemical functionalization, co-polymerization, molecule immobilization) and pharmacological properties influence or enhance their blending/bioactivity capacity. Through the finetuning of these parameters, we can modulate cell behavior and mimic the target tissue microenvironment, enhancing the material's biocompatibility and easing its clinical applicability to stem cell therapy [42,73,74]. Biocompatibility can be screened using both in vitro (cultured cell lines and stem cells) and in vivo (biological samples, animal models) assays, with experimental timespans depending on the effects under observation [74].

Electroconductive polymers were shown to be capable of sustaining cell growth [75] and conduct electrical stimuli to cultured cells. Electroconductive polymers are composed by a sequence of monomers that form a conjugated single–double bond system, that is, an alternation of π-bonds. Through doping, either by charge or proton transfer, it is possible to induce charged defects in this regular bond alternation that can be propagated through the polymer chains. This delocalization capacity can be quantified through the determination of a polymer's electroconductivity (σ), usually expressed in S m^{-1} (SI) or S cm^{-1} (the most-used in scaffold design). These polymers are also particularly advantageous for other applications such as bioelectronics, due to their combined ionic/electronic conductivities, which decreases their interfacial impedance in biologic fluids, and improves the performance of signal transmission between sensors and cells/tissues [76,77]. When compared to metals, electroconductive polymers are easily processed into 3D structures at relatively low temperatures, have more favorable mechanical properties [32] and can be easy and cheap to produce [78]. These features justify the wide range of additive manufacturing techniques that can be employed in their processing and also make these ideal candidates for designing scaffolds for the potential therapy of neurological diseases [79,80]. Moreover, electroconductive polymers can slightly improve NSC differentiation even in the absence of electrical stimulation [42,81].

Some examples of electroconductive polymers include PANI, PPY and PEDOT [82]. All these polymers require doping agents to create and stabilize their most conductive form, by either charge transfer doping or proton transfer. While PANI is mostly doped via proton transfer, using (mild-) strong acids (e.g., CSA, hydrochloric acid, phosphoric acid, sulfuric acid), PEDOT is stabilized in the oxidized form during its chemical synthesis, using for instance hyaluronic acid or poly(styrene sulfonic acid) [83–85]. Usually, the doping agent's proportion to the conductive polymer must be optimized to obtain the best electric conductivity values. After a certain value (usually 50–60%), electroconductivity starts to decrease with increasing amounts of doping agents.

3. PANI:CSA, a Versatile Electroconductive System Suitable for Neural Tissue Engineering

The use of PANI as an electroconductive substrate for electric stimulation of cells in biomaterials has already been reviewed by Arteshi and colleagues and Qazi and colleagues [74,86]. Moreover, more specialized reviews, such as that by Wei and colleagues [87] summarize the design of conductive fibers using PANI, and other conductive polymers, for diverse medical and soft-electronics applications. In this section, we intend to complement such reviews by exploring the physico-chemical determinants necessary to improve PANI:CSA electroconductivity and the design of platforms/devices for neural applications. We propose to explore these in more detail and explore their influence on the biological performance of platforms for the electrical stimulation of neural cells. We will also place a special focus on in vitro assays due to their wide availability in the literature, while complementing this focus with in vivo assays when possible.

PANI is regarded as a good electroconductive and inexpensive material candidate for the design of electroconductive scaffolds for tissue engineering [88]. When doped, PANI undergoes a change in its resonance structure, switching from the non-conductive emeraldine base (blue) form to the highly conductive emeraldine salt (green) form. This stable salt is the most conductive form of PANI [89,90]. PANI can be doped with several doping agents with positive effects on the polymer's electroconductivity, including CSA, formic acid, hydrochloric acid, picric acid, and phosphoric acid [91–93]. CSA is the most widely used doping agent for PANI (Figure 2), and an optimal proportion between the two is necessary to obtain the highest conductivity value. Such optimization studies were already performed by Holland and colleagues. The authors tested different doping levels of PANI:CSA (10–90%) and concluded that a doping level of 60% resulted in the highest electroconductivity at RT with metallic behavior, even at low temperatures [88].

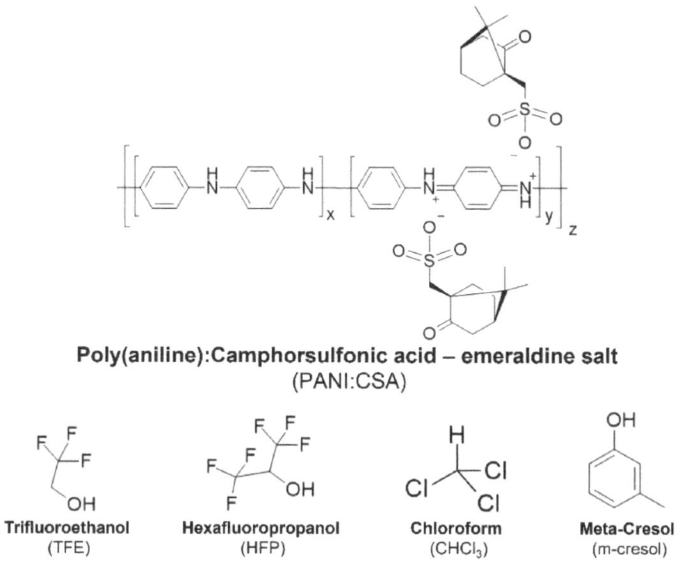

Poly(aniline):Camphorsulfonic acid – emeraldine salt
(PANI:CSA)

Trifluoroethanol
(TFE)

Hexafluoropropanol
(HFP)

Chloroform
(CHCl₃)

Meta-Cresol
(m-cresol)

Figure 2. Chemical structure of poly(aniline) doped with camphorsulfonic acid (PANI:CSA) and solvents suitable for its processability.

3.1. PANI:CSA–Processing Methods, Solvent Systems and Stability

The molecular organization and electroconductivity of PANI:CSA are influenced by the solvent(s) used in its processing, such as *m*-cresol, hexafluoropropanol (HFP) and trifluoroethanol (TFE). These consequences are described in terms of a pseudo-doping effect of the solvents used. A pseudo-doping agent is capable of inducing permanent structural and electroconductive changes to PANI:CSA, in combination with, or even without, a primary doping agent (e.g., CSA). It chemically interacts with PANI:CSA and promote the uncoiling of PANI:CSA aggregates that are naturally formed when dispersed [94,95]. The positive consequences of pseudo-doping on the electroconductivity of PANI:CSA can be observed in numerous studies. Cao and colleagues [96] studied the effect of different organic solvents on the electroconductivity of cast PANI:CSA films, using as reference the conductivity of pressed films (10 S cm^{-1}). They found that solvents such as *m*-cresol were able to solvate PANI:CSA (300 S cm^{-1}), stabilizing it in a more rigid conformation, leading to an increase in conductivity. According to the authors, this arises from the hydrogen-bonding ability of *m*-cresol, allowing chemical interaction between the hydroxyl group of *m*-cresol and the amine groups of PANI. When hydrogen-bond acceptors such as dimethylsulfoxide (DMSO) or chloroform ($CHCl_3$) were used, either alone or mixed with *m*-cresol, the conductivity decreased dramatically (10^{-2} S cm^{-1}). The authors also describe the effect of other bulkier yet weaker hydrogen-bond donor solvents, such as 2-ethylphenol (228 S cm^{-1}), trifluoroacetic acid (100 S cm^{-1}) and hexafluoropropanol (60 S cm^{-1}), which had a similar, but less pronounced effect, when compared to *m*-cresol. In a different work by Hopkins and colleagues [91], HFP was found to pseudo-dope PANI, allowing it to adopt an expanded conformation that increased electroconductivity and decreased chain aggregation. In contrast to *m*-cresol, HFP could be completely removed from PANI:CSA without compromising its electroconductivity. Fryczkowski and colleagues [97] also describe the positive effect of TFE on PANI:CSA electroconductivity when used for processing.

Changes in electroconductivity induced by pseudo-doping agents are concomitant to structural changes on PANI:CSA, which are readily visible through different spectroscopy/diffraction methods. Xia and colleagues [98] reported that processing PANI:CSA with *m*-cresol yields changes in the respective ultraviolet-visible (UV/Vis) and near-infrared (NIR) spectra, including: (1) reduced intensity of the band at 350 nm (π-π^* transition); (2) increased intensity of the band at 440 nm (polaron-π^* transition); (3) reduced intensity of the band at 780 nm (π-polaron transition); and (4) the formation of a wide and intense free-carrier-tail/conduction band in the region of 1000–2600 nm. These changes are explained by an expansion of the coil-like conformation of PANI:CSA aggregates, which improves electroconductivity by an increase in interchain contacts and the spatial extension of the conjugated chain segments. Similar results were observed by other works in the literature [99,100]. Yao and colleagues [99] describe an increased crystallinity of PANI:CSA when processed using solvent systems with increasing concentrations of *m*-cresol. Such effect was attributed to *m*-cresol inducing the formation of more frequent ordered regions in PANI:CSA samples, as suggested by increasing intensity of x-ray diffraction (XRD) peaks at $2\theta = 20°$ and $25°$. Such change is compatible with the observations of Lee and colleagues [101] which associate these changes to an increase in π-π stacking of PANI:CSA chains and a more planar and rigid chain conformation. Similar observations were also made by Chaudari and colleagues [92] for acid-doped PANI and by Hopkins and colleagues [91] for HFP-processed PANI:CSA. A planar and more rigid conformation of PANI:CSA has positive consequences for both electroconductivity, ionic conductivity and capacitance. This justifies the use of such systems in the design of supercapacitors [102,103] and presents an opportunity for the design of devices suitable for brain (cell) stimulation. Otherwise, electroconductivity will remain low even when high amounts of PANI are used to fabricate the films [104].

PANI:CSA properties can change with environmental conditions. PANI:CSA has a polyelectrolyte structure, with a high ability to absorb water. In the short term, this is beneficial for the material's electroconductivity by favoring proton mobility through the polymer chain. In the long term, this effect promotes de-doping of PANI:CSA, compromising its electroconductivity [89,105]. An example of this can be found in the work of Hobaica and colleagues [106], where it was observed that the doping extension of PANI with hydrochloric acid decreased when samples were immersed in water at a pH of 1.18 for approximately 140 h. This led to a decrease in the material's electroconductivity (6.25 to 1.94 S cm^{-1}), which was not observed with prolonged storage in air (five years at room temperature). A similar effect was observed by Bidez and colleagues [107], whose PANI film's resistivity increased by 1,000 times when incubated in Dulbecco's modified eagle's media (DMEM) (37 °C and 5% CO_2) for 100 h. The best option available to overcome this limitation is the use of a carrier polymer to shield PANI from water contact. In fact, an improvement can occur, as was observed in the work of Qazi and colleagues [75], where the electroconductivity of PANI:CSA/poly(glycerol sebacate) (PGS) composites decreased by only 10 times (1.77×10^{-2} to 1.03×10^{-3} S cm^{-1} for PAN/PGS 30%) when immersed in phosphate-buffered saline (PBS) for four days. Another option is the design of coaxial electrospun fibers, where blended PANI:CSA is partly shielded from the environment, as we described in a previous study by our group [38].

The degradation profile of PANI was previously assessed. Using bidimensional spectrochemistry, Lopez-Palácios and colleagues [108] correlated the effect of the number of voltametric cycles with PANI weight loss. This was associated, in an initial stage, with the loss of synthesis byproducts entrapped in the polymers, and in a later stage with degradation and release of soluble products. This confirms the observations made by Kobayahi and colleagues [109] describing PANI's oxidative degradation mechanism on electrodes. These results are important to fully understand the biocompatibility profile of PANI and the potential risks of using PANI in scaffolds for electrical stimulation.

3.2. Biocompatibility for Mammalian Cells

The biocompatibility of PANI has always been a subject of intense debate in the literature. Numerous in vitro and in vivo studies have demonstrated that PANI is biocompatible, either alone or when blended with other materials. For example, PANI is capable of the adsorption of proteins at its surface, avoiding its aggregation and making it a good substrate for cell adhesion and extension [62,71,110]. One of the most robust biocompatibility studies of PANI alone was performed by Humpolicek and colleagues [111]. In a first approach, biocompatibility tests following ISO-10993 were performed on PANI following several un-doping/re-doping cycles, showing that these operations had a positive effect on biocompatibility when human cancer cell lines (HepG2 and HaCaT) were used. Toxicity was therefore attributed to the presence of low-molecular-weight impurities arising from its synthesis. Interestingly, the base form of PANI was observed to be more biocompatible than its salt form [112]. A follow-up study by the same author [113] showed that, among all the possible impurities, ammonium persulfate, an oxidizing agent necessary for the synthesis of PANI, was found to be the most toxic species present. The toxicity of aniline monomers was also evaluated separately, and toxicity was observed above 0.25 mg mL^{-1} for aniline (An) and 0.75 mg mL^{-1} for aniline hydrochloride (AnH). When the monomers were combined with ammonium persulfate (as low as 0.1 mg mL^{-1}) a synergistic toxic effect occurred for every combination of the concentrations tested. The biocompatibility of PANI doped with different acids was also analyzed. Its toxicity was again attributed to the presence of impurities. These results clearly indicate that PANI is a biocompatible polymer and that possible toxic effects on cells can be overcome by improving the material-purification steps after synthesis.

Stem cells are essential for building trustworthy in vitro disease models and tissue engineering strategies due to their ability to differentiate into mature cells. Since stem cells are extremely sensitive to the surrounding microenvironment, it is therefore paramount to determine their compatibility with poly(aniline) in the construction of conductive scaffolds. Initial work from Humpolicek and colleagues [112] showed that extracts from PANI emeraldine salt, but not emeraldine base, could impact the differentiation of R1 mouse embryonic stem cells into erythroid cells and cardiomyocytes (embryotoxicity). No impairments were observed when cardiomyocyte adhesion was evaluated on PANI films (base and salt forms). Likewise, with cell differentiation, both in cardiomyogenesis from mouse ESCs and neurogenesis from mouse neural progenitors, no negative effects were observed. In another study conducted by Xu and colleagues [48], coating poly(2-vinyl-4,6-diamino-1,3,5-triazine)-*co*-1-vinylimidazole) copolymer (PVV)-based hydrogels with PANI did not affect the material's biocompatibility with hNSCs.

In vivo biocompatibility studies for PANI are also available in the literature. Zhang and colleagues [114,115] found that the oral LD50 (lethal dose of a substance that kills 50% of the animals in a test) of pure, in-lab-synthesized PANI nanospheres and nanofibers (40.9–100 mg mL^{-1}) was higher than 3000 mg Kg^{-1}. Interestingly, they also reported that, apart from the lethargy and weight loss observed when only PANI fibers were administered at concentrations higher than 64 mg mL^{-1}, no other toxicity symptoms were observed. Kidney and liver pathohistological analysis revealed no changes in tissue architecture, except for the fatty degeneration of liver cells when mice were administered PANI fibers at 100 mg mL^{-1}. The authors attribute this effect to the lower molecular weight of PANI fibers when compared with PANI nanospheres, which reportedly tend to aggregate and become toxic. Zhou and colleagues [116] showed that, also in mice, the intra-tumoral administration of colloidal PANI-poly(glutamic acid) nanogels was associated with normal tissue histology of heart, liver, spleen lung and kidneys 20 days post-injection. This suggests that leaking and circulating nanogel debris are safe. Regarding the local effects of implanted PANI, Qu and colleagues [117] observed that PANI limited inflammatory response and biodegradation after 28 days implantation of chitosan-PANI/dextran hydrogels in mice. Similar observations were made by Das and colleagues [118], whose silk fibroin (SF)-PANI electrospun nerve conduit enabled successful myelinization of damaged nerves in mice when transplanted for 12 months, especially when Schwann cells were co-transplanted, with no inflammatory response observed. Wang and colleagues [63] observed similar beneficial results in sciatic nerve regeneration of rats for their PANI-coated zein microtubes when transplanted for two months. The local inflammatory response was observed after four months due to PANI debris remaining in the tissue. All in all, PANI can be regarded as a biocompatible material for tissue-engineering applications, with both in vitro and in vivo data supporting this conclusion.

3.3. PANI:CSA and Blend Preparation for Scaffold Design

Due to its poor mechanical properties and inability to be processed in 3D structures alone, the use of PANI for tissue-engineering applications requires the use of carrier polymers. Different polymers can be used, including PCL, PGS, and PLCL. The most widely studied and promising scaffolds for neural tissue engineering include electrospun fibers [70,119] and hydrogels [120,121], where the authors make use of different PANI blends to fully take advantage of its electroconductive properties and positive effects on neural cells.

PGS is a polymer with important applications in neural tissue engineering due to its biocompatibility, biodegradability, and tunable mechanical properties. Conductive PGS composites were obtained by Qazi and colleagues [75] after blending with PANI:CSA (15–30%). The incorporation of PANI directly increased the system's electroconductivity (maximum of $\approx 1.8 \times 10^{-2}$ S cm^{-1}), Young's elastic modulus (0.32 MPa for pure PGS vs. 9.2 MPa for PANI:PGS 30%) and contact angle (86° for pure PGS vs. 107° for PANI:PGS 30%). In vitro stability/degradation of this scaffold was also studied and a constant decrease in 10-times the conductivity value was observed in just four days when incubated with PGS. In fact, the authors reported changes in the FTIR spectra of the composite correlated with the de-doping of PANI and degradation of the PANI:CSA:PGS composite after 15 and 45 days in PBS, which might explain these changes. Finally, the composite was proved to be biocompatible with C2C12 myoblast cells. According to the literature, this PANI:CSA:PGS composite has not been tested with neural cells to date.

PANI can be blended with PCL to produce conductive fibers suitable for biomedical applications. Previous studies have already shown the potential application of conductive fibers made of PCL:PANI blend as gas sensors, where high conductivities were obtained (8×10^{-2} S cm^{-1}) [122]. Our group was the first to systematically study the effect of a PANI:CSA concentration in PCL using TFE as the solvent for dispersion [123]. Our physico-chemical analysis of the PCL:PANI systems showed a phase segregation between PCL and the PANI:CSA system, suggesting that the electroconductivity of PANI:CSA would not be significantly affected by PCL. We also showed that the obtained PCL:PANI fibers are soft and more fragile due to PANI:CSA disruption of the PCL packaging. We observed a steady electroconductivity increase until the PANI content totaled 5% (sample composition 95:5, $\sigma = 4.28 \times 10^{-2}$ S cm^{-1}), after which a plateau was reached until the PANI content was raised to 12% of the fiber content (sample 88:12, $\sigma = 7.7 \times 10^{-2}$ S cm^{-1}). The plateau was attributed to the presence of segregated microscopic PANI:CSA aggregates that disrupted the charge transfer. In a second work [37], we optimized the solvent system used for dispersing PCL:PANI by mixing equal parts of TFE and HFP (TH55). Therefore, the electroconductivity of samples produced in a 50% humidity atmosphere increased by 100× from $8.4 \times 10^{-4}\,\mathrm{S\,cm^{-1}}$ (sample TFE) to $1.9 \times 10^{-1}\,\mathrm{S\,cm^{-1}}$ (sample TH55), in a pair with $1.9 \times 10^{-1}\,\mathrm{S\,cm^{-1}}$ for the sample HFP. Moreover, the obtained PCL:PANI fibers had a larger diameter (373 nm for TH55 vs. 190 nm for TFE), were softer ($\varepsilon = 1.6$ MPa for TH55 vs. $\varepsilon = 4.8$ MPa for TFE and $\varepsilon \approx 10$ MPa for PCL fibers), and were more hydrophilic ($\theta = 45°$ for TH55 vs. $\theta = 87°$ for TFE). We attribute this modification of the properties to the pseudo-doping properties of HFP, and we hypothesized that the PANI chains (in PANI:CSA) adopted an expanded coil conformation, which favored chain–chain contact and facilitated electron transfer through hopping. This strategy was employed by our group in the design of biodegradable and electroconductive coaxial fibers [38]. The coaxial structure was obtained using PGS as the core polymer and the PC:-PANI blend (TH55) as the shell/cladding. The obtained coaxial structure prolonged the stability of the electroconductivity of the fibers for up to 21 days in PBS. PANI:CSA was found to retard fiber degradation by lipase, potentially due to enzyme adsorption by electrostatic interaction [124,125]. Other groups have since developed further work on PCL:PANI blends for tissue-engineering applications, including the development of topographic structures [126] and, in the work of Licciardello and colleagues [127] highly hydrophilic and biocompatible fibers after plasma treatment.

The use of PCL:PANI blend for neural cell cultures was first reported by Yang and colleagues using PC12 cells. They tested two concentrations (1% and 3%) and observed constant proliferation for seven days, with normal cell morphology for both samples. More-over, PC12 neural differentiation was enhanced in PCL:PANI fibers. Our group was the first to test the performance of PCL:PANI fibers with the human-derived neural stem cell lin-eage ReNCell-VM. Increasing PANI concentration on the fibers did not affect the scaffolds' biocompatibility, since NSCs were able to attach and interact with the substrate, proliferate, and keep their spindle/stretched morphology. What is more, NSCs on PCL:PANI samples presented higher growth rates than on pristine PCL fibers and acquired a stellar/spindle morphology [123] In a following work, we also demonstrated that PCL:PANI fibers are suitable for the neural differentiation of both ReNCell-VM [37] (monoaxial fibers) and neural progenitors derived from patient-derived iPSCs (coaxial fibers) [38], attesting their suitability for clinical applications.

Other studies addressed the use of PCL:PANI fibers for other relevant tissue-engineering applications [128,129]. One of the first studies addressing the biocompatibility of the PCL:PANI blend, using the C2C12 myoblast cell line, was published by Ku and col-leagues [128]. PCL and PANI were blended before electrospinning. Although electro-conductivity values were not reported, cyclic-voltammetry measurements showed the redox potential of the fibers and an increasing intensity response with PANI concentra-tions. Both random and aligned PCL:PANI fibers supported proliferation and stimulated the differentiation of C2C12 cells, and the PANI concentration was correlated with the increasing number of myotubes formed and the higher expression of final differentiation genes such as Tropoin T and MHC. In another work, Li and colleagues [129] synthetized PANI directly on PCL 7% fibers, and a high conductivity value for the fibers was obtained ($6.71 \times 10^{-3}\,\mathrm{S\,cm^{-1}}$). The initial biocompatibility study with human umbilical-vein en-dothelial cells (HUVECs) showed successful adhesion of the cells to the substrate after 24 h of culture, showing that the material had good biocompatibility. Further studies with electrical stimulation for five days showed an increase in the proliferation performance of the cells on PCL:PANI fibers when compared with pristine fibers, with an increase in the voltage values applied (200, 300 and $400\,\mathrm{mV\,cm^{-1}}$). Interestingly, HUVECs were better able to spread on PCL:PANI fibers than on pristine PCL fibers, and this effect was enhanced by electrical stimulation. Finally, Wibowo and colleagues [65] reported the use of a PCL:PANI blend, with up to 2% of PANI content, for the 3D bioprinting of scaffolds for potential bone-tissue-engineering applications.

PANI can also be used to create blends for hydrogels. PANI is not water soluble and therefore its incorporation in hydrogels depends on the direct coating or dispersion of PANI nanoparticles into the hydrogel, impairing current distribution through the material. Nevertheless, works in the literature describe the development of PANI-based hydrogels with favorable mechanical properties for neural tissue engineering. Stejskal and colleagues [130] developed a PANI cryogel where poly(aniline) was polymerized in situ inside a frozen ($-20\ ^\circ$C) PVA (poly(vinyl alcohol) or poly(vinylacetate) solution. The result is a highly porous PANI:PVA cryogel, which is hydrophilic, soft (9.7 kPa) and electroconductive ($\approx 10^{-3}$ S cm^{-1}). Humpolicek and colleagues [68] successfully tested the biocompatibility of PANI:PVA cryogels using ESCs, ES-derived cardiomyocytes, and mice neural progenitors. Neural progenitor cell adhesion and spreading onto the material was possible, despite the low adhesion values obtained. This new materials presentation will surely enable a more homogeneous current distribution, suitable for the electrical stimulation of neural cells.

3.4. Electrical Stimulation of Neural Cells on PANI:CSA-Based Platforms

The conductivity and biocompatibility of PANI justify its use in the electrical stimulation of neural cells. The first electrical stimulation studies using PANI-based scaffolds were performed by Ghasemi-Mobarakeh and colleagues [47]. The authors designed electrospun PCL:gelatin:PANI-blend fibers produced for use as nerve guides. The obtained fibers were electroconductive (conductance of 2×10^{-8} S), biodegradable, and biocompatible. Finally, NSCs were cultured under electrical stimulation (DC, 1 V cm^{-1}) for 60 min and their growth profile and morphology post-stimulation (at one, three and five days) were evaluated. Cell numbers through five days of cell culture post-stimulation were consistently higher and neurite length was higher (30 μm vs. 22 μm) than the non-stimulated samples. According to the authors, electrical stimulation caused depolarization of the cell membrane and induced the effects. Additionally, PANI incorporation did not induce visible effects on NSC proliferation or compromise cell viability after electrical stimulation.

Wang and colleagues [104] reported the use of a PAN:PANI:nickel nanoparticles for the development of an electroconductive hydrogel for the electrical stimulation of Schwann cells. The authors observed an increase in cell proliferation when an electrical field of 100 mV cm^{-1} was applied for one hour per day for five days. A similar influence of the low-intensity electrical fields on NSC proliferation on PANI-based hydrogels was observed by Xu and colleagues [48]. From the three values of voltage tested (15, 35 and 75 mV), NSC proliferation peaked when 15 mV were applied for six hours a day and for seven days. Nevertheless, all three stimulation conditions improved NSC proliferation when compared to the non-stimulated hydrogel.

Zhang and colleagues [70] show the potential of using a dual stimulation system to enhance neural cell differentiation of PC12 cells using PANI-based scaffolds. The authors developed electroconductive poly(L-lactic) acid (PLLA).SF.PANI-blend monoaxial fibers and succeeded in increasing both the length (0.04 vs. 13.9 μm) and number of neurite-positive cells (1.5 vs. 29.7%) by using an electrical field of 100 mV cm^{-1} (one hour per day, five days). Cell alignment with the electrical field was also observed. The blend was used to prepare coaxial fibers for the encapsulation and controlled release of NGF under electrical stimulation. PC12 neurite extension and numbers per cell were further enhanced when the co-axial fibers were used concomitant with electrical stimulation and NGF release.

Our group has also developed PANI-based electrospun fibers for neural-differentiation studies under electrical stimulation. In a first approach [37], ReNCell-VM (hNSCs) were cultured and left to proliferate for four days on randomly oriented PCL:PANI monoaxial fibers, followed by another four days of differentiation. In both stages, electrical stimulation of 1 V cm^{-1}, 100 Hz, for 12 h a day (AC stimulation) was applied. The high-intensity electrical field induced cell alignment at the end of both the proliferation and differentiation stages. Moreover, it was observed that electrical stimulation induced a statistically significant upregulation of the neural markers *Dcx*, *Map2* and S100 calcium-binding protein B (*S100β*). In a second study [38], electroconductive and biodegradable PGS/PCL:PANI fibers were used for the electrical stimulation of hiNPCs, obtained after the neural induction of patient-derived iPSCs. The electrical stimulation protocol used was 1 V cm^{-1}, 100 Hz for 12 h a day, but instead used pDC stimulation to maximize neuronal differentiation [30]. Electrical stimulation was found to induce an increase in the axon diameter and alignment, as well as increased expression of specific markers associated with (1) cytoskeleton remodeling; (2) general/functional maturation; and (3) a switch in the neural population present from dominant-inhibitory (GABAergic +) to dominant-stimulatory (Glutamatergic +). These results open doors to new strategies for treating neurological diseases.

PANI-based scaffolds have also been tested for in vivo regeneration of peripheral nerves. Wu and colleagues [131] developed a sensor device made of hydroxyethyl cellulose:soy protein isolate:PANI sponge, suitable for man–machine interfaces due to the high electrical signal sensitivity and low inflammatory response after implantation. Electroconductivity of the best scaffolds gradually decreased for 40 days, but these presented favorable mechanical behavior, even after 14 days in PBS, that were deemed suitable for use as a pressure sensor. This sponge was later processed in a nerve conduit used for in vivo sciatic nerve regeneration under electrical stimulation (60 min every 2 days, 14 days) [132]. Rats treated with a combination of the scaffold and electrical stimulation scored higher, similar to the autograph on the sciatic functional index, indicating functional recovery. Three months after surgery, the regenerated nerve was found to be electrophysiologically functional, presented a higher degree of myelinization and a larger axon diameter. Overall, this study demonstrates the potential clinical applications of PANI-based scaffolds.

PANI can also be used in the design of nanoparticles suitable for performing wireless electrical stimulation for cells that internalize them. Kim and colleagues [133] developed gold-doped PANI nanoparticles suitable for internalization by human mesenchymal/stromal stem cells (hMSCs). Nanoparticle uptake allowed for higher calcium internalization by the cells and maximized the response to electrical stimulation (one pulse, 500 V, three days). hMSCs subjected to the electrical stimulation protocol evidenced a change in phenotype to a neuron-like morphology. The phenomenon was accompanied by (1) increased neurite length; (2) higher expression of neural markers (e.g., TUJ1, MAP2); and (3) upregulation of genes associated with neurogenesis, cell proliferation, apoptosis, DNA repair and neural differentiation. Despite the short duration of the assay and the absence of functionality tests of the cells, this work indicated a positive synergy between the electroconductive materials and wireless electrical stimulation for neural applications.

3.5. Improving the Bioactivity of the PANI:CSA System

The functionalization of biomaterials to increase their bioactivity is another strategy to improve their applications in tissue engineering. This opens the way for the material's surface properties to be tailored to the target tissue, allowing cell recognition, boosting biocompatibility, and allowing for a more optimized cell metabolism. Various strategies can be employed to improve cell responses to the electrical field, including: the (1) immobilization of active molecules on a materials surface; (2) the modulation of substrate topography; and (3) the encapsulation and release of small/large electro-facilitator molecules, such as BDNF or ion-channel modulators.

For the design of neural-friendly biomaterials, functionalization with glycosaminoglycans (GAGs) can potentiate the effect of electrical stimulation on neural cells. Erskine and McCaig [72] studied the effect of GAGs on neurite growth of *Xenopus* sp. nerve cells under electrical stimulation. Cells were cultured for two to four hours for neurites to begin to sprout, and the supplementation of culture media with GAGs enriched in chondroitin-4-sulfate (CS-4S), chondroitin-6-sulfate (CS-6S), dermatan sulfate (DS) or keratan sulfate (KS) was performed during the five-hour stimulation (50–133 mV mm^{-1}). Differences were found for neurite growth. Namely, CS-6S (10 µg mL^{-1}) promoted the neurites to turn to the cathode, whereas CS-4S and DS did not have any effect and KS reduced neurite turning. When higher electrical fields were used, CS-6S continued to promote neurite turning to the cathode but CS-4S decreased neurite turning instead. Concerning neurite growth rates, no changes were observed with GAG supplementation. Interestingly, disaccharide units of CS-4S, CS-6S, DS or KS did not produce any effect on neurite turning. This work suggests that GAGs have a positive effect on neurite directionality, possibly through a direct interaction with the growth cone of the cells. As such, chondroitin-sulfate moieties, or those from other GAGs, could be used to modulate neural cell morphology and growth [134]. Recently, Thompson and colleagues demonstrated the different effects of extracellular-matrix (ECM) molecules isolated from differentiated protoplastamic (grey matter) and fibrous (white matter) astrocytes on axonal growth and guidance [135]. This strategy can be used together with electrical stimulation to improve differentiation or to guide neural cell neurites to specific target regions.

Topography influences neural cell differentiation by influencing cytoskeleton organization and by concentrating growth/differentiation factors near the differentiating cells. Topographic effects can be achieved when, for example, ridges and grooves are produced (by lithography, two-photon printing) in the flat film of a biomaterial. In their work, Tan and colleagues [136] focused on testing the influence of three kinds of gratings on a dopaminergic differentiation protocol for murine NPCs, these included biomaterials with 2 μm, 250 nm and $2 \times 2 \times 2$ μm with interior $250 \times 250 \times 250$ nm (hierarchical grafting). In initial studies, the immunofluorescence of TUJ1 (early neural marker), MAP2 (intermediate neural marker) and glial fibrillary acidic protein (GFAP) (early astrocyte marker) were analyzed. All three gratings showed higher expressions of TUJ1 protein when compared to GFAP, suggesting that more cells were differentiating into neurons, and MAP2 expression was higher in the 2 μm gratings. Elongation and cell alignment was also higher for the 2 μm and hierarchical gratings. When tyrosine hydroxylase (TH), an enzyme involved in dopamine synthesis, was targeted (dopaminergic neurons), high expression values were also found in the same structures, which were not significantly different even after normalization with TUJ1 values. This suggests that the three major tested structures enhanced dopaminergic neuron differentiation (at the protein level) compared to an unpatterned control. However, gene expression analysis revealed significant differences: tyrosine hydroxylase cDNA (*Th*) expression was higher in the hierarchical samples and the expression of pituitary homebox 3 (PITX3), a transcription factor important for dopaminergic neuron survival during development, was higher in the 2 μm gratings. Overall, the work shows the positive effect of topography on neural cell differentiation, especially for constructs with higher spacings (2 μm and hierarchical). Not only do these patterned materials provide more surface area for cells to adhere to, but they also initiate mechano-sensing mechanisms that can potentially lead to changes in the expression of the genes involved in cell differentiation.

Yang and colleagues [137] investigated the additive effect of nanopatterning and electroconductive substrates on myogenic and neural cell differentiation. In one study, gold (2.0×10^5 S m^{-1}) or titanium (1.2×10^3 S m^{-1}) were electro-evaporated on nanopatterned (800 nm width × 600 nm height) poly(urethane acrylate) (PUA), dimensioned to match the naturally aligned collagen fibers in the muscle tissue. The results they obtained indicate that C2C12 cells align with these designed nanopatterns and myotube length and width, and that the fusion index of myocytes and the expression of differentiation genes increase with the increasing substrate electroconductivity. Both these effects were found to act synergically in promoting cell differentiation in myocytes.

In another study by Yang and colleagues [42] the separate and combined effects of nanopatterning, the electroconductive layer (titanium), and electrical stimulation in neural stem cell differentiation were examined. A positive effect of electroconductivity on cell differentiation was observed. This was attributed to the ability of the titanium layer to short-circuit natural electrical signals across it (passive stimulation), leading to increased protein adsorption, thus affording a more cell-friendly scaffold. Different groove–ridge distances were tested (150, 200, 250 and 300 nm) and only 150 and 200 nm distances were found to increase the overall neural differentiation phenotype of the cells (increasing TUJ1 expression, increasing body length and increasing neurite formation). The process was mediated by mechano-transduction and was associated with:

- More contact points for the local adhesion of the cells inducing integrin-binding and clustering, with the consequential increase in integrin-binding domains causing focal adhesion kinase (FAK) activation;
- Actin rearrangement and the modulation of cell contractibility by actomyosin, mediated by the rho-associated protein kinase (ROCK) pathway. The expression of MAP2 levels in hESCs was also important;
- Consequential activation of the mitogen-activated protein kinase (MEK)-ERK pathway, and the enhancement of neuronal differentiation.

They concluded that cytoskeletal organization and actomyosin contractibility were paramount and enhanced hNSC differentiation. This can be promoted when cells are cultured on nanopatterned substrates, and further enhanced when the substrate is electroconductive. Similar results concerning patterned scaffolds and their influence on cell differentiation were found in other studies [138–140].

Patterning results from fiber diameter and alignment using electrospun fibers of poly(ethersulfone) (PES) by Christopherson and colleagues [141] showed the effect of fiber diameter on rat neural stem cell differentiation. The obtained PES fibers had average diameters of 273 ± 45, 749 ± 153 and 1452 ± 312 nm. One main change observed in cell growth was a higher proliferation on the tissue culture plate and fibers with smaller diameters. Fiber diameter did not affect the amount of adsorbed laminin, and nestin expression was observed to be equal for all samples. Therefore, the most plausible factor responsible for these cell growth properties was topography. Moreover, when the differentiation profile was analyzed, the cells were found to express higher levels of TUJ1 (neurons) and lower levels of nestin (neural stem cells) and GFAP (astrocytes) on fibers with a larger diameter (749 nm) than on thinner ones. Wang and colleagues [73] established a relation between fiber diameter and neurite growth and migration of chick dorsal root ganglia-derived Schwann cells. For the three types of PLLA fibers tested (293 ± 65, 759 ± 169 and 1325 ± 383 nm), a higher diameter was associated with a higher neurite length and a greater migration distance of Schwann cells.

Similar results for cell alignment were obtained by Johnson and colleagues [142] for aligned PLLA fibers when rat astrocytes and neurons were co-cultured. Interestingly, it was the topography induced by the presence of electrospun fibers and not the fiber diameters alone that was a critical factor in increasing the neuroprotective properties of astrocytes towards co-cultured neurons. This phenomenon was mediated by the increased expression of the glutamate receptor GLUT-1 in astrocytes, responsible for glutamate removal from the extracellular media. Overall, aligned fibers enhance cell elongation and promote faster regeneration of peripheral nerve cells. This was also shown by Chang and colleagues [143] who found a spiral scaffold to be more effective for nerve repair/regeneration of rats when aligned fibers were used as guiding cues.

4. Challenges, Opportunities and Conclusions

In the final section of this review, we discuss how PANI:CSA platforms can be improved and provide possible future development avenues. The first is by increasing the chemically stability of the doping. PANI:CSA platforms that are suitable for ES must also be stable in aqueous media at physiological conditions (e.g., 37 °C, 5% CO_2, pH = 7.4; ionic strength of 90 mM), without, and/or in the presence of relevant enzymes. In most of the studies reviewed here, such screening is mostly overlooked in favor of the maximization of electroconductivity. It is arguable that the mechanism of transduction of ES into a biological response is solely dependent on the electron or ionic conductivity of an electroconductive polymers [144,145], thus potentially making this a non-issue. However, the natural de-doping of PANI:CSA that occurs in aqueous solutions [38,106] limits the reliability and applicability of PANI:CSA-based platforms for the long-term ES of neural cells (e.g., 28 days). In our past work, we addressed this issue by developing coaxial PGS/PCL:PANI fibers with prolonged stability of electroconductivity in a saline solution and in the presence of the enzyme lipase [38]. We attribute this improved stability to PGS, which has previously been shown to improve the stability of PANI:CSA, as studied by Qazi and colleagues [75]. However, we believe this is not enough. We believe that the PANI:CSA system should be chemically modified to improve its stability in aqueous media, as its instability results from CSA dissolution in aqueous media, especially at pH values close to 7. We propose that CSA and/or other relevant doping agents (e.g., hyaluronic acid) should be chemically linked to a support material (e.g., PCL, PLLA). Since PANI is not water soluble, such a strategy would allow for the immobilization of PANI and improve the stability of PANI:CSA, while increasing its versatility for additive manufacturing.

Some groups have reported the development of water-soluble and self-doped PANI, which might be advantageous for processing in adequate ES platforms [146,147]. However, the production of PANI derivates involves complex synthesis and purification steps that are not easily reproducible. As such, another, easier, strategy to improve the stability of PANI:CSA is to change the doping agent used. Other groups developed composites with PANI doped with water-soluble molecules/polymers (e.g., hyaluronic acid, PLLA) with the ability to dope PANI through free carboxylic groups (pKa = 4–5) [148,149]. CSA has a pKa $\approx$ 1.2 [150], and potential candidates for replacement should have a similar or lower pKa value, while also allowing for an easy processability into scaffolds. An interesting candidate class for replacing CSA would be polyelectrolytes. Among these, PSS (pKa = 1.0) [151] is the most interesting due to its wide reported use with PEDOT, another biocompatible electroconductive polymer. In the PEDOT:PSS system, PSS is also responsible for improving the water dispersibility of PEDOT. It is possible that a similar PANI:PSS system would improve the water dispersibility of PANI and greatly reduce the use of organic solvents. Similar to PEDOT:PSS, cross-linking with (3-glycidyloxypropyl)trimethoxysilane (GOPS) [152] and divinyl sulfone (DVS) [153], or even sulfuric acid treatment [154], could also be employed to improve water resistance and improve the electrical properties of PANI. Other polyelectrolytes are also possible, including biological-derived ones such as polyglutamic acid (pKa = 4.86) [155], hyaluronic acid (pKa = 3.0) [149], or even desoxyribonucleic acid (DNA) (pKa $\approx$ 2.0) [156].

Finally, if de-doping of PANI:CSA cannot be avoided, it can be harnessed for other applications. In the work of Bhattacharya and colleagues [157], PANI was doped with the antibiotic chloroxylenol and used for the production of poly(ethylene oxide) (PEO):PANI:chloraroxylenol electrospun fibers (3.2 S cm^{-1}). The released chloroxylenol greatly improved the antibacterial properties of PANI:PEO fibers against Gram-positive and Gram-negative bacteria. Other, more sophisticated, systems could be developed for the controlled release of immobilized drugs through the application of specific voltage values. This is similar to work described by Sun and colleagues [44].

In conclusion, the manipulation of bioelectricity through electrical stimulation has dramatic effects on the phenotype of neural cells. When physiologically relevant electrical fields are applied, several biomolecular cascades are activated, resulting in an improvement of NSC differentiation. Electroconductive materials, such as PANI:CSA, can be used in the development of bioelectroactive composites and scaffolds suitable for directing electrical stimulation to the cultured cells. Pseudo-doping agents and smart composite design can be employed to maximize electroconductivity and stability. The biocompatibility of PANI:CSA ensures its potential clinical use in long-term applications, including transplantation. The electrical stimulation of neural cells on PANI-based scaffolds improves differentiation as well as when other electroactive materials are used, including PEDOT:PSS and PPY. Improving bioactivity through the immobilization of biomolecules and/or processing PANI-based constructs into topographic/3D scaffolds is possible and desirable for a more efficient neural cell differentiation. Overall, PANI:CSA continues to be a relevant system for the design of platforms/scaffolds for the therapy of neurological diseases using electrical stimulation and tissue engineering. Further studies on PANI:CSA should focus on improving doping stability, potentially through cross-linking and/or a change of CSA for a bulkier doping agent. This would also increase the diversity of scaffolds that could be obtained.

Author Contributions: Conceptualization: F.F.F.G. Supervision: All authors. Project administration: F.F.F.G. Methodology, Investigation and Validation: F.F.F.G. Formal Analysis: F.F.F.G. Data Curation and Visualization: F.F.F.G. Funding Acquisition and Resources: F.F.F.G., R.J.L., F.C.F. and J.M. Writing—Original Draft: F.F.F.G. Writing—Review and Editing: All authors. All authors have read and agreed to the published version of the manuscript.

Funding: This research was funded by dedicated funding from FCT—Portuguese Foundation for Science and Technology (FCT/MCTES) through project BioMaterARISES (EXPL/CTM-CTM/0995/2021). The authors also thank FCT/MCTES for funding through IT (UIDB/50008/2020), iBB (UID/BIO/04565/2022, UIDB/04565/2020 and UIDP/04565/2020), and Laboratório-Associado I4HB (LA/P/0140/2020).

Institutional Review Board Statement: Not applicable.

Data Availability Statement: Not applicable.

Conflicts of Interest: The authors declare no conflict of interest.

References

1. Sankar, T.; Chakravarty, M.; Bescos, A.; Lara, M.; Obuchi, T.; Laxton, A.; McAndrews, M.; Tang-Wai, D.; Workman, C.; Smith, G.; et al. Deep Brain Stimulation Influences Brain Structure in Alzheimer's Disease. *Brain Stimul.* **2015**, *8*, 645–654. [CrossRef]

2. Mann, A.; Gondard, E.; Tampellini, D.; Milsted, J.; Marillac, D.; Hamani, C.; Kalia, S.; Lozano, A. Chronic deep brain stimulation in an Alzheimer's disease mouse model enhances memory and reduces pathological hallmarks. *Brain Stimul.* **2018**, *11*, 435–444. [CrossRef]

3. Little, S.; Pogosyan, A.; Neal, S.; Zavala, B.; Zrinzo, L.; Hariz, M.; Foltynie, T.; Limousin, P.; Ashkan, K.; FitzGerald, J.; et al. Adaptive deep brain stimulation in advanced Parkinson disease. *Ann. Neurol.* **2013**, *74*, 449–457. [CrossRef]

4. Khabarova, E.A.; Denisova, N.P.; Dmitriev, A.B.; Slavin, K.V.; Verhagen Metman, L. Deep Brain Stimulation of the Subthalamic Nucleus in Patients with Parkinson Disease with Prior Pallidotomy or Thalamotomy. *Brain Sci.* **2018**, *8*, 66. [CrossRef]

5. Raza, C.; Anjum, R.; Shakeel, N.U. Parkinson's disease: Mechanisms, translational models and management strategies. *Life Sci.* **2019**, *226*, 77–90. [CrossRef] [PubMed]

6. Yeh, E.; Dao, D.; Wu, Z.; Kandalam, S.; Camacho, F.; Tom, C.; Zhang, W.; Krencik, R.; Rauen, K.; Ullian, E.; et al. Patient-derived iPSCs show premature neural differentiation and neuron type-specific phenotypes relevant to neurodevelopment. *Mol. Psychiatr.* **2018**, *23*, 1687–1698. [CrossRef] [PubMed]

7. Hernández, R.; Jiménez-Luna, C.; Perales-Adán, J.; Perazzoli, G.; Melguizo, C.; Prados, J. Differentiation of Human Mesenchymal Stem Cells towards Neuronal Lineage: Clinical Trials in Nervous System Disorders. *Biomol. Ther.* **2020**, *28*, 34–44. [CrossRef]

8. Ma, Y.; Tang, C.; Chaly, T.; Greene, P.; Breeze, R.; Fahn, S.; Freed, C.; Dhawan, V.; Eidelberg, D. Dopamine Cell Implantation in Parkinson's Disease: Long-Term Clinical and 18F-FDOPA PET Outcomes. *J. Nucl. Med.* **2010**, *51*, 7–15. [CrossRef]

9. Radoszkiewicz, K.; Hribljan, V.; Isakovic, J.; Mitrecic, D.; Sarnowska, A. Critical points for optimizing long-term culture and neural differentiation capacity of rodent and human neural stem cells to facilitate translation into clinical settings. *Exp. Neurol.* **2023**, *363*, 114353. [CrossRef] [PubMed]

10. Kolagar, T.; Farzaneh, M.; Nikkar, N.; Anbiyaiee, A.; Heydari, E.; Khoshnam, S. Human Pluripotent Stem Cells in Neurodegenerative Diseases: Potentials, Advances, and Limitations. *Curr. Stem. Cell Res. Ther.* **2020**, *14*, 102–110. [CrossRef]

11. Mathews, J.; Levin, M. The body electric 2.0: Recent advances in developmental bioelectricity for regenerative and synthetic bioengineering. *Curr. Opin. Biotech.* **2018**, *52*, 134–144. [CrossRef] [PubMed]

12. Clark, M.; Reguera, G. Biology and biotechnology of microbial pilus nanowires. *J. Ind. Microbiol. Biot.* **2020**, *47*, 897–907. [CrossRef]

13. Pfeffer, C.; Larsen, S.; Song, J.; Dong, M.; Besenbacher, F.; Meyer, R.; Kjeldsen, K.; Schreiber, L.; Gorby, Y.; El-Naggar, M.; et al. Filamentous bacteria transport electrons over centimetre distances. *Nature* **2012**, *491*, 218. [CrossRef] [PubMed]

14. Levin, M.; Pezzulo, G.; Finkelstein, J.M. Endogenous bioelectric signaling networks: Exploiting voltage gradients for control of growth and form. *Annu. Rev. Biomed. Eng.* **2017**, *19*, 353–387. [CrossRef] [PubMed]

15. McLaughlin, K.A.; Levin, M. Bioelectric signaling in regeneration: Mechanisms of ionic controls of growth and form. *Dev. Biol.* **2018**, *433*, 177–189. [CrossRef]

16. Vertkin, I.; Styr, B.; Slomowitz, E.; Ofir, N.; Shapira, I.; Berner, D.; Fedorova, T.; Laviv, T.; Barak-Broner, N.; Greitzer-Antes, D.; et al. GABAB receptor deficiency causes failure of neuronal homeostasis in hippocampal networks. *Proc. Natl. Acad. Sci. USA* **2015**, *112*, E3291–E3299. [CrossRef]

17. Stevens, B. Neuron-Astrocyte Signaling in the Development and Plasticity of Neural Circuits. *Neurosignals* **2008**, *16*, 278–288. [CrossRef]

18. Levine, M.; Stevenson, C.G. Regulation of Cell Behavior and Tissue Patterning by Bioelectrical Signals: Challenges and Opportunities for Biomedical Engineering. *Annu. Rev. Biomed. Eng.* **2012**, *14*, 295–332. [CrossRef]

19. Drew, D.; Boudker, O. Shared Molecular Mechanisms of Membrane Transporters. *Annu. Rev. Biochem.* **2016**, *85*, 543–572. [CrossRef]

20. Counillon, L.; Bouret, Y.; Marchiq, I.; Pouysségur, J. Na^+/H^+ antiporter (NHE1) and lactate/H^+ symporters (MCTs) in pH homeostasis and cancer metabolism. *Biochim. Biophys. Acta (BBA) Mol. Cell Res.* **2016**, *1863*, 2465–2480. [CrossRef]
21. Nguyen, H.; Bursac, N. Ion channel engineering for modulation and de novo generation of electrical excitability. *Curr. Opin. Biotechnol.* **2019**, *58*, 100–107. [CrossRef] [PubMed]
22. Cao, L.; Liu, J.; Pu, J.; Collinson, J.; Forrester, J.; McCaig, C. Endogenous bioelectric currents promote differentiation of the mammalian lens. *J. Cell Physiol.* **2018**, *233*, 2202–2212. [CrossRef] [PubMed]
23. Nagy, J.I.; Pereda, A.E.; Rash, J.E. Electrical Synapses in Mammalian CNS: Past Eras, Present Focus and Future Directions. *Biochim. Biophys. Acta* **2018**, *1860*, 102–123. [CrossRef] [PubMed]
24. Bhattacharya, A.; Aghayeva, U.; Berghoff, E.G.; Hobert, O. Plasticity of the electrical connectome of *C. elegans*. *Cell* **2019**, *176*, 1174–1189. [CrossRef] [PubMed]
25. Liu, P.; Chen, B.; Mailler, R.; Wang, Z.-W. Antidromic-rectifying gap junctions amplify chemical transmission at functionally mixed electrical-chemical synapses. *Nat. Commun.* **2017**, *8*, 14818. [CrossRef]
26. McCaig, C.; Rajnicek, A.; Song, B.; Zhao, M. Controlling cell behavior electrically: Current views and future potential. *Physiol. Rev.* **2005**, *85*, 943–978. [CrossRef]
27. Heubach, J.; Graf, E.; Leutheuser, J.; Bock, M.; Balana, B.; Zahanich, I.; Christ, T.; Boxberger, S.; Wettwer, E.; Ravens, U. Electrophysiological properties of human mesenchymal stem cells. *J. Physiol.* **2004**, *554*, 659–672. [CrossRef]
28. Wang, K.; Xue, T.; Tsang, S.; Huizen, R.; Wong, C.; Lai, K.; Ye, Z.; Cheng, L.; Au, K.; Zhang, J.; et al. Electrophysiological Properties of Pluripotent Human and Mouse Embryonic Stem Cells. *Stem. Cells* **2005**, *23*, 1526–1534. [CrossRef]
29. Jiang, P.; Rushing, S.; Kong, C.; Fu, J.; Lieu, D.; Chan, C.; Deng, W.; Li, R. Electrophysiological properties of human induced pluripotent stem cells. *Am. J. Physiol. Cell Physiol.* **2010**, *298*, C486–C495. [CrossRef]
30. Sordini, L.; Garrudo, F.; Rodrigues, C.; Linhardt, R.; Cabral, J.; Ferreira, F.; Morgado, J. Effect of Electrical Stimulation Conditions on Neural Stem Cells Differentiation on Cross-Linked PEDOT:PSS Films. *Front. Bioeng. Biotechnol.* **2021**, *9*, 591838. [CrossRef]
31. Freeman, D.; Eddington, D.; Rizzo, J.; Fried, S. Selective Activation of Neuronal Targets With Sinusoidal Electric Stimulation. *J. Neurophysiol.* **2010**, *104*, 2778–2791. [CrossRef]
32. Pires, F.; Ferreira, Q.; Rodrigues, C.; Morgado, J.; Ferreira, F. Neural stem cell differentiation by electrical stimulation using a cross-linked PEDOT substrate: Expanding the use of biocompatible conjugated conductive polymers for neural tissue engineering. *Biochim. Biophys. Acta Gen. Subj.* **2015**, *1850*, 1158–1168. [CrossRef]
33. Zhu, W.; Ye, T.; Lee, S.-J.; Cui, H.; Miao, S.; Zhou, X.; Shuai, D.; Zhang, L. Enhanced Neural Stem Cell Functions in Conductive Annealed Carbon Nanofibrous Scaffolds with Electrical Stimulation. *Nanomedicine* **2017**, *14*, 2485–2494. [CrossRef]
34. Derhambakhsh, S.; Mohammadi, J.; Shokrgozar, M.; Rabbani, H.; Sadeghi, N.; Nekounam, H.; Mohammadi, S.; Lee, K.-B.; Khakbiz, M. Investigation of electrical stimulation on phenotypic vascular smooth muscle cells differentiation in tissue-engineered small-diameter vascular graft. *Tissue Cell* **2023**, *81*, 101996. [CrossRef] [PubMed]
35. Zhao, Y.; Liang, Y.; Ding, S.; Zhang, K.; Mao, H.; Yang, Y. Application of conductive PPy/SF composite scaffold and electrical stimulation for neural tissue engineering. *Biomaterials* **2020**, *255*, 120164. [CrossRef] [PubMed]
36. Xu, Q.; Jin, L.; Li, C.; Kuddannayai, S.; Zhang, Y. The effect of electrical stimulation on cortical cells in 3D nanofibrous scaffolds. *RSC Adv.* **2018**, *8*, 11027–11035. [CrossRef]
37. Garrudo, F.; Mikael, P.; Rodrigues, C.; Udangawa, R.; Paradiso, P.; Chapman, C.; Hoffman, P.; Colaço, R.; Cabral, J.; Morgado, J.; et al. Polyaniline-polycaprolactone fibers for neural applications: Electroconductivity enhanced by pseudo-doping. *Mater. Sci. Eng. C* **2021**, *120*, 111680. [CrossRef]
38. Garrudo, F.F.; Nogueira, D.E.S.E.; Rodrigues, C.A.V.A.; Ferreira, F.A.A.; Paradiso, P.; Colaço, R.; Marques, A.C.; Cabral, J.M.S.M.; Morgado, J.; Linhardt, R.J.; et al. Electrical stimulation of neural-differentiating iPSCs on novel coaxial electroconductive nanofibers. *Biomater. Sci.* **2021**, *9*, 5359–5382. [CrossRef]
39. Chen, C.; Chen, X.; Zhang, H.; Zhang, Q.; Wang, L.; Li, C.; Dai, B.; Yang, J.; Liu, J.; Sun, D. Electrically-responsive core-shell hybrid microfibers for controlled drug release and cell culture. *Acta Biomater.* **2017**, *55*, 434–442. [CrossRef]
40. Koppes, A.N.; Zaccor, N.W.; Rivet, C.J.; Williams, L.A.; Piselli, J.M.; Gilbert, R.J.; Thompson, D.M. Neurite outgrowth on electrospun PLLA fibers is enhanced by exogenous electrical stimulation. *J. Neural Eng.* **2014**, *11*, 046002. [CrossRef] [PubMed]
41. Zhou, X.; Yang, A.; Huang, Z.; Yin, G.; Pu, X.; Jin, J. Enhancement of neurite adhesion. alignment and elongation on conductive polypyrrole-poly(lactide acid) fibers with cell-derived extracellular matrix. *Colloids Surf. B* **2017**, *149*, 217–225. [CrossRef] [PubMed]
42. Yang, K.; Yu, S.; Lee, J.; Lee, H.-R.; Chang, G.-E.; Seo, J.; Lee, T.; Cheong, E.; Im, S.G.; Cho, S.W. Electroconductive nanoscale topography for enhanced neuronal differentiation and electrophysiological maturation of human neural stem cells. *Nanoscale* **2017**, *9*, 18737–18752. [CrossRef] [PubMed]
43. Koppes, A.; Seggio, A.; Thompson, D. Neurite outgrowth is significantly increased by the simultaneous presentation of Schwann cells and moderate exogenous electric fields. *J. Neural. Eng.* **2011**, *8*, 046023. [CrossRef] [PubMed]
44. Sun, X.; Li, L.; Tan, Z.; Li, J.; Hou, Y.; Wang, X.; Liu, B.; Xing, X.; Rong, L.; He, L. On-demand release of the small-molecule TrkB agonist improves neuron-Schwann cell interactions. *J. Control Release* **2022**, *343*, 482–491. [CrossRef]
45. Zhang, Z.; Liu, X.; Schroeder, J.; Chan, C.-B.; Song, M.; Yu, S.; Weinshenker, D.; Ye, K. 7,8-Dihydroxyflavone Prevents Synaptic Loss and Memory Deficits in a Mouse Model of Alzheimer's Disease. *Neuropsychopharmacology* **2014**, *39*, 638–650. [CrossRef]

46. Jeong, S.; Jun, S.; Song, J.; Kim, S. Activity-dependent neuronal cell migration induced by electrical stimulation. *Med. Biol. Eng. Comput.* **2009**, *47*, 93–99. [CrossRef]

47. Ghasemi-Mobarakeh, L.; Prabhakaran, M.P.; Morshed, M.; Nasr-Esfahani, M.H.; Ramakrishna, S. Electrical stimulation of nerve cells using conductive nanofibrous scaffolds for nerve tissue engineering. *Tissue Eng. Part A* **2009**, *15*, 3605–3619. [CrossRef]

48. Xu, B.; Bai, T.; Sinclair, A.; Wang, W.; Wu, Q.; Gao, F.; Jia, H.; Jiang, S.; Liu, W. Directed neural stem cell differentiation on polyaniline-coated high strength hydrogels. *Mater. Today Chem.* **2016**, *1*, 15–22. [CrossRef]

49. Song, S.; Amores, D.; Chen, C.; McConnell, K.; Oh, B.; Poon, A.; George, P. Controlling properties of human neural progenitor cells using 2D and 3D conductive polymer scaffolds. *Sci. Rep.* **2019**, *9*, 19565. [CrossRef]

50. Borah, R.; Ingavle, G.; Sandeman, S.; Kumar, A.; Mikhalovsky, S. Electrically conductive MEH-PPV:PCL electrospun nanofibres for electrical stimulation of rat PC12 pheochromocytoma cells. *Biomater. Sci.* **2018**, *6*, 2342–2359. [CrossRef]

51. Blanquie, O.; Kilb, W.; Sinning, A.; Luhmann, H. Homeostatic interplay between electrical activity and neuronal apoptosis in the developing neocortex. *Neuroscience* **2017**, *358*, 190–200. [CrossRef] [PubMed]

52. Isensee, J.; Schild, C.; Schwede, F.; Hucho, T. Crosstalk from cAMP to ERK1/2 emerges during postnatal maturation of nociceptive neurons and is maintained during aging. *J. Cell Sci.* **2017**, *130*, 2134–2146. [PubMed]

53. Meissner, A.; Noack, T. Proliferation of human lens epithelial cells (HLE-B3) is inhibited by blocking of voltage-gated calcium channels. *Pflügers. Arch. Eur. J. Physiol.* **2008**, *457*, 47–59. [CrossRef] [PubMed]

54. Liu, J.; Dong, Y.; Wen, Y.; Shi, L.; Zhu, Z.; Ke, G.; Gu, Y. LncRNA KCNQ1OT1 knockdown inhibits viability, migration and epithelial-mesenchymal transition in human lens epithelial cells via miR-26a-5p/ITGAV/TGF-beta/Smad3 axis. *Exp. Eye Res.* **2020**, *200*, 108251. [CrossRef] [PubMed]

55. Wang, S.; Guan, S.; Sun, C.; Liu, H.; Liu, T.; Ma, X. Electrical stimulation enhances the neuronal differentiation of neural stem cells in three-dimensional conductive scaffolds through the voltage-gated calcium ion channel. *Brain Res.* **2023**, *1798*, 148163. [CrossRef]

56. Zhu, R.; Sun, Z.; Li, C.; Ramakrishna, S.; Chiu, K.; He, L. Electrical stimulation affects neural stem cell fate and function in vitro. *Exp. Neurol.* **2019**, *319*, 112963. [CrossRef]

57. Balint, R.; Cassidy, N.; Cartmell, S. Electrical Stimulation: A Novel Tool for Tissue Engineering. *Tissue Eng. Part B Rev.* **2013**, *19*, 48–57. [CrossRef]

58. Pascoal-Faria, P.; Ferreira, P.; Datta, A.; Amado, S.; Moura, C.; Alves, N. Electrical Stimulation Optimization in Bioreactors for Tissue Engineering Applications. *Appl. Mech. Mater.* **2019**, *890*, 314–323. [CrossRef]

59. Bodamyali, T.; Kanczler, J.; Simon, B.; Blake, D.; Stevens, C. Effect of Faradic Products on Direct Current-Stimulated Calvarial Organ Culture Calcium Levels. *Biochem. Bioph. Res. Commun.* **1999**, *264*, 657–661. [CrossRef] [PubMed]

60. Rossmeisl, J.; Dimitrievski, K.; Siegbahn, P.; Nørskov, J. Comparing Electrochemical and Biological Water Splitting. *J. Phys. Chem. C* **2007**, *111*, 18821–18823. [CrossRef]

61. Meneses, J.; Fernandes, S.; Alves, N.; Pascoal-Faria, P.; Miranda, P. Effects of Scaffold Electrical Properties on Electric Field Delivery in Bioreactors. In Proceedings of the 2021 43rd Annual International Conference of the IEEE Engineering in Medicine & Biology Society (EMBC), Mexico, 1–5 November 2021.

62. Wang, L.-P.; Wang, W.; Di, L.; Lu, Y.-N.; Wang, J.-Y. Protein adsorption under electrical stimulation of neural probe coated with polyaniline. *Colloids Surf. B Biointerfaces* **2010**, *80*, 72–78. [CrossRef] [PubMed]

63. Wang, G.; Wu, W.; Yang, H.; Zhang, P.; Wang, J. Intact polyaniline coating as a conductive guidance is beneficial to repairing sciatic nerve injury. *J. Biomed. Mater. Res. Part B Appl. Biomater.* **2020**, *108*, 128–142. [CrossRef] [PubMed]

64. Khan, M.; Cantù, E.; Tonello, S.; Serpelloni, M.; Lopomo, N.; Sardini, E. A Review on Biomaterials for 3D Conductive Scaffolds for Stimulating and Monitoring Cellular Activities. *Appl. Sci.* **2019**, *9*, 961. [CrossRef]

65. Wibowo, A.; Vyas, C.; Cooper, G.; Qulub, F.; Suratman, R.; Mahyuddin, A.; Dirgantara, T.; Bartolo, P. 3D Printing of Polycaprolactone–Polyaniline Electroactive Scaffolds for Bone Tissue Engineering. *Materials* **2020**, *13*, 512. [CrossRef]

66. Menzel, V.; Tudela, I. Additive manufacturing of polyaniline-based materials: An opportunity for new designs and applications in energy and biotechnology. *Curr. Opin. Chem. Eng.* **2022**, *35*, 100742. [CrossRef]

67. Distler, T.; Boccaccini, A. 3D Printing of Electrically Conductive Hydrogels for Tissue Engineering and Biosensors—A Review. *Acta Biomater.* **2020**, *101*, 1–13. [CrossRef] [PubMed]

68. Humpolíček, P.; Radaszkiewicz, K.; Capáková, Z.; Pacherník, J.; Bober, P.; Kašpárková, V.; Rejmontová, P.; Lehocký, M.; Ponížil, P.; Stejskal, J. Polyaniline cryogels: Biocompatibility of novel conducting macroporous material. *Sci. Rep.* **2018**, *8*, 135. [CrossRef]

69. Xu, J.; Wong, C.-W.; Hsu, S. An Injectable. Electroconductive Hydrogel/Scaffold for Neural Repair and Motion Sensing. *Chem. Mater.* **2020**, *32*, 10407–10422. [CrossRef]

70. Zhang, J.; Qiu, K.; Sun, B.; Fang, J.; Zhang, K.; EI-Hamshary, H.; Al-Deyab, S.S.; Mo, X. The aligned core–sheath nanofibers with electrical conductivity for neural tissue engineering. *J. Mater. Chem. B* **2014**, *2*, 7945–7954. [CrossRef]

71. Wang, L.; Huang, Q.; Wang, J.-Y. Nanostructured Polyaniline Coating on ITO Glass Promotes the Neurite Outgrowth of PC 12 Cells by Electrical Stimulation. *Langmuir* **2015**, *31*, 12315–12322. [CrossRef]

72. Erskine, L.; McCaig, C.D. Integrated interactions between chondroitin sulphate proteoglycans and weak dc electric fields regulate nerve growth cone guidance in vitro. *J. Cell Sci.* **1997**, *110*, 1957–1965. [CrossRef]

73. Wang, H.; Mullins, M.; Cregg, J.; McCarthy, C.; Gilbert, R. Varying the diameter of aligned electrospun fibers alters neurite outgrowth and Schwann cell migration. *Acta Biomater.* **2010**, *6*, 2970–2978. [CrossRef]

74. Arteshi, Y.; Aghanejad, A.; Davaran, S.; Omidi, Y. Biocompatible and electroconductive polyaniline-based biomaterials for electrical stimulation. *Eur. Pol. J.* **2018**, *108*, 150–170. [CrossRef]

75. Qazi, T.H.; Rai, R.; Dippold, D.; Roether, J.E.; Schubert, D.W.; Rosellini, E.; Barbani, N.; Boccaccini, A.R. Development and characterization of novel electrically conductive PANI–PGS composites for cardiac tissue engineering applications. *Acta Biomater.* **2014**, *10*, 2434–2445. [CrossRef] [PubMed]

76. Kim, S.; Kim, K.-M.; Hoffman-Kim, D.; Song, H.-K.; Palmore, G. Quantitative Control of Neuron Adhesion at a Neural Interface Using a Conducting Polymer Composite with Low Electrical Impedance. *ACS Appl. Mater. Interfaces* **2010**, *3*, 16–21. [CrossRef] [PubMed]

77. Alba, N.; Du, Z.; Catt, K.; Kozai, T.; Cui, X. In Vivo Electrochemical Analysis of a PEDOT/MWCNT Neural Electrode Coating. *Biosensors* **2015**, *5*, 618–646. [CrossRef] [PubMed]

78. Guimard, N.; Gomez, N.; Schmidt, C. Conducting polymers in biomedical engineering. *Prog. Polym. Sci.* **2007**, *32*, 876–921. [CrossRef]

79. Balint, R.; Cassidy, N.; Cartmell, S. Conductive polymers: Towards a smart biomaterial for tissue engineering. *Acta Biomater.* **2014**, *10*, 2341–2353. [CrossRef]

80. Sensharma, P.; Madhumathi, G.; Jayant, R.D.; Jaiswal, A.K. Biomaterials and cells for neural tissue engineering: Current choices. *Mat. Sci. Eng. C* **2017**, *77*, 1302–1315. [CrossRef]

81. Shin, J.; Choi, E.; Cho, J.; Cho, A.-N.; Jin, Y.; Yang, K.; Song, C.; Cho, S.-W. Three-Dimensional Electroconductive Hyaluronic Acid Hydrogels Incorporated with Carbon Nanotubes and Polypyrrole by Catechol-Mediated Dispersion Enhance Neurogenesis of Human Neural Stem Cells. *Biomacromolecules* **2017**, *18*, 3060–3072. [CrossRef]

82. Rocha, I.; Cerqueira, G.; Penteado, F.; de Torresi, S.I.C. Electrical Stimulation and Conductive Polymers as a Powerful Toolbox for Tailoring Cell Behaviour in vitro. *Front. Med. Technol.* **2021**, *3*, 670274. [CrossRef] [PubMed]

83. Neoh, K.G.; Pun, M.Y.; Kang, E.T.; Tan, K.L. Polyaniline treated with organic acids: Doping characteristics and stability. *Synth. Met.* **1995**, *73*, 209–215. [CrossRef]

84. Dimitriev, O.P.; Grinko, D.A.; Noskov, Y.V.; Ogurtsov, N.A.; Pud, A.A. PEDOT:PSS films—Effect of organic solvent additives and annealing on the film conductivity. *Synth. Met.* **2009**, *159*, 2237–2239. [CrossRef]

85. Glipa, X.; Bonnet, B.; Mula, B.; Deborah, J.; Rozière, J. Investigation of the conduction properties of phosphoric and sulfuric acid doped polybenzimidazole. *J. Mater. Chem.* **1999**, *9*, 3045–3049. [CrossRef]

86. Qazi, T.H.; Rai, R.; Boccaccini, A.R. Tissue engineering of electrically responsive tissues using polyaniline based polymers: A review. *Biomaterials* **2014**, *35*, 9068–9086. [CrossRef]

87. Wei, L.; Wang, S.; Shan, M.; Li, Y.; Wang, Y.; Wang, F.; Wang, L.; Mao, J. Conductive fibers for biomedical applications. *Bioact. Mater.* **2023**, *22*, 343–364. [CrossRef]

88. Holland, E.R.; Pomfret, S.J.; Adams, P.N.; Monkman, A.P. Conductivity studies of polyaniline doped with CSA. *J. Phys. Condens. Matter.* **1996**, *8*, 2991–3002. [CrossRef]

89. Focke, W.; Wnek, G. Conduction mechanisms in polyaniline (emeraldine salt). *J. Electroanal. Chem. Interfacial Electrochem.* **1988**, *256*, 343–352. [CrossRef]

90. Li, M.; Guo, Y.; Wei, Y.; MacDiarmid, A.; Lelkes, P. Electrospinning polyaniline-contained gelatin nanofibers for tissue engineering applications. *Biomaterials* **2006**, *27*, 2705–2715. [CrossRef]

91. Hopkins, A.R.; Rasmussen, P.G. Characterization of solution and solid state properties of undoped and doped polyanilines processed from hexafluoro-2-propanol. *Macromolecules* **1996**, *29*, 7838–7846. [CrossRef]

92. Chaudhari, H.; Kelkar, D. Investigation of Structure and Electrical Conductivity in Doped Polyaniline. *Polym. Int.* **1997**, *42*, 380–384. [CrossRef]

93. Blinova, N.; Stejskal, J.; Trchová, M.; Prokeš, J. Control of polyaniline conductivity and contact angles by partial protonation. *Polym. Int.* **2008**, *57*, 66–69. [CrossRef]

94. Leite, F.; Neto, M.; Paterno, L.; Ballestero, M.; Polikarpov, I.; Mascarenhas, Y.; Herrmann, P.; Mattoso, L.; Oliveira, O. Nanoscale conformational ordering in polyanilines investigated by SAXS and AFM. *J. Colloid Interf. Sci.* **2007**, *316*, 376–387. [CrossRef]

95. Lobov, I.; Davletkildeev, N.; Sokolov, D. AFM study of the supramolecular transformation of polyaniline and polyaniline/carbon nanotubes composite upon doping with dodecylbenzenesulfonic acid in the presence of a solvent. *IOP Conf. Ser. Mater. Sci. Eng.* **2017**, *256*, 012017. [CrossRef]

96. Cao, Y.; Qiu, J.; Smith, P. Effect of solvents and co-solvents on the processibility of polyaniline: I. Solubility and conductivity studies. *Synth. Met.* **1995**, *69*, 187–190. [CrossRef]

97. Fryczkowski, R.; Gorczowska, M.; Fryczkowska, B.; Janicki, J. The effect of solvent on the properties of nanofibres obtained by electrospinning from a mixture of poly(3-hydroxybutyrate) and polyaniline. *Synth. Met.* **2013**, *166*, 14–21. [CrossRef]

98. Xia, Y.; Wiesinger, J.M.; MacDiarmid, A.G. Camphorsulfonic acid fully doped polyaniline emeraldine salt: Conformations in different solvents studied by an ultraviolet/visible/near-infrared spectroscopic method. *Chem. Mater.* **1995**, *7*, 443–445. [CrossRef]

99. Yao, Q.; Wang, Q.; Wang, L.; Wang, Y.; Sun, J.; Zeng, H.; Jin, Z.; Huang, X.; Chen, L. The synergic regulation of conductivity and Seebeck coefficient in pure polyaniline by chemically changing the ordered degree of molecular chains. *J. Mater. Chem. A* **2014**, *2*, 2634. [CrossRef]

100. Lenin, R.; Singh, A.; Bera, C. Effect of dopants and morphology on the electrical properties of polyaniline for various applications. *J. Mater. Sci. Mater. Electron.* **2021**, *32*, 24710–24725. [CrossRef]

101. Lee, K.; Cho, S.; Park, S.; Heeger, A.; Lee, C.-W.; Lee, S.-H. Metallic transport in polyaniline. *Nature* **2006**, *441*, 65–68. [CrossRef]
102. Bhattacharya, S.; Roy, I.; Tice, A.; Chapman, C.; Udangawa, R.; Chakrapani, V.; Plawsky, J.L.; Linhardt, R.J. High Conductivity and High Capacitance Electrospun Fibers for Supercapacitor Applications. *ACS Appl. Mater. Interfaces* **2020**, *12*, 19369–19376. [CrossRef]
103. Chen, X.; Liu, P.; Liu, C.; Liu, G.; Wei, J.; Xu, J.; Jiang, Q.; Liu, X.; Jiang, F. Microstructure control for high-capacitance polyaniline. *Electrochim. Acta* **2021**, *391*, 138977. [CrossRef]
104. Wang, M.; Tremblay, P.-L.; Zhang, T. Optimizing the electrical conductivity of polyacrylonitrile/polyaniline with nickel nanoparticles for the enhanced electrostimulation of Schwann cells proliferation. *Bioelectrochemistry* **2021**, *140*, 107750. [CrossRef]
105. Ayad, M.; Zaki, E. Doping of polyaniline films with organic sulfonic acids in aqueous media and the effect of water on these doped films. *Eur. Polym. J.* **2008**, *44*, 3741–3747. [CrossRef]
106. Hobaica, S. Stability of polyaniline in air and acidic water. *J. Polym. Sci. Part B Polym. Phys.* **2003**, *41*, 807–822. [CrossRef]
107. Bidez, P.; Li, S.; Macdiarmid, A.; Venancio, E.; Wei, Y.; Lelkes, P. Polyaniline, an electroactive polymer. supports adhesion and proliferation of cardiac myoblasts. *J. Biomater. Sci. Polym. Ed.* **2006**, *17*, 199–212. [CrossRef]
108. López-Palacios, J.; Muñoz, E.; Heras, A.; Colina, Á.; Ruiz, V. Study of polyaniline films degradation by thin-layer bidimensional spectroelectrochemistry. *Electrochim. Acta* **2006**, *52*, 234–239. [CrossRef]
109. Kobayashi, T.; Yoneyama, H.; Tamura, H. Oxidative degradation pathway of polyaniline film electrodes. *J. Electroanal. Chem. Interfacial Electrochem.* **1984**, *177*, 293–297. [CrossRef]
110. Di, L.; Wang, L.-P.; Lu, Y.-N.; He, L.; Lin, Z.-X.; Wu, K.-J.; Ren, Q.-S.; Wang, J.-Y. Protein adsorption and peroxidation of rat retinas under stimulation of a neural probe coated with polyaniline. *Acta Biomater.* **2011**, *7*, 3738–3745. [CrossRef] [PubMed]
111. Humpolicek, P.; Kasparkova, V.; Saha, P.; Stejskal, J. Biocompatibility of polyaniline. *Synth. Met.* **2012**, *162*, 722–727. [CrossRef]
112. Humpolíček, P.; Kašpárková, V.; Pacherník, J.; Stejskal, J.; Bober, P.; Capáková, Z.; Radaszkiewicz, K.; Junkar, I.; Lehocký, M. The biocompatibility of polyaniline and polypyrrole: A comparative study of their cytotoxicity. embryotoxicity and impurity profile. *Mater. Sci. Eng. C* **2018**, *91*, 303–310. [CrossRef]
113. Kašpárková, V.; Humpolíček, P.; Stejskal, J.; Capáková, Z.; Bober, P.; Skopalová, K.; Lehocký, M. Exploring the Critical Factors Limiting Polyaniline Biocompatibility. *Polymers* **2019**, *11*, 362. [CrossRef]
114. Zhang, Y.; Zhou, M.; Dou, C.; Ma, G.; Wang, Y.; Feng, N.; Wang, W.; Fang, L. Synthesis and biocompatibility assessment of polyaniline nanomaterials. *J. Bioact. Compat. Polym.* **2019**, *34*, 16–24. [CrossRef]
115. Ibarra, L.; Tarres, L.; Bongiovanni, S.; Barbero, C.; Kogan, M.; Rivarola, V.; Bertuzzi, M.; Yslas, E. Assessment of polyaniline nanoparticles toxicity and teratogenicity in aquatic environment using Rhinella arenarum model. *Ecotox. Environ. Safe* **2015**, *114*, 84–92. [CrossRef]
116. Zhou, Y.; Hu, Y.; Sun, W.; Zhou, B.; Zhu, J.; Peng, C.; Shen, M.; Shi, X. Polyaniline-loaded γ-polyglutamic acid nanogels as a platform for photoacoustic imaging-guided tumor photothermal therapy. *Nanoscale* **2017**, *9*, 12746–12754. [CrossRef] [PubMed]
117. Qu, J.; Zhao, X.; Ma, P.; Guo, B. Injectable antibacterial conductive hydrogels with dual response to an electric field and pH for localized "smart" drug release. *Acta Biomater.* **2018**, *72*, 55–69. [CrossRef]
118. Das, S.; Sharma, M.; Saharia, D.; Sarma, K.; Muir, E.; Bora, U. Electrospun silk-polyaniline conduits for functional nerve regeneration in rat sciatic nerve injury model. *Biomed. Mater.* **2017**, *12*, 045025. [CrossRef] [PubMed]
119. Bhang, S.; Jeong, S.; Lee, T.; Jun, I.; Lee, Y.; Kim, B.; Shin, H. Electroactive Electrospun Polyaniline/Poly[(L-lactide)-co-(ε-caprolactone)] Fibers for Control of Neural Cell Function. *Macromol. Biosci.* **2012**, *12*, 402–411. [CrossRef]
120. Mawad, D.; Stewart, E.; Officer, D.; Romeo, T.; Wagner, P.; Wagner, K.; Wallace, G.A. A Single Component Conducting Polymer Hydrogel as a Scaffold for Tissue Engineering. *Adv. Funct. Mater.* **2012**, *22*, 2692–2699. [CrossRef]
121. Guarino, V.; Alvarez-Perez, M.; Borriello, A.; Napolitano, T.; Ambrosio, L. Conductive PANi/PEGDA Macroporous Hydrogels For Nerve Regeneration. *Adv. Healthc. Mater.* **2013**, *2*, 218–227. [CrossRef]
122. Low, K.; Chartuprayoon, N.; Echeverria, C.; Li, C.; Bosze, W.; Myung, N.; Nam, J. Polyaniline/poly(ε-caprolactone) composite electrospun nanofiber-based gas sensors: Optimization of sensing properties by dopants and doping concentration. *Nanotechnology* **2014**, *25*, 115501. [CrossRef] [PubMed]
123. Garrudo, F.F.F.; Chapman, C.A.; Hoffman, P.R.; Udangawa, R.W.; Silva, J.C.; Mikael, P.E.; Rodrigues, C.A.V.; Cabral, J.M.S.; Morgado, J.M.F.; Ferreira, F.C.; et al. Polyaniline-Polycaprolactone Blended Nanofibers for Neural Cell Culture. *Eur. Polym. J.* **2019**, *117*, 28–37. [CrossRef]
124. Hong, S.-G.; Kim, H.; Kim, J. Highly Stabilized Lipase in Polyaniline Nanofibers for Surfactant-Mediated Esterification of Ibuprofen. *Langmuir* **2014**, *30*, 911–915. [CrossRef] [PubMed]
125. Soni, S.; Dwivedee, B.; Banerjee, U. Tailoring a stable and recyclable nanobiocatalyst by immobilization of surfactant treated Burkholderia cepacia lipase on polyaniline nanofibers for biocatalytic application. *Int. J. Biol. Macromol.* **2020**, *161*, 573–586. [CrossRef]
126. Hanumantharao, S.; Que, C.; Rao, S. Self-Assembly of 3D Nanostructures in Electrospun Polycaprolactone-Polyaniline Fibers and their Application as Scaffolds for Tissue Engineering. *Materialia* **2019**, *6*, 100296. [CrossRef]
127. Licciardello, M.; Ciardelli, G.; Tonda-Turo, C. Biocompatible Electrospun Polycaprolactone-Polyaniline Scaffold Treated with Atmospheric Plasma to Improve Hydrophilicity. *Bioengineering* **2021**, *8*, 24. [CrossRef]
128. Ku, S.; Lee, S.; Park, C. Synergic effects of nanofiber alignment and electroactivity on myoblast differentiation. *Biomaterials* **2012**, *33*, 6098–6104. [CrossRef]

129. Li, Y.; Li, X.; Zhao, R.; Wang, C.; Qiu, F.; Sun, B.; Ji, H.; Qiu, J.; Wang, C. Enhanced adhesion and proliferation of human umbilical vein endothelial cells on conductive PANI-PCL fiber scaffold by electrical stimulation. *Mater. Sci. Eng. C* **2017**, *72*, 106–112. [CrossRef]
130. Stejskal, J.; Bober, P.; Trchová, M.; Kovalcik, A.; Hodan, J.; Hromádková, J.; Prokeš, J. Polyaniline Cryogels Supported with Poly(vinyl alcohol): Soft and Conducting. *Macromolecules* **2017**, *50*, 972–978. [CrossRef]
131. Wu, P.; Xiao, A.; Zhao, Y.; Chen, F.; Ke, M.; Zhang, Q.; Zhang, J.; Shi, X.; He, X.; Chen, Y. An implantable and versatile piezoresistive sensor for the monitoring of human–machine interface interactions and the dynamical process of nerve repair. *Nanoscale* **2019**, *11*, 21103–21118. [CrossRef]
132. Wu, P.; Zhao, Y.; Chen, F.; Xiao, A.; Du, Q.; Dong, Q.; Ke, M.; Liang, X.; Zhou, Q.; Chen, Y. Conductive Hydroxyethyl Cellulose/Soy Protein Isolate/Polyaniline Conduits for Enhancing Peripheral Nerve Regeneration via Electrical Stimulation. *Front. Bioeng. Biotechnol.* **2020**, *8*, 709. [CrossRef]
133. Kim, H.; Lee, J.; Park, J.; Lee, S.; Hong, S.; Park, J.; Park, K.-H. Fabrication of Nanocomposites Complexed with Gold Nanoparticles on Polyaniline and Application to Their Nerve Regeneration. *Acs Appl. Mater. Inter.* **2020**, *12*, 30750–30760. [CrossRef]
134. Song, I.; Dityatev, A. Crosstalk between glia. extracellular matrix and neurons. *Brain Res. Bull.* **2018**, *136*, 101–108. [CrossRef]
135. Thompson, R.; Pardieck, J.; Smith, L.; Kenny, P.; Crawford, L.; Shoichet, M.; Sakiyama-Elbert, S. Effect of hyaluronic acid hydrogels containing astrocyte-derived extracellular matrix and/or V2a interneurons on histologic outcomes following spinal cord injury. *Biomaterials* **2018**, *162*, 208–223. [CrossRef]
136. Tan, K.; Tann, J.; Sathe, S.; Goh, S.; Ma, D.; Goh, E.; Ma, D.; Goh, E.; Yim, E. Enhanced differentiation of neural progenitor cells into neurons of the mesencephalic dopaminergic subtype on topographical patterns. *Biomaterials* **2015**, *43*, 32–43. [CrossRef]
137. Yang, H.; Lee, B.; Tsui, J.; Macadangdang, J.; Jang, S.; Im, S.; Kim, D. Electroconductive Nanopatterned Substrates for Enhanced Myogenic Differentiation and Maturation. *Adv. Healthc. Mater.* **2016**, *5*, 137–145. [CrossRef] [PubMed]
138. Ankam, S.; Suryana, M.; Chan, L.; Moe, A.; Teo, B.; Law, J.; Sheetz, M.; Low, H.; Yim, E. Substrate topography and size determine the fate of human embryonic stem cells to neuronal or glial lineage. *Acta Biomater.* **2013**, *9*, 4535–4545. [CrossRef]
139. Ankam, S.; Lim, C.; Yim, E. Actomyosin contractility plays a role in MAP2 expression during nanotopography-directed neuronal differentiation of human embryonic stem cells. *Biomaterials* **2015**, *47*, 20–28. [CrossRef]
140. Yang, K.; Park, E.; Lee, J.; Kim, I.; Hong, K.; Park, K.; Cho, S.; Yang, H. Biodegradable Nanotopography Combined with Neurotrophic Signals Enhances Contact Guidance and Neuronal Differentiation of Human Neural Stem Cells. *Macromol. Biosci.* **2015**, *15*, 1348–1356. [CrossRef] [PubMed]
141. Christopherson, G.; Song, H.; Mao, H.-Q. The influence of fiber diameter of electrospun substrates on neural stem cell differentiation and proliferation. *Biomaterials* **2009**, *30*, 556–564. [CrossRef] [PubMed]
142. Johnson, C.D.L.; Zuidema, J.M.; Kearns, K.R.; Maguire, A.B.; Desmond, G.P.; Thompson, D.M.; Gilbert, R.J. The Effect of Electrospun Fiber Diameter on Astrocyte-Mediated Neurite Guidance and Protection. *ACS Appl. Bio. Mater.* **2019**, *2*, 104–117. [CrossRef]
143. Chang, W.; Shah, M.; Lee, P.; Yu, X. Tissue-Engineered Spiral Nerve Guidance Conduit for Peripheral Nerve Regeneration. *Acta Biomater.* **2018**, *73*, 302–311. [CrossRef]
144. Morgado, J. Modulation of the electrical double layer in metals and conducting polymers. *Sci. Rep.* **2022**, *12*, 307. [CrossRef] [PubMed]
145. Morgado, J.; Sordini, L.; Ferreira, F. Electronic to ionic transduction of the electric field applied to PEDOT:PSS substrates to the cell cultures on top. *Bioelectrochemistry* **2022**, *145*, 108099. [CrossRef] [PubMed]
146. Chan, H.; Ho, P.; Ng, S.; Tan, B.; Tan, K. A new water-soluble, self-doping conducting polyaniline from poly (o-aminobenzylphosphonic acid) and its sodium salts: Synthesis and characterization. *J. Am. Chem. Soc.* **1995**, *117*, 8517–8523. [CrossRef]
147. Alva, K.S.; Kumar, J.; Marx, K.A.; Tripathy, S.K. Enzymatic synthesis and characterization of a novel water-soluble polyaniline: Poly (2, 5-diaminobenzenesulfonate). *Macromolecules* **1997**, *30*, 4024–4029. [CrossRef]
148. Wang, S.; Guan, S.; Wang, J.; Liu, H.; Liu, T.; Ma, X.; Cui, Z. Fabrication and characterization of conductive poly (3,4-ethylenedioxythiophene) doped with hyaluronic acid/poly (l-lactic acid) composite film for biomedical application. *J. Biosci. Bioeng.* **2017**, *123*, 116–125. [CrossRef]
149. Zheng, T.; Wang, X.; Liu, Y.; Bayaniahangar, R.; Li, H.; Lu, C.; Xu, N.; Yao, Z.; Qiao, Y.; Zhang, D.; et al. Polyaniline-decorated hyaluronic acid-carbon nanotube hybrid microfiber as a flexible supercapacitor electrode material. *Carbon* **2020**, *159*, 65–73. [CrossRef]
150. Garrudo, F.F.F.; Udangawa, R.N.; Hoffman, P.R.; Sordini, L.; Chapman, C.A.; Mikael, P.E.; Ferreira, F.A.; Silva, J.C.; Rodrigues, C.A.V.; Cabral, J.M.S.; et al. Polybenzimidazole nanofibers for neural stem cell culture. *Mater. Today Chem.* **2019**, *14*, 100185. [CrossRef]
151. Dickhaus, B.; Priefer, R. Determination of polyelectrolyte pKa values using surface-to-air tension measurements. *Colloids Surf. Physicochem. Eng. Asp.* **2016**, *488*, 15–19. [CrossRef]
152. Håkansson, A.; Han, S.; Wang, S.; Lu, J.; Braun, S.; Fahlman, M.; Berggren, M.; Crispin, X.; Fabiano, S. Effect of (3-glycidyloxypropyl)trimethoxysilane (GOPS) on the electrical properties of PEDOT:PSS films. *J. Polym. Sci. Part B Polym. Phys.* **2017**, *55*, 814–820. [CrossRef]

153. Mantione, D.; del Agua, I.; Schaafsma, W.; ElMahmoudy, M.; Uguz, I.; Sanchez-Sanchez, A.; Sardon, H.; Castro, B.; Malliaras, G.; Mecerreyes, D. Low-Temperature Cross-Linking of PEDOT:PSS Films Using Divinylsulfone. *ACS Appl. Mater. Interfaces* **2017**, *9*, 18254–18262. [CrossRef] [PubMed]

154. Alesary, H.; Ismail, H.; Khudhair, A.; Mohammed, M. Effects of Dopant Ions on the Properties of Polyaniline Conducting Polymer. *Orient. J. Chem.* **2018**, *34*, 2525–2533. [CrossRef]

155. Wang, L.-L.; Chen, J.-T.; Wang, L.-F.; Wu, S.; Zhang, G.; Yu, H.-Q.; Ye, X.; Shi, Q.-S. Conformations and molecular interactions of poly-γ-glutamic acid as a soluble microbial product in aqueous solutions. *Sci. Rep.* **2017**, *7*, 12787. [CrossRef]

156. Thaplyal, P.; Bevilacqua, P.C. Experimental Approaches for Measuring pKa's in RNA and DNA. *Methods Enzymol.* **2014**, *549*, 189–219.

157. Bhattacharya, S.; Kim, D.; Gopal, S.; Tice, A.; Lang, K.; Dordick, J.; Plawsky, J.; Linhardt, R. Antimicrobial effects of positively charged. conductive electrospun polymer fibers. *Mater. Sci. Eng. C* **2020**, *116*, 111247. [CrossRef]

Review

A Comprehensive Review of Electrospun Fibers, 3D-Printed Scaffolds, and Hydrogels for Cancer Therapies

Angelika Zaszczyńska , Beata Niemczyk-Soczynska and Paweł Sajkiewicz *

Laboratory of Polymers & Biomaterials, Institute of Fundamental Technological Research, Polish Academy of Sciences, Pawińskiego 5B, 02-106 Warsaw, Poland
* Correspondence: psajk@ippt.pan.pl

Abstract: Anticancer therapies and regenerative medicine are being developed to destroy tumor cells, as well as remodel, replace, and support injured organs and tissues. Nowadays, a suitable three-dimensional structure of the scaffold and the type of cells used are crucial for creating bio-inspired organs and tissues. The materials used in medicine are made of non-degradable and degradable biomaterials and can serve as drug carriers. Developing flexible and properly targeted drug carrier systems is crucial for tissue engineering, regenerative medicine, and novel cancer treatment strategies. This review is focused on presenting innovative biomaterials, i.e., electrospun nanofibers, 3D-printed scaffolds, and hydrogels as a novel approach for anticancer treatments which are still under development and awaiting thorough optimization.

Keywords: scaffolds; hydrogels; tissue engineering; polymers; anticancer treatments; cancer therapy; regenerative medicine

Citation: Zaszczyńska, A.; Niemczyk-Soczynska, B.; Sajkiewicz, P. A Comprehensive Review of Electrospun Fibers, 3D-Printed Scaffolds, and Hydrogels for Cancer Therapies. *Polymers* **2022**, *14*, 5278. https://doi.org/10.3390/polym14235278

Academic Editors: João Carlos Silva and Frederico Castelo Ferreira

Received: 12 October 2022
Accepted: 28 November 2022
Published: 2 December 2022

Publisher's Note: MDPI stays neutral with regard to jurisdictional claims in published maps and institutional affiliations.

1. Introduction

The human body has limited regenerative abilities. Thus, it cannot heal itself from severe damage and significant defects due to congenital abnormalities and diseases [1]. Consequently, tissue engineering strategies, including stromal cell-based therapies, are being developed. The main goal is to regenerate, remodel, replace, or support damaged organs and tissues or to induce a healing process by activating the body's self-healing capacity [2,3].

Cancer is a serious socio-economic problem affecting people of all genders and ages [4]. In 2022, 609,360 cancer deaths in the US are predicted, while in Europe, cancer deaths are estimated at 1,269,200 [5,6]. Although it is expected that lung, stomach, colorectal, breast, uterine, prostate cancer, and leukemia mortality will decrease both in the US and Europe [4,5], there are still types of cancer, such as pancreatic cancer and glioblastoma, that lead to deaths of a large number of people and are extremely challenging for oncology [3,4,7–9].

Current anticancer treatments are based on surgical treatments followed by radiotherapy, chemotherapy, high-intensity focused ultrasound (HIFU), and systemic therapies, i.e., chemotherapy and hormonal therapies [10–12]. In some cases of anti-cancer therapies, e.g., in breast cancer therapies, three main aspects are taken under consideration, namely, treatment, regeneration of the tissue, and restoration of the physical appearance. However, their implementation usually carries many side effects [13,14]. For instance, chemotherapy causes the nonspecific distribution of chemotherapeutics, resulting in decreased anticancer effects as well as systemic toxicity [15]. In addition, long-term anticancer drug intake very often leads to tumor resistance, which makes it difficult to cure cancer even with alternative treatment methods [16].

A relatively novel, promising, and still developing approach is near-infrared light (NIR)-induced photothermal therapy (PTT) as a local, minimally invasive anticancer therapy that destroys cancer cells via hyperthermia [17,18]. The PTT method is based on controlling NIR and using appropriate photothermal conversion agents that prevent damage to surrounding tissues [19,20].

Moreover, an innovative method of treatment used in the prevention, diagnostics, and cancer therapy is personalized medicine (PM, Figure 1) [21]. PM selects a specifically matched treatment based on the patient's genome, lifestyle, and medical test results. In the case of cancer, tumor heterogeneity significantly decreases the efficacy of conventional treatments, thus requiring therapies with a higher level of personalization. Additionally, PM can adjust therapy to reduce side effects and provide efficient therapeutic delivery [13]. Implementation of such treatment requires using technologies that allow the designing and formation of biomaterials providing effective therapeutics delivery systems or tissue regeneration. In this respect, the use of various materials in regenerative medicine has been expanding rapidly in the last few years [22]. The evolution of modern biomaterials would be impossible without the special parameters of certain materials. Originally, these materials were used in tissue engineering to improve the living standards of patients. Personalized medicine can help increase the effectiveness of the cancer treatment process because the methods and active substances used in it are aimed at learning about the abnormal mechanisms underlying the disease, and then finding the right drug. Personalized medicine has many advantages because it entails the greater effectiveness of individually selected drugs, the ability to prevent the development of the disease and cure it, knowledge about the patient's health condition, the predisposition of his body, and fewer side effects. These advantages are very important from the perspective of a patient with a weakened organism [23].

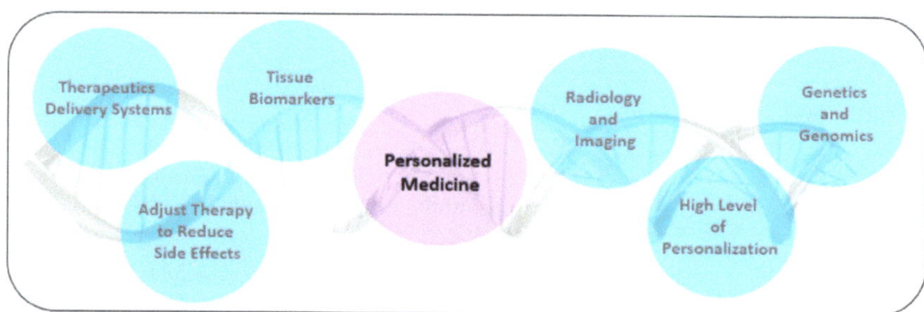

Figure 1. Scheme of personalized medical treatment.

Human tissues have different biological, physical, and biochemical properties. To achieve those requirements, the choice of appropriate material is essential for effective treatment. Degradation kinetics, design aspects, and cell-material interaction are crucial in developing biomaterials. The U.S. Food and Drug Administration (FDA) has approved many materials and technologies for medical purposes, such as scaffolds, drug-delivery treatments, fabrication techniques, and implants. This multidisciplinary area's rapid progress also allows for improving modern medical applications [24].

Ideal biomaterials must create conditions similar to those occurring in the natural extracellular matrix (ECM), that provide adequate cell adhesion, proliferation, and differentiation. An ECM is a dynamic structure that occurs between neighboring cells, which is made from water, structural proteins, minerals, proteoglycans, and specialized proteins (Figure 2) [25]. The knowledge of the composition of various tissues, ECMs, and tumor physiology plays a crucial role in understanding their function–structure relationship, and thus in anticancer and regenerative medicine [26,27].

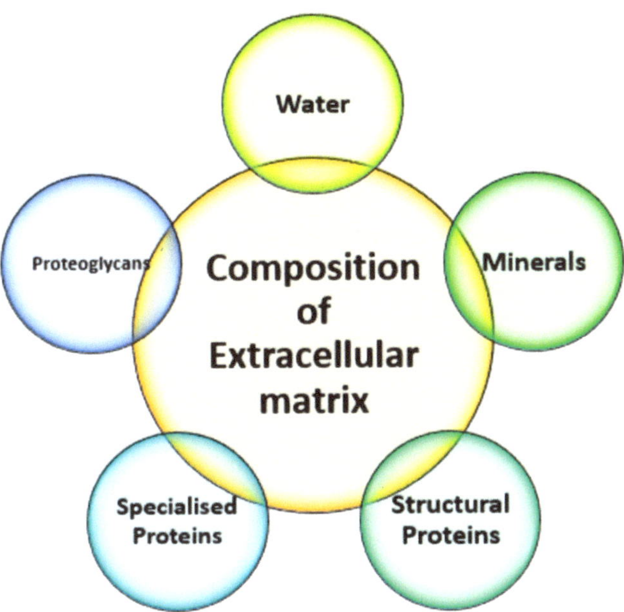

Figure 2. Composition of the ECM.

Currently, for anticancer treatments and subsequent regenerative medicine, various polymeric materials are used due to their favorable structure, chemical composition, and molecular weight. They can be generally categorized as naturally occurring or synthetic polymers. The most important natural materials used as cell scaffolds include proteins (collagen and gelatin, fibrin, fibrinogen, silk, etc.), polysaccharides (cellulose, chitosan, dextran, hyaluronic acid, agarose, etc.), and polynucleotides (RNA, DNA). Synthetic polymers are being widely investigated, with their advantages and disadvantages compared to natural polymers. In this respect, poly (methyl methacrylate) (PMMA) [28], polycaprolactone (PCL) [29], poly (L-lactide) (PLLA) [30], poly (lactic-co-glycolic acid) (PLGA), polyvinylidene difluoride (PVDF) [31], polyglycolic acid (PGA), polyhydroxyalkanoates (PHA), poly (3-hydroxybutyrate-co-3-hydroxyvalerate) PHBV [32], poly (hydroxybutyrate) (PHB) [27], etc. have been distinguished. In addition, many scientists test a combination of polymers and additives to obtain better biological and mechanical properties [33].

The main goal of this paper was to present a review of the published research concerning materials for anticancer therapies and tissue regeneration after resection. Current approaches to biomaterials, i.e., electrospun nanofibers, 3D-printed scaffolds, and hydrogels in the field of modern medicine are reported. We further discuss the novelties in the field of new technologies, cell, and drug delivery systems, and their use as approaches for cancer therapies.

2. Fibrous Scaffolds for Anticancer Treatments

Progress in materials science and human expectations have caused the development of biodegradable and non-biodegradable scaffolds for restoring the function of damaged tissues [34]. Some of them have found applications in cancer therapy. Materials and methods for forming highly functional scaffolds are discussed below.

There exist several types of nanofibers scaffolds manufacturing dedicated to tissue engineering and anticancer treatment. The most widely investigated is the electrospinning process, but there are also other methods, such as bicomponent fiber spinning, solution blowing, melt blowing, phase separation, drawing, and others [35].

Electrospinning (Figure 3.) is a method of manufacturing nanofibers with the use of a high electric voltage. In the electrospinning technique, a strong electric field is applied to the liquid polymer (or melt-melt electrospinning), resulting in the distribution of electric charges on the surface of the polymer droplet coming from the needle. An electrode with positive potential is connected to the capillary with the polymer solution, and another, with negative potential, to the collector. It is worth noting that manufactured nanofiber mats have a high surface area and controllable pores and can potentially diagnose and treat cancer cells [36]. Nanofibers, with the addition of anticancer drugs, can give sustained release after cancer tumor removal [37]; however, there is a need to produce more advanced materials which will support the complicated treatment after tumor removal [38].

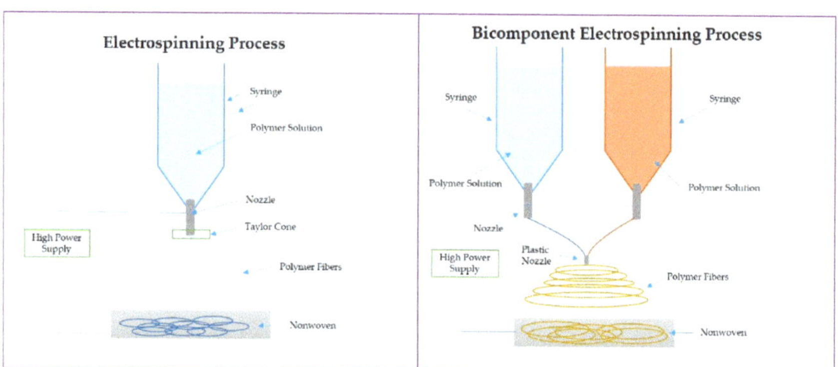

Figure 3. Scheme of the electrospinning process (**left**) and bicomponent electrospinning process (**right**).

Thus, based on the electrospinning technique, various methods of complex nanofiber production have been developed. Bicomponent spinning (BCS, Figure 3 (right)) allows the manufacturing of nanofibers in a two-step process. Two polymers can be produced by splitting or removing components [39] and, with the dual release, can be used in anticancer therapy [40]. Another type, coaxial electrospinning (CEs), allows the production of core-shell fibers and offers an advanced system of drug carrier options for multiple drug delivery [41]. Multi-jet electrospinning (MNEs) allows the production of nanofibers using several jets, which improves productivity [42], although it can result in poor fiber quality. Emulsion electrospinning (EEs) produces fibers from two immiscible solutions; nevertheless, due to the high number of variables, the fibers are difficult to produce [43].

Aside from electrospinning methods, several non-electrospinning techniques have been developed for improving nanofiber production; these include phase separation, solution blowing, template synthesis, freeze/drying synthesis, interfacial polymerization, drawing techniques, and splitting [44]. These techniques use centrifugal force or gases instead of electricity to produce nanofibers, but they yield low-quality fibers, and thus their application in tissue engineering is limited [45]. Depending on the type of process and parameters, different types and shapes of materials can be manufactured, such as nanowires, nano-webs, porous fibers, random and aligned nanofibers, ribbons, and others [46]. Figure 4 presents the classification and examples of materials used in medicine.

Biomaterials in Medicine

Natural biomaterials		**Synthetic biomaterials**

	Ceramic	**Polymers**
• Alginate		• Polycaprolactone (PCL)
• Hyaluronan		• Polylactic acid (PLA)
• Agarose	• Hydroxyapatite	• Poly(l-lactide-co-glycolide) (PLGA)
• Silk	• Zirconia	• Polyethylenoxide (PEG)
• Collagen	• Bioactive glass	• Poly(vinylidene difluoride –
• Cgitosan	• Calcium phosphate	Trifluoroethylene) (PVDF-TrFE)
• Gelatin		

Figure 4. Classification of materials used in medicine.

2.1. Degradable Scaffolds

Degradable polymers, the most preferable candidates as scaffolds for tissue engineering, can be further classified in terms of their origin as natural or synthetic. In anticancer treatments, degradable polymers, e.g., poly (vinyl alcohol) (PVA), poly (β-hydroxybutyrate-β-hydroxyvalerate) (PHBV), poly (lactic-co-glycolic acid) (PLGA), polycaprolactone (PCL), and polylactic acid (PLA) nanofibers, serve as scaffolds and site-specific drug delivery systems [47–50].

2.1.1. Natural Degradable Polymers

Natural degradable polymers are incessantly investigated in anticancer treatment. Nowadays, 3D silk is gaining interest, because 3D silk scaffolds are biodegradable and easy to form, and have excellent mechanical properties [51]. Electrospinning gives the opportunity to obtain very thin silk fibers (SF) with a diameter of <500 nm [52]. Recently, to develop scaffolds with good mechanical properties, scientists used SF for reinforcing hydrogels [53]. SF and silk-sericin (SS) are perfect materials for use as a biotemplate to develop lung anticancer system drug delivery systems [54]. They are widely used as chemotherapeutics delivery systems. For instance, SF/PCL fibers have been used to release titanocene, a drug for breast cancer therapies resulting in the promotion of MCF-7 breast cancer cell apoptosis [55]. Especially important here were interactions between SF amine groups and titanocene, which increased the rate of cancer cell apoptosis. In another study reported by Li et al. [56], silk regenerated SF loaded with curcumin (CUR, hydrophobic drug) and doxorubicin hydrochloride (DOX·HCl, hydrophilic drug) were colloid-electrospun to provide the dual drug in a sustained way for breast and skin anticancer therapies. TEM microscopy has confirmed the effectiveness of the electrospinning by the presence of both drugs. Additionally, both drugs showed controlled release according to the Fickian model. Such an approach was a preliminary study, and, in the future, it should be tested on breast or skin tumors to evaluate its usefulness for anticancer treatments.

Another natural material, collagen, can serve as an excellent scaffold for supporting cell–matrix interactions and cell proliferation, adhesion, and migration [57]. To achieve high mechanical strength, collagen can be mixed with synthetic polymers such as PGA, PLA, or P (LLA-CL) [58,59]. Collagen has a high degradation rate, biocompatibility, and minimum immune response [60]. Collagen scaffolds are widely investigated in medicine, including for wound healing [61], tissue engineering, drug delivery systems [62], and other tissue engineering applications [63]. Many experiments have shown the influence of specific collagen on anticancer treatments. Typical matrix collagen COLI takes place in ECM remodeling, which in effect induces metastasis or invasion of epithelial ovarian cancer cells; after remodeling, it affects the ovarian cancer cells by signaling pathways [64]. These studies have shown that ECM remodeling increases the invasion of aggressive ovarian cells by adequate signaling and the further blocking of these signaling pathways could result in the successful destruction of ovarian cell proliferation and further invasion.

In other studies, PCL/Collagen and piperine served as anti-breast cancer drug delivery systems. Based on the results, such combinations showed good mechanical properties, influenced sustained piperine release, decreased tumor size, and induced apoptosis in MCF-7 and 4T1 breast tumor cell lines. On the one hand, collagen is a material that brings many advantages, but on the other, collagen is an expensive material [52], and there are plenty of polymers that are cheaper alternatives showing comparable properties. One example is gelatin. Gelatin has found many applications in industry as well as in biomedicine. Further, after chemical modification or blending with different polymers it can be used as a biomaterial in long-term applications, such as brain hydrogel [65]. The anticancer use of gelatin in modern anticancer therapies additionally provides PTT effect. Such a combination is described in detail below, in Section 2.2.

Another natural material, alginate, can be used as a biomedical scaffold due to its viscoelastic properties [66]. Alginate shows good biocompatibility, nontoxicity, and low price, and can be crosslinked with Ca^{2+} ions. Chen et al. [67] formed an interesting composite consisting of freeze-dried alginate/gelatin sponge and curcumin-loaded electrospun fibers (CFAGS) for wound healing after tumor resection. The results showed that CFAGS released curcumin stably during the entire study (15 days). In vitro and in vivo studies showed that stable curcumin release destroyed MCF-7 tumor cells and prevented tumor recurrence after surgery.

Chitosan is a biomaterial with high solubility properties and cationic nature [68]. It can be applied to organs, nerves, and cardiac tissue due to its similarity to the extracellular matrix (ECM) [69]. Chitosan combinations with poly (vinyl pyrrolidone) (PVP) [70], poly (vinyl alcohol) (PVA) [71], or natural materials gelatin and collagen are usually formed via electrospinning as anticancer drug delivery systems. Similarly to SF, the free amine groups that are present in chitosan can interact with hydrophilic anticancer drugs. This feature makes it favorable from the anticancer drug delivery perspective. Shafabakhsh et al. [72] described that the interactions between chitosan and anticancer drugs resulted in an effective decrease in tumor cell proliferation, higher apoptosis, as well as decreased metastasis.

2.1.2. Synthetic Degradable Polymers

The most common synthetic degradable polymers dedicated to anticancer therapies are listed below.

PLA is widely used in resorbable medical sutures due to its non-toxic nature [73]. In tissue engineering, various materials are mixed with PLA polymer to achieve higher mechanical properties. In bone tissue engineering, PLA-based composites, such as PLA-hydroxyapatite, PLA-chitosan, PLA-PLDA, and PLA-phosphates have been created [74,75]. To achieve a composite with high stiffness, PLA with tricalcium phosphate was mixed and applied as a bone implant [76]. PLA biomaterials can be manufactured using different technologies, such as film casting, thermoforming, electrospinning, and nano- and micro-methods, and can be formed into various sizes and shapes to be used as a medical scaffold [77].

In oncology, PLA-based scaffolds can be used in drug delivery due to the incorporation of chemotherapeutic agents inside fibers [78]. For instance, Yuan et al. [79] formed doxorubicin (DOX)-loaded mesoporous silica nano-particle with PLLA nanofibers composite showing early phase and sustained drug release for breast cancer therapies. The results showed successful sustained DOX release from 50 to 120 days, which is extremely long in comparison to other drug delivery systems. Additionally, the composite induced MDA-MB-231 breast tumor cell with decreased Bcl-2 and TNF-α gene expression. Considering controlled prolonged drug release and positive anticancer response, such fibrous composite has huge potential for future anticancer treatments.

Another example of a synthetic polymer is PLGA. The degradation of PLGA can result in an acidic environment. Tumor environments are also usually acidic, so PLGA could be useful or not depending on its precise application [80]. PLGA is widely examined in drug delivery systems [81] and as a nanoparticle-based system in cancer tumor treatment [82]. An interesting approach is a PLGA electrospun mat loaded with Ag nanoparticles [83] for liver anticancer therapies. In vitro studies on Hep-G2 cell lines showed the anticancer character of PLGA fibers and the anticancer effect increased with the Ag nanoparticles contribution in the fibers. In other studies, PLGA [84] nanofibers were loaded with metformin to study the anticancer effect on A549 human lung adenocarcinoma cells. The results showed sustained drug release over two weeks and subsequent polymer degradation after ca. 24 days. The controlled release of metformin resulted in a cytotoxic effect on A549 tumor cells after two days, which resulted in their apoptosis.

PCL is another polymer commonly used in electrospinning for fibrous mat formation dedicated to anticancer therapies. PCL is one of the most promising biodegradable biomaterials approved by the Food and Drug Administration (FDA) as sutures, drug delivery devices, and adhesion barriers. In cancer treatment, nanoparticulated PCL-based drug delivery systems have much to offer [85]. Additional advantages of these drug systems include drug transport to the injured tissue, protection of the drug from disintegration, and stability in biological fluids compared to conventional chemotherapy [86]. Various drug delivery systems based on PCL have been investigated in glioblastoma, a primary brain tumor. For instance, [87] core-shell electrospun PCL/mycophenolic acid (MPA) was studied by Han et al. The results showed an effective anticancer effect of such an approach through hampering glioblastoma U-87 MG multiforme cell growth. The drug delivery system provided sustained drug release after 100 h. The authors understand the need to release the drugs over a longer time (ca. 49 days) and characterized the polymers more to provide longer drug release to fulfill anticancer treatment needs. In other studies, [88] Irani et al. formed PCL-Diol-b-PU/Au/temozolomide (TMZ) nanofibers for glioma tumor treatments. On the one hand, the results showed a cytotoxic effect of the composite against U-87 human glioblastoma, confirming the chemical composition is adequate for further, more advanced anti-glioblastoma therapy studies. On the other hand, however, the study showed drug release for 24 h only, which is too short for the actual needs of anti-cancer therapies. The prolonged TMZ release studies should be carried out in the future or the chemical composition of such an approach should be adjusted more thoroughly.

PVA is another polymer widely used in tissue engineering and anticancer therapies [89]. To achieve desired properties, such as high mechanical strength, PVA, similarly to other polymers, could be blended with various types of materials. For instance, PCL/PVA core-shell electrospun fibers [90] were loaded with the anticancer drug paclitaxel (PTX). PCL was a core, while PVA served as a shell, and combining both of these polymers led to obtaining a biocompatible and biodegradable composite, while PTX is an anticancer drug that additionally shows pH sensitivity. The drug delivery system showed the sustained release of PTX under acidic pH after over 25 days. This indicates that PCL/PVA/PTX could be useful in anticancer drug delivery. Additionally, the composite, due to its sustained drug release, inhibited the proliferation and growth of LoVo colon cancer cells, resulting in their destruction. Those results indicated that such an approach has great application potential in the chemotherapy of some solid tumors in the clinical setting. An overview of electrospun degradable and non-degradable polymers and their biomedical applications is given in Table 1.

Table 1. Overview of degradable and non-degradable polymers and their biomedical applications in cancer treatment and regenerative medicine.

Polymer	Method of Fabrication/Type of Material	Application	Ref.
Polycaprolactone (PCL)	Electrospinning	Skin tissue regeneration, wound healing, skin cancer	[91]
Polyethylenoxide (PEG)	Electrospinning	Tissue regeneration, especially soft tissue regeneration	[92]
Polycaprolactone (PCL) -gelatin	Electrospinning	Tracheal tissue engineering applications	[93]
Poly(l-lactide-co-glycolide) (PLGA)	Electrospinning	Neural tissue engineering applications (axons stretching)	[94]
Poly(lactic-co-glycolicacid) (PLGA)	Electrospinning	Skeletal muscle regeneration after muscle tumor	[95]
Polymeric/solid/hydrogels	Nanoparticles, microparticles	Ocular drug delivery, eye cancer, implants	[96]
Polycaprolactone-Polydimethyl Siloxane (PCL-PDMS)	Electrospinning	Bone tissue engineering, bone cancer	[97]
Polycaprolactone-Polylactic acid PCL/PLA	Coaxial Electrospinning	Bone tissue engineering, bone cancer	[98]
Polycaprolactone-poly(l-lactide-co-glycolide) PCL/PLGA	3D printing/Electrospinning	Bone tissue engineering, bone cancer	[99]
Polylactic acid (PLA)	Electrospinning	Tendon/joint stability	[100]
Polylactic acid (PLA)	Electrospinning	Soft tissue engineering/repair and regeneration of tendon defects and injuries	[101]
Polycaprolactone (PCL)	Electrospinning	Vascular graft	[102]
Poly(vinylidene difluoride-Trifluoroethylene) (PVDF-TrFE)	Electrospinning	Neural tissue engineering/regenerative medicine	[103]
Poly(glycerol sebacate)	Electrospinning	Tissue engineering/soft tissue	[104]
PLGA/collagen		Skin	[105]
Gelatin/cellulose acetate/elastin	Electrospinning	Skin/biomimicking	[106]
Laminin-functionalized PDLLA	Electrospinning	Skin cancer/skin healing	[107]

2.2. Electrospun Scaffolds Combined with PTT Effect

An interesting approach that is currently studied is broad of elecrospun fibers functionalities via for instance conjugation of electrospun mats with PTT effect. This conjugation involves both natural and synthetic degradable polymers.

A great representative of natural polymers is gelatin. There are many scientific reports describing its positive effect with PTT for anticancer treatments. For instance, a gelatin scaffold with gold nanoparticles can be used in photothermal cancer therapy [65]. Zhang et al. [65] adopted such an approach, which besides good biocompatibility, upon NIR laser irradiation at 805 nm heated up to 50–60 °C. The photothermal effect resulted in the effective destruction of HeLa cancer cells. A similar study used a folic acid/gelatin/gold nanoparticles composite scaffold for the photothermal ablation of breast cancer cells [66]. The gold nanoparticles generated a photothermal effect under an NIR laser of 805 nm, while folic acid served as a targeting ligand in nanoparticles. Additionally, folic acid could recognize and bind the specific receptors overexpressed by breast cancer cells. The results showed the material heated up under an NIR laser up to 35.1 °C, which resulted in the local destruction of MDA-MB231 breast cancer cells.

The incorporation of the PTT effect with synthetic electrospun fibers is currently being studied as well. For instance, Cheng et al. [108] formed gold nanorods (GNR) polyethylenoxide (PEG) and PLGA composite fibers which served as membranes as a photothermal platform for anticancer therapy. Not only did such an approach show an excellent photothermal effect resulting in effective tumor cell destruction but also the biodegradability of the mat provided tissue regeneration after resection. The presence of GNRs allowed the material to produce heat up to 42 °C at NIR light of 850 nm, which effectively destroyed HeLa and breast cancer MCF-7 cells.

Another trend in anticancer approach designing is the combination of the PTT effect and chemotherapy. Such a combination allows adequate anticancer drug delivery via electro-spun mats and at the same time provides additional therapy that will strengthen cancer cell destruction. Chen et al. [109] combined electrospun fibers to release in a controlled way chemotherapeutics with PTT effect to effectively destroy cancer stem cells (CSCs). In this respect, PCL nanofibers released all-trans retinoic acid (ATRA) drugs, while multi-walled carbon nanotubes (MWCNTs-OH) served as PTT agents. The ATRA and MWCNTs-OH were placed in PCL solution and electrospun. The obtained composite fibers were capable of sustaining drug release for a long time (more than 15 days). The presence of MWCNTs-OH increased the mechanical properties of the composite materials and created heat in the range of 42–47 °C under NIR light (808 nm) to destroy CSCs. Additionally, ATRA stimulated CSC differentiation improving the PTT effect. Treatment with composite fibers with the PTT effect showed a decrease in the mice's tumor size.

Another interesting approach was reported by Azerbaijan et al. [110], where pH-sensitive core-shell nanofibers with PTT effect were produced of molecules doped with poly (tetramethylene ether) glycol-based polyurethane (PTMG-PU) which served as a core, paclitaxel, graphene oxide/gold (GO/Au) nanorods loaded into PTMG-PU, and chitosan. Chitosan served as a shell. In this composite, paclitaxel was chemotherapeutic and the GO/Au nanorods were PTT agents while chitosan performed pH sensitivity promoting the acceleration of anticancer drug release in an acidic (cancer) environment.

The composite showed an excellent PTT effect, where the material heated up to 51–55 °C under 808 nm laser irradiation. Additionally, chemotherapeutic was released in a controllable manner near the tumor area due to the PTT effect and pH sensitivity. The results showed an increase in apoptotic nuclei in A549 lung cancer cell lines indicating the huge potential of such an approach for future anticancer treatments. Other examples of electrospun scaffolds combined with PTT effect were shown in Table 2.

Table 2. Electrospun scaffolds combined with PTT effect.

Material	Multifunctional Effect	Light WaveLength	Application	Ref.
PLA/PCL with Cu_2S	- high mortality (>90%) of skin tumor cells - inhibited tumor growth	808 nm	Skin cancer therapy and wound healing	[111]
PCL microfiber/GO scaffold	- cancer cell in situ ($\approx$98%) removal	810 nm	Breast cancer therapy/adipose tissue repair	[112]
PCL/Gelatin scaffold with BP nanosheets	- create a tumor-suppressive microenvironment - increased tissue repair ability	808 nm	Skin cancer treatments	[113]
Gelatin/Au	- PTT effect supports cell adhesion and proliferation	805 nm	Breast tumor cells	[114]
PLA/PCL composite scaffolds	- inhibit in vivo tumor growth - advance healing of cancer-surgery caused wounds	808 nm	Skin cancer cells	[115]
PCL/Gelatin	- control of drug release by PTT effect	808 nm	Inhibition of cancer tumor growth	[116]
PCL	- Control of drug release	808 nm	- chemotherapy with PTT effect	[117]

3. 3D Printed Scaffolds and Hydrogels for Anticancer Treatments

A 3D-printing method allows many types of scaffold manufacturing. The 3D printing might be classified in terms of the techniques used e.g., fused deposition modeling (FDM, Figure 5, left), inkjet printing, laser beam melting, selective laser sintering (SLS, Figure 5, right), bioprinting, extrusion, digital laser printing (DLP), polyjet, stereolithography, and electron beam melting [118–120]. Additionally, the method could be divided into classical printing [121] and bioink, i.e., 3D printing with cells or bioactive substances [122,123].

The 3D printers currently in use are based on three principles: liquid solidification, powder solidification, and extrusion. The most popular method of 3D printing is SLA. This method uses the phenomenon of photopolymerization, which causes the solidification of the liquid. The process is repeated layer by layer until an object is created. In the SLS method, the laser heats and melts the powder, which creates a 3D object. The first step is to distribute the powder on the platform evenly, and then the roller aligns the surface of the object. Another method of inkjet 3D printing is to print objects drop by drop. The droplets sprayed from the nozzle are applied in thin layers and cured with high-energy light or cool air. FDM is a method based on the extrusion of material from a nozzle and distribution in layers on the worktable. Each method has its pros and cons, which are detailed in Table 3. This does not change the fact that 3D printing is a promising tool in personalized medicine, which tries to treat cancer patients [124].

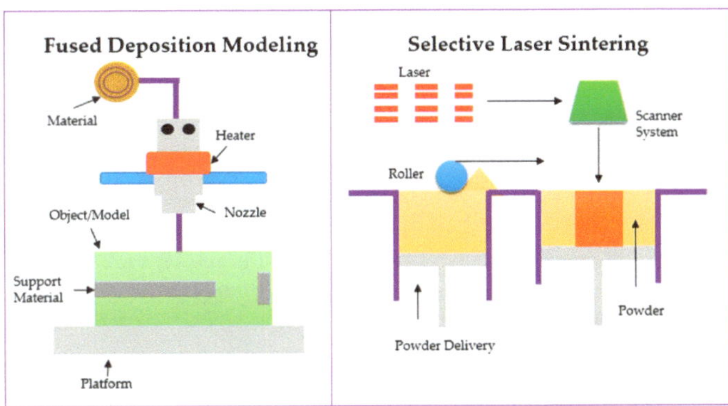

Figure 5. Scheme of the FDM (**left**) and SLS (**right**) process.

Table 3. The three-dimensional printing methods as a promising tool in anticancer therapies and PM.

Method	Advantages	Disadvantages	Ref.
Stereolithography: photopolymerization of liquid material	- High quality of printed objects - Low thermal stress	- Cytotoxicity of the products - Post-curing processes - Material must have a photo-curable properties	[125]
Selective laser sintering (SLS): laser energy absorbing material (powder)	- Solvent-free and fast printing - high quality	- High energy may cause anticancer drug degradation - Limited number of materials with laser absorbing properties	[126]
Inkjet 3D printing: ink (drug solution) and substrate (polymer-based material)	- Continuous printing prevents clogging of the needle - Low cost - High precision	- Low resolution - High cost of production	[127]
Fused deposition modelling (FDM): printing using filaments	- Low cost - No post-production - Solvent-free process	- Only thermoplastic polymers - Preparation of filaments - Heat can cause the degradation	[128]

Like all scaffold manufacturing methods, this one also has its pros and cons. On the one hand, the 3D printing method brings great advantages, i.e., easily accessible, easy processing, variety of bioprinter types, relatively low costs, generates less waste, and most importantly, the ability to form precisely controlled structures [129,130]. On the other, depending on the chosen material, bioprinter, and fabrication process still, there are some problems with the restricted built size, not-fully crosslinked material, or cell aggregation that leads to the nozzle tip clogging [131].

Since studies of anticancer therapies involve a wide spectrum of diagnosis, treatments, prognosis, and metastasis, 3D printing is a technique that could play a role in all these areas. The 3D-printed materials might be applied as tumor models (3D cell culture models), personalized drug delivery systems, bioprinted organs, organ-on-chip models, and devices used in diagnosis [132]. Three-dimensional-bioprinted scaffolds, including aliphatic polyester or hydrogels for cancer therapies, are also formed mainly for treating residual cancer after the resection of laminin-functionalized PDLLA surgery but also serve as a 3D cell culture model [133,134].

Currently, there is a clear trend of using 3D-printed materials that provide the synergistic effects of using current anti-cancer therapies (chemotherapy, radiotherapy, PTT, and others), promoting tissue regeneration after resection, and avoiding cancer relapse. On the one hand, such approaches increase the scope of their applications and face most of the requirements imposed by cancer therapies; on the other, the fabrication of such biomaterial is challenging and complex, and requires laborious and multi-step functionalization [135]. Moreover, functionalization could change the scaffold properties such as biodegradation or mechanical properties.

3.1. Three-Dimensional-Printed Scaffolds Combined with PTT Effect

Three-dimensional-printed scaffolds provide many benefits from the tissue engineering point of view. One of them is interconnected pores and controlled porosity, which could provide a decent environment for cellular activities as well as nutrient and gas transportation [136]. Another is customized parameters of the scaffolds, i.e., architecture, composition, and stiffness. However, for anticancer treatment providing all of the above-mentioned parameters is still insufficient. Combining 3D printing as a scaffold preparation method triggers the ablation of tumors through hyperthermia and the simultaneous promotion of tissue regeneration [137]. PTT could promote immunoadjuvant-like effects to generate the immunity responsible for tumor-attacking [138]. The photothermal agents are usually metals, metal oxides, NIR dyes, graphenes, and others [139,140]. The photothermal surfaces are fabricated via firm immobilization or deposition of photothermal agents or via direct fabrication of materials showing the photothermal effect [141]. The greatest advantage of photothermal material is providing precise control of time, heat area, and intensity by adjusting light irradiation. The wavelength that activates the photothermal effect is in the range of 700–1400 nm [142]. By combining such approaches with ceramics or polymers, novel multifunctional scaffolds are formed and presented in Table 2 [143–145].

Most of the current activity is focused on the 3D printing of ceramic scaffolds providing the PTT effect. For instance, He et al. [146] formed a thermoactive biodegradable 3D-printed scaffold consisting of immune adjuvant (R837)-loaded and niobium carbide (Nb_2C) MXene-modified bioglass (BG@NbSiR) as bone metastasis of breast cancer therapy. This strategy was based on the PTT and immune-activation properties of NS Nb_2C@Si loaded with R837. This multifunctional scaffold stimulated long-term immune memory, providing adequate protection against breast cancer metastasis. Additionally, the unique properties of the material and the biodegradation of BG@NbSiR provided more efficient bone regeneration. Similar studies were conducted by Ma et al. [147], who designed the Fe-$CaSiO_3$ scaffold using a facile ball-milling and 3D printing technique as a bone cancer treatment. The scaffolds combined high compressive strength, photothermal effect, and reactive oxygen species (ROS) production that is harmful to tumor cells, and provided effective bone regeneration. The high compressive strength (up to 126 MPa) allowed it to fill, withstand, and mechanically support bone cortical defects. It is reported that the compressive modulus of human bone is in the range of 90–170 MPa. Additionally, such scaffolds provide excellent photothermal effects, anticancer therapeutic effects as a result of Fe ions, and sustained release. Fe ions catalyze the degradation of H_2O_2 in tumor cells resulting in ROS production. In vitro studies have shown improved rat bone mesenchymal stem cells (rBMSCs) adhesion and their osteogenic differentiation, while in vivo tests on New Zealand rabbits confirmed the scaffold's bone regenerative effect.

An interesting approach is the development of polymer/ceramic-based scaffolds for bone cancer treatment. One example is an approach developed by Yang et al. [140]. The 3D-printed $SrCuSi_4O_{10}$/PCL scaffold was developed for inducing osteosarcoma ablation and providing effective bone vascularization. The composite scaffold provided an excellent PTT effect for osteosarcoma without side effects. Additionally, sustained Sr, Si, and Cu ions release increased rBMSCs adhesion, proliferation, and osteogenic differentiation. In other studies, such as that by Wang et al. [145], the 3D-printed borosilicate bioactive glass (BG) was functionalized with MoS_2-PLGA film to provide a PTT effect. Such composite not only decreased the viability of MNNG/HOS osteosarcoma cells in vitro and stopped tumor development in mice in vivo but also stimulated the proliferation and osteogenic differentiation of rBMSCs. The 3D-printed polymeric/ceramic multifunctional scaffolds with the PTT effect are still relatively new approaches, but the authors believe that the development of this particular field of materials science will provide many benefits in future anticancer treatments. The 3D-printed scaffolds combined with the PTT effect are summarized in Table 4.

Table 4. Three-dimensional printed scaffolds combined with PTT effect.

Material	Multifunctional Effect	Light Wavelength	Application	Ref.
carbon-embedding larnite (larnite/C) /$CaCO_3$	- destroying human osteosarcoma cells (MNNG/HOS) - increased bone regeneration ability PTT effect	808 nm	Bone cancer treatment	[148]
Niobium carbide (Nb_2C) MXene-modified 3D-printing biodegradable bioglass (BG@NbSiR)	- increased bone regeneration ability - providing long-term immune therapy PTT effect	808 nm	Breast cancer/bone metastasis therapy	[149]
Fe-$CaSiO_3$	- excellent mechanical properties - ROS production - increased bone regeneration ability PTT effect	808 nm	Bone cancer treatments	[150]
$SrCuSi_4O_{10}$/PCL	- increased bone regeneration ability - enhanced vascularized bone regeneration PTT effect	650–1000 nm	Bone cancer treatments (osteosarcoma, chondrosarcoma, fibrosarcoma, etc.)	[151]
MoS_2/PLGA/BG		808 nm	Bone cancer treatments	[152]

Three-Dimensional-Printed Hydrogels Combined with PTT Effect

Hydrogels are highly hydrated, three-dimensional polymers that mimic native ECM and, at the same time, might serve as drug/cell/growth factor delivery systems and scaffolds. The tunable mechanical and biochemical properties are another advantage of these materials [153]. Similar to 3D-printed scaffolds, due to the complex and dynamic tumor microenvironment, anticancer approaches based on hydrogels must be multifunctional and targeted. They should provide more features than a conventional hydrogel scaffold, e.g., the PTT effect [154]. One such hydrogel-based therapy is gelatin (Gel), sodium alginate (Alg) hydrogel system loaded with CuO nanoparticles obtained by Dang et al. [155]. This approach combined the photothermal effect and provided biochemical cues resulting in tumor recurrence. Gel and Alg provided good biocompatibility, biodegradability, controlled release of CuO nanoparticles, and 3D printability, while CuO nanoparticles served as photothermal agents and released Cu ions in a controllable way, resulting in ROS production.

In other studies, conducted by Lia et al. [133], methacrylated gelatin (GelMA) and methacrylated chondroitin sulfate (CSMA) hydrogels were loaded with gold nanorods (GNRs) and nanohydroxyapatite (nHA) for bone cancer treatments. Hydrogel components provided biocompatibility and 3D printability, the GNRs provided a PTT effect at 808 nm, while nHA mimicked the native bone ECM, and promoted osteogenic differentiation of MSCs and bone mineralization, which led to bone regeneration in defect areas previously invaded by a tumor.

The conjugation of scaffolding, local chemotherapy, and the PTT effect is also an interesting therapeutic direction. Xu et al. [134] designed 3D-printed sodium alginate (SA), gellan gum (GG), and polydopamine nanoparticles (PDA NPs) loaded with the chemotherapeutic of doxorubicin (DOX). In this composite, SA-GG provided biocompatibility, thermal sensitivity, and 3D printability, while PDA NPs provided a great PTT effect at 808 nm, ca. 1.5-fold increased mechanical properties of the scaffold, and provided increased wound healing ability after surgery. Combining PTT and chemotherapy provided significantly lower viability of $B_{16}F_{10}$ tumor cells and prevented tumor recurrence.

In recent times, a significant trend observed during scaffold design has been to conjugate current anti-cancer therapies and diagnostic methods. Such a smart scaffold could provide tissue reconstruction after breast cancer resection, photothermal conversion required for PTT, as well as a scaffold imaging possibility via photoacoustic imaging (PAI) or magnetic resonance imaging (MRI) [135]. In this regard, Luo et al. designed 3D-printed dopamine (DA)-modified sodium alginate/polydopamine (Alg-PDA) scaffold for breast cancer treatment. In this system, Alg provided desirable biocompatibility and safe crosslinking using Mn^{2+} cations. Not only did Mn^{2+} play a crucial role as a crosslinking agent and chemically coordinated PDA but also served as a contrast agent used in MRI. While DA and PDA increased cellular response due to the presence of catechol moieties, provided a photothermal effect for PTT, and allowed diagnosis via PAI and MRI. Such a 3D scaffold was characterized by increased porosity, with the pore size ranging from 1 μm to 1 mm, which provided adequate vascularization and cell infiltration. Additionally, the Alg-PDA scaffold showed mechanical properties of ca. 2 kPa, which were comparable to the human breast tissue (ca. 3 kPa), and an excellent photothermal effect of 2-fold increased temperature after MRC laser exposure. Most of all, multiple exposures by laser did not weaken the material mechanically. Conjugating the scaffolding with laser exposure showed over 60% decreased viability of 4T1 breast cancer cells. Simultaneously, the scaffolds provided good proliferation of MCF-10A breast cells showing good potential for anti-cancer treatments and further tissue reconstruction.

3.2. Hydrogels as 3D Cell Culture Models for Cancer Therapies

Cell culture models are biomedical approaches that monitor tumor cell behavior and signaling pathways, which is important information from the perspective of anti-cancer treatments. Currently, two types of cell culture models can be distinguished, 2D and 3D. The 2D cell culture models are simple and inexpensive [149]. However, they do not reflect complex 3D tumor environments [9]. The cell-cell and cell-ECM interactions strongly influence cell behavior. In 2D models, those interactions are damped, which most likely leads to cell dysfunction, and this has been observed in many cell types, such as hepatocytes or colorectal cancer cells [149–151]. Additionally, cancer cells seeded on 2D models do not show the real drug resistance observed in in vivo conditions.

A promising alternative that overcomes those limitations is the 3D cell culture model. The 3D cell culture models are more biochemically diverse and structurally complex than 2D models, enabling cell-cell and cell-ECM interactions and providing the same environmental conditions as in vivo [9,149]. Cells growing in 3D scaffolds show a slower proliferation rate and restore the histological differentiation characteristic of primary tumors that were not present in 2D cultures. Cells seeded on various 3D models also show varying cell morphology, various gene expressions, as well as various levels of drug resistance [9,151,156–159]. Ruedinger et al. reported cells seeded on 3D models to show a slower proliferation rate and present histological differentiation characteristics for primary tumors. Such a phenomenon is not observed for 2D cell culture models [160]. Hence, many types of 3D culture models are currently available. In this instance, spheroids [161], organoids [162], microfluidic devices [163], microfibers [164], and hydrogels [165] can be distinguished.

Hydrogels dedicated to cancer therapies very often serve as biomedical 3D approaches which accurately reflect the tumor microenvironments. To fairly mimic the native tumor microenvironment, such hydrogels should correspond mechanically and physiologically to the tumor but also to native healthy tissue [166]. As with scaffolds for tissue engineering, cells are grown on 3D scaffolds in in vitro conditions that induce adequate regulation of specific gene markers of tumor cells. In this section, hydrogels as an attractive approach applied to 3D cell culture models will be discussed. Hydrogels are especially interesting candidates for monitoring soft tissue cancers, e.g., glioma or glioblastoma, and breast cancer cells. Hydrogels well mimic the aqueous environment of human tissues, and their biological and mechanical properties, porosity, gasses, and nutrient diffusion can be easily adjusted.

3.2.1. Natural Hydrogels

Since cancer cells mostly respond to the biochemical and mechanical cues of the tissue, a huge impact is imposed on adequate adjustment of those properties during 3D model designing. Additionally culturing of tumor cells within 3D allows for acquiring phenotypes and response to stimuli that are observed in vivo conditions. [167]. Thus, cancer drug delivery systems can be optimized using 3D platforms like hydrogels [168]. Due to the ability to mimic key characteristics of in vivo tumor progression, there are many examples of hydrogels that serve as 3D models for cancer therapies. The most common natural hydrogels used as 3D models are mostly proteins and polysaccharides [169,170].

Collagen I is a protein hydrogel that shows great biocompatibility and could easily mimic the native environment of a tumor. The most important feature of this hydrogel is the presence of tripeptide RGD (Arg-Gly-Asp) which effectively binds the cell surface receptors [167]. Additionally, Collagen I promotes glioma cell growth, forming spheroid aggregates, and allows us to understand the tumor's properties such as growth, proliferation, and invasion [171]. In studies conducted by Szot et al., the 3D collagen I cell culture model provided fast growth of MDA-MB-231 breast cancer cells with the parallel capability of necrotic and hypoxic areas development. Additionally, cell-cell and cell-ECM interactions resulted in cell signaling and their phenotype observed in 3D models reflected in vivo conditions reality. Despite simplification of the tumor's complex nature, these studies allowed the study and reproduction of tumor necrosis, hypoxia, or gene expression that natively occur in vivo conditions.

In other studies, Jia et al. [172] studied the changes and key genes and miRNA affecting stemness functions and anticancer drug sensitivity of U87, U251, and HS683 glioma cells cultured in 3D collagen I culture model. Collagen I enabled glioma colony formation and evaluated their drug resistance. Studies on 3D models have allowed the investigation of 77 genes of glioma as well as signaling paths of protein network interactions as a response to stresses, DNA damage/repair, and drug metabolism. In these studies, it was discovered that AKT1, ATM, CASP3, CCND1, EGFR, PARP1, and TP53 genes and miRNA in glioma cells showed related pathways increasing the stemness and decreasing drug sensitivity of glioma, which suggests their crucial role in future diagnosis and further potential treatments.

Hyaluronans, i.e., hyaluronic acid or its salts, are polysaccharides that represent natural hydrogels [173]. The molecules of HA natively occur in native ECM, and HA has easily tunable mechanical properties through adjustment of concentration and crosslinking degree. The great advantage of HA is the ability to modulate healthy and tumor cells' fate [174].

Additionally, HA content, especially synthase gene expression in HA, is upregulated in many cancers, such as breast, pancreatic, colon, lung, and prostate cancer, and thus is a perfect material for 3D tumor cell culturing [175–179]. It is reported that a high content of synthase gene expression correlates with higher mortality in patients. Thus, the investigation of tumor cells on HA-based 3D platforms is fully justified. On the other hand, it is reported that HA alone, i.e., Matrigel® is overexpressed in breast cancer, providing increased response, and influencing the polarity of macrophages in many types of tumors [166]. To overcome that problem, many HA-composite 3D cell culture models have been formed. For instance, Baker et al. [160] formed an HA derivative crosslinked with matrix metalloproteinase (MMP)/oxime (HA-MMPx) modified with collagen and laminin. Such hydrogel composition allowed made it possible to obtain a 3D cell culture platform that was stable in physiological temperature, adequate mechanically, without distorted macrophage polarization and excessive response. The addition of laminin increased the hydrogel's stress relaxation, enabling the native moves of MCF-10A mammary epithelial cells. This property of HA-MMPx significantly increased healthy tissue regeneration in comparison to Matrigel®. Moreover, the well-defined composition of HA-MMPx provided the highest invasion of immune cells, i.e., host natural killer (NK) cell infiltration in comparison to other cell culture platforms. The increased NK infiltration in breast cancer correlates with a better prognosis for patients. Thus, more advanced studies on this HA-MMPx platform could be a breakthrough in future anticancer therapies' design.

Natural hydrogels seem to be perfect 3D tumor culture models; however, their mechanical properties range 0.1–0.4 kPa, which is insufficient from the perspective of tumors, whose mechanical properties depending on tumor type range ca. 1–6 kPa [180–182]. To overcome these problems, natural hydrogels are usually reinforced with other hydrogels showing increased mechanical properties or various types of nanomaterials.

Cellulose-based materials can help increase the mechanical properties of natural hydrogels. They can serve as hydrogel additives or hydrogels, and in the latest times, these materials have gained a lot of interest among both engineers and biologists [183–186]. For instance, collagen–nanocellulose hydrogel fairly mimics the ECM of pancreatic ductal adenocarcinoma (PDAC) [186]. The hydrogel consisted of short nanocrystals or elongated nanofibers (so-called nanocellulose) and cell adhesive proteins that provided adequate cell–materials surface interactions. Such composition not only provided stiffness in the range of 0.6–1.2 kPa that might be easily adjusted to mimic the lower profile of PDAC's ECM but also decent proliferation and morphology of MIA PaCa-2 and PANC-1 tumor cells in 3D culture over 14 days.

In other studies, Shokri et al. [187] combined methylcellulose (MC), HA, and silk fibroin (SF) to obtain a thermosensitive physically crosslinked hydrogel platform (MCHASF) that mimics the native breast tumor ECM. The obtained model showed 6 weeks of physiological stability, similar mechanical properties (elastic modulus of ca. 1 kPa) to native breast tumor (1–4 kPa), and provided an environment that allowed the mimicking of the morphology of human breast cancer (MDA-MB-231) cells characteristic for native malignancy. This model also kept the native drug resistance and metastatic potential of MDA-MB-231. Additionally, the migration rate and overexpression of MMP2, MMP9, and VEGF proteins were increased, which proves the metastatic potential of the tumor that resembles native in vivo conditions.

3.2.2. Synthetic Hydrogels

Although synthetic hydrogels make it possible to control many of the hydrogel parameters such as crosslinking rate or mechanical properties, hydrogels consisting of synthetic hydrogels are biochemically inert. To overcome that limitation, they can be modified with bioactive macromolecules or natural polymers [29,176,183].

Hybrid hydrogels are also beneficial approaches that serve as 3D cell culture models. Such combinations allow for overcoming the limitations of both, natural hydrogels, i.e., poor mechanical properties, and stability in physiological conditions, and synthetic hydrogels, i.e., biochemical inertia.

One of the examples of synthetic/natural hydrogel composites is gelatin methacrylate (GelMA) [9]. Shah et al. studied the GelMA 3D model which was seeded with patient neurospheres and U251 glioblastoma tumor cells. In this study, GelMa stiffeners ranged from ca. 5–19 kPa, which corresponds to the stiffness that is provided in natural glioblastomas environment, i.e., ca., 1–13 kPa. The results showed that 3D models provided a 3.5-fold increased invasive potential of U251 cells than those seeded on 2D models.

The 3D hydrogel models for tumor cell culturing are interesting approaches that allow a thorough investigation of the physiology of tumors, i.e., cellular morphology, phenotype, gene expression pattern, tumor–immune cell interactions, and drug resistance. Nevertheless, it should be noted that animal models based on rodents still are one of the best approaches to imitating tumor physiology. On the other hand, it should be noted that the preclinical testing results of drug efficacy on animal models rarely correspond to human clinical trials [167,188]. This is a result of differences in animal and human size, cells, and genes. Additionally, current ethical issues limit animal and human preclinical and clinical trials. Given the current state of knowledge, 3D cell culture models and animal models should be combined and used in preclinical trials. Although 3D cell culture models are promising approaches, at this time they will not replace small animal culture models. That is why there is still a huge demand for studies on designing and validating 3D cell culture models to provide an adequate physiologically, chemically, and mechanically complete environment for tumor testing that will reflect the physiology of the human organism [160].

4. Conclusions and Future Directions in Designing Biomaterials and 3D Cell Culture Models in Cancer Therapy

Anticancer therapies are a very challenging and complex field with many obstacles to overcome [189]. Many methods of forming biomaterials dedicated to anticancer therapies are being widely investigated [190]. The variations of materials manufacturing techniques could be observed, including co-axial electrospinning, emulsion electrospinning, bio-electrospraying, solution blow spinning, and various types of 3D printing and bioprinting. Scientists try to manipulate the parameters of production methods to receive a high-quality biomaterial with appropriate properties for anticancer therapy and regeneration of organs or tissues. Currently, most of these materials and methods are in the preclinical stage, although it is clear that their clinical applications are not far off [191].

The current trends in anticancer therapies are focused on combining biomaterials with chemotherapies, radiotherapies, and PTT to provide a synergistic effect that will result in winning the battle against cancer, promoting tissue regeneration, and preventing cancer relapse. Such an approach might help to face most of the anticancer therapies' requirements, but it should also be remembered that the fabrication and implementation of such biomaterial is challenging and involves many laborious functionalizations and optimizations. It should be noted that biological systems are dynamic, and under NIR, properties such as stiffness can change, and in the effect can influence cell adhesion.

Another approach that could help in anticancer therapies is 3D cell culturing models. The idea behind using such approaches is a thorough investigation of tumor physiology. Many scientists are working on a new type of 3D cell culture model [192]. Despite 3D cell culture models offering many advantages such as the examination and understanding of gene expression patterns, tumor–immune cell interactions, and tumor resistance to anticancer drugs, there is still plenty to do in this field. However, the future perspectives of 3D cell culture models and scaffolds assume their functionalization by incorporating endothelial and stromal cells to induce the vascularization process. Such an approach will be more functional and approximate these materials to the animal models [167].

There are still many problems to resolve in the modern anticancer drug delivery systems. In the field of drug delivery systems there is a need for improvements in targeted and controlled rate and dose of drug release in a desired localization, for example, in the vicinity of the cancer tumor. Numerous research teams [193] are still working on more realistic solutions. The long-term goal is a high-standard drug delivery and release system. Very small but targeted doses in targeted therapy can change patients' quality of life. This can eliminate side effects and increase tissues' and whole organs' regenerative capacity.

To summarize, many proposals and applications of electrospun nanofibers, 3D-printed materials, and hydrogels discussed in this paper concern the stage of in vivo tests or proof-of-principle studies. Scientists from various universities are trying to translate these ideas and accomplishments into clinical trials. The long-term goal is to overcome the current problems of implants and transplants, drug delivery, and cell culture strategies, resulting in designing more successful cancer therapies.

Author Contributions: Conceptualization, A.Z. and B.N.-S.; validation, A.Z., B.N.-S. and P.S.; formal analysis, P.S.; investigation, A.Z. and B.N.-S.; writing—original draft preparation, A.Z. and B.N.-S.; writing—review and editing, A.Z., B.N.-S. and P.S.; visualization, A.Z., B.N.-S. and P.S.; supervision, P.S. All authors have read and agreed to the published version of the manuscript.

Funding: This research received no external funding.

Institutional Review Board Statement: Not applicable.

Data Availability Statement: Not applicable.

References

1. Wei, H.; Cui, J.; Lin, K.; Xie, J.; Wang, X. Recent advances in smart stimuli-responsive biomaterials for bone therapeutics and regeneration. *Bone Res.* **2022**, *10*, 17. [CrossRef] [PubMed]
2. Charbe, N.B.; Tambuwala, M.; Palakurthi, S.S.; Warokar, A.; Hromić-Jahjefendić, A.; Bakshi, H.; Zacconi, F.; Mishra, V.; Khadse, S.; Aljabali, A.A.; et al. Biomedical applications of three-dimensional bioprinted craniofacial tissue engineering. *Bioeng. Transl. Med.* **2022**, e10333. [CrossRef]
3. Zaszczyńska, A.; Gradys, A.; Sajkiewicz, P. Progress in the Applications of Smart Piezoelectric Materials for Medical Devices. *Polymers* **2020**, *12*, 2754. [CrossRef] [PubMed]
4. Dorak, M.T.; Karpuzoglu, E. Gender Differences in Cancer Susceptibility: An Inadequately Addressed Issue. *Front. Genet.* **2012**, *3*, 268. [CrossRef]
5. Siegel, R.L.; Miller, K.D.; Fuchs, H.E.; Jemal, A. Cancer statistics, 2022. *CA Cancer J. Clin.* **2022**, *72*, 7–33. [CrossRef] [PubMed]
6. Dalmartello, M.; La Vecchia, C.; Bertuccio, P.; Boffetta, P.; Levi, F.; Negri, E.; Malvezzi, M. European cancer mortality predictions for the year 2022 with focus on ovarian cancer. *Ann. Oncol.* **2022**, *33*, 330–339. [CrossRef] [PubMed]
7. Bruns, J.; Egan, T.; Mercier, P.; Zustiak, S.P. Glioblastoma spheroid growth and chemotherapeutic responses in single and dual-stiffness hydrogels. *Acta Biomater.* **2022**. [CrossRef]
8. Zaszczynska, A.; Sajkiewicz, P.; Gradys, A. Piezoelectric Scaffolds as Smart Materials for Neural Tissue Engineering. *Polymers* **2020**, *12*, 161. [CrossRef]
9. Shah, N.; Hallur, P.M.; Ganesh, R.A.; Sonpatki, P.; Naik, D.; Chandrachari, K.P.; Puchalski, R.B.; Chaubey, A. Gelatin methacrylate hydrogels culture model for glioblastoma cells enriches for mesenchymal-like state and models interactions with immune cells. *Sci. Rep.* **2021**, *11*, 17727. [CrossRef]
10. Zhou, Y.; Chen, J.; Li, Q.; Huang, W.; Lan, H.; Jiang, H. Association between breastfeeding and breast cancer risk: Evidence from a meta-analysis. *Breastfeed Med.* **2015**, *10*, 175–182. [CrossRef]
11. Borasi, G.; Russo, G.; Nahum, A.; Zhang, Q.; Gilardi, M. HIFU generated fast hyperthermia as a new option for glioblastoma treatment. *Phys. Med.* **2016**, *32*, 5. [CrossRef]
12. Wang, X.; Chen, H.; Chen, Y.; Ma, M.; Zhang, K.; Li, F.; Zheng, Y.; Zeng, D.; Wang, Q.; Shi, J. Perfluorohexane-Encapsulated Mesoporous Silica Nanocapsules as Enhancement Agents for Highly Efficient High Intensity Focused Ultrasound (HIFU). *Adv. Mater.* **2012**, *24*, 785–791. [CrossRef] [PubMed]

13. Fan, W.; Yung, B.; Huang, P.; Chen, X. Nanotechnology for Multimodal Synergistic Cancer Therapy. *Chem. Rev.* **2017**, *117*, 13566–13638. [CrossRef] [PubMed]

14. Wei, X.; Liu, C.; Wang, Z.; Luo, Y. 3D printed core-shell hydrogel fiber scaffolds with NIR-triggered drug release for localized therapy of breast cancer. *Int. J. Pharm.* **2020**, *580*, 119219. [CrossRef] [PubMed]

15. Li, L.; Tang, F.; Liu, H.; Liu, T.; Hao, N.; Chen, D.; Teng, X.; He, J. In Vivo Delivery of Silica Nanorattle Encapsulated Docetaxel for Liver Cancer Therapy with Low Toxicity and High Efficacy. *ACS Nano* **2010**, *4*, 6874–6882. [CrossRef]

16. Ling, D.; Park, W.; Park, S.-J.; Lu, Y.; Kim, K.S.; Hackett, M.J.; Kim, B.H.; Yim, H.; Jeon, Y.S.; Na, K.; et al. Multifunctional Tumor pH- Sensitive Self-Assembled Nanoparticles for Bimodal Imaging and Treatment of Resistant Heterogeneous Tumors. *J. Am. Chem. Soc.* **2014**, *136*, 5647–5655. [CrossRef]

17. Jung, H.S.; Verwilst, P.; Sharma, A.; Shin, J.; Sessler, J.L.; Kim, J.S. Organic molecule-based photothermal agents: An expanding photothermal therapy universe. *Chem. Soc. Rev.* **2018**, *47*, 2280–2297. [CrossRef]

18. Lin, J.; Wang, M.; Hu, H.; Yang, X.; Wen, B.; Wang, Z.; Jacobson, O.; Song, J.; Zhang, G.; Niu, G.; et al. Multimodal-Imaging-Guided Cancer Phototherapy by Versatile Biomimetic Theranostics with UV and γ-Irradiation Protection. *Adv. Mater.* **2016**, *28*, 3273–3279. [CrossRef]

19. Liu, Y.; Bhattarai, P.; Dai, Z.; Chen, X. Photothermal therapy and photoacoustic imaging *via* nanotheranostics in fighting cancer. *Chem. Soc. Rev.* **2019**, *48*, 2053–2108. [CrossRef]

20. Xiang, H.; Chen, Y. Energy-converting nanomedicine. *Small* **2019**, *15*, 1805339. [CrossRef]

21. Verma, M. Personalized Medicine and Cancer. *J. Pers. Med.* **2012**, *2*, 1–14. [CrossRef] [PubMed]

22. Stock, U.A.; Vacanti, J.P. Tissue Engineering: Current State and Prospects. *Annu. Rev. Med.* **2001**, *52*, 443–451. [CrossRef] [PubMed]

23. Tu, Z.; Zhong, Y.; Hu, H.; Shao, D.; Haag, R.; Schirner, M.; Lee, J.; Sullenger, B.; Leong, K.W. Design of therapeutic biomaterials to control inflammation. *Nat. Rev. Mater.* **2022**, *7*, 557–574. [CrossRef] [PubMed]

24. Kharbikar, B.N.; Mohindra, P.; Desai, T.A. Biomaterials to enhance stem cell transplantation. *Cell Stem Cell* **2022**, *29*, 692–721. [CrossRef]

25. Schultz, G.S.; Ladwig, G.; Wysocki, A. Extracellular matrix: Review of its roles in acute and chronic wounds. *World Wide Wounds* **2005**, *2005*, 1–18.

26. Alberts, B. Cell junctions, cell adhesion and the extracellular matrix. In *Molecular Biology of the Cell*, 2nd ed.; Alberts, B., Ed.; Garland Scienc: New York, NY, USA, 1994; Volume 19, pp. 949–1009.

27. Dias, Y.J.; Robles, J.R.; Sinha-Ray, S.; Abiade, J.; Pourdeyhimi, B.; Niemczyk-Soczynska, B.; Kolbuk, D.; Sajkiewicz, P.; Yarin, A.L. Solution-blown poly (hydroxybutyrate) and ε-poly-l-lysine submicro- and microfiber-based sustainable nonwovens with antimicrobial activity for single-use applications. *ACS Biomater. Sci. Eng.* **2021**, *7*, 3980–3992. [CrossRef]

28. Ura, D.P.; Rosell-Llompart, J.; Zaszczyńska, A.; Vasilyev, G.; Gradys, A.; Szewczyk, P.K.; Stachewicz, U. The role of electrical polarity in electrospinning and on the mechanical and structural properties of as-spun fibers. *Materials* **2020**, *13*, 4169. [CrossRef]

29. Niemczyk-Soczynska, B.; Gradys, A.; Sajkiewicz, P. Hydrophilic surface functionalization of electrospun nanofibrous scaffolds in tissue engineering. *Polymers* **2020**, *12*, 2636. [CrossRef]

30. Niemczyk-Soczyńska, B.; Dulnik, J.; Jeznach, O.; Kołbuk, D.; Sajkiewicz, P. Shortening of electrospun PLLA fibers by ultrasonication. *Micron* **2021**, *145*, 103066. [CrossRef]

31. Zaszczyńska, A.; Sajkiewicz, P.Ł.; Gradys, A.; Tymkiewicz, R.; Urbanek, O.; Kołbuk, D. Influence of process-material conditions on the structure and biological properties of electrospun polyvinylidene fluoride fibers. *Bull. Pol. Acad. Sci. Tech. Sci.* **2020**, *68*, 627–633.

32. Kaniuk, Ł.; Ferraris, S.; Spriano, S.; Luxbacher, T.; Krysiak, Z.; Berniak, K.; Zaszczyńska, A.; Marzec, M.; Bernasik, A.; Sajkiewicz, P.; et al. Time-dependent effects on physicochemical and surface properties of PHBV fibers and films in relation to their interactions with fibroblasts. *Appl. Surf. Sci.* **2021**, *545*, 148983. [CrossRef]

33. Mandal, S.; Mohanto, S.; Mantry, S.; Das, S.; Datta, S.; Ghosh, M. A novel approach on microencapsulation in drug delivery system for cancer diagnosis. *IJPAR J.* **2017**, *6*, 327–334.

34. Ghosal, K.; Augustine, R.; Zaszczynska, A.; Barman, M.; Jain, A.; Hasan, A.; Kalarikkal, N.; Sajkiewicz, P.; Thomas, S. Novel drug delivery systems based on triaxial electrospinning based nanofibers. *React. Funct. Polym.* **2021**, *163*, 104895. [CrossRef]

35. Shafiee, A. Design and Fabrication of Three-Dimensional Printed Scaffolds for Cancer Precision Medicine. *Tissue Eng. Part A* **2020**, *26*, 305–317. [CrossRef]

36. Rabionet, M.; Yeste, M.; Puig, T.; Ciurana, J. Electrospinning PCL Scaffolds Manufacture for Three-Dimensional Breast Cancer Cell Culture. *Polymers* **2017**, *9*, 328. [CrossRef]

37. Cheng, S.; Du, Y.; Ma, B.; Tan, D. Total synthesis of a furostan saponin, timosaponin BII. *Org. Biomol. Chem.* **2009**, *7*, 3112–3118. [CrossRef]

38. Sridhar, R.; Sundarrajan, S.; Venugopal, J.R.; Ravichandran, R.; Ramakrishna, S. Electrospun inorganic and polymer composite nanofibers for biomedical applications. *J. Biomater.Sci. Polym. Ed.* **2013**, *24*, 365–385. [CrossRef]

39. Harish, V.; Tewari, D.; Gaur, M.; Yadav, A.B.; Swaroop, S.; Bechelany, M.; Barhoum, A. Review on Nanoparticles and Nanostructured Materials: Bioimaging, Biosensing, Drug Delivery, Tissue Engineering, Antimicrobial, and Agro-Food Applications. *Nanomaterials* **2022**, *12*, 457. [CrossRef]

40. Zhou, Y.; Zhao, K.Q.; Tsai, N.L.Y.; Wang, M. Bicomponent nanofibrous scaffolds with dual release of anticancer drugs and biomacromolecules. *MRS Commun.* **2019**, *9*, 413–420. [CrossRef]

41. Chen, S.; Boda, S.K.; Batra, S.K.; Li, X.; Xie, J. Emerging Roles of Electrospun Nanofibers in Cancer Research. *Adv. Healthc. Mater.* **2018**, *7*, e1701024. [CrossRef]

42. Thoppey, N.M.; Bochinski, J.R.; Clarke, L.I.; Gorga, R.E. Unconfined fluid electrospun into high quality nanofibers from a plate edge. *Polymer* **2010**, *51*, 4928–4936. [CrossRef]

43. Gugulothu, D.; Barhoum, A.; Nerella, R.; Ajmer, R.; Bechlany, M. Fabrication of Nanofibers: Electrospinning and Non-electrospinning Techniques. In *Barhoum*, 2nd ed.; Barhoum, A., Bechelany, M., Makhlouf, A., Eds.; Springer: New York, NY, USA, 2019; pp. 45–77. [CrossRef]

44. Stojanovska, E.; Canbay, E.; Pampal, E.S.; Calisir, M.D.; Ağma, O.; Polat, Y.; Simsek, R.; Gundogdu, N.A.S.; Akgul, Y.; Kilic, A. A review on non-electro nanofibre spinning techniques. *RSC Adv.* **2016**, *6*, 83783–83801. [CrossRef]

45. Liu, F.; Li, S.; Fang, Y.; Zheng, F.; Li, J.; He, J. Fabrication of highly oriented nanoporous fibers via airflow bubble-spinning. *Appl. Surf. Sci.* **2017**, *421*, 61. [CrossRef]

46. Sahay, R.; Thavasi, V.; Ramakrishna, S. Design modifications in electrospinning setup for advanced applications. *J. Nanomater.* **2011**, *2011*, 17. [CrossRef]

47. Czwartos, J.; Zaszczyńska, A.; Nowak-Stępniowska, A.; Fok, T.; Budner, B.; Bartnik, A.; Fiedorowicz, H. The novel approach to physico-chemical modification and cytocompatibility enhancement of fibrous polycaprolactone (PCL) scaffolds using soft X-ray/extreme ultraviolet (SXR/EUV) radiation and low-temperature, SXR/EUV induced, nitrogen and oxygen plasmas. *Appl. Surf. Sci.* **2022**, *606*, 154779. [CrossRef]

48. Khan, I.; Pandit, J.; Ahmed, S.; Zameer, S.; Nikita; Ahmad, S.; Bano, S.; Ansari, M.D.; Solanki, P.; Jahan, R.N.; et al. Development and evaluation of biodegradable polymeric lomustine nanofibres for the efficient tumor targeting: In vitro characterization, ex vivo permeation and degradation study. *J. Drug Deliv. Sci. Technol.* **2022**, *75*, 103685. [CrossRef]

49. Mayilswamy, N.; Prakash, N.J.; Kandasubramanian, B. Design and fabrication of biodegradable electrospun nanofibers loaded with biocidal agents. *Int. J. Polym. Mater. Polym. Biomater.* **2022**, 1–27. [CrossRef]

50. Wang, Z.; Zhao, Y.; Shen, M.; Tomás, H.; Zhou, B.; Shi, X. Antitumor Efficacy of Doxorubicin-Loaded Electrospun Attapulgite–Poly (lactic-co-glycolic acid) Composite Nanofibers. *J. Funct. Biomater.* **2022**, *13*, 55. [CrossRef]

51. Sutherland, T.D.; Young, J.H.; Weisman, S.; Hayashi, C.Y.; Merritt, D.J. Insect Silk: One Name, Many Materials. *Annu. Rev. Entomol.* **2010**, *55*, 171–188. [CrossRef]

52. Ramakrishna, S. *An Introduction to Electrospinning and Nanofibers*, 3rd ed.; World Scientific: Central Area, Singapore, 2005. [CrossRef]

53. Rockwood, D.N.; Gil, E.S.; Park, S.-H.; Kluge, J.A.; Grayson, W.; Bhumiratana, S.; Rajkhowa, R.; Wang, X.; Kim, S.J.; Vunjak-Novakovic, G.; et al. Ingrowth of human mesenchymal stem cells into porous silk particle reinforced silk composite scaffolds: An in vitro study. *Acta Biomater.* **2011**, *7*, 144–151. [CrossRef]

54. Mottaghitalab, F.; Farokhi, M.; Fatahi, Y.; Atyabi, F.; Dinarvand, R. New insights into designing hybrid nanoparticles for lung cancer: Diagnosis and treatment. *J. Control. Release* **2019**, *295*, 250–267. [CrossRef]

55. Laiva, A.L.; Venugopal, J.R.; Karuppuswamy, P.; Navaneethan, B.; Gora, A.; Ramakrishna, S. Controlled release of titanocene into the hybrid nanofibrous scaffolds to prevent the proliferation of breast cancer cells. *Int. J. Pharm.* **2015**, *483*, 115–123. [CrossRef] [PubMed]

56. Li, H.; Zhu, J.; Chen, S.; Jia, L.; Ma, Y. Fabrication of aqueous-based dual drug loaded silk fibroin electrospun nanofibers embedded with curcumin-loaded RSF nanospheres for drugs controlled release. *RSC Adv.* **2017**, *7*, 56550–56558. [CrossRef]

57. Yang, J.; Zhang, Y.S.; Yue, K.; Khademhosseini, A. Cell-laden hydrogels for osteochondral and cartilage tissue engineering. *Acta Biomater.* **2017**, *57*, 1–25. [CrossRef] [PubMed]

58. Park, T.G. Degradation of poly(lactic-co-glycolic acid) microspheres: Effect of copolymer composition. *Biomaterials* **1995**, *16*, 1123–1130. [CrossRef]

59. Mehaffey, J.H.; Charles, E.J.; Sharma, A.K.; Money, D.T.; Zhao, Y.; Stoler, M.H.; Lau, C.L.; Tribble, C.G.; Laubach, V.E.; Roeser, M.E.; et al. Airway pressure release ventilation during ex vivo lung perfusion attenuates injury. *J. Thorac. Cardiovasc. Surg.* **2017**, *153*, 197–204. [CrossRef]

60. Lynn, A.; Yannas, I.; Bonfield, W. Antigenicity and immunogenicity of collagen. *J. Biomed. Mater. Res.–Part B Appl. Biomater.* **2004**, *71*, 343–354. [CrossRef]

61. Liu, X.; Zheng, C.; Luo, X.; Wang, X.; Jiang, H. Recent advances of collagen-based biomaterials: Multi-hierarchical structure, modification and biomedical applications. *Mater. Sci. Eng. C* **2019**, *99*, 1509–1522. [CrossRef]

62. Cen, L.; Liu, W.; Cui, L.; Zhang, W.; Cao, Y. Collagen Tissue Engineering: Development of Novel Biomaterials and Applications. *Pediatr. Res.* **2008**, *63*, 492–496. [CrossRef]

63. Irawan, V.; Sung, T.-C.; Higuchi, A.; Ikoma, T. Collagen Scaffolds in Cartilage Tissue Engineering and Relevant Approaches for Future Development. *Tissue Eng. Regen. Med.* **2018**, *15*, 673–697. [CrossRef]

64. Shen, Y.; Shen, R.; Ge, L.; Zhu, Q.; Li, F. Fibrillar type I collagen matrices enhance metastasis/invasion of ovarian epithelial cancer via beta1 integrin and PTEN signals. *Int. J. Gynecol. Cancer* **2012**, *22*, 1316–1324. [CrossRef] [PubMed]

65. Zhang, T.; Yan, Y.; Wang, X.; Xiong, Z.; Lin, F.; Wu, R.; Zhang, R. Three-dimensional Gelatin and Gelatin/Hyaluronan Hydrogel Structures for Traumatic Brain Injury. *J. Bioact. Compat. Polym.* **2007**, *22*, 19–29. [CrossRef]

66. Zhang, J.; Li, J.; Kawazoe, N.; Chen, G. Composite scaffolds of gelatin and gold nanoparticles with tunable size and shape for photothermal cancer therapy. *J. Mater. Chem. B.* **2017**, *5*, 245–253. [CrossRef] [PubMed]

67. Chen, K.; Pan, H.; Yan, Z.; Li, Y.; Ji, D.; Yun, K.; Su, Y.; Liu, D.; Pan, W. A novel alginate/gelatin sponge combined with cur-cumin-loaded electrospun fibers for postoperative rapid hemostasis and prevention of tumor recurrence. *Int. J. Biol. Macromol.* **2021**, *182*, 1339–1350. [CrossRef] [PubMed]

68. Baharlouei, P.; Rahman, A. Chitin and Chitosan: Prospective Biomedical Applications in Drug Delivery, Cancer Treatment, and Wound Healing. *Mar. Drugs.* **2022**, *20*, 460. [CrossRef] [PubMed]

69. Zakhireh, S.; Barar, J.; Adibkia, K.; Beygi-Khosrowshahi, Y.; Fathi, M.; Omidain, H.; Omidi, Y. Bioactive Chitosan-Based Organometallic Scaffolds for Tissue Engineering and Regeneration. *Top. Curr. Chem.* **2022**, *380*, 13. [CrossRef]

70. Mangla, D.; Sharma, A.; Ikram, S. Synthesis of ecological chitosan/PVP magnetic composite: Remediation of amoxicillin trihydrate from its aqueous solution, isotherm modelling, thermodynamic, and kinetic studies. *React. Funct. Polym.* **2022**, *175*, 105261. [CrossRef]

71. Chopra, H.; Bibi, S.; Kumar, S.; Khan, M.S.; Kumar, P.; Singh, I. Preparation and Evaluation of Chitosan/PVA Based Hydrogel Films Loaded with Honey for Wound Healing Application. *Gels* **2022**, *8*, 111. [CrossRef]

72. Shafabakhsh, R.; Yousefi, B.; Asemi, Z.; Nikfar, B.; Mansournia, M.A.; Hallajzadeh, J. Chitosan: A compound for drug delivery system in gastric cancer-a review. *Carbohydr. Polym.* **2020**, *242*, 116403. [CrossRef]

73. Ranjbar-Mohammadi, M.; Sa'Di, V.; Moezzi, M.; Saghafi, R. Fabrication and Characterization of Antibacterial Suture Yarns Containing PLA/Tetracycline Hydrochloride-PVA/Chitosan Nanofibers. *Fibers Polym.* **2022**, *23*, 1538–1547. [CrossRef]

74. Nikcevic, I.; Maravic, D.; Ignjatovic, N.; Mitric, M.; Makovec, D.; Uskokovic, D. The Formation and Characterization of Nanocrystalline Phases by Mechanical Milling of Biphasic Calcium Phosphate/Poly-L-Lactide Biocomposite. *Mater. Trans.* **2006**, *47*, 2980–2986. [CrossRef]

75. Davachi, S.M.; Kaffashi, B. Polylactic Acid in Medicine. *Polym. Technol. Eng.* **2015**, *54*, 944–967. [CrossRef]

76. Szaraniec, B. Durability of Biodegradable Internal Fixation Plates. In *Materials Science Forum*, 2nd ed.; Pinto, A.M.P., Pouzada, A.S., Eds.; Trans Tech Publications: Zurich, Switzerland, 2013; Volume 730, pp. 15–19. [CrossRef]

77. Tyler, B.; Gullotti, D.; Mangraviti, A.; Utsuki, T.; Brem, H. Polylactic acid (PLA) controlled delivery carriers for biomedical applications. *Adv. Drug Deliv. Rev.* **2016**, *107*, 163–175. [CrossRef] [PubMed]

78. Hu, J.; Sun, X.; Ma, H.; Xie, C.; Chen, Y.E.; Ma, P.X. Porous nanofibrous PLLA scaffolds for vascular tissue engineering. *Biomaterials* **2010**, *31*, 7971–7977. [CrossRef] [PubMed]

79. Yuan, Z.; Pan, Y.; Cheng, R.; Sheng, L.; Wu, W.; Pan, G.; Feng, Q.; Cui, W. Doxorubicin-loaded mesoporous silica nanoparticle composite nanofibers for long-term adjustments of tumor apoptosis. *Nanotechnology* **2016**, *27*, 245101. [CrossRef] [PubMed]

80. Boedtkjer, E.; Pedersen, S.F. The Acidic Tumor Microenvironment as a Driver of Cancer. *Annu. Rev. Physiol.* **2020**, *82*, 103–126. [CrossRef]

81. Mir, M.; Ahmed, N.; Rehman, A.U. Recent applications of PLGA based nanostructures in drug delivery. *Colloids Surf. B Biointerfaces* **2017**, *159*, 217–231. [CrossRef]

82. Parveen, S.; Sahoo, S.K. Long circulating chitosan/PEG blended PLGA nanoparticle for tumor drug delivery. *Eur. J. Pharmacol.* **2011**, *670*, 372–383. [CrossRef]

83. Almajhdi, F.N.; Fouad, H.; Khalil, K.A.; Awad, H.M.; Mohamed, S.H.S.; Elsarnagawy, T.; Albarrag, A.M.; Al-Jassir, F.F.; Abdo, H.S. In-vitro anticancer and antimicrobial activities of PLGA/silver nanofiber composites prepared by electrospinning. *J. Mater. Sci. Mater. Med.* **2014**, *25*, 1045–1053. [CrossRef]

84. Samadzadeh, S.; Mousazadeh, H.; Ghareghomi, S.; Dadashpour, M.; Babazadeh, M.; Zarghami, N. In vitro anticancer efficacy of Metformin-loaded PLGA nanofibers towards the post-surgical therapy of lung cancer. *J. Drug Deliv. Sci. Technol.* **2021**, *61*, 102318. [CrossRef]

85. Espinoza, S.M.; Patil, H.I.; San Martin Martinez, E.; Casañas Pimentel, R.; Ige, P.P. Poly-ε-caprolactone (PCL), a promising polymer for pharmaceutical and biomedical applications: Focus on nanomedicine in cancer. *Int. J. Polym. Mater. Polym. Biomater.* **2020**, *69*, 85–126. [CrossRef]

86. Shang, Q.; Dong, Y.; Su, Y.; Leslie, F.; Sun, M.; Wang, F. Local scaffold-assisted delivery of immunotherapeutic agents for improved cancer immunotherapy. *Adv. Drug Deliv. Rev.* **2022**, *185*, 114308. [CrossRef] [PubMed]

87. Han, D.; Sasaki, M.; Yoshino, H.; Kofuji, S.; Sasaki, A.T.; Steckl, A.J. In-vitro evaluation of MPA-loaded electrospun coaxial fiber membranes for local treatment of glioblastoma tumor cells. *J. Drug Deliv. Sci. Technol.* **2017**, *40*, 45–50. [CrossRef]

88. Irani, M.; Sadeghi, G.M.M.; Haririan, I. The sustained delivery of temozolomide from electrospun PCL-Diol-b-PU/gold nanocomposite nanofibers to treat glioblastoma tumors. *Mater. Sci. Eng. C* **2017**, *75*, 165–174. [CrossRef]

89. Yap, J.X.; Leo, C.; Yasin, N.H.M.; Show, P.L.; Chu, D.-T.; Singh, V.; Derek, C. Recent advances of natural biopolymeric culture scaffold: Synthesis and modification. *Bioengineered* **2022**, *13*, 2226–2247. [CrossRef]

90. Yan, E.; Jiang, J.; Ren, X.; Gao, J.; Zhang, X.; Li, S.; Chen, S.; Li, Y. Polycaprolactone/polyvinyl alcohol core-shell nanofibers as a pH-responsive drug carrier for the potential application in chemotherapy against colon cancer. *Mater. Lett.* **2021**, *291*, 129516. [CrossRef]

91. Asran, A.S.; Razghandi, K.; Aggarwal, N.; Michler, G.H.; Groth, T. Nanofibers from blends of polyvinyl alcohol and polyhydroxy butyrate as potential scaffold material for tissue engineering of skin. *Biomacromolecules* **2010**, *11*, 3413–3421. [CrossRef]

92. Avci, H.; Monticello, R.; Kotek, R. Preparation of antibacterial PVA and PEO nanofibers containing Lawsonia Inermis (henna) leaf extracts. *J. Biomater. Sci. Polym. Ed.* **2013**, *24*, 1815–1830. [CrossRef]

93. Liu, H.; Zuo, B. Sound absorption property of PVA/PEO/GO nanofiber membrane and non-woven composite material. *J. Ind. Text.* **2020**, *50*, 512–525. [CrossRef]

94. Linh, N.T.B.; Lee, B.-T. Electrospinning of polyvinyl alcohol/gelatin nanofiber composites and cross-linking for bone tissue engineering application. *J. Biomater. Appl.* **2012**, *27*, 255–266. [CrossRef]

95. Mohan, N.; Nair, P.D. Polyvinyl alcohol-poly (caprolactone) Semi IPN scaffold with implication for cartilage tissue engineering. *J. Biomed. Mater. Res. Part B Appl. Biomater.* **2008**, *84*, 584–594. [CrossRef] [PubMed]

96. Yang, C.; Wu, X.; Zhao, Y.; Xu, L.; Wei, S. Nanofibrous scaffold prepared by electrospinning of poly (vinyl alcohol)/gelatin aqueous solutions. *J. Appl. Polym. Sci.* **2011**, *121*, 3047–3055. [CrossRef]

97. Hasbum, A.; Karabulut, O.; Reyes, R.E.; Ricci, C.; Franchi, A.; Danti, S.; Chew, S.A. Combined Application of Patient-Derived Cells and Biomaterials as 3D In Vitro Tumor Models. *Cancers* **2022**, *14*, 2503. [CrossRef] [PubMed]

98. Qian, Y.; Li, L.; Jiang, C.; Xu, W.; Lv, Y.; Zhong, L.; Cai, K.; Yang, L. The effect of hyaluronan on the motility of skin dermal fibroblasts in nanofibrous scaffolds. *Int. J. Biol. Macromol.* **2015**, *79*, 133–143. [CrossRef] [PubMed]

99. Zhong, S.; Teo, W.E.; Zhu, X.; Beuerman, R.; Ramakrishna, S.; Yung, L.Y.L. Formation of Collagen-Glycosaminoglycan Blended Nanofibrous Scaffolds and Their Biological Properties. *Biomacromolecules* **2005**, *6*, 2998–3004. [CrossRef]

100. Hinderer, S.; Schesny, M.; Bayrak, A.; Ibold, B.; Hampel, M.; Walles, T.; Stock, U.A.; Seifert, M.; Schenke-Layland, K. Engineering of fibrillar decorin matrices for a tissue-engineered trachea. *Biomaterials* **2012**, *33*, 5259–5266. [CrossRef]

101. Wen, X.; Wang, Y.; Guo, Z.; Meng, H.; Huang, J.; Zhang, L.; Zhao, B.; Zhao, Q.; Zheng, Y.; Peng, J. Cauda Equina-Derived Extracellular Matrix for Fabrication of Nanostructured Hybrid Scaffolds Applied to Neural Tissue Engineering. *Tissue Eng. Part A* **2015**, *21*, 1095–1105. [CrossRef]

102. Shin, Y.C.; Lee, J.H.; Jin, L.; Kim, M.J.; Kim, Y.-J.; Hyun, J.K.; Jung, T.-G.; Hong, S.W.; Han, D.-W. Stimulated myoblast differentiation on graphene oxide-impregnated PLGA-collagen hybrid fibre matrices. *J. Nanobiotechnol.* **2015**, *13*, 21. [CrossRef]

103. Kim, H.; Woo, S. Ocular Drug Delivery to the Retina: Current Innovations and Future Perspectives. *Pharmaceutics* **2021**, *13*, 108. [CrossRef]

104. Sun, L.; Gao, X.; Wu, D.; Guo, Q. Advances in Physiologically Relevant Actuation of Shape Memory Polymers for Biomedical Applications. *Polym. Rev.* **2021**, *61*, 280–318. [CrossRef]

105. Baudequin, T.; Gaut, L.; Mueller, M.; Huepkes, A.; Glasmacher, B.; Duprez, D.; Bedoui, F.; Legallais, C. The osteogenic and tenogenic differentiation potential of C3H10T1/2 (mesenchymal stem cell model) cultured on PCL/PLA electrospun scaffolds in the absence of specific differentiation medium. *Materials* **2017**, *10*, 1387. [CrossRef] [PubMed]

106. Maurmann, N.; Pereira, D.P.; Burguez, D.; de S Pereira, F.D.; Neto, P.I.; Rezende, R.A.; Gamba, D.; da Silva, J.V.L.; Pranke, P. Mesenchymal stem cells cultivated on scaffolds formed by 3D printed PCL matrices, coated with PLGA electrospun nanofibers for use in tissue engineering. *Biomed. Phys. Eng. Express.* **2017**, *3*, 045005. [CrossRef]

107. Zhang, C.; Wang, X.; Zhang, E.; Yang, L.; Yuan, H.; Tu, W.; Zhang, H.; Yin, Z.; Shen, W.; Chen, X.; et al. An epigenetic bioactive composite scaffold with well-aligned nanofibers for functional tendon tissue engineering. *Acta Biomater.* **2018**, *66*, 141–156. [CrossRef] [PubMed]

108. Cheng, M.; Wang, H.; Zhang, Z.; Li, N.; Fang, X.; Xu, S. Gold Nanorod-Embedded Electrospun Fibrous Membrane as a Photothermal Therapy Platform. *ACS Appl. Mater. Interfaces* **2014**, *6*, 1569–1575. [CrossRef]

109. Chen, H.; Shi, Y.; Sun, L.; Ni, S. Electrospun composite nanofibers with all-trans retinoic acid and MWCNTs-OH against cancer stem cells. *Life Sci.* **2020**, *258*, 118152. [CrossRef]

110. Azerbaijan, M.H.; Bahmani, E.; Jouybari, M.H.; Hassaniazardaryani, A.; Goleij, P.; Akrami, M.; Irani, M. Electrospun gold nanorods/graphene oxide loaded-core-shell nanofibers for local delivery of paclitaxel against lung cancer during pho-to-chemotherapy method. *Eur. J. Pharm. Sci.* **2021**, *164*, 105914. [CrossRef]

111. Wang, X.; Lv, F.; Li, T.; Han, Y.; Yi, Z.; Liu, M.; Chang, J.; Wu, C. Electrospun Micropatterned Nanocomposites Incorporated with Cu_2S Nanoflowers for Skin Tumor Therapy and Wound Healing. *ACS Nano* **2017**, *11*, 11337–11349. [CrossRef]

112. Mauro, N.; Scialabba, C.; Pitarresi, G.; Giammona, G. Enhanced adhesion and in situ photothermal ablation of cancer cells in surface-functionalized electrospun microfiber scaffold with grapheme oxide. *Int. J. Pharm.* **2017**, *526*, 167–177. [CrossRef]

113. Xue, C.; Sutrisno, L.; Li, M.; Zhu, W.; Fei, Y.; Liu, C.; Wang, X.; Cai, K.; Hu, Y.; Luo, Z. Implantable multifunctional black phosphorus nanoformulation-deposited biodegradable scaffold for combinational photothermal/ chemotherapy and wound healing. *Biomaterials* **2021**, *269*, 120623. [CrossRef]

114. Available online: https://pubs.rsc.org/en/content/articlehtml/2018/tb/c8tb02325e (accessed on 1 September 2022).

115. Available online: https://www.sciencedirect.com/science/article/pii/S0142961218308342 (accessed on 1 September 2022).

116. Cen, D.; Wan, Z.; Fu, Y.; Pan, H.; Xu, J.; Wang, Y.; Wu, Y.; Li, X.; Cai, X. Implantable fibrous 'patch' enabling preclinical chemo-photothermal tumor therapy. *Colloids Surf. B Biointerfaces* **2020**, *192*, 111005. [CrossRef]

117. Tan, Z.; Gao, X.; Liu, T.; Yang, Y.; Zhong, J.; Tong, C.; Tan, Y. Electrospun vein grafts with high cell infiltration for vascular tissue engineering. *Mater. Sci. Eng. C* **2017**, *81*, 407–415. [CrossRef] [PubMed]

118. Niemczyk-Soczynska, B.; Zaszczyńska, A.; Zabielski, K.; Sajkiewicz, P. Hydrogel, Electrospun and Composite Materials for Bone/Cartilage and Neural Tissue Engineering. *Materials* **2021**, *14*, 6899. [CrossRef] [PubMed]

119. Jammalamadaka, U.; Tappa, K. Recent Advances in Biomaterials for 3D Printing and Tissue Engineering. *J. Funct. Biomater.* **2018**, *9*, 22. [CrossRef] [PubMed]

120. Zaszczyńska, A.; Moczulska-Heljak, M.; Gradys, A.; Sajkiewicz, P. Advances in 3D Printing for Tissue Engineering. *Materials* **2021**, *14*, 3149. [CrossRef] [PubMed]

121. Sigaux, N.; Pourchet, L.; Breton, P.; Brosset, S.; Louvrier, A.; Marquette, C.A. 3D Bioprinting: Principles, fantasies and prospects. *J. Stomatol. Oral Maxillofac. Surg.* **2019**, *120*, 128–132. [CrossRef]

122. Hospodiuk, M.; Dey, M.; Sosnoski, D.; Ozbolat, I.T. The bioink: A comprehensive review on bioprintable materials. *Biotechnol. Adv.* **2017**, *35*, 217–239. [CrossRef]

123. Gillispie, G.J.; Prim, P.; Copus, J.S.; Fisher, J.P.; Mikos, A.G.; Yoo, J.J.; Atala, A.; Lee, S.J.J. Assessment methodologies for extrusion-based bioink printability. *Biofabrication* **2020**, *12*, 022003. [CrossRef]

124. Sears, N.A.; Seshadri, D.R.; Dhavalikar, P.S.; Cosgriff-Hernandez, E. A Review of Three-Dimensional Printing in Tissue Engineering. *Tissue Eng. Part B Rev.* **2016**, *22*, 298–310. [CrossRef]

125. Vanderburgh, J.; Sterling, J.A.; Guelcher, S.A. 3D Printing of Tissue Engineered Constructs for In Vitro Modeling of Disease Progression and Drug Screening. *Ann. Biomed. Eng.* **2017**, *45*, 164–179. [CrossRef]

126. Lan, W.; Huang, X.; Huang, D.; Wei, X.; Chen, W. Progress in 3D printing for bone tissue engineering: A review. *J. Mater. Sci.* **2022**, *57*, 12685–12709. [CrossRef]

127. Boland, T.; Xu, T.; Damon, B.; Cui, X. Application of inkjet printing to tissue engineering. *Biotechnol. J. Healthc. Nutr. Technol.* **2006**, *1*, 910–917. [CrossRef] [PubMed]

128. Suffo, M.; López-Marín, C.J. A Comparative Study of Turbulence Methods Applied to the Design of a 3D-Printed Scaffold and the Selection of the Appropriate Numerical Scheme to Simulate the Scaffold for Tissue Engineering. *Appl. Sci.* **2022**, *12*, 191. [CrossRef]

129. Luo, Y.; Li, Y.; Qin, X.; Wa, Q. 3D printing of concentrated alginate/gelatin scaffolds with homogeneous nano apatite coating for bone tissue engineering. *Mater. Des.* **2018**, *146*, 12–19. [CrossRef]

130. Moroni, S.; Casettari, L.; Lamprou, D.A. 3D and 4D Printing in the Fight against Breast Cancer. *Biosensors* **2022**, *12*, 568. [CrossRef] [PubMed]

131. Bhuskute, H.; Shende, P.; Prabhakar, B. 3D Printed Personalized Medicine for Cancer: Applications for Betterment of Diagnosis, Prognosis and Treatment. *AAPS PharmSciTech* **2022**, *23*, 8. [CrossRef]

132. Xiang, H.; Yang, Q.; Gao, Y.; Zhu, D.; Pan, S.; Xu, T.; Chen, Y. Cocrystal Strategy toward Multifunctional 3D-Printing Scaffolds Enables NIR-Activated Photonic Osteosarcoma Hyperthermia and Enhanced Bone Defect Regeneration. *Adv. Funct. Mater.* **2020**, *30*, 1909938. [CrossRef]

133. Niemczyk, B.; Sajkiewicz, P.; Kolbuk, D. Injectable hydrogels as novel materials for central nervous system regeneration. *J. Neural Eng.* **2018**, *15*, 051002. [CrossRef]

134. Huang, Z.; Tian, Z.; Zhu, M.; Wu, C.; Zhu, Y. Recent Advances in Biomaterial Scaffolds for Integrative Tumor Therapy and Bone Regeneration. *Adv. Ther.* **2021**, *4*, 2000212. [CrossRef]

135. Alamdari, S.G.; Amini, M.; Jalilzadeh, N.; Baradaran, B.; Mohammadzadeh, R.; Mokhtarzadeh, A.; Oroojalian, F. Recent advances in nanoparticle-based photothermal therapy for breast cancer. *J. Control. Release* **2022**, *349*, 269–303. [CrossRef]

136. Wang, S.; Wang, F.; Zhao, X.; Yang, F.; Xu, Y.; Yan, F.; Xia, D.; Liu, Y. The effect of near-infrared light-assisted photothermal therapy combined with polymer materials on promoting bone regeneration: A systematic review. *Mater. Des.* **2022**, *217*, 110621. [CrossRef]

137. Hossain, I.; Nanda, S.S.; Selvan, S.T.; Yi, D.K. Recent Insights into NIR-Light-Responsive Materials for Photothermal Cell Treatments. *Nanomaterials* **2022**, *12*, 3318. [CrossRef] [PubMed]

138. Han, S.; Wu, J. Three-dimensional (3D) scaffolds as powerful weapons for tumor immunotherapy. *Bioact. Mater.* **2022**, *17*, 300–319. [CrossRef] [PubMed]

139. Xu, X.; Chen, X.; Wang, H.; Mei, X.; Chen, B.; Li, R.; Qin, Y. Balancing the toxicity, photothermal effect, and promotion of osteogenesis: Photothermal scaffolds for malignant bone tumor therapy. *Mater. Today Adv.* **2022**, *13*, 100209. [CrossRef]

140. Yang, C.; Ma, H.; Wang, Z.; Younis, M.R.; Liu, C.; Wu, C.; Luo, Y.; Huang, P. 3D Printed Wesselsite Nanosheets Functionalized Scaffold Facilitates NIR-II Photothermal Therapy and Vascularized Bone Regeneration. *Adv. Sci.* **2021**, *8*, 2100894. [CrossRef] [PubMed]

141. Zhao, J.; Xu, W.; Zhao, Z.; Ling, G.; Zhang, P. Intelligent nanocomposite hydrogels with simultaneous photothermal antitumor and antibacterial efficacy for cutaneous melanoma treatment. *Compos. Part B Eng.* **2022**, *243*, 110130. [CrossRef]

142. Qu, Y.; Lu, K.; Zheng, Y.; Huang, C.; Wang, G.; Zhang, Y.; Yu, Q. Photothermal scaffolds/surfaces for regulation of cell behaviors. *Bioact. Mater.* **2022**, *8*, 449–477. [CrossRef]

143. Fu, S.; Hu, H.; Chen, J.; Zhu, Y.; Zhao, S. Silicone resin derived larnite/C scaffolds via 3D printing for potential tumor therapy and bone regeneration. *Chem. Eng. J.* **2020**, *382*, 122928. [CrossRef]

144. Zhang, X.; Li, Y.; Dong, X.; Wang, H.; Chen, B.; Li, R.; Qin, Y.; Ivasishin, O. 3D-printed bioactive ceramic scaffolds with MoSe$_2$ nanocrystals as photothermal agents for bone tumor therapy. *RSC Adv.* **2022**, *12*, 30588–30597. [CrossRef]

145. Wang, H.; Zeng, X.; Pang, L.; Wang, H.; Lin, B.; Deng, Z.; Qi, E.L.X.; Miao, N.; Wang, D.; Huang, P.; et al. Integrative treatment of anti-tumor/bone repair by combination of MoS2 nanosheets with 3D printed bioactive borosilicate glass scaffolds. *Chem. Eng. J.* **2020**, *396*, 125081. [CrossRef]

146. He, C.; Yu, L.; Yao, H.; Chen, Y.; Hao, Y. Combinatorial Photothermal 3D-Printing Scaffold and Checkpoint Blockade Inhibits Growth/Metastasis of Breast Cancer to Bone and Accelerates Osteogenesis. *Adv. Funct. Mater.* **2020**, *31*, 2006214. [CrossRef]

147. Ma, H.; Li, T.; Huan, Z.; Zhang, M.; Yang, Z.; Wang, J.; Chang, J.; Wu, C. 3D printing of high-strength bioscaffolds for the synergistic treatment of bone cancer. *NPG Asia Mater.* **2018**, *10*, 31–44. [CrossRef]

148. Wilson, J.L.; McDevitt, T.C. Biofunctional Hydrogels for Three-Dimensional Stem Cell Culture. In *Biology and Engineering of Stem Cell Niches*; Academic Press: Cambridge, MA, USA, 2017; pp. 345–362. [CrossRef]

149. Sun, Z.; Song, C.; Wang, C.; Hu, Y.; Wu, J. Hydrogel-based controlled drug delivery for cancer treatment: A review. *Mol Pharm.* **2019**, *17*, 373–391. [CrossRef] [PubMed]

150. Ma, H.; He, C.; Cheng, Y.; Yang, Z.; Zang, J.; Liu, J.; Chen, X. Localized co-delivery of doxorubicin, cisplatin, and methotrexate by thermosensitive hydrogels for enhanced osteosarcoma treatment. *ACS Appl. Mater. Interfaces* **2015**, *7*, 27040–27048. [CrossRef]

151. Xu, L.; Chen, Y.; Zhang, P.; Tang, J.; Xue, Y.; Luo, H.; Liu, J. 3D printed heterogeneous hybrid hydrogel scaffolds for sequential tumor photothermal-chemotherapy and wound healing. *Biomater. Sci.* **2022**, *10*, 5648–5661. [CrossRef] [PubMed]

152. Askari, E.; Seyfoori, A.; Amereh, M.; Gharaie, S.S.; Ghazali, H.S.; Ghazali, Z.S.; Akbari, M. Stimuli-responsive hydrogels for local post-surgical drug delivery. *Gels* **2020**, *6*, 14. [CrossRef]

153. Aisenbrey, E.A.; Murphy, W.L. Synthetic alternatives to matrigel. *Nat. Rev. Mater.* **2020**, *5*, 539–551. [CrossRef] [PubMed]

154. Ma, H.; He, C.; Cheng, Y.; Li, D.; Gong, Y.; Liu, J.; Tian, H.; Chen, X. PLK1shRNA and doxorubicin co-loaded thermosensitive PLGA-PEG-PLGA hydrogels for osteosarcoma treatment. *Biomaterials* **2014**, *35*, 8723–8734. [CrossRef] [PubMed]

155. Dang, W.; Chen, W.-C.; Ju, E.; Xu, Y.; Li, K.; Wang, H.; Wang, K.; Lv, S.; Shao, D.; Tao, Y.; et al. 3D printed hydrogel scaffolds combining glutathione depletion-induced ferroptosis and photothermia-augmented chemodynamic therapy for efficiently inhibiting postoperative tumor recurrence. *J. Nanobiotechnol.* **2022**, *20*, 266. [CrossRef]

156. Liao, J.; Shi, K.; Jia, Y.; Wu, Y.; Qian, Z. Gold nanorods and nanohydroxyapatite hybrid hydrogel for preventing bone tumor recurrence via postoperative photothermal therapy and bone regeneration promotion. *Bioact. Mater.* **2021**, *6*, 2221–2230. [CrossRef]

157. Luo, Y.; Wei, X.; Wan, Y.; Lin, X.; Wang, Z.; Huang, P. 3D printing of hydrogel scaffolds for future application in photothermal therapy of breast cancer and tissue repair. *Acta Biomater.* **2019**, *92*, 37–47. [CrossRef]

158. Guguen-Guillouzo, C.; Guillouzo, A. Modulation of functional activities in cultured rat hepatocytes. In *Enzym Induction Modul*; Springer: New York, NY, USA, 1983; pp. 35–56. [CrossRef]

159. Luca, A.C.; Mersch, S.; Deenen, R.; Schmidt, S.; Messner, I.; Schäfer, K.-L.; Baldus, S.E.; Huckenbeck, W.; Piekorz, R.P.; Knoefel, W.T.; et al. Impact of the 3D Microenvironment on Phenotype, Gene Expression, and EGFR Inhibition of Colorectal Cancer Cell Lines. *PLoS ONE* **2013**, *8*, e59689. [CrossRef] [PubMed]

160. Ruedinger, F.; Lavrentieva, A.; Blume, C.; Pepelanova, I.; Scheper, T. Hydrogels for 3D mammalian cell culture: A starting guide for laboratory practice. *Appl. Microbiol. Biotechnol.* **2014**, *99*, 623–636. [CrossRef] [PubMed]

161. Filipiak-Duliban, A.; Brodaczewska, K.; Majewska, A.; Kieda, C. Spheroid culture models adequately imitate distinctive features of the renal cancer or melanoma microenvironment. *Vitr. Cell. Dev. Biol.-Anim.* **2022**, *58*, 349–364. [CrossRef] [PubMed]

162. Gunti, S.; Hoke, A.T.; Vu, K.P.; London, N.R., Jr. Organoid and spheroid tumor models: Techniques and applications. *Cancers* **2021**, *13*, 874. [CrossRef] [PubMed]

163. Chaw, K.C.; Manimaran, M.; Tay, E.H.; Swaminathan, S. Multi-step microfluidic device for studying cancer metastasis. *Lab Chip* **2007**, *7*, 1041–1047. [CrossRef]

164. Dragoj, M.; Stojkovska, J.; Stanković, T.; Dinić, J.; Podolski-Renić, A.; Obradović, B.; Pešić, M. Development and Validation of a Long-Term 3D Glioblastoma Cell Culture in Alginate Microfibers as a Novel Bio-Mimicking Model System for Preclinical Drug Testing. *Brain Sci.* **2021**, *11*, 1025. [CrossRef]

165. Jensen, C.; Shay, C.; Teng, Y. The New Frontier of Three-Dimensional Culture Models to Scale-Up Cancer Research. In *Methods in Molecular Biology*, 2nd ed.; Guest, P.C., Ed.; Springer: New York, NY, USA, 2022; Volume 2343, pp. 3–18.

166. Baker, A.E.; Bahlmann, L.C.; Xue, C.; Lu, Y.H.J.; Chin, A.A.; Cruickshank, J.; Cescon, D.W.; Shoichet, M.S. Chemically and mechanically defined hyaluronan hydrogels emulate the extracellular matrix for unbiased in vivo and in vitro organoid formation and drug testing in cancer. *Mater. Today* **2022**, *56*, 96–113. [CrossRef]

167. Szot, C.S.; Buchanan, C.F.; Freeman, J.W.; Rylander, M.N. 3D in vitro bioengineered tumors based on collagen I hydrogels. *Biomaterials* **2011**, *32*, 7905–7912. [CrossRef]

168. Conde, J.; Oliva-Jorge, N.; Atilano, M.; Song, H.S.; Artzi, N. Self-assembled RNA-triple-helix hydrogel scaffold for microRNA modulation in the tumour microenvironment. *Nat. Mater.* **2016**, *15*, 353–363. [CrossRef]

169. Sepantafar, M.; Maheronnaghsh, R.; Mohammadi, H.; Rajabi-Zeleti, S.; Annabi, N.; Aghdami, N.; Baharvand, H. Stem cells and injectable hydrogels: Synergistic therapeutics in myocardial repair. *Biotechnol. Adv.* **2016**, *34*, 362–379. [CrossRef]

170. Shen, H.; Li, F.; Wang, D.; Yang, Z.; Yao, C.; Ye, Y.; Wang, X. Chitosan–alginate BSA-gel-capsules for local chemotherapy against drug-resistant breast cancer. *Drug Des. Dev. Ther.* **2018**, *12*, 921–934. [CrossRef] [PubMed]

171. Calori, I.R.; Alves, S.R.; Bi, H.; Tedesco, A.C. Type-I Collagen/Collagenase Modulates the 3D Structure and Behavior of Glioblastoma Spheroid Models. *ACS Appl. Bio Mater.* **2022**, *5*, 723–733. [CrossRef] [PubMed]

172. Jia, W.; Zhu, H.; Zhao, M.; Zhou, Q.; Yin, W.; Liu, W.; Wang, L.; Xiao, Z.; Jiang, X.; Dai, J.; et al. Potential mechanisms underlying the promoting effects of 3D collagen scaffold culture on stemness and drug resistance of glioma cells. *Biochim. Biophys. Acta (BBA)-Mol. Basis Dis.* **2022**, *1868*, 166522. [CrossRef] [PubMed]

173. McCleave, M.J. Is breast augmentation using hyaluronic acid safe? *Aesthetic Plast. Surg.* **2010**, *34*, 65–70. [CrossRef]

174. Spinelli, F.M.; Rosales, P.; Pluda, S.; Vitale, D.L.; Icardi, A.; Guarise, C.; Reszegi, A.; Kovalszky, I.; García, M.; Sevic, I.; et al. The effects of sulfated hyaluronan in breast, lung and colorectal carcinoma and monocytes/macrophages cells: Its role in angiogenesis and tumor progression. *IUBMB Life* **2022**, *74*, 927–942. [CrossRef]

175. Schmid, K.; Grundboeck-Jusco, J.; Kimura, A.; Tschopp, F.A.; Zollinger, R.; Binette, J.; Lewis, W.; Hayashi, S. The distribution of the glycosaminoglycans in the anatomic components of the lung and the changes in concentration of these macromolecules during development and aging. *Biochim. Biophys. Acta (BBA)-Gen. Subj.* **1982**, *716*, 178–187. [CrossRef]

176. Waisberg, J.; Marolla, A.P.C.; Saba, G.T.; Germini, D.; Pinhal, M.A.D.S. Hyaluronic Acid in Normal and Neoplastic Colorectal Tissue: Electrospray Ionization Mass Spectrometric and Fluor Metric Analysis. *Cancer Transl. Med.* **2016**, *2*, 79. [CrossRef]

177. Cheng, X.-B.; Sato, N.; Kohi, S.; Yamaguchi, K. Prognostic Impact of Hyaluronan and Its Regulators in Pancreatic Ductal Adenocarcinoma. *PLoS ONE* **2013**, *8*, e80765. [CrossRef]

178. Tolg, C.; Yuan, H.; Flynn, S.M.; Basu, K.; Ma, J.; Tse, K.C.K.; Kowalska, B.; Vulkanesku, D.; Cowman, M.K.; McCarthy, J.B.; et al. Hyaluronan modulates growth factor induced mammary gland branching in a size dependent manner. *Matrix Biol.* **2017**, *63*, 117–132. [CrossRef]

179. Lokeshwar, V.B.; Rubinowicz, D.; Schroeder, G.L.; Forgacs, E.; Minna, J.D.; Block, N.L.; Nadji, M.; Lokeshwar, B.L. Stromal and Epithelial Expression of Tumor Markers Hyaluronic Acid and HYAL1 Hyaluronidase in Prostate Cancer. *J. Biol. Chem.* **2001**, *276*, 11922–11932. [CrossRef]

180. Soofi, S.S.; Last, J.A.; Liliensiek, S.J.; Nealey, P.F.; Murphy, C.J. The elastic modulus of Matrigel™ as determined by atomic force microscopy. *J. Struct. Biol.* **2009**, *167*, 216–219. [CrossRef] [PubMed]

181. Rubiano, A.; Delitto, D.; Han, S.; Gerber, M.; Galitz, C.; Trevino, J.; Thomas, R.M.; Hughes, S.J.; Simmons, C.S. Viscoelastic properties of human pancreatic tumors and in vitro constructs to mimic mechanical properties. *Acta Biomater.* **2018**, *67*, 331–340. [CrossRef] [PubMed]

182. Laklai, H.; Miroshnikova, Y.A.; Pickup, M.W.; Collisson, E.A.; Kim, G.E.; Barrett, A.S.; Hill, R.C.; Lakins, J.N.; Schlaepfer, D.D.; Mouw, J.K.; et al. Genotype tunes pancreatic ductal adenocarcinoma tissue tension to induce matricellular fibrosis and tumor progression. *Nat. Med.* **2016**, *22*, 497–505. [CrossRef]

183. Curvello, R.; Raghuwanshi, V.S.; Garnier, G. Engineering nanocellulose hydrogels for biomedical applications. *Adv. Colloid Interface Sci.* **2019**, *267*, 47–61. [CrossRef]

184. Niemczyk-Soczynska, B.; Gradys, A.; Kolbuk, D.; Krzton-Maziopa, A.; Sajkiewicz, P. Crosslinking kinetics of methylcellulose qqueous solution and its potential as a scaffold for tissue engineering. *Polymers* **2019**, *11*, 1772. [CrossRef] [PubMed]

185. Niemczyk-Soczynska, B.; Sajkiewicz, P.; Gradys, A. Toward a Better Understanding of the Gelation Mechanism of Methylcellulose via Systematic DSC Studies. *Polymers* **2022**, *14*, 1810. [CrossRef] [PubMed]

186. Curvello, R.; Kast, V.; Abuwarwar, M.H.; Fletcher, A.L.; Garnier, G.; Loessner, D. 3D Collagen-Nanocellulose Matrices Model the Tumour Microenvironment of Pancreatic Cancer. *Front. Digit. Health* **2021**, *3*. [CrossRef]

187. Shokri, R.; Fuentes-Chandía, M.; Ai, J.; Roudkenar, M.H.; Mahboubian, A.R.; Malekshahi, M.R.; Ostad, S.N. A thermo-sensitive hydrogel composed of methylcellulose/hyaluronic acid/silk fibrin as a biomimetic extracellular matrix to simulate breast cancer malignancy. *Eur. Polym. J.* **2022**, *176*, 111421. [CrossRef]

188. Sharpless, N.E.; Depinho, R.A. The mighty mouse: Genetically engineered mouse models in cancer drug development. *Nat. Rev. Drug Discov.* **2006**, *5*, 741–754. [CrossRef]

189. Natarajan, D.; Ye, Z.; Wang, L.; Ge, L.; Pathak, J.L. Rare earth smart nanomaterials for bone tissue engineering and implantology: Advances, challenges, and prospects. *Bioeng. Transl. Med.* **2022**, *7*, e10262. [CrossRef]

190. Li, L.; Zhang, X.; Zhou, J.; Zhang, L.; Xue, J.; Tao, W. Non-Invasive Thermal Therapy for Tissue Engineering and Regenerative Medicine. *Small* **2022**, *18*, 2107705. [CrossRef] [PubMed]

191. Maia, F.R.; Bastos, A.R.; Oliveira, J.M.; Correlo, V.M.; Reis, R.L. Recent approaches towards bone tissue engineering. *Bone* **2022**, *154*, 116256. [CrossRef] [PubMed]

192. Amann, A.; Zwierzina, M.; Gamerith, G.; Bitsche, M.; Huber, J.M.; Vogel, G.F.; Blumer, M.; Koeck, S.; Pechriggl, E.J.; Kelm, J.M.; et al. Development of an Innovative 3D Cell Culture System to Study Tumour–Stroma Interactions in Non-Small Cell Lung Cancer Cells. *PLoS ONE* **2014**, *9*, e92511. [CrossRef] [PubMed]

193. Baião, A.; Dias, S.; Soares, A.F.; Pereira, C.L.; Oliveira, C.; Sarmento, B. Advances in the use of 3D colorectal cancer models for novel drug discovery. *Expert Opin. Drug Discov.* **2022**, *17*, 569–580. [CrossRef] [PubMed]

Article

Poly(ε-caprolactone)-poly(ethylene glycol) Tri-Block Copolymer as Quercetin Delivery System for Human Colorectal Carcinoma Cells: Synthesis, Characterization and In Vitro Study

Nancy Ferrentino [1,†], Maria Preziosa Romano [1,2,†], Silvia Zappavigna [3], Marianna Abate [3], Vitale Del Vecchio [4], Dario Romano [5], Chiara Germinario [1], Celestino Grifa [1], Rosanna Filosa [1,6] and Daniela Pappalardo [1,*]

[1]   Dipartimento di Scienze e Tecnologie, Università del Sannio, Via de Sanctis snc, 82100 Benevento, Italy
[2]   Advanced Medical Pharma, (AMP-BIOTEC) Healthcare Research and Innovation Center, 82030 San Salvatore Telesino, Italy
[3]   Department of Precision Medicine, University of Campania "Luigi Vanvitelli", Via Santa Maria di Costantinopoli, 16, 80138 Naples, Italy
[4]   Department of Experimental Medicine, University of Campania "Luigi Vanvitelli", Via Santa Maria di Costantinopoli, 16, 80138 Naples, Italy
[5]   Physical Science and Engineering Division, King Abdullah University of Science and Technology (KAUST), Thuwal 23955-6900, Saudi Arabia
[6]   Istituti Clinici Scientifici Maugeri IRCCS, Cardiac Rehabilitation Unit of Telese Terme Institute, 82037 Telese Terme, Italy
*   Correspondence: pappalardo@unisannio.it
†   These authors contributed equally to this work.

Citation: Ferrentino, N.; Romano, M.P.; Zappavigna, S.; Abate, M.; Del Vecchio, V.; Romano, D.; Germinario, C.; Grifa, C.; Filosa, R.; Pappalardo, D. Poly(ε-caprolactone)-Poly(ethylene glycol) Tri-Block Copolymer as Quercetin Delivery System for Human Colorectal Carcinoma Cells: Synthesis, Characterization and In Vitro Study. *Polymers* **2023**, *15*, 1179. https://doi.org/10.3390/polym15051179

Academic Editors: João Carlos Silva and Frederico Castelo Ferreira

Received: 28 January 2023
Revised: 16 February 2023
Accepted: 23 February 2023
Published: 26 February 2023

Abstract: Quercetin is a hydrophobic molecule with short blood circulation times and instability. The development of a nano-delivery system formulation of quercetin may increase its bioavailability, resulting in greater tumor suppressing effects. Triblock ABA type polycaprolactone-polyethylenglycol-polycaprolactone (PCL-PEG-PCL) copolymers have been synthetized using ring-opening polymerization of caprolactone from PEG diol. The copolymers were characterized by nuclear magnetic resonance (NMR), diffusion-ordered NMR spectroscopy (DOSY), and gel permeation chromatography (GPC). The triblock copolymers self-assembled in water forming micelles consisting of a core of biodegradable polycaprolactone (PCL) and a corona of polyethylenglycol (PEG). The core-shell PCL-PEG-PCL nanoparticles were able to incorporate quercetin into the core. They were characterized by dynamic light scattering (DLS) and NMR. The cellular uptake efficiency of human colorectal carcinoma cells was quantitatively determined by flow cytometry using nanoparticles loaded with Nile Red as hydrophobic model drug. The cytotoxic effect of quercetin-loaded nanoparticles was evaluated on HCT 116 cells, showing promising results.

Keywords: polycaprolactone-*block*-polyethylenglycol; copolymers; quercetin; drug delivery system; colorectal carcinoma cells

1. Introduction

The development of delivery systems for anticancer drugs into tumor tissues to avoid systemic toxicity is a crucial challenge in cancer therapy. Drug delivery systems (DDSs) could increase the therapeutic effect, the bioavailability, circulation time, and stability of drugs [1–6]. Several DDSs have been investigated including polymers, liposomes, and block copolymer micelles [7]. Compared to liposomes, polymeric nanoparticles, have a greater stability in vivo and they are more versatile, since polymer structure, molecular weights, and composition can be changed to meet the requirements of a particular application [8]. Polymeric nanoparticles based on amphiphilic block copolymers have aroused significant interest because of their ability to self-assemble into a micellar form in water phase, with a hydrophilic outer shell or corona and a hydrophobic core. The hydrophilic shell of the micelle prevents steric recognition by the immune system and its removal from the bloodstream by the macrophage system. At the same time, the hydrophobic core can encapsulate fragile hydrophobic anticancer therapeutics [9,10]. The encapsulation of anticancer drugs in nanocarriers enhances drug accumulation in the tumor tissues through the enhanced permeability and retention (EPR) effect [1].

Polymeric nanoparticles derived from biodegradable and biocompatible polyesters have been largely used as a DDS. Many studies have been conducted on micelles consisting of a core of biodegradable polycaprolactone (PCL) and a corona of polyethylenglycol (PEG). PEG is hydrophilic and it exhibits excellent biocompatibility and lack of toxicity. The presence of PEG as an outer shell in DDSs reduces opsonization and the aggregation of DDSs in vivo [11]. PCL is non-toxic and biodegradable; these characteristics are important for drug delivery as they help maintain clearance of the nanocarrier after injection. PCL-based systems can be completely degraded within the body due to the presence of ester bonds that can easily be subjected to hydrolysis by esterases [12]. The main degradation product of PCL is 6-hydroxycaproic acid, that can be eliminated by urinary excretion or pulmonary elimination as carbon dioxide [13]. Due to the high degree of crystallinity and hydrophobicity, PCL is less biocompatible with soft tissues limiting its clinical application. Therefore, the addition of PEG to the PCL chain allows improvements to the hydrophilicity, biodegradability, and performance in the cell culture studies when compared with the PCL homopolymer, thus allowing the use of such systems in different biomedical applications [14].

The synthesis of PCL–PEG block copolymers can be achieved by the ring-opening polymerization of ε-caprolactone (ε-CL) using PEG as initiator [10,15]. The molecular weight of the PCL block is primarily controlled by the molar ratio of ε-CL to the initiator PEG. Copolymers of various block lengths of PEG and PCL have been described in the literature. Moreover, diblock (AB) or triblock (ABA) copolymers have been prepared using a mono-hydroxy or α, ω-dihydroxy PEG as initiator for the polymerization of lactone monomers [14,16–19].

PEG-PCL-based DDSs have been used as carriers of diverse hydrophobic drugs and they have been shown to induce therapeutic effects both in vitro and in vivo. Many in vitro toxicity studies have been performed on PEG-PCL nanoparticle systems. These systems proved to be stable and non-toxic, allowing them to protect the drug and to administer it in a controlled way. To mention some of them, nanoparticles constituted by diblock PEG-PCL were tested in vitro and in vivo for the delivery of paclitaxel with high encapsulation efficiency for glioblastoma multiform treatment [20]. Doxorubicin-loaded nanoparticles made of triblock PCL-PEG-PCL were proposed as promising DDSs for breast cancer therapy [21]. Moreover, nanoparticles based on the same PCL-PEG-PCL triblock copolymer were able to simultaneously encapsulate two drugs, doxorubicin and quercetin [22]. High drug-loading efficiency and efficient cellular uptake were also recognized for the PTX-loaded PCL-PEG-PCL nanoparticles in EMT-6 breast cancer cells [23].

PEG-PCL drug delivery systems have also been tested on colon cancer cells. Nanoparticles based on the PEG-PCL-PEG triblock copolymer conjugated with folic acid and loaded with 5-fluorouracil and magnetite increased the inhibitory effect against several colon cancer cells when compared with free drug applications [24]. Moreover, PCL-PEG diblock copolymers have been coupled to tyrosine and angiopep-2 to obtain docetaxel-loaded PEG-PCL-dual-modified-nanoparticles for the treatment of colorectal cancer [25]. Di-block PEG-PCL copolymers were an important DDS for tetradrine delivery [26], while triblock PCL-PEG-PCL copolymers, loaded with auraptene, increased drug biodistribution and bioavailability into colon cancer cells [27].

All these results indicated that PEG-PCL nanoparticles had great potential as optimal delivery systems for hydrophobic molecules and to increase the pharmacokinetic profile of drug but also to decrease the metabolism of active compound. Taking this paradigm into account, the combination of natural products with anticancer properties and chemotherapy can be a novel strategy for both the possible improvement of therapeutic efficacy of drugs and in achieving better pharmacokinetic profiles and avoiding the side effects of drug toxicity. For several decades, flavonoids played an essential role in the human diet as antioxidants, antinflammatories, and cancer/chemo preventive agents [28]. Among those known for chemo-preventive efficacy, quercetin, as a member of the flavonols subgroup, is one of the main compounds that has been extensively described in several in vitro and in vivo studies [29]. In this regard, its beneficial anti-mutagenic and anti-proliferative effects, its antioxidant properties, and its role in the regulation of cell signalling, cell cycle, and apoptosis, assert its significant pharmacological potential against cancer. Unfortunately, the bioavailability of quercetin is well known to be poor in limited low plasma concentrations. It is reported that the application of a nanosystem formulation of quercetin results in an increased water solubility and thus enhancement of its bioavailability, resulting in greater tumor suppressing effects [30,31].

In the framework of our interest in block copolymers based on PEG and resorbable aliphatic polyesters [32–35], we have synthetized and characterized amphiphilic PCL-PEG-PCL triblock copolymers and obtained micelle nanoparticles. The micelles were loaded with the hydrophobic molecule Nile Red (NR) to characterize and evaluate the loading capacity of the nanoparticles and monitoring the entry and loading of the fluorescent substance. Subsequently, the micelles were loaded with quercetin to confirm their capability to maintain the anticancer property of the natural compound, increasing its water solubility.

2. Materials and Methods

2.1. Instruments and Measurements

2.1.1. Nuclear Magnetic Resonance (NMR)

NMR spectra were recorded on a Bruker AM300 (^{1}H, 300 MHz; ^{13}C, 75 MHz) and a Bruker Advance 400 (^{1}H, 400 MHz; ^{13}C, 100 MHz). Chemical shifts (^{1}H and ^{13}C NMR) were assigned using tetramethylsilane (TMS) as an internal reference and were expressed as parts per million. Coupling constants (J) were expressed in Hertz. NMR spectra were referenced using the residual solvent peak at δ = 7.27 (^{1}H) for CDCl$_3$ and at δ = 2.50 (^{1}H) for DMSO. NMR signals were reported as follows: chemical shift (δ ppm), relative integral, and multiplicity (s = singlet, d = doublet, t = triplet, q = quartet, m = multiplet, dd = doublet of doublet, br = broad). The NMR samples were prepared by dissolving about 10 mg of the compound in 0.5 mL of the deuterated solvent. Spectra were recorded using Bruker TopSpin v3.2 software. Data processing was performed using TopSpin v3.2 or MestReNova v12.0.2 software.

2D Diffusion-Ordered (DOSY) PGSE NMR spectra of the block copolymers were recorded on a Bruker Avance 400 spectrometer 10 mg of the copolymer were dissolved in 0.5 mL of CDCl$_3$ and the spectra were recorded at room temperature without spinning; the parameters δ and Δ were kept constant during the experiments, whereas G was varied from 2 to 95 % in 25 steps, 64 scans per step.

2.1.2. Gel Permeation Chromatography (GPC)

Gel permeation chromatography was performed using a GPC Agilent with a refractive detector and a PLgel 5 μm Mixed-C column. The samples were dissolved and eluted in tetrahydrofuran. After complete dissolution, samples were filtered through PTFE membranes (0.22 μm). The injection volume was 20.00 μL and the flow rate was 1.00 mL/min. The chosen method of analysis was universal calibration based on polystyrene standards with a narrow molecular weight distribution. The measurements were performed at 35 °C according to the temperatures of the columns and detectors.

2.1.3. Fourier Transform Infrared Spectroscopy (FTIR)

FTIR analyses were carried out with the diamond crystal Attenuated Total Reflectance module (ATR) of a Bruker ALPHA spectrometer, equipped with ROCKSOLID™ interferometer and a ZnSe/KBr beam splitter with a DTGS detector. The spectra, processed using Bruker OPUS 7.2 software, were acquired in the mid-infrared spectral range (4000–400 cm^{-1}) with a resolution 4 cm^{-1} and 64 scans per minute.

2.1.4. Dinamic Light Scattering (DLS)

DLS measurements were performed using a Malvern Zetasizer Nano Z. Measurements were performed at 25 °C and at 37 °C after an equilibration time of 60 s. The angle of detection was 173° using the non-invasive back scatter system. The measurements were performed in deionized water; the viscosity (η) and the refractive index (RI) of the dispersant at 25 °C (η_{25} = 0.8960 mPa s; RI_{25} = 1.393) and 37 °C (η_{37} = 0.7424 mPa s; RI_{37} = 1.391) were set accordingly.

The intensity-weighted mean values of the diameter and polydispersity index (PDI) were reported as the average of three measurements. For each measurement, 10 scans were recorded and the reported error represents the average error incurred during the three measurements.

The quercetin-loaded nanoparticles (0.78 mg) were suspended in deionized water (2 mL), sonicated for 30 min, and then placed in a glass cuvette and analyzed.

2.2. Chemicals

Moisture and air-sensitive materials were manipulated under nitrogen using Schlenk techniques. The chemicals anhydrous toluene (99.8%), hexane ($\geq$95%), stannous octoate (92.5–100.0%) (SnOct$_2$), methanol (99.8%), dimethylformamide ($\geq$99.8%) (DMF), chloroform ($\geq$99.5%), tetrahydrofuran ($\geq$99.9%) (THF), Nile Red ($\geq$97.0%) (NR), quercetin ($\geq$95%) (Q), and 3-(4,5-dimethylthiazol-2-yl)-2,5-diphenyltetrazolium bromide ($\geq$97.5%) (MTT) were acquired from Merck and used as received. The ε-caprolactone (97%) (ε-CL), acquired from Merck, was distilled in vacuo from CaH$_2$ prior to use. Polyethylene glycol (PEG-2000 MW), acquired from Merck, was dried in vacuo over phosphorus pentoxide (P$_2$O$_5$) for 72 h.

2.3. Cell Cultures

HCT-116 and LoVo cells, both human colon carcinoma cells (American Type Tissue culture Collection, Rockville, MD, USA), were cultured in Dulbecco's Modified Eagle Medium (DMEM) and F-12K, respectively, and both supplemented with 10% heat inactivated Fetal Bovine Serum (FBS), 100 μg/mL penicillin, 100 μg/mL streptomycin, 1% L-glutamine, and 1% sodium pyruvate. Cells were grown at 37 °C in a humidified atmosphere of 95% air/5% CO$_2$.

2.4. Synthetic Procedures

2.4.1. Synthesis of Triblock PCL-PEG-PCL Copolymer

The PCL-PEG-PCL triblock copolymer was synthesized by ring-opening polymerization (ROP) of ε-CL on PEG and in the presence of stannous octanoate as catalyst under nitrogen atmosphere. In detail, 0.5 g of telechelic PEG-2000 (0.25 mmol) and ε-CL (2.2 mL, 19.8 mmol) were introduced in a three neck flask and then 6.1 mL of a 9 mM solution of Sn(Oct)$_2$ in toluene was added. The resulting mixture was placed in an oil bath at 130 °C and stirred for 44 h. After that time, the mixture was cooled to room temperature, dissolved in chloroform, and the resulting mixture was poured in hexane. The precipitated triblock copolymer was recovered by filtration and dried in vacuo. Yield: 96%.

^{1}H NMR (400 MHz, CDCl$_3$): δ = 4.04 (t, 2H, -(CH$_2$)$_4$CH$_2$O-), 3.65 (s, 4H,-OCH$_2$CH$_2$O-), 2.29 (t, 2H, -CH$_2$C(O)O-), 1.64–1.60 (m, 4H, -CH$_2$-), 1.37 (m, 2H, -CH$_2$-).

$M_{n\,NMR}$ = 11,020 g/mol; GPC analysis (THF): M_n = 19,420 g/mol; $Đ$ = 1.9

2.4.2. Preparation of Triblock PCL-PEG-PCL Copolymer Nanoparticles (NP)

The PCL-PEG-PCL copolymer (10 mg) was dissolved in 1.0 mL of DMF then transferred into 10 mL of stirred deionized water at room temperature. After 2 h, the solution was added to a dialysis membrane with a molecular weight cutoff between 6000 and 8000 g/mol and a volume per length ratio of 1.7 mL cm^{-1} then dialyzed against water for 72 h. The dialysis solution was replaced with fresh deionized water every 3 h; then, dialysis was allowed to occur overnight. The solution of the purified nanoparticles was then lyophilized.

2.4.3. Preparation of Triblock PCL-PEG-PCL Copolymer Nanoparticles Loaded with Nile Red (NP+NR)

The PCL-PEG-PCL copolymer (50 mg) was dissolved in 5 mL of DMF and a solution of Nile Red (1.5 mg, 4.7 μmol) in 1.5 mL of DMF was slowly added. The obtained solution was transferred into 10 mL of deionized water and stirred for 2 h at room temperature. After this time, the solution was dialyzed against 3 L of deionized water for 72 h with water changes every 3 h. The NR-loaded micelles were dried by using lyophilization.

The loading of NR was evaluated by ^{1}H NMR measuring the molar amount of NR in respect to the molar amount of the copolymer using the following formula:

$$\text{NR content } (\% \ w/w) = MW_{NR} \times mol_{NR} / (MW_{copolymer} \times mol_{copolymer} + MW_{NR} \times mol_{NR}) \times 100 = 1.6\%$$

The loading efficiency was evaluated by ^{1}H NMR using the following formula: % loading efficiency $= [w_{(NR\ in\ NP)}] / [w_{(initial\ NR\ added)}] \times 100 = 53\%$

2.4.4. Preparation of Triblock PCL-PEG-PCL Copolymer Nanoparticles Loaded with Quercetin (NP+Q)

The PCL-PEG-PCL copolymer (10 mg) was dissolved in 1.5 mL of DMF and a solution of quercetin (3 mg, 10 μmol) in 0.5 mL of DMF was slowly added. The obtained solution was transferred into 10 mL of deionized water and stirred for 2 h at room temperature. After this time, the solution was dialyzed against 3 L of deionized water for 72 h. The dialysis solution was replaced with fresh deionized water every 3 h. The quercetin-loaded nanoparticles were dried by using lyophilization.

The loading of quercetin into the nanoparticles was evaluated by ^{1}H NMR measuring the molar amount of quercetin respect to the molar amount of the copolymer using the following formula:

$$\text{Quercetin content } (\% \ w/w) = MW_{quercetin} \times mol_{quercetin} / (MW_{copolymer} \times mol_{copolymer} + MW_{quercetin} \times mol_{quercetin}) \times 100 = 1.6\%$$

The loading efficiency was evaluated by ^{1}H NMR by using the following formula: % loading efficiency $= [w_{(quercetin\ in\ NP)}] / [w_{(initial\ quercetin\ added)}] \times 100 = 5.3\%$

2.5. In Vitro Cells Studies

2.5.1. Analysis of Cell Uptake of Nile Red-Loaded Nanoparticles (NP+NR) by Fluorescence Microscopy and Fluorescence-Activated Cell Sorting (FACS)

Nile Red (NR)-loaded nanoparticles PCL-PEG-PCL (NP+NR) were prepared using the dialysis method. HCT 116 e LoVo cells were seeded in 12-well plates at a density of 1×10^5/well and incubated at 37 °C for 24 h. The culture medium containing free NR or NP+NR (10 μM of NR) was added to the cells and incubated for 1 h, 3 h, and 5 h. After incubation, the cells were washed with phosphate-buffered saline (PBS) at pH 7.4 for 1 min and observed under the fluorescence microscope. Red fluorescence was observed by an inverted digital fluorescence microscope (EVOS M5000 Imaging System, Thermo Fisher Invitrogen) using a 20× magnification.

A BD FACSCanto™fluorescent-activated flow cytometer and the BD FACSCanto™ Clinical software (BD Biosciences, San Jose, CA, USA) were used to perform flow cytometry analysis. In detail, the cells were detached with 1X trypsin, centrifuged for 5 min, washed with PBS, centrifuged for 5 min, and fixed with formaldehyde solution (4% v/v) for 15 min at 4 °C. Then, the cells were washed again with PBS at pH 7.4 and, for each tube, 20,000 cells were measured on a FACSCanto™ flow cytometer (Becton Dickinson, BD Biosciences, San Jose, CA, USA) using FACSCanto™ Clinicalsoftware (Becton Dickinson, BD Biosciences, San Jose, CA, USA). The detection wavelengths were 546 nm and 590 nm for excitation and emission, respectively; the emission spectrum was collected by FLH-2 channel of the FACSCanto™.

2.5.2. Cell Proliferation Assay

Cell proliferation of HCT 116 was evaluated in the presence of quercetin-loaded nanoparticles (NP+Q). After trypsinization, cells were seeded in 96-well plates at the density of 3×10^3 cells/well in 100 µL of medium DMEM 10%. After 24 h incubation at 37 °C, the medium was removed and replaced with a fresh one containing increasing concentrations of NP+Q (0.6 µM, 1.2 µM, 2.5 µM, 5 µM). Cells were incubated under these conditions for a time course spanning 24 h, 48 h, and 72 h. The cell viability was assessed by MTT assay. The MTT solution (5 mg/mL in PBS) was added (10 µL/well) and the plates were incubated for 3 h at 37 °C. Then, the MTT-formazan salts were dissolved with an isopropanol/hydrochloric acid 10% solution for 20 min in constant agitation. Then, the absorbance values of the solution were measured at 570 nm using a BioRad 550 microplate reader (BioRad Laboratories, Milan, Italy). Combination index (CI) between NPs and quercetin was calculated by the dedicated software Compusyn, using dose-effect curves of the different drugs. CI values lower than 1, equal to 1, and higher than 1 indicate synergy, additivity, and antagonism, respectively. Data are as mean ± standard deviation (SD). Each value was obtained from three independent experiments.

2.5.3. Statistical Analysis

All data were expressed as mean values ± SD of three independent experiments. Statistical evaluation of the data was performed by one-way ANOVA compared to the control, followed by Tukey–Kramer. Significant differences were determined at p value < 0.05 (*), $p < 0.005$ (**); $p < 0.001$ (***); $p < 0.0001$ (****). Graphs were obtained using GraphPad Prism Version 6 software (San Diego, CA, USA).

3. Results

3.1. Synthesis and Characterization of PCL-PEG-PCL Copolymer

PCL and PEG are the most popular materials for the preparation of DDSs [18]. The synthesis of PCL–PEG block copolymers can be achieved by the ROP of ε-CL using a telechelic PEG diol as initiator. The amphiphilic triblock PCL-PEG-PCL copolymer was thus obtained by ROP of ε-CL with PEG diol (2000 g/mol) as initiator and Sn(Oct)$_2$ as catalyst under nitrogen atmosphere (Scheme 1). The ^{1}H NMR spectrum in CDCl$_3$ (Figure 1) confirmed that the triblock PCL-PEG-PCL copolymer was successfully synthesized due to the presence of characteristic signals corresponding to both PCL (1.36, 1.61–1.65, 2.29, and 4.04 ppm) and PEG (3.65 ppm). In the copolymer sample, the molar ratio between the ε-CL units and the ethylene glycol units, evaluated by ^{1}H NMR analysis, was 79:45. To determine the molecular weight of the triblock PCL-PEG-PCL copolymer, ^{1}H NMR and GPC analysis were performed. In particular, the ^{1}H NMR in CDCl$_3$ showed a molecular weight of 11,020 g/mol, evaluated by integrating the -CH$_2$O- peak at 4.04 ppm of PCL versus the -OCH$_2$CH$_2$O- peak at 3.65 of PEG. In Figure S1 the ^{1}H NMR of PEG diol is reported. This value was in perfect agreement with the theoretical molecular weight (M_n = 11,030 g/mol). The GPC analysis, performed using THF, indicated a molecular weight of 19,420 g/mol and a dispersity $Đ$ = 1.9 (Figure S2). The GPC-chromatogram of the copolymer showed monomodal distribution of the molecular weight, thus indicating that no transesterification reaction has occurred during the ring-opening polymerization (Figure S2); moreover the molecular weight increased respect to the PEG diol (Figure S3).

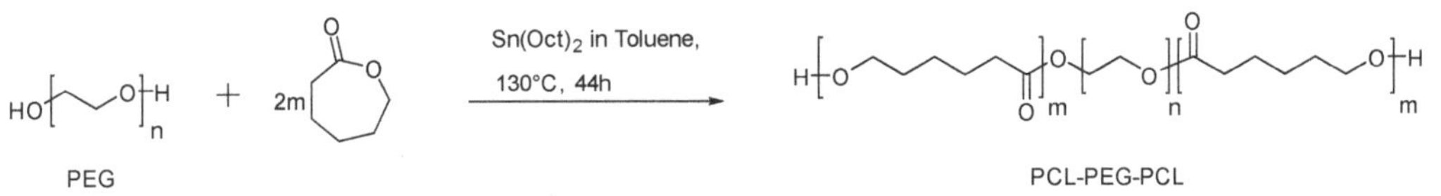

Scheme 1. Synthesis of triblock PCL-PEG-PCL copolymer.

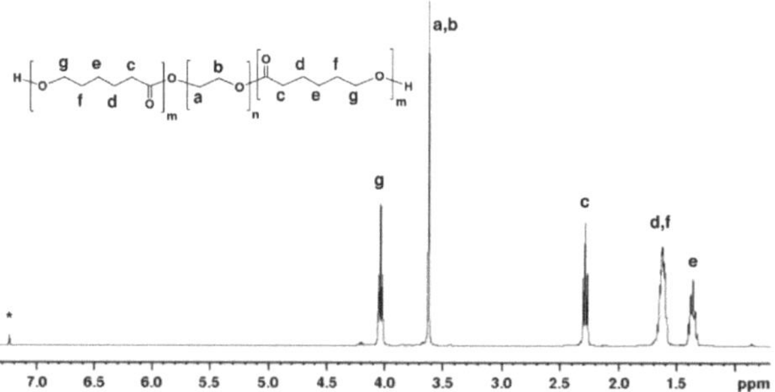

Figure 1. ^{1}H NMR spectrum (400 MHz, CDCl$_3$, 298 K) of triblock PCL-PEG-PCL copolymer.

In the Fourier Transform Infrared (FTIR) spectrum of PCL-PEG-PCL, shown in Figure S4, a strong C=O peak appeared at 1721 cm^{-1}. In addition, the CH stretching band of caprolactone monomeric unit shifted to 2943 cm^{-1}, in respect to the one of the ε-CL monomers, while the band of CH stretching vibration of PEG at 2894 cm^{-1}. Furthermore, the spectrum showed a decrease in the intensity of the –OH peak (3560 cm^{-1}) compared with peaks of individual PEGs and PCLs, respectively [22]. This indicated the formation of triblock PCL-PEG-PCL copolymer.

Finally, the Diffusion Ordered Spectroscopy (DOSY) NMR analysis showed that the PEG and PCL portion had the same diffusion coefficient and that they therefore belonged to the same macromolecule (Figure 2). This result clearly demonstrated that the obtained product was the desired PCL-PEG-PCL triblock copolymer and not simply the mixture of the two homo-polymers. As a reviewer suggested, for comparison purposes, the DOSY spectrum of a simple PCL sample has been registered as well (Figure S5).

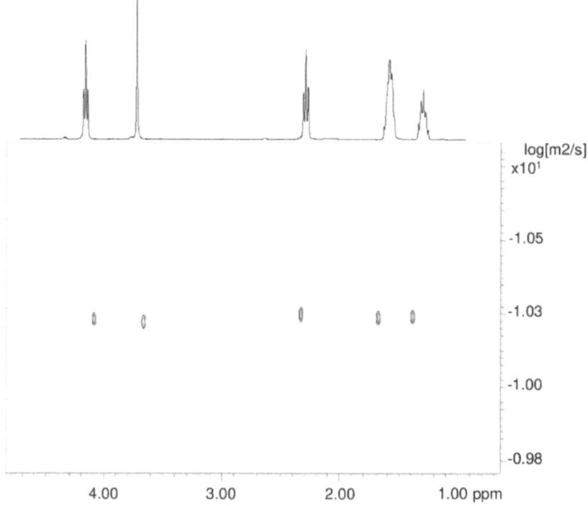

Figure 2. DOSY NMR spectrum (400 MHz, CDCl$_3$, 298 K) of triblock PCL-PEG-PCL copolymer.

3.2. Self-Assembly of Triblock PCL-PEG-PCL Copolymers to Form Nanoparticles

Due to its amphiphilic character, the PCL-PEG-PCL copolymers in water phase could self-assemble to form micelles, having the hydrophilic PEG as a corona and the hydrophobic PCL as a core [18]. In the hydrophobic core, a lipophilic molecule can be loaded, and the micelles may be used as carrier for, e.g., poorly water-soluble drugs.

The micelle solutions were prepared using the dialysis method. The PCL-PEG-PCL copolymers were first dissolved in a water miscible organic solvent, such as DMF. Then the DMF copolymer solution was added to the water and spontaneous formation of micelles occurred due to reduced polymer solubility and interfacial hydrodynamic phenomena. The organic solvent was removed by dialyzing against water, and the nanoparticles (NP) were recovered by freeze-drying.

With the same approach, nanoparticles loaded with NR (NP+NR) and quercetin (NP+Q) were also prepared. The NR is a hydrophobic, fluorescent, poorly water-soluble dye, and it was used as a model guest molecule for initial cellular study. Subsequently, quercetin was loaded into the micelles to evaluate its antioxidant and antiproliferative activity on tumor cells.

To form drug-loaded micelles, the PCL-PEG-PCL copolymer and drug were dissolved in minimum amount of DMF. This solution was added slowly into water, transferred into the dialysis bag (MWCO: 6000–8000 g/mol), and dialyzed against water for 3 days. The dialysis solution was replaced with fresh deionized water every 3 h. The so-formed micelles were dried using lyophilization. The loading of the drugs (NR or quercetin) was evaluated by ^{1}H NMR analysis in DMSO (Figures S6–S8). The loading of quercetin into the nanoparticles was 1.6% w/w; the same value was obtained for the NR.

The quercetin-loaded nanoparticle (NP+Q) dimensions were evaluated by dynamic light scattering (DLS) at 25 °C and 37 °C (Figure 3). The data are summarized in Table 1. The average diameter of the nanoparticles was 152.5 and 128.5 nm with a PDI of 0.26 and 0.21 at 25 and 37 °C, respectively. Additional peaks (around 0.7, 4–11, and 5000 nm) are attributed to contaminants in the water and observed also in the absence of quercetin loaded nanoparticles (See Figure S9). Interestingly, at higher temperatures, narrower size distributions were observed, and the average values determined for the three measurements were much closer to each other, though within the experimental error.

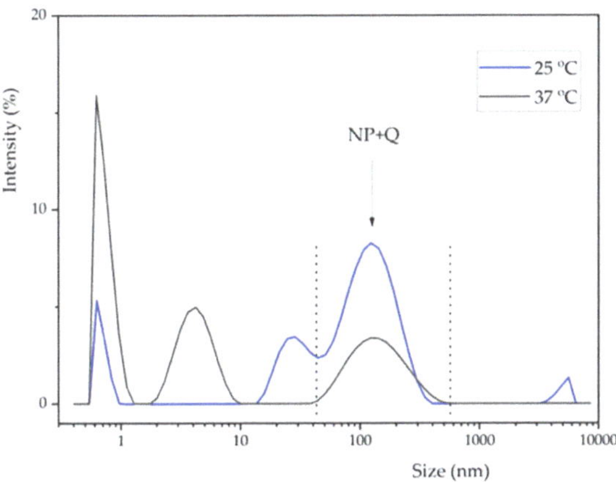

Figure 3. Quercetin loaded nanoparticles (NP+Q) size distribution obtained in water by DLS analysis at 25 and at 37 °C.

Table 1. DLS Data of the quercetin loaded nanoparticles (NP+Q).

Sample	Average Diameter (nm) [1]	
NP+Q	25 °C	37 °C
	152.5 ± 77.3	128.5 ± 59.0

[1] Determined by DLS in water.

The ^{1}H NMR spectrum in DMSO of the PCL-PEG-PCL nanoparticles (NP+Q) loaded with quercetin (Figures S8 and S10) showed the characteristic signals of quercetin (6.15, 6.37, 6.85, 7.51, 7.64, and 12.47 ppm), PEG (3.48 ppm), and PCL (1.28, 1.54–1.47, 2.22, and 3.95 ppm), while the NMR spectrum in D_2O of the same micelles showed only the signal of the PEG (3.48 ppm) (Figure 4). This confirmed that, in an aqueous environment, PEG arranges itself as a "corona" around the PCL and the hydrophobic drug. The formation of flower-like core-shell structures for PCL-PEG-PCL copolymers was hypothesized. The core of the micelles accommodated the hydrophobic drugs via hydrophobic interactions, while the "shell" was a hydrophilic corona, made of PEG, which allowed the micelles to be dispersible in water.

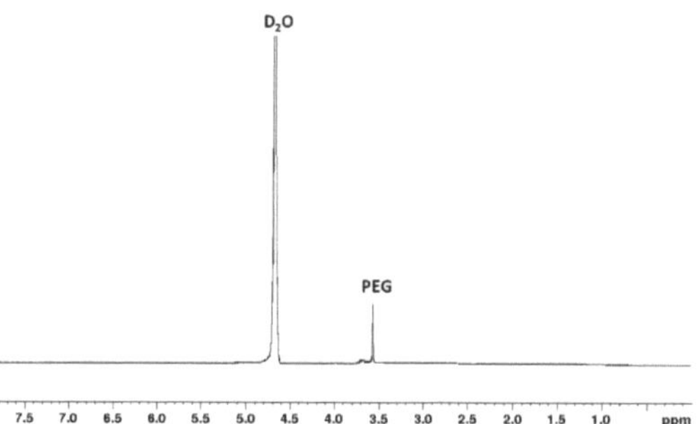

Figure 4. ^{1}H NMR spectrum (400 MHz, D_2O, 298 K) of PCL-PEG-PCL micelles loaded with quercetin.

3.3. Cellular Studies

3.3.1. Nanoparticle Uptake in HCT 116 and LoVo Cells

To evaluate the cellular uptake of the nanoparticles and their intracellular distribution, fluorescence microscopy and FACS analysis were performed on HCT 116 and LoVo cells treated with Nile Red-loaded nanoparticles (NP+NR). NR is a lipophilic red dye that has been widely used to visualize and localize drug vectors [36]. The intracellular distribution of NP+NR after 1 h, 3 h, and 5 h of incubation was studied using fluorescence microscopy in both cell lines and compared to free (not loaded) NR. However, HCT 116 treated with NP+NR showed more diffuse and intense red fluorescence intensity after as early as 1 h of incubation when compared to the resulting cells incubated with free NR, and red fluorescence remained almost constant even after 3 h and 5 h of incubation (Figure 5A).

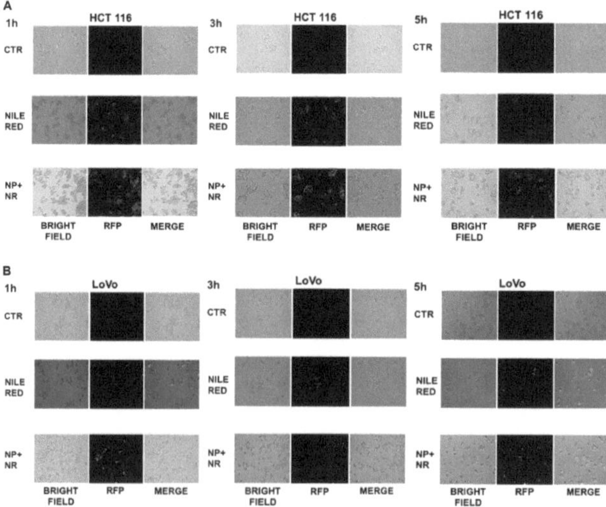

Figure 5. Fluorescence microscopy images of HCT 116 and LoVo, reported in panel (**A**,**B**) respectively, after 1, 3 and 5 h of incubation at 37 °C with NR and NP+NR. For each cell line, the bright field image, the red fluorescence image (RFP), and the overlap of the two images (MERGE) are reported. The cells were visualized under a fluorescence microscopy at 20× magnification.

In addition, LoVo showed a more widespread and stronger red fluorescence intensity after 1 h of incubation than cells incubated with free NR, but this slightly decreased after 3 and 5 h of incubation (Figure 5B). Furthermore, the cellular uptake efficiency of HCT 116 and LoVo cells was measured by flow cytometry. After 1 h, 3 h, and 5 h of cell incubation with NR and NP+NR, the median X of the fluorescence intensity compared to the negative control (CTR) was detected by FACS and assumed to be 100%. In detail, as shown in Figure 6A, in HCT 116 cells, the nanoparticle uptake reached a maximum peak after 1 h and it kept constant after 3 and 5 h; moreover, a 6-fold increase in fluorescence intensity was observed when cells were treated with NP+NR compared to cells treated with free NR (NP+NR 24,814.6% vs. NR 3831.9%). Similarly, as shown in Figure 6B, the highest cell uptake efficiency was also observed after 1 h for LoVo, but it slightly decreased after 3 h and 5 h; however, a fluorescence intensity rate of 9055.8% was observed for NP+NR treated cells compared to a rate of 426.9% for cells treated with free NR. These findings confirmed that PEG-PCL-PEG-based nanoparticles were able to increase and speed up NR uptake inside the cells by improving drug delivery and stability.

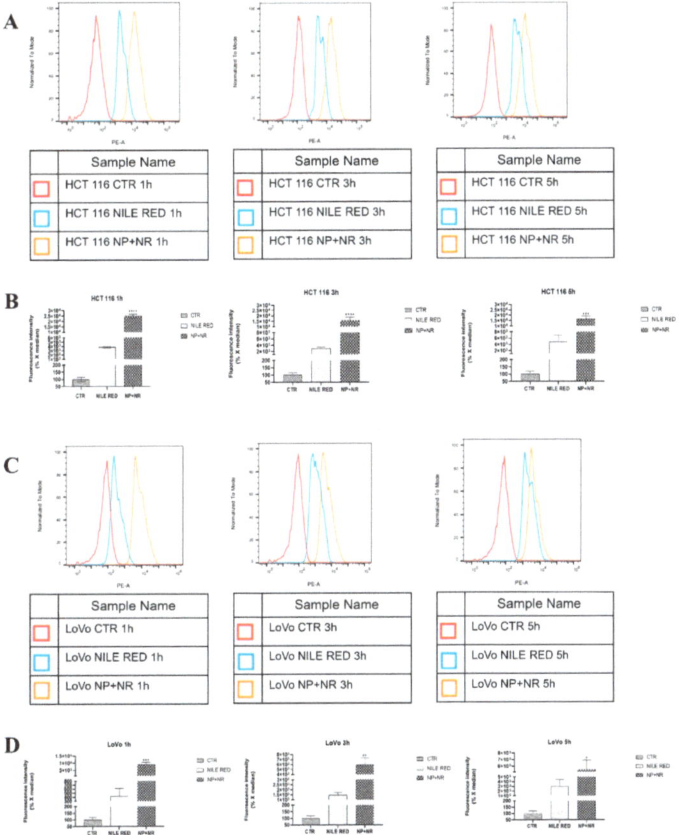

Figure 6. FACS analysis of nanoparticle uptake in colorectal cancer cell lines. Graphical representation of cellular NR uptake evaluated by flow cytometry in HCT 116 (**A**) and LoVo (**C**) cells after 1 h, 3 h, and 5 h of incubation, at the absorption peak at 590 nm wavelength (PE-A). Bar graphs reported the median X% of NR fluorescence intensity in HCT 116 (**B**) and LoVo (**D**) cell lines. The experiments were performed at least three times and the results were always similar. Bars, SD. (* $p < 0.05$; ** $p < 0.005$; *** $p < 0.001$; **** $p < 0.0001$ vs. negative control).

3.3.2. Effects of Quercetin-Loaded Nanoparticles (NP+Q) on Colon Cancer Cell Line HCT 116

The cytotoxic effect of quercetin-loaded nanoparticles (NP+Q) was evaluated on HCT 116 cells; these cells showed a higher cellular uptake efficiency compared to LoVo by MTT assay as described in "Experimental Section". Many studies have described the strict correlation between the dietary flavonoid intake and cancer risk; in particular, they showed that high flavonoid intake, including quercetin, could reduce the risk of colon cancer. In our colon cell model, NP+Q induced 50% growth inhibition at a concentration of 3.7 μM and 2.1 μM (IC50 values) after 48 h and 72 h, respectively (see Table 2). Interestingly, NP+Q induced a dose-dependent growth inhibition in this cell line more evident after 72 h (Figure 7). Quercetin has low stability and easily undergoes degradation; for these reasons, its antiproliferative effects are limited, with IC50 values ranging from 100 to 150 μM, as reported in previous works [37]. Our results showed that quercetin loading in PCL-PEG-PCL nanoparticles improved drug stability and efficiency; in fact, IC50 values of NP+Q indicated that the drug is already effective at low concentrations. To additionally confirm this, we compared the antiproliferative effects of NP+Q to those induced by empty NPs and free quercetin and calculated the combination index (CI) using the dedicated software Compusyn. CI values lower than 1, equal to 1, and higher than 1 indicate synergy, additivity, and antagonism, respectively. We found that CI was much lower than 1 for each tested concentration, suggesting a strong synergism between NPs and quercetin (see Figure S11).

Table 2. IC50 (μM) of quercetin-loaded nanoparticles on HCT 116, after 24 h, 48 h, and 72 h of treatment.

HCT 116	**IC$_{50}$**
24 h	>5 μM (low toxicity)
48 h	3.7 μM
72 h	2.1 μM

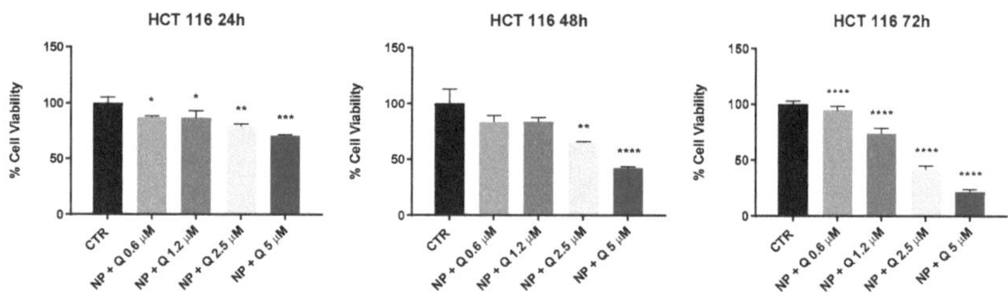

Figure 7. Effect of NP+Q on HCT 116 (colon cancer cell line) after 24 h, 48 h and 72 h, evaluated by MTT assay and expressed as percentage of cell viability (compared to untreated cells). Data are reported as mean of three independent experiments $\pm$ SD. (* $p < 0.05$; ** $p < 0.005$; *** $p < 0.001$; **** $p < 0.0001$ vs. negative control).

Moreover, NP+Q were able to induce a dose dependent growth inhibition by maintaining more stable the levels of quercetin inside the nanoparticles. Quercetin is easily degraded in cell medium, as recently reported [38], and its loading in PCL-PEG-PCL nanoparticles probably enhance its antiproliferative effects by improving its stability.

4. Conclusions

Quercetin is a hydrophobic molecule with very poor water solubility and short blood circulation times. A critical aspect for the successful application of quercetin in cancer treatment is its encapsulation in proper delivery systems. In the realm of nanoparticles for drug delivery systems, PEG-PCL based amphiphilic copolymers are the most used and applied for in vitro cancer cells studies and for the delivery of hydrophobic drug molecules [20–27]. To the best of our knowledge, they have not been used for the delivery of quercetin into human colorectal carcinoma cells.

We have synthetized and characterized triblock ABA type PCL-PEG-PCL copolymers and used them for the preparation of nanoparticles. The triblock copolymers self-assembled in water forming micelles consisting of a core of biodegradable PCL and a corona of PEG. The copolymers formed core-shell nanoparticles and were able to incorporate quercetin into the core.

The cellular uptake efficiency of HCT 116 and LoVo cells was quantitatively determined by flow cytometry using nanoparticles loaded with NR as a model of a hydrophobic drug. The cytotoxic effect of quercetin-loaded nanoparticles (NP+Q) was evaluated on HCT 116 cells and showed promising results. Indeed, by loading the quercetin in PCL-PEG-PCL nanoparticles, its antiproliferative effects were likely enhanced by improving its stability.

Our results indicated that the prepared PCL-PEG-PCL nanoparticles are a promising DDS for the delivery of quercetin into human colorectal carcinoma cells. Further studies could be addressed to prepare different PEG/PCL nanoparticles with diverse composition [14,16–19], to evaluate the difference in nanoparticle size, loading efficiency, nanoparticle uptake, and to understand the biochemical mechanisms of quercetin uptake in carcinoma cells. Our results, moreover, encourage research in the use of PEG-PCL as a DDS not only for quercetin but also for other flavonoids, for the possible improvement of therapeutic efficacy and pharmacokinetic profiles.

Supplementary Materials: The following supporting information can be downloaded at: https://www.mdpi.com/article/10.3390/polym15051179/s1, Figure S1: ^{1}H NMR spectrum (300 MHz, $CDCl_3$, 298 K) of PEG diol; Figure S2: GPC curve of triblock PCL-PEG-PCL copolymer; Figure S3: GPC curve of PEG diol; Figure S4: FT-IR spectrum of triblock PCL–PEG–PCL copolymer; Figure S5: DOSY NMR spectrum (400 MHz, $CDCl_3$, 298 K) of PCL; Figure S6: ^{1}H NMR spectrum (400 MHz, DMSO, 298 K) of triblock PCL-PEG-PCL copolymer; Figure S7: ^{1}H NMR spectrum (300 MHz, DMSO, 298 K) of PCL-PEG-PCL micelles loaded with Nile Red; Figure S8: ^{1}H NMR spectrum (400 MHz, DMSO, 298 K) of PCL-PEG-PCL micelles loaded with quercetin; Figure S9: DLS measurement of aqueous solution in the absence of quercetin loaded nanoparticles; Figure S10: ^{1}H NMR spectrum (400 MHz, DMSO, 298 K) of (a) unloaded polymer, (b) quercetin and (c) polymer loaded with quercetin; Figure S11: Isobologram analysis of the effects of NP+Q combination at different concentrations.

Author Contributions: Conceptualization, D.P. and R.F.; methodology, D.P. and S.Z.; software, N.F., C.G. (Chiara Germinario), C.G. (Celestino Grifa), D.R. and M.P.R.; validation, N.F., M.P.R. and M.A.; formal analysis, C.G. (Chiara Germinario), C.G. (Celestino Grifa), M.A.; investigation, D.P., R.F., D.R. and S.Z.; resources, D.P., R.F., S.Z., V.D.V., C.G. (Celestino Grifa); data curation, R.F., S.Z., D.P., C.G. (Chiara Germinario), D.R., writing—original draft preparation, N.F. and M.P.R.; writing—review and editing, D.P.; visualization, D.P. and R.F.; supervision, D.P., R.F. and S.Z.; project administration, D.P., R.F.; funding acquisition, D.P. and R.F. All authors have read and agreed to the published version of the manuscript.

Funding: This research received no external funding.

Institutional Review Board Statement: Not applicable.

Data Availability Statement: The data presented in this study are available on request from the authors.

Acknowledgments: The authors gratefully acknowledged the FAR Università del Sannio and the PON "Ricerca e Innovazione" 2014–2020, azione 4, dottorati su tematiche dell'innovazione.

Conflicts of Interest: The authors declare no conflict of interest.

References

1. Nichols, J.W.; Bae, Y.H. EPR: Evidence and Fallacy. *J. Control. Release* **2014**, *190*, 451–464. [CrossRef]
2. Senapati, S.; Mahanta, A.K.; Kumar, S.; Maiti, P. Controlled Drug Delivery Vehicles for Cancer Treatment and Their Performance. *Signal Transduct. Target. Ther.* **2018**, *3*, 7. [CrossRef]

3. Miyata, T.; Namera, T.; Liu, Y.; Kawamura, A.; Yamaoka, T. Photoresponsive Behaviour of Zwitterionic Polymer Parti-cles with Photodimerizable Groups on Their Surfaces. *J. Mater. Chem. B* **2022**, *10*, 2637–2648. [CrossRef]
4. Qin, L.; Liang, F.; Li, Y.; Wu, J.; Guan, S.; Wu, M.; Xie, S.; Luo, M.; Ma, D. A 2D Porous Zinc-Organic Framework Platform for Loading of 5-Fluorouracil. *Inorganics* **2022**, *10*, 202. [CrossRef]
5. Zheng, R.; Guo, J.; Cai, X.; Bin, L.; Lu, C.; Singh, A.; Trivedi, M.; Kumar, A.; Liu, J. Manganese Complexes and Manga-nese-Based Metal-Organic Frameworks as Contrast Agents in MRI and Chemotherapeutics Agents: Applications and Prospects. *Colloids Surf. B Biointerfaces* **2022**, *213*, 112432. [CrossRef]
6. Rao, C.; Liao, D.; Pan, Y.; Zhong, Y.; Zhang, W.; Ouyang, Q.; Nezamzadeh-Ejhieh, A.; Liu, J. Novel formulations of metal-organic frameworks for controlled drug delivery. *Expert Opin. Drug Deliv.* **2022**, *19*, 1183–1202. [CrossRef]
7. Tiwari, G.; Tiwari, R.; Bannerjee, S.; Bhati, L.; Pandey, S.; Pandey, P.; Sriwastawa, B. Drug Delivery Systems: An Updated Review. *Int. J. Pharm. Investig.* **2012**, *2*, 2. [CrossRef]
8. Buishvili, L.L.; Khalvashi, E.K. The Theory of Nonstationary Dynamic Polarization of Nuclei. *Radiophys. Quantum Electron.* **1971**, *14*, 1143–1144. [CrossRef]
9. Huang, J.; Zhang, H.; Yu, Y.; Chen, Y.; Wang, D.; Zhang, G.; Zhou, G.; Liu, J.; Sun, Z.; Sun, D.; et al. Biodegradable Self-Assembled Nanoparticles of Poly (d,l-Lactide-Co-Glycolide)/Hyaluronic Acid Block Copolymers for Target Delivery of Docetaxel to Breast Cancer. *Biomaterials* **2014**, *35*, 550–566. [CrossRef] [PubMed]
10. Shuai, X.; Ai, H.; Nasongkla, N.; Kim, S.; Gao, J. Micellar Carriers Based on Block Copolymers of Poly(ε-Caprolactone) and Poly(Ethylene Glycol) for Doxorubicin Delivery. *J. Control. Release* **2004**, *98*, 415–426. [CrossRef] [PubMed]
11. Knop, K.; Hoogenboom, R.; Fischer, D.; Schubert, U.S. Poly(Ethylene Glycol) in Drug Delivery: Pros and Cons as Well as Potential Alternatives. *Angew. Chem.–Int. Ed.* **2010**, *49*, 6288–6308. [CrossRef]
12. Sisson, A.L.; Ekinci, D.; Lendlein, A. The Contemporary Role of ε-Caprolactone Chemistry to Create Advanced Polymer Architectures. *Polymer* **2013**, *54*, 4333–4350. [CrossRef]
13. Pappalardo, D.; Mathisen, T.; Finne-Wistrand, A. Biocompatibility of Resorbable Polymers: A Historical Perspective and Framework for the Future. *Biomacromolecules* **2019**, *20*, 1465–1477. [CrossRef] [PubMed]
14. Wei, X.W.; Gong, C.Y.; Gou, M.L.; Fu, S.Z.; Guo, Q.F.; Shi, S.; Luo, F.; Guo, G.; Qiu, L.Y.; Qian, Z.Y. Biodegradable Poly(ε-Caprolactone)-Poly(Ethylene Glycol) Copolymers as Drug Delivery System. *Int. J. Pharm.* **2009**, *381*, 1–18. [CrossRef] [PubMed]
15. Sana, B.; Ferrentino, N.; Kohlan, T.B.; Liu, Y.; Pasiskevicius, V.; Finne-Wistrand, A.; Pappalardo, D. Coumarin End-Capped Poly(ε-Caprolactone)-Poly(Ethylene Glycol) Tri-Block Copolymer: Synthesis, Characterization and Light-Response Behavior. *Eur. Polym. J.* **2023**, *183*, 111760. [CrossRef]
16. Youxin, L.; Kissel, T. Synthesis and Properties of Biodegradable ABA Triblock Copolymers Consisting of Poly(l-Lactic Acid) or Poly (l-Lactic-Co-Glycolic Acid) A-Blocks Attached to Central Poly (Oxyethylene) B-Blocks. *J. Control. Release* **1993**, *27*, 247–257. [CrossRef]
17. Gan, Z.; Zhang, J.; Jiang, B. Poly(ε-Caprolactone)/Poly(Ethylene Oxide) Diblock Copolymer II. Nonisothermal Crystallization and Melting Behavior. *J. Appl. Polym. Sci.* **1997**, *63*, 1793–1804. [CrossRef]
18. Zamani, S.; Khoee, S. Preparation of Core-Shell Chitosan/PCL-PEG Triblock Copolymer Nanoparticles with ABA and BAB Morphologies: Effect of Intraparticle Interactions on Physicochemical Properties. *Polymer* **2012**, *53*, 5723–5736. [CrossRef]
19. Qi, W.; Ghoroghchian, P.; Li, G.; Hammer, D.; Therien, M. Aqueous Self-Assembly of Poly(ethylene oxide)-block-Poly(ε-caprolactone) (PEO-b-PCL) Copolymers: Disparate Diblock Copolymer Compositions Give Rise to Nano- and Meso-Scale Bilayered Vesicles. *Nanoscale* **2013**, *5*, 10908–10915. [CrossRef]
20. Xin, H.; Chen, L.; Gu, J.; Ren, X.; Wei, Z.; Luo, J.; Chen, Y.; Jiang, X.; Sha, X.; Fang, X. Enhanced Anti-Glioblastoma Efficacy by PTX-Loaded PEGylated Poly(ε-Caprolactone) Nanoparticles: In Vitro and in Vivo Evaluation. *Int. J. Pharm.* **2010**, *402*, 238–247. [CrossRef]
21. Zhang, L.; Chen, Z.; Wang, H.; Wu, S.; Zhao, K.; Sun, H.; Kong, D.; Wang, C.; Leng, X.; Zhu, D. Preparation and Evaluation of PCL-PEG-PCL Polymeric Nanoparticles for Doxorubicin Delivery against Breast Cancer. *RSC Adv.* **2016**, *6*, 54727–54737. [CrossRef]
22. Hakemi, P.; Ghadi, A.; Mahjoub, S.; Zabihi, E.; Tashakkorian, H. Ratio Design of Docetaxel/Quercetin Co-Loading-to-Nanocarrier: Synthesis of PCL–PEG–PCL Copolymer, Study of Drug Release Kinetic and Growth Inhibition of Human Breast Cancer (MCF-7) Cell Line. *Russ. J. Appl. Chem.* **2021**, *94*, 388–401. [CrossRef]
23. Hu, C.; Chen, Z.; Wu, S.; Han, Y.; Wang, H.; Sun, H.; Kong, D.; Leng, X.; Wang, C.; Zhang, L.; et al. Micelle or Polymersome Formation by PCL-PEG-PCL Copolymers as Drug Delivery Systems. *Chin. Chem. Lett.* **2017**, *28*, 1905–1909. [CrossRef]
24. Mirzaghavami, P.S.; Khoei, S.; Khoee, S.; Shirvalilou, S. Folic Acid-Conjugated Magnetic Triblock Copolymer Nanoparticles for Dual Targeted Delivery of 5-Fluorouracil to Colon Cancer Cells. *Cancer Nanotechnol.* **2022**, *13*, 1–18. [CrossRef]
25. Ni, R.; Duan, D.; Li, B.; Li, Z.; Li, L.; Ming, Y.; Wang, X.; Chen, J. Dual-Modified PCL-PEG Nanoparticles for Improved Targeting and Therapeutic Efficacy of Docetaxel against Colorectal Cancer. *Pharm. Dev. Technol.* **2021**, *26*, 910–921. [CrossRef] [PubMed]
26. Li, R.; Li, X.; Xie, L.; Ding, D.; Hu, Y.; Qian, X.; Yu, L.; Ding, Y.; Jiang, X.; Liu, B. Preparation and Evaluation of PEG-PCL Nanoparticles for Local Tetradrine Delivery. *Int. J. Pharm.* **2009**, *379*, 158–166. [CrossRef]
27. Jalilzadeh, N.; Samadi, N.; Salehi, R.; Dehghan, G.; Iranshahi, M.; Dadpour, M.R.; Hamishehkar, H. Novel Nano-Vehicle for Delivery and Efficiency of Anticancer Auraptene against Colon Cancer Cells. *Sci. Rep.* **2020**, *10*, 1606. [CrossRef]

28. Chen, L.; Li, K.; Liu, Q.; Quiles, J.L.; Filosa, R.; Kamal, M.A.; Wang, F.; Kai, G.; Zou, X.; Teng, H.; et al. Protective Effects of Raspberry on the Oxidative Damage in HepG2 Cells through Keap1/Nrf2-Dependent Signaling Pathway. *Food Chem. Toxicol.* **2019**, *133*, 110781. [CrossRef]

29. Kim, D.H.; Khan, H.; Ullah, H.; Hassan, S.T.S.; Šmejkal, K.; Efferth, T.; Mahomoodally, M.F.; Xu, S.; Habtemariam, S.; Filosa, R.; et al. MicroRNA Targeting by Quercetin in Cancer Treatment and Chemoprotection. *Pharmacol. Res.* **2019**, *147*, 104346. [CrossRef]

30. Tan, B.J.; Liu, Y.; Chang, K.L.; Lim, B.K.W.; Chiu, G.N.C. Perorally Active Nanomicellar Formulation of Quercetin in the Treatment of Lung Cancer. *Int. J. Nanomed.* **2012**, *7*, 651–661. [CrossRef]

31. Xu, G.Y.; Shi, H.S.; Ren, L.B.; Gou, H.F.; Gong, D.Y.; Gao, X.; Huang, N. Enhancing the Anti-Colon Cancer Activity of Quercetin by Self-Assembled Micelles. *Int. J. Nanomed.* **2015**, *10*, 2051–2063. [CrossRef]

32. Garofalo, C.; Capuano, G.; Sottile, R.; Tallerico, R.; Adami, R.; Reverchon, E.; Carbone, E.; Izzo, L.; Pappalardo, D. Different Insight into Amphiphilic PEG-PLA Copolymers: Influence of Macromolecular Architecture on the Micelle Formation and Cellular Uptake. *Biomacromolecules* **2014**, *15*, 403–415. [CrossRef] [PubMed]

33. Fuoco, T.; Pappalardo, D.; Finne-Wistrand, A. Redox-Responsive Disulfide Cross-Linked PLA-PEG Nanoparticles. *Macromolecules* **2017**, *50*, 7052–7061. [CrossRef]

34. Caruso, E.; Orlandi, V.T.; Malacarne, M.C.; Martegani, E.; Scanferla, C.; Pappalardo, D.; Vigliotta, G.; Izzo, L. Bodipy-Loaded Micelles Based on Polylactide as Surface Coating for Photodynamic Control of Staphylococcus Aureus. *Coatings* **2021**, *11*, 223. [CrossRef]

35. Adami, R.; Liparoti, S.; Izzo, L.; Pappalardo, D.; Reverchon, E. PLA-PEG Copolymers Micronization by Supercritical Assisted Atomization. *J. Supercrit. Fluids* **2012**, *72*, 15–21. [CrossRef]

36. Ostacolo, L.; Marra, M.; Ungaro, F.; Zappavigna, S.; Maglio, G.; Quaglia, F.; Abbruzzese, A.; Caraglia, M. In Vitro Anticancer Activity of Docetaxel-Loaded Micelles Based on Poly(Ethylene Oxide)-Poly(Epsilon-Caprolactone) Block Copolymers: Do Nanocarrier Properties Have a Role? *J. Control. Release* **2010**, *148*, 255–263. [CrossRef]

37. Zhang, Z.; Li, B.; Xu, P.; Yang, B. Integrated Whole Transcriptome Profiling and Bioinformatics Analysis for Revealing Regulatory Pathways Associated With Quercetin-Induced Apoptosis in HCT-116 Cells. *Front. Pharmacol.* **2019**, *10*, 798. [CrossRef]

38. Cao, H.; Högger, P.; Prieto, M.; Simal-Gandara, J.; Xiao, J. Stability of Quercetin in DMEM and Cell Culture with A549 Cells. *eFood* **2022**, *3*, e13. [CrossRef]

Article

Selection and Optimization of a Bioink Based on PANC-1-Plasma/Alginate/Methylcellulose for Pancreatic Tumour Modelling

Cristina Banda Sánchez [1], Nieves Cubo Mateo [2,3,†], Laura Saldaña [4,5], Alba Valdivieso [3], Julie Earl [6,7], Itziar González Gómez [3] and Luis M. Rodríguez-Lorenzo [1,*]

[1] Institute of Science and Technology of Polymers (ICTP-CSIC), 28006 Madrid, Spain
[2] Nebrija Research Group ARIES, Higher Polytechnic School, Antonio de Nebrija University, 28015 Madrid, Spain
[3] Institute for Physical and Information Technologies (ITEFI-CSIC), Sensors and Ultrasonic Systems, 28006 Madrid, Spain
[4] IdiPAZ, Hospital Universitario La Paz, 28046 Madrid, Spain
[5] Biomedical Research Networking Center in Bioengineering, Biomaterials, and Nanomedicine, CIBER-BBN, 28029 Madrid, Spain
[6] Ramón y Cajal Health Research Institute (IRYCIS), Molecular Epidemiology and Predictive Tumour Markers, 28034 Madrid, Spain
[7] Biomedical Research Network in Cancer (CIBERONC), 28034 Madrid, Spain
* Correspondence: luis.rodriguez-lorenzo@ictp.csic.es
† Formerly at Institute for Physical and Information Technologies (ITEFI-CSIC), Sensors and Ultrasonic Systems, 28006 Madrid, Spain.

Abstract: 3D bioprinting involves using bioinks that combine biological and synthetic materials. The selection of the most appropriate cell-material combination for a specific application is complex, and there is a lack of consensus on the optimal conditions required. Plasma-loaded alginate and alginate/methylcellulose (Alg/MC) inks were chosen to study their viscoelastic behaviour, degree of recovery, gelation kinetics, and cell survival after printing. Selected inks showed a shear thinning behavior from shear rates as low as 0.2 s^{-1}, and the ink composed of 3% *w/v* SA and 9% *w/v* MC was the only one showing a successful stacking and 96% recovery capacity. A 0.5×10^6 PANC-1 cell-laden bioink was extruded with an Inkredible 3D printer (Cellink) through a D = 410 μm tip conical nozzle into 6-well culture plates. Cylindrical constructs were printed and crosslinked with $CaCl_2$. Bioinks suffered a 1.845 Pa maximum pressure at the tip that was not deleterious for cellular viability. Cell aggregates can be appreciated for the cut total length observed in confocal microscopy, indicating a good proliferation rate at different heights of the construct, and suggesting the viability of the selected bioink PANC-1/P-Alg$_3$/MC$_9$ for building up three-dimensional bioprinted pancreatic tumor constructs.

Keywords: 3D bioprinting; bioinks; plasma; alginate; methylcellulose; PANC-1; tumour modelling

1. Introduction

A major hurdle in cancer research is the pre-clinical modelling strategies that have been conventionally applied to evaluate the efficacy of new cancer therapies. Traditionally, new anticancer drugs have been evaluated in two-dimensional (2D) cell culture platforms. However, 2D cultured cancer cells cannot mimic the complexity and heterogeneity of in vivo tumours, which usually grow in a three-dimensional (3D) conformation [1]. During the last few years, several 3D cell culturing techniques have emerged to overcome the observed gap between in vivo and in vitro experiments in cancer research [2]. Three-dimensional cancer models are anticipated to mimic the in vivo tumour microenvironment in human patients by recapitulating the proper tumour cell/matrix composition and yielding properties that match the type and stage of the intended disease. Therefore, it would be possible to perform accurate mechanistic studies on these in vitro models [3]. Many of these techniques base

---

Citation: Banda Sánchez, C.; Cubo Mateo, N.; Saldaña, L.; Valdivieso, A.; Earl, J.; González Gómez, I.; Rodríguez-Lorenzo, L.M. Selection and Optimization of a Bioink Based on PANC-1- Plasma/Alginate/ Methylcellulose for Pancreatic Tumour Modelling. *Polymers* **2023**, *15*, 3196. https://doi.org/10.3390/polym15153196

Academic Editors: Swee Leong Sing, Frederico Castelo Ferreira and João Carlos Silva

Received: 4 June 2023
Revised: 18 July 2023
Accepted: 25 July 2023
Published: 27 July 2023

their performance on the generation of multicellular tumour spheroids (MCTS) [4], hydrogel embedding [5], cell patterning [6], or microfluidic chips [7]. These techniques are helpful for many studies. i.e., multicellular tumour spheroids have been used to study fundamental cancer biology and drug screening [8,9]. However, the tumour microenvironment, including both chemical cues (growth factors and cytokines) and biophysical cues (interstitial pressure and matrix mechanics), is extremely complex, and most of those models lack well-organized spatial distribution of tumour cells and ECM compositions [3]. Therefore, there are still significant challenges that need to be overcome by 3D in vitro models on the road to using them for therapeutic drug development. These include batch-to-batch variability [10], limited control over cell patterning [11], low throughput, oversimplified structures [3] and limited vascularization potential [12].

3D bioprinting combines the ability to design geometrical parameters for constructs such as pore size, pore strut thickness, pore interconnectivity, and pores morphology with the capacity of depositing several types of co-cultured cells in a single spatial arrangement matching the natural architecture of native tissues [13,14]. This tissue engineering approach may be reproduced to engineer preclinical tumour models resulting in spatiotemporal control of physical and biological elements [15]. Mechanical forces, different cell populations [16] and bioactive signals can be incorporated into biological environments [17] to drive cell phenotypes that resemble tumour microenvironments [18]. 3D bioprinting involves using bioinks that combine biological and hosting synthetic materials. The selection of the most appropriate cell-material combination for a specific application, the printing conditions (printer type, temperature oxygen rate, speed of deposition), and the maturation procedure (signals and bioreactors) are so complex that there is a lack of consensus in the optimal conditions required for each specific case [19]. In addition, when tumour modelling is considered, there is high variability between cancer types and stages, making the selection of conditions even more specific.

Developing suitable cell-material combinations that can be used as bioinks for each specific application is the bottleneck hindering the advance of bioprinting. In addition to the already strict conditions that made a material suitable for building up a cytocompatibility scaffold, there are very specific conditions that materials should fulfil to be used as components in bioinks. Alginate has been selected as the first component of the intended bioink due to its gelling capacity, low toxicity, high availability, and low cost. Alginate hydrogels are formed by ionic crosslinking in the presence of calcium ions, and they reach similar mechanical properties to extracellular matrices, which explains why they are frequently used for studying cell response in tissue engineering testing. In addition, alginates have been declared safe by the United States Food and Drug Administration (FDA) [20] for application in humans [21]. Literature search results in terms of cell viability for different alginate concentrations [22] and previous works from team members [23] were used to select a 3% w/v alginate concentration as starting reference. The combination of viscoelastic values in the ECM range, potential cell viability and ease of preparing/handling was the criteria used. However, building up 3D printed construct in the Z direction with alginate hydrogels has been described as non-successful [24], whereas blending with a second polysaccharide, such as methylcellulose (MC), makes it easier to modulate the viscoelastic properties of the material and enables the printing of strands in the Z direction without fusing [25]. MC is a water-soluble cellulose derivative, cytocompatible and approved for food and drug administration [24].

This manuscript intends to describe the selection of a bioink (cell-material combination) that can be used in the future in modelling pancreatic tumours. Plasma was incorporated not only because plasma-based bioinks have demonstrated that they allow cell spreading, colony formation and angiogenic cues even for long-term in vitro tests [23]. In addition, plasma contains fibrinogen that should help induce fibroblast to generate the thick fibrous walls characteristic of pancreatic tumours when multicellular models are assayed [26]. Pancreatic tumour cells, PANC-1, have been shown to grow well in vitro 2D as attached cells [27] and were selected for this proof-of-concept study.

This paper deals with the translation and validation of a recently proposed synthetic material composition in the field of bone engineering [23] for preparing a PANC-1-based bioink for future modelling of primary pancreas tumours and with the establishing of the procedure and conditions required for the validation of a bioink. The states that must be established include viscoelasticity properties, gelation characteristics, geometrical fidelity and cellular viability after the stress suffered by the selected cell line within the tip during the bioprinting process.

2. Materials and Methods

2.1. Materials and Cells

2.1.1. Polymers

Alginic acid sodium salt from brown algae (SA) (Mw: 8945 g/mol, mannuronate/guluronate ratio of 0.63, Sigma, Saint-Quentin-Fallavier, France) [28] and methylcellulose (MC) (viscosity 4000 cP, M0512-500G, Sigma were purchased by Sigma-Aldrich (Madrid, Spain). Fresh frozen human plasma was donated by Hospital Ciudad de Coria. In this study, fibrinogen concentration was not measured, but coagulation was tested through a tube inverted test, obtaining a coagulation time below 10 min. For the inverted tube test, three vials were filled with 1 mL of plasma, and then 0.2 mL of $CaCl_2$, 1.5% w/v, was added while a timer was started for each one. Tube inversion was performed every 30 s to visually inspect the sol-gel transition until the gel remained attached to the top. To have a reference, as previously reported in the literature, fibrinogen's normal concentration range is 200 to 400 mg/dL (2.0 to 4.0 g/L), which may vary slightly among different laboratories and patients [29].

2.1.2. Cells

PANC-1 cells, an epithelioid carcinoma cell line derived from the human pancreas [30], were provided by ATCC (frozen, CRL-1469, tissue: Pancreas; Duct) from the human cell culture collection (https://www.atcc.org/ accessed on 1 September 2020).

2.2. Inks and Bioink Preparation

Fresh frozen human plasma was slowly thawed at 37 °C and mixed with SA under stirring. Then, MC was added and mixed with a spatula. Three different inks were prepared using the following concentrations: plasma containing alginate 3% w/v (P-Alg$_3$/MC$_0$), plasma containing alginate 3% w/v, methylcellulose 3% w/v (P-Alg$_3$/MC$_3$) and plasma containing alginate 3% w/v, methylcellulose 9% w/v (P-Alg$_3$/MC$_9$).

For the bioink (PANC-1/P-Alg$_3$MC$_9$), standard cell culture protocols for PANC-1 cell lines were applied for cell sample preparation [30,31]. PANC-1 cell lines were cultured in RPMI (Gibco/Invitrogen Bleiswijk, Netherlands) supplemented with 10% fetal bovine serum (FBS; Invitrogen Bleiswijk, Netherlands) and 50 units/mL penicillin/streptomycin (Invitrogen) and kept in an incubator at 5% CO_2 and 37 °C. Then cells were resuspended and embedded into the P-Alg$_3$/MC$_9$ ink, slowly mixing with a spatula, obtaining a final concentration of 5×10^5 cells/mL. Every procedure is accomplished under sterile conditions inside a laminar flow cabinet.

2.3. Optimization & Characterization of the Inks

Inks printability was assessed by a layer stacking test, printing 20 mm successive layers with a pneumatic extrusion Inkredible 3D printer (Cellink) through a conical nozzle (Nordson, SmoothFlow Tapered Dispense Tips, ref 7018298, ID = 0.41 mm/0.016″) and visually analyzing the merge of the layers. Also, rheological tests were performed on an ARG2 (TA instruments, New Castle, DE, USA) rheometer with a sandblasted parallel plate D = 25 mm geometry at 37 °C before, during, and after the gelation event upon the addition of 1.5% w/v $CaCl_2$ crosslinker [32]. From a parallel plate geometry, the Power law index (η) (Equation (1)) is obtained as previously described [33]:

$$\eta = k \dot{\Upsilon}^{n-1} \tag{1}$$

where η is viscosity, k is the consistency and $\ddot{Y}$ the shear rate. Oscillatory frequency sweeps evaluate the viscosity behaviour at the shear thinning region. The linear viscoelastic region had a torque sweep between 0.5 and 10^3 s^{-1}. The critical yield point, $\gamma c = 154$ μNm, determined from the LVR test, delimits the maximum value of stress that the ink can support. Frequency tests were made below that value. Frequency sweeps, in the range from 10^{-2} to 10^2 Hz, were performed at a torque of 50 μNm. Thixotropy and degree of recovery after a structural disruption are computed by applying a low shear rate (0.5 s^{-1}) for 60 s and then a 250 s^{-1} shear for 30 s. Gelation kinetics are measured in a time sweep for 30 min, where crosslinker (1.5% CaCl$_2$) is added 1 min after starting recording. All rheological treatments have been replicated three times.

2.4. 3D Bioprinting

Cell-laden bioink solution is loaded into a 5 mL cartridge and then stored at RT to stabilise bioink's rheological properties. Pneumatic extrusion Inkredible 3D printer (Cellink) dispenses the hydrogel bioink through a conical nozzle (D = 410 μm). Cylindrical constructs of D = 15 mm are printed into 6-well cell culture plates with a strand width of 1.8 mm, creating 3D stacking pores of 1.3 mm (with a 0.48 mm infill extrusion width). Printed hydrogels are crosslinked for 20 min under a bath with 1.5% *w/v* CaCl$_2$; then, the crosslinker is removed and replaced with fresh medium. Finally, constructs are incubated at 37 °C, 5% CO$_2$.

Shear stress suffered by the bioinks in the syringe is calculated with the equation.

$$\dot{\gamma}_1 = \left(\frac{V_1 R_1^2}{\left(\frac{n}{3n+1}\right)\left(R_1^{\frac{3n+1}{n}}\right)} \right) \sqrt[n]{r} \tag{2}$$

And the shear stress suffered at the tip is calculated as shown in Equation (3):

$$\dot{\gamma}_2 = \left(\frac{V_2 R_2^2}{\left(\frac{n}{3n+1}\right)\left(R_2^{\frac{3n+1}{n}}\right)} \right) \sqrt[n]{r} \tag{3}$$

where $\dot{\gamma}$ is the shear rate, V is the dispensing speed, R is the maximum ratio, and r is the ratio at the calculation point.

2.5. Viability Assay on Cell-Laden Constructs

Cellular survival and proliferation are assessed with a Live/Dead viability kit, calcein AM (0.5 mL) and ethidium homodimer-1 (2.0 mL) were dissolved in 997.5 mL PBS, added to the samples, and incubated for 30 min while protected from light at 37 °C in a humidified atmosphere with 5% CO$_2$. (EthD-1) (ThermoFisher Scientific #L3224, Karlsruhe, Germany) according to the manufacturer's protocol. Afterwards, constructs are imaged in a fluorescence microscope (Smart Cell Imager PAULA, Leica Microsystems, Wetzlar, Germany) for green (live) and red (dead) spectra. Viability is screened at 1-, 6-, and 14-day post-printing with a Leica TCS SPE, Wetzlar, Germany, from MNCN-CSIC, Madrid. For the visualization of nuclear morphology and proliferation within the construct, cells were fixed with 4% *w/v* formaldehyde in PBS and stained with PBS containing 3×10^{-6} M, 4,6-diamidino-2-phenylindole DAPI (D1306, Invitrogen) and Actin (Alexa Fluor 488 phalloidin, A12379, Invitrogen) and by confocal microscopy using a confocal microscope (Leica TCS SPE, Wetzlar, Germany) from idiPAZ-Madrid.

2.6. Statistics

Three repetitions have been conducted to verify the reproducibility of the results obtained in the rheology and cytocompatibility test. A One-way ANOVA was performed.

Pairwise comparisons of means with equal variances have been implemented using Tukey's post hoc analysis. Statistical differences were assumed at $p < 0.05$. All analyses were performed using STATA/SE, StataCorp LLC Statistics/Data Analysis (Special Edition College Station, TX 77845, USA).

3. Results

3.1. Ink Selection, Printability and Printing Fidelity

Plasma-based bioinks containing an alginate concentration of 3% and Methyl-cellulose (MC) to refine printability was selected based on published works for applications in bone engineering [23]. Three different concentrations of MC were tested: 0, 3 and 9%. A layer stacking test was used for the initial qualitative screening. It could be observed that a successful stacking without merging of the layers is obtained only for the composition P-Alg$_3$/MC$_9$, as displayed in Figure 1.

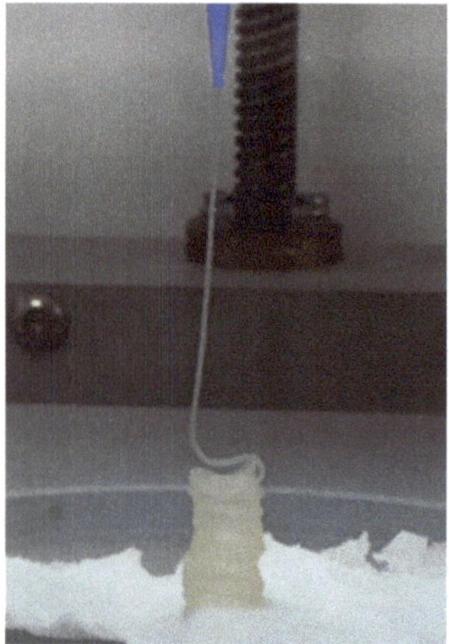

Figure 1. Layer stacking test. Ink containing plasma, 3% alginate and 9% methylcellulose (P-Alg$_3$/MC$_9$) is the only one where successive layers did not merge with the layers below.

Printability was evaluated, then, through rheology tests. Viscosity versus shear rate curves are shown in Figure 2. They are obtained using the Cox-Merz transformation. The two MC-containing inks, P-Alg$_3$/MC$_3$ and P-Alg$_3$/MC$_9$ show a shear thinning behaviour from a yield stress point as low as 0.2 s^{-1}. Whereas the control ink, P-Alg$_3$/MC$_0$, shows a shear thinning behaviour from a yield stress point of 2 s^{-1}, indicating that the inclusion of the MC expands the range of shear rate that can be applied when printing. Also, a greater viscosity is obtained for P-Alg$_3$/MC$_9$. The n Index, between 0.2–0.3, for P-Alg$_3$/MC$_3$ and P-Alg$_3$/MC$_9$ indicates a rapid decrease of viscosity with the increase of shear rate that it should favour not only the printability but also the recovery of the viscosity of the ink when the printing process is over.

The recovery experiments for the three tested inks are shown in Figure 3. P-Alg$_3$/MC$_9$ ink eventually recovers 96% of the viscosity after the period of high shear rate, whereas P-Alg$_3$/MC$_3$ recovers 85% and P-Alg$_3$/MC$_0$ recovers 89% in the first cycle, but it drops in the second. Also, P-Alg$_3$/MC$_9$ displays a quicker recovery in the first 60 s after disruptive shear stress application than P-Alg$_3$/MC$_3$.

The frequency sweeps made on the inks are shown in Figure 4. Similar rubbery behaviour is observed for the three inks along the tested range of frequencies where the elastic behaviour prevails. P-Alg$_3$/MC$_0$ displays a fall in tan(δ) for higher frequencies. Similar behaviour is also observed for P-Alg$_3$/MC$_3$. However, this ink is more stable at low frequencies, and P-Alg$_3$/MC$_9$ is stable over the whole range of frequencies, also showing the lower value for tan(δ), indicating the highest elastic vs. viscous behaviour of the three inks. Thus, the highest capacity for absorbing impacts with no deformation.

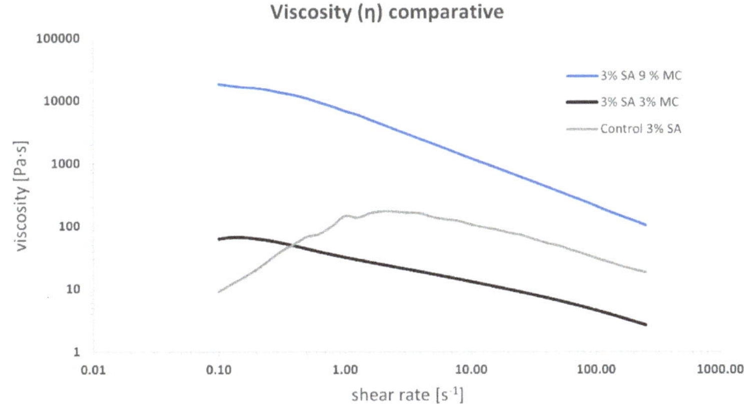

Figure 2. Viscosity vs. shear rate for inks P-Alg$_3$/MC$_0$, P-Alg$_3$/MC$_3$ and P-Alg$_3$/MC$_9$. An extended shear thinning region can be observed for MC-containing inks.

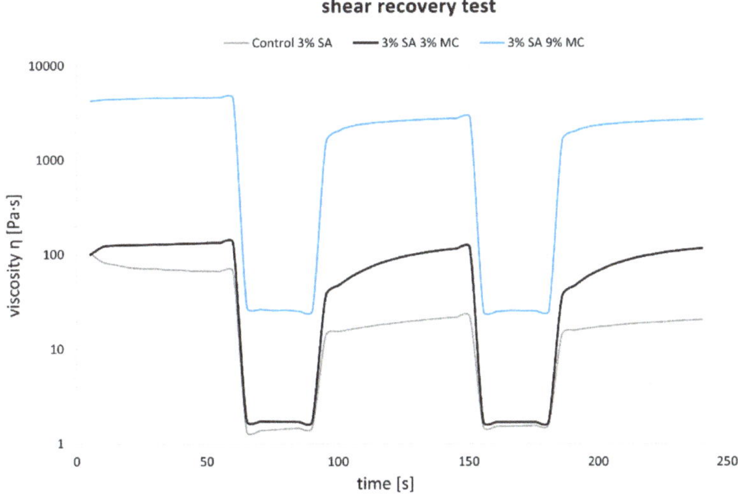

Figure 3. Recovery test experiments. P-Alg$_3$/MC$_9$ displays the highest and quickest recovery percentage out of the three tested inks.

From these results, P-Alg$_3$/MC$_9$ was chosen for printing PANC-1-loaded models. Printing fidelity was then evaluated to learn the capacity of this ink to reproduce accurately the design. The CAD design and the printed result with P-Alg$_3$/MC$_9$ ink can be shown in Figure 5. At the time of printing, the ink offers great printability in terms of layer stacking, shape consistency and CAD fidelity, no ink clumping at the tip of the nozzle, and homogeneous qualitative ink distribution. The external skirt has been used to determine the average required printing pressure of 78 ± 1 kPa.

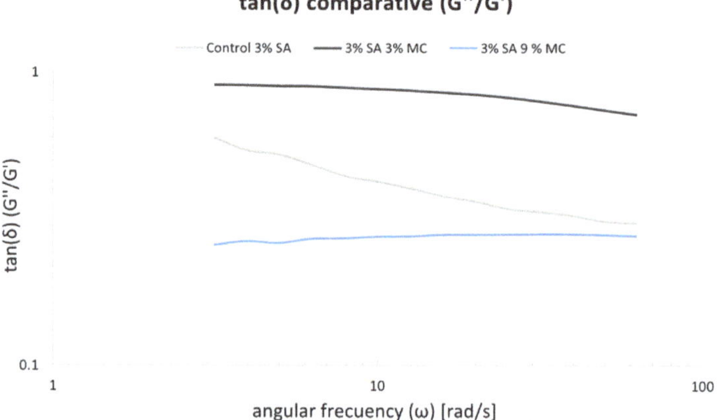

Figure 4. tan(δ) results obtained with the frequency swaps for the three inks. P-Alg$_3$/MC$_9$ is stable over the whole range of frequencies.

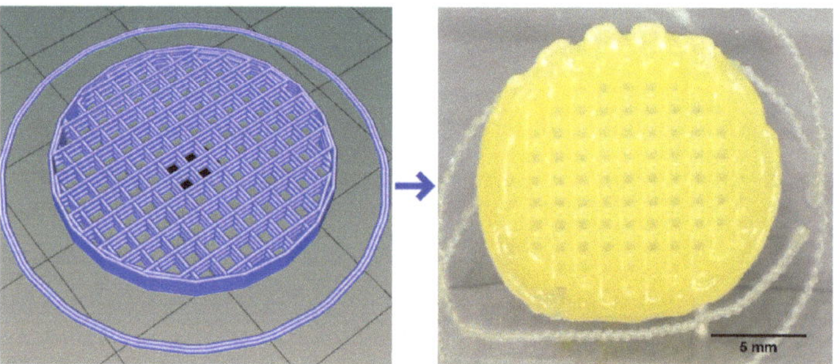

Figure 5. CAD design on the left and printed device on the right reproducing the design.

The evolution of the tan(δ) after adding the ionic crosslinker is displayed in Figure 6. The ink gels quickly during the first five minutes, then slowly gains elastic prevalence, thus decreasing tan(δ). Past 20 min from the crosslinker solution addition, the constructs have swelled slightly, doubling the filament diameter yet not clogging the printed grids.

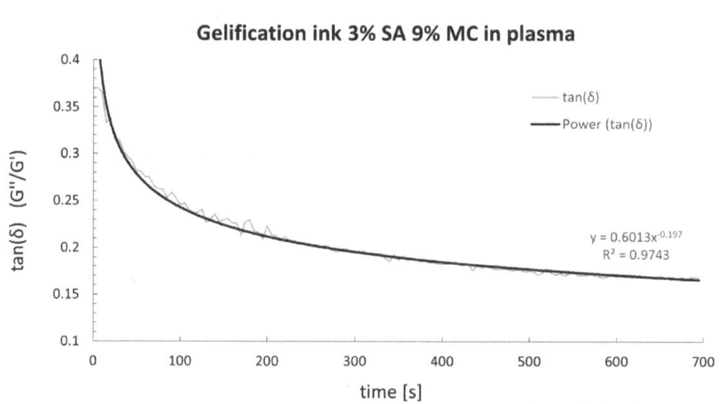

Figure 6. tan(δ) evolution of P-Alg$_3$/MC$_9$ ink after adding 1.5 mL of CaCl$_2$ 1.5%.

A representation of the syringe used for the process of printing and the shear rate calculations made to evaluate the shear rate that cells will suffer when printing (and thus to evaluate their potential for survival) is shown in Figure 7.

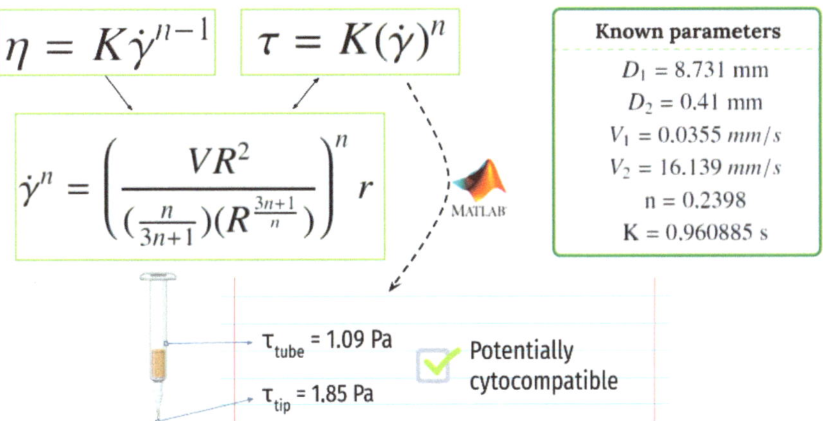

Figure 7. The shear rate that the cells will suffer when printing at the syringe and at the needle tip.

With a printing pressure of 78 $\pm$ 1 kPa, the required time to dispense 3 mL of ink has been measured to be 23′28″. Using the equation shown in Figure 7, a V_2 = 16.139 mm/s speed at the tip of the dispensing needle and a V_1 = 0.0355 mm/s speed at the core of the syringe have been determined. Equation (2) can now be used to determine the shear rate the cells will suffer when printing and to evaluate the potential for survival in the printing process. Calculated values at the syringe and the tip of the needle are τ_{max1} = 1.088 Pa and τ_{max1} = 1.845 Pa, respectively.

3.2. Viability Assays on Cell-Laden Constructs

The top row in Figure 8 displays a macroscopic picture of the printed constructs on days 1, 6 and 14. The image after six days shows that the construct maintains its structural integrity and still can be handled. The image after 14 days shows a swollen construct with diffuse edges indicating that the degradation process has started to affect the integrity of the construct. Figure 8, bottom row, collects the fluorescent channels red and green at days 1 and 6 taken with the Smart Cell Imager PAULA and the live/dead ratio at day 14 taken with a Leica TCS SPE confocal microscopy. Some red dots can be observed on day one, suggesting that the printing process was barely detrimental for cells. Green fluorescence occupies a larger surface at day six, indicating a good proliferation rate after six days of culture. Calculations based on fluorescence image yield cell viability of 74 $\pm$ 4% at day 1, 81 $\pm$ 4% at day 6 and 77 $\pm$ 1 at day 14.

Figure 9 collects confocal microscopy images of the constructs. The left image, a depth map, shows how the cells are distributed in different z-levels along the construct, and the colour corresponds to cells at different depths. Cell aggregates for the total length of the cut can be appreciated, suggesting a good proliferation rate to varying heights of the construct. In the right image, a blue DAPI staining for the nucleus and a green actin staining for the cytoplasm enable us to appreciate the formation of cell aggregates and the maintenance of the characteristic prismatic forms of the PANC-1 cells selected.

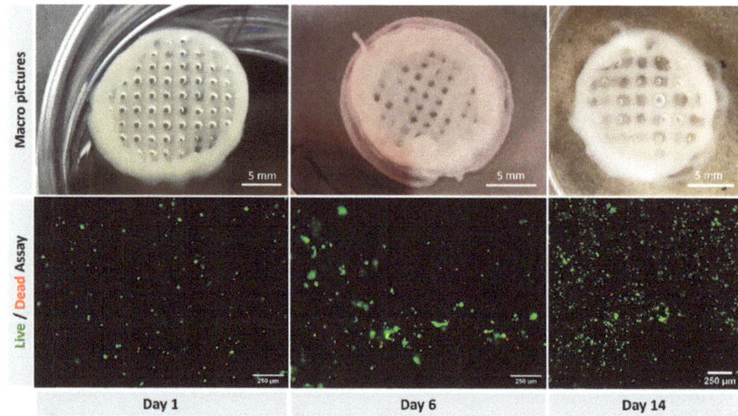

Figure 8. (**Top**): macroscopic view of printed constructs at days 1, 6 and 14. (**Bottom**): fluorescence with green (calcein staining) for live cells and red (EthD-1) for dead, at days 1, 6 and 14.

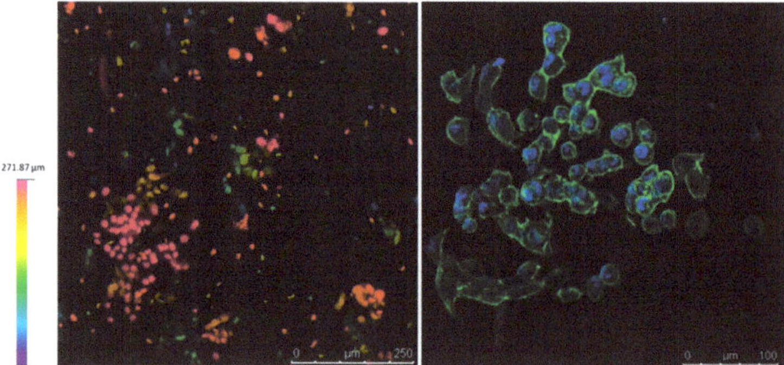

Figure 9. Confocal microscopy on day 9. (**Left**), depth map showing that cells survive at different levels within the construct. (**Right**), Dapi/Actin staining for the nucleus and cytoplasm shows the PANC-1 phenotype within the construct.

4. Discussion

4.1. Selection of Bioink Composition, Printability and Viscoelastic Properties

Plasma would be an ideal vehicle for cell-laden bioinks because it contains over 700 proteins such as fibrinogen, fibronectin, or albumin as significant components and growth factors, hormones, and cytokines that would assist cell growth and survival [23]. Plasma enabled the development of patient-specific bioinks suitable for the bioprinting of constructs with patient-specific shapes, material composition, and cells [23]. This is particularly relevant for cancer modelling since anticancer drugs have been shown to vary substantially among patients [34], and there is a heightened initiative to develop in vitro tumour models that could be patient-derived and able to phenocopy as much as possible from the original tumour [35]. However, plasma consistency is similar to water and requires supplementation to reach the viscoelastic properties needed for bioprinting [36]. Some studies have shown that mixing plasma with polysaccharides such as cellulose nanocrystals or alginate proves the compatibility among these components and shows great potential for building bioinks with regenerative abilities that could be used in personalized medicine [13,37].

Alginate-based bioinks are among the most used and successful materials in bioprinting due to their shear thinning character, the ability for rapid crosslinking, which provides good shape fidelity to the printed construct, and the feasibility of printing viable cells [38]. Since the viscosity of alginate bioinks depends on the alginate concentration, the alginate molecular weight and the cell density loading, printability could be promoted by controlling these parameters and blending with different saccharides. Using different alginate concentrations have been discharged because it may affect cellular viability, as will be discussed later. The addition of methylcellulose (MC) to a low-concentrated alginate solution has previously been shown to strongly improve the printability of alginate solutions, enabling 3D plotting of centimetre-scaled 3D constructs with tailored architecture and with high shape fidelity [24,25]. Based on these arguments, three different plasma-based inks were proposed for this project, low-concentrated alginate (P-Alg$_3$) and two inks modified with MC, P-Alg$_3$/MC$_3$ and P-Alg$_3$/MC$_9$. Printability was tested with these three inks. The three inks present a shear thinning behaviour that should reduce the shear stress of cells during printing. It can be noticed that inks modified with MC expand the range of shear thinning behaviour compared to the P-Alg$_3$/MC$_0$ ink, as seen in Figure 2. This would also expand the range of conditions and printing techniques; i.e., inkjet techniques require lower viscosities than extrusion techniques [39], which could be used with these inks, though this is not the subject of this work.

In a first analysis, viscosity during printing and, consequently, the shear stress suffered by cells should be higher in p-Alg$_3$/MC$_9$ than in the other two inks. However, rheological properties must also be adjusted to generate constructs with high shape fidelity. And, as can be seen in Figure 4, P-Alg$_3$/MC$_9$ ink is the most stable over the whole range of frequencies, also showing the lower value for tan(δ), indicating the highest elastic behaviour out of the three inks. Thus, the highest capacity for absorbing impacts with no deformation. Also, only P-Alg$_3$/MC$_9$ was consistent enough to pile layers with no merging, as displayed in Figure 1. Thus it would be the only of the three proposed inks capable of printing multilayer constructs. In addition, P-Alg$_3$/MC$_9$ is the ink that eventually recovers most of the viscosity that it had before printing (Figure 3). This result indicates that P-Alg$_3$/MC$_9$ is the ink that suffers lower damage during the printing process and suggests that higher printing fidelity would be achievable. When the ionic crosslinker is added immediately after printing, the printed ink displays a quick gelling behaviour that provides further consistency, as deduced from the quick decrease of the tan(δ) value (Figure 7). Eventually, the quick "solidification" produces an acceptable printing fidelity, as can be observed in Figure 5, comparing the device designed and the device printed. However, even this ink displays some swelling that somehow limits the fidelity of the design and suggests that a minimum separation distance between printed lines would be required.

The suitability of our bioink for modelling pancreas tumours can be considered by comparison with natural tissues. The elastic modulus of diseased tissue is frequently higher than that of healthy tissue. In particular, for the pancreas, human pancreatitis tissue possesses a modulus two times higher than healthy tissue. In contrast, tumour tissues are five times higher than healthy tissue [40]. In rheology experiments, pancreas tissue models have been reported to be in the range from 120 to 180 Pa with a weakly frequency-dependent dynamic modulus from 0.1 to 0.8 Hz [41], whereas the reported values from our bioink are in the range of 400 to 1100 Pa, thus roughly 4 to 6 times higher than the healthier tissue model in the literature.

4.2. Crosslinking Behaviour

Only the P-Alg$_3$/MC$_9$ ink displayed a printing fidelity appropriate for building up 3-dimensional constructs. Thus, only this ink was selected for testing the gelation process and cellular viability. In addition, extrusion bioprinting requires quick gelation of the bioink after the printing process to maintain the structure long enough for the cells to proliferate [42]. Alginate hydrogels offer a rapid gelation ability by ionic crosslinking with divalent cations. The rheology test of P-Alg$_3$/MC$_9$ showed a reduction of the tanδ value from 0.4 to 0.2 in five minutes after adding CaCl$_2$, followed by a slow stabilization later. The prevalence of the elastic component over the viscous component of the material should help maintain the construct's morphology for the required period for cell proliferation. (Figure 6). However, it has to be clear that this behaviour is due to the specific characteristics of the selected alginate, and then it is specific to the prepared ink. Ionic crosslinking of alginate is a complex process that depends on the M and G blocks content of the corresponding alginate. G blocks form ionic bridges with divalent cations and increase the gel-forming ability, whereas M blocks form weak junctions with divalent cations, and MG and M blocks increase flexibility [38]. It has to be considered that the biological response depends on this ratio since it has been described that a high amount of M blocks could cause immunogenicity [43]. The crosslinking mechanism involves the coordination of divalent ions with four-carboxyl groups to form an egg-box arrangement, as described in previous work [38]. Our ink was prepared with the alginic acid sodium salt of Mw: 8945 g/mol and a M/G ratio of 0.63. Full characterization of the alginate used is described in a former paper [28], and the elasticity of the printed construct, crosslinking ability and biological response are only reproducible with this alginate Mw and M/G ratios. The batch-to-batch variability reported as one of the problems with 3D models in tissue engineering or cancer research [44] may be related to the variability of the alginates used or the lack of specifications of the polysaccharides used.

4.3. PANC-1 Viability on 3D Bioprinted Constructs

As stated above, the printability of alginate-based bioinks can be promoted by modifying the alginate concentration or the alginate molecular weight or by blending with different saccharides. Since low-concentrated alginate solutions have been shown to improve the printability of the material strongly [38] and also high concentrations can negatively influence long-term biological performance [45], it was decided that the incorporation of an additional saccharide to be a more convenient approach to combine printability and cellular viability. The composition P-Alg$_3$/MC$_9$ is the most convenient prepared ink from the printability side. It is necessary to check whether it is also suitable for our particular cell line survival as well. Firstly, it was calculated that the maximum shear stress at the syringe would be τ_{max1} = 1088 Pa, whereas, at the tip, it would be τ_{max1} = 1845 Pa, and since a conical tip is being used, the cell will suffer this stress for about 2 milliseconds only [46]. However, short-time exposure to high levels of shear stress has been shown that affects cell viability and can induce long-term alterations in the proliferation potential of the cells that have survived the printing process [47,48]. The literature describes that fibroblast viability dropped below 80% for shear stress higher than 5000 Pa for 30 ms [46] and a maximum shear stress value of 160 Pa was detrimental to chondrocyte viability [49]. These results st then that each cell type has a different resistance. Nor and no studies have been found so far about the PANC-1 cell line needed for building pancreatic tumor models and cell viability needs to be checked after printing. Cell morphology, proliferation and phenotype have been examined after printing and after two weeks of printing. Results are collected in Figures 8 and 9. It can be observed that viability is not significantly affected, and phenotypic expressions are preserved [50,51], thus, pending further examination. They are expected to behave similarly as they would in their native niches. Also, good control of the spatial repartition of the cells and cell density can be observed (Figure 9 left), enabling a reproducible cell patterning that can be used to develop a controlled tumour microenvironment.

In summary, the proper selection of the bioprinting parameters, starting with the correct selection of the bioink components, the viscosity achieved, the capacity for printing recovery, and the exposure to the shear stress during the bioprinting process are major challenges that need to be overcome to facilitate any application of 3D bioprinting. Gelling kinetics is the key parameter to obtain constructs that maintain the designed features, and it is also of great importance to preserve the cell bioactivity and phenotypic expressions in the way they behave in their native niches. Our results show for the first time that the combination of a PANC-1/P-alg$_3$/MC$_9$ bioink and a pneumatic extrusion configuration with a 0.4 mm end conical nozzle can be used to build up a pancreatic tumour 3D biomodel.

Polymers **2023**, *15*, 3196

Although these results are specific only for the characteristics of the alginate selected and for the chosen cell line and thus, they cannot be directly applied to other hydrogels, printing techniques or cell lines, these results are of great value for other biofabrication modelling attempts as they give a tip on which parameters should be examined, how they should be examined and what should be expected from the cells respond.

5. Conclusions

Our results show that a PANC-1, plasma-loaded bioink with a 3% alginate concentration and alginate: methylcellulose ratio 1:3 provides a viscosity and printing recovery features suitable for bioprinting. A 3D bioprinting configuration based on pneumatically pressure extrusion with a conical 0.4 mm width end nozzle induces a $\tau max1 = 1845$ Pa low enough to maintain PANC-1 phenotype, viability over 70% and proliferation in suitable levels for building up PANC-1-based 3D printed biomodels. The bioink composition and printing conditions did not inhibit a homogeneous proliferation of PANC-1 cells at different heights, suggesting the viability of building three-dimensional bioprinted pancreas tumor models based on PANC-1/P-Alg$_3$/Mc$_9$ bioinks.

Author Contributions: Conceptualization, N.C.M. and L.M.R.-L.; Data curation, C.B.S.; Formal analysis, C.B.S.; Funding acquisition, J.E. and I.G.G.; Investigation, C.B.S., N.C.M. and L.M.R.-L.; Methodology, C.B.S., L.S., A.V. and J.E.; Project administration, I.G.G.; Resources, J.E. and L.M.R.-L.; Writing—original draft, C.B.S.; Writing—review & editing, L.M.R.-L. All authors have read and agreed to the published version of the manuscript.

Funding: This research was funded by project PID2021-128985OB-I00 funded by MCIN/AEI/10.13039/ 501100011033/FEDER Una manera de hacer Europa. Grant PI22/00939 from ISCIII-FEDER, Ministerio de Ciencia, Innovación y Universidades (MICINN)-AES.

Institutional Review Board Statement: Not applicable.

Data Availability Statement: The data that support the findings of this study are available on request from the corresponding author.

Acknowledgments: Hospital de Coria (Spain) is acknowledged for supplying the plasma.

Conflicts of Interest: The authors declare no conflict of interest.

References

1. Gebeyehu, A.; Surapaneni, S.K.; Huang, J.; Mondal, A.; Wang, V.Z.; Haruna, N.F.; Bagde, A.; Arthur, P.; Kutlehria, S.; Patel, N.; et al. Polysaccharide Hydrogel Based 3d Printed Tumor Models for Chemotherapeutic Drug Screening. *Sci. Rep.* **2021**, *11*, 372. [CrossRef] [PubMed]
2. Peela, N.; Truong, D.; Saini, H.; Chu, H.H.; Mashaghi, S.; Ham, S.L.; Singh, S.; Tavana, H.; Mosadegh, B.; Nikkhah, M. Advanced Biomaterials and Microengineering Technologies to Recapitulate the Stepwise Process of Cancer Metastasis. *Biomaterials* **2017**, *133*, 176–207. [CrossRef] [PubMed]
3. Zhang, Y.S.; Duchamp, M.; Oklu, R.; Ellisen, L.W.; Langer, R.; Khademhosseini, A. Bioprinting the Cancer Microenvironment. *ACS Biomater. Sci. Eng.* **2016**, *2*, 1710–1721. [CrossRef]
4. Poggi, A.; Villa, F.; Fernadez, J.L.C.; Costa, D.; Zocchi, M.R.; Benelli, R. Three-Dimensional Culture Models to Study Innate Anti-Tumor Immune Response: Advantages and Disadvantages. *Cancers* **2021**, *13*, 3417.
5. Chaicharoenaudomrung, N.; Kunhorm, P.; Noisa, P. Three-Dimensional Cell Culture Systems as an In Vitro Platform for Cancer and Stem Cell Modeling. *World J. Stem Cells* **2019**, *11*, 1065–1083. [CrossRef]
6. Tian, C.; Tu, Q.; Liu, W.; Wang, J. Recent Advances in Microfluidic Technologies for Organ-on-a-Chip. *Trac-Trends Anal. Chem.* **2019**, *117*, 146–156. [CrossRef]
7. Rodrigues, T.; Kundu, B.; Silva-Correia, J.; Kundu, S.C.; Oliveir, J.M.; Reis, R.L.; Correlo, V.M. Emerging Tumor Spheroids Technologies for 3d in Vitro Cancer Modeling. *Pharmacol. Ther.* **2018**, *184*, 201–211. [CrossRef]
8. Gunti, S.; Hoke, A.T.K.; Vu, K.P.; London, N.R., Jr. Organoid and Spheroid Tumor Models: Techniques and Applications. *Cancers* **2021**, *13*, 874.
9. Vinci, M.; Gowan, S.; Boxall, F.; Patterson, L.; Zimmermann, M.; Court, W.; Lomas, C.; Mendiola, M.; Hardisson, D.; Eccles, S.A. Advances in Establishment and Analysis of Three-Dimensional Tumor Spheroid-Based Functional Assays for Target Validation and Drug Evaluation. *BMC Biol.* **2012**, *10*, 29. [CrossRef]
10. Loessner, D.; Stok, K.S.; Lutolf, M.P.; Hutmacher, D.W.; Clements, J.A.; Rizzi, S.C. Bioengineered 3d Platform to Explore Cell–Ecm Interactions and Drug Resistance of Epithelial Ovarian Cancer Cells. *Biomaterials* **2010**, *31*, 8494–8506. [CrossRef]

11. Sung, K.E.; Beebe, D.J. Microfluidic 3d Models of Cancer. *Adv. Drug Deliv. Rev.* **2014**, *79–80*, 68–78. [CrossRef]
12. Nishiguchi, A.; Matsusaki, M.; Asano, Y.; Shimoda, H.; Akashi, M. Effects of Angiogenic Factors and 3d-Microenvironments on Vascularization within Sandwich Cultures. *Biomaterials* **2014**, *35*, 4739–4748. [CrossRef]
13. Cubo, N.; Garcia, M.; del Cañizo, J.F.; Velasco, D.; Jorcano, J.L. 3d Bioprinting of Functional Human Skin: Production and In Vivo Analysis. *Biofabrication* **2016**, *9*, 015006. [CrossRef]
14. Cortes, E.D.; Molina, C.M.; Rodriguez-Lorenzo, L.; Cubo-Mateo, N. Generation of Controlled Micrometric Fibers inside Printed Scaffolds Using Standard Fdm 3d Printers. *Polymers* **2023**, *15*, 96. [CrossRef]
15. Hughes, A.M.; Kolb, A.D.; Shupp, A.B.; Shine, K.M.; Bussard, K.M. Printing the Pathway Forward in Bone Metastatic Cancer Research: Applications of 3d Engineered Models and Bioprinted Scaffolds to Recapitulate the Bone–Tumor Niche. *Cancers* **2021**, *13*, 507.
16. Pati, F.; Gantelius, J.; Svahn, H.A. 3d Bioprinting of Tissue/Organ Models. *Angew. Chem. Int. Ed.* **2016**, *55*, 4650–4665. [CrossRef]
17. Sánchez-Salazar, M.G.; Álvarez, M.M.; Santiago, G.T.-D. Advances in 3d Bioprinting for the Biofabrication of Tumor Models. *Bioprinting* **2021**, *21*, e00120. [CrossRef]
18. Herrada-Manchón, H.; Celada, L.; Rodríguez-González, D.; Fernández, M.A.; Aguilar, E.; Chiara, M.D. Three-Dimensional Bioprinted Cancer Models: A Powerful Platform for Investigating Tunneling Nanotube-Like Cell Structures in Complex Microenvironments. *Mater. Sci. Eng. C* **2021**, *128*, 112357. [CrossRef]
19. Sigaux, N.; Pourchet, L.; Breton, P.; Brosset, S.; Louvrier, A.; Marquette, C.A. 3D Bioprinting: Principles, fantasies and prospects. *Oral Maxillofac. Surg.* **2019**, *120*, 128–132.
20. Xu, Z.; Lam, M.T. Alginate Application for Heart and Cardiovascular Diseases. In *Alginates and Their Biomedical Applications*; Rehm, B.H.A., Moradali, M.F., Eds.; Springer: Singapore, 2018; pp. 185–212.
21. de Vos, P.; Lazarjani, H.A.; Poncelet, D.; Faas, M.M. Polymers in Cell Encapsulation from an Enveloped Cell Perspective. *Adv. Drug Deliv. Rev.* **2014**, *67–68*, 15–34. [CrossRef]
22. Dani, S.; Ahlfeld, T.; Albrecht, F.; Duin, S.; Kluger, P.; Lode, A.; Gelinsky, M. Homogeneous and Reproducible Mixing of Highly Viscous Biomaterial Inks and Cell Suspensions to Create Bioinks. *Gels* **2021**, *7*, 227. [CrossRef]
23. Ahlfeld, T.; Cubo-Mateo, N.; Cometta, S.; Guduric, V.; Vater, C.; Bernhardt, A.; Akkineni, A.R.; Lode, A.; Gelinsky, M. A Novel Plasma-Based Bioink Stimulates Cell Proliferation and Differentiation in Bioprinted, Mineralized Constructs. *ACS Appl. Mater. Interfaces* **2020**, *12*, 12557–12572. [CrossRef] [PubMed]
24. Ahlfeld, T.; Guduric, V.; Duin, S.; Akkineni, A.R.; Schütz, K.; Kilian, D.; Emmermacher, J.; Cubo-Mateo, N.; Dani, S.; Witzleben, M.V.; et al. Methylcellulose-a Versatile Printing Material That Enables Biofabrication of Tissue Equivalents with High Shape Fidelity. *Biomater. Sci.* **2020**, *8*, 2102–2110. [CrossRef]
25. Schütz, K.; Placht, A.-M.; Paul, B.; Brüggemeier, S.; Gelinsky, M.; Lode, A. Three-Dimensional Plotting of a Cell-Laden Alginate/Methylcellulose Blend: Towards Biofabrication of Tissue Engineering Constructs with Clinically Relevant Dimensions. *J. Tissue Eng. Regen. Med.* **2017**, *11*, 1574–1587.
26. Rasheed, Z.A.; Matsui, W.; Maitra, A. Pathology of Pancreatic Stroma in Pdac. In *Pancreatic Cancer and Tumor Microenvironment*; Munshi, H.G., Grippo, P.J., Eds.; Transworld Research Network: Trivandrum, India, 2012.
27. Gonzalez, I.; Luzuriaga, J.; Valdivieso, A.; Candil, M.; Frutos, J.; Lopez, J.; Hernandez, L.; Rodriguez-Lorenzo, L.; Yaguee, V.; Blanco, J.L.; et al. Low-Intensity Continuous Ultrasound to Inhibit Cancer Cell Migration. *Front. Cell Dev. Biol.* **2023**, *10*, 842965. [CrossRef] [PubMed]
28. Hernández-González, A.C.; Téllez-Jurado, L.; Rodríguez-Lorenzo, L.M. Synthesis of in-Situ Silica-Alginate Hybrid Hydrogels by a Sol-Gel Route. *Carbohydr. Polym.* **2020**, *250*, 116877. [CrossRef]
29. Chernecky, C.C.; Berger, B.J. *Laboratory Tests & Diagnostic Procedures*; Elsevier/Saunders: St. Louis, MO, USA, 2013.
30. Deer, E.L.; Gonzalez-Hernandez, J.; Coursen, J.D.; Shea, J.E.; Ngatia, J.; Scaife, C.L.; Firpo, M.A.; Mulvihill, S.J. Phenotype and Genotype of Pancreatic Cancer Cell Lines. *Pancreas* **2010**, *39*, 425–435. [CrossRef]
31. Cukjati, D.; Rebersek, S.; Miklavcic, D. A Reliable Method of Determining Wound Healing Rate. *Med. Biol. Eng. Comput.* **2001**, *39*, 263–271. [CrossRef]
32. Martínez Ávila, H.; Schwarz, S.; Rotter, N.; Gatenholm, P. 3d Bioprinting of Human Chondrocyte-Laden Nanocellulose Hydrogels for Patient-Specific Auricular Cartilage Regeneration. *Bioprinting* **2016**, *1–2*, 22–35. [CrossRef]
33. Li, H.; Liu, S.; Li, L. Rheological Study on 3d Printability of Alginate Hydrogel and Effect of Graphene Oxide. *Int. J. Bioprint.* **2016**, *2*, 54–66. [CrossRef]
34. Verma, M. Personalized Medicine and Cancer. *J. Pers. Med.* **2012**, *2*, 1–14. [CrossRef] [PubMed]
35. Rizzo, G.; Bertotti, A.; Leto, S.M.; Vetrano, S. Patient-Derived Tumor Models: A More Suitable Tool for Pre-Clinical Studies in Colorectal Cancer. *J. Exp. Clin. Cancer Res.* **2021**, *40*. [CrossRef]
36. Li, M.; Tian, X.; Kozinski, J.A.; Chen, X.; Hwang, D.K. Modeling Mechanical Cell Damage in the Bioprinting Process Employing a Conical Needle. *J. Mech. Med. Biol.* **2015**, *15*, 1550073. [CrossRef]
37. Mendes, B.B.; Gómez-Florit, M.; Hamilton, A.G.; Detamore, M.S.; Domingues, R.M.A.; Reis, R.L.; Gomes, M.E. Human Platelet Lysate-Based Nanocomposite Bioink for Bioprinting Hierarchical Fibrillar Structures. *Biofabrication* **2019**, *12*, 015012. [CrossRef]
38. Hernández-González, A.C.; Téllez-Jurado, L.; Rodríguez-Lorenzo, L.M. Alginate Hydrogels for Bone Tissue Engineering, from Injectables to Bioprinting: A Review. *Carbohydr. Polym.* **2020**, *229*, 115514.

39. Hölzl, K.; Lin, S.; Tytgat, L.; Van Vlierberghe, S.; Gu, L.; Ovsianikov, A. Bioink Properties before, During and after 3d Bioprinting. *Biofabrication* **2016**, *8*, 032002. [CrossRef]
40. Rubiano, A.; Delitto, D.; Han, S.; Gerber, M.; Galitz, C.; Trevino, J.; Thomas, R.M.; Hughes, S.J.; Simmons, C.S. Viscoelastic Properties of Human Pancreatic Tumors and in Vitro Constructs to Mimic Mechanical Properties. *Acta Biomater.* **2018**, *67*, 331–340. [CrossRef]
41. Nicolle, S.; Noguer, L.; Palierne, J.F. Shear Mechanical Properties of the Porcine Pancreas: Experiment and Analytical Modelling. *J. Mech. Behav. Biomed. Mater.* **2013**, *26*, 90–97. [CrossRef]
42. Axpe, E.; Oyen, M.L. Applications of Alginate-Based Bioinks in 3d Bioprinting. *Int. J. Mol. Sci.* **2016**, *17*, 1976. [CrossRef]
43. Kulseng, B.; Skjak-Braek, G.; Ryan, L.; Andersson, A.; King, A.; Faxvaag, A.; Espevik, T. Transplantation of Alginate Microcapsules-Generation of Antibodies against Alginates and Encapsulated Porcine Islet-Like Cell Clusters. *Transplantation* **1999**, *67*, 978–984. [CrossRef]
44. Ashworth, J.C.; Thompson, J.L.; James, J.R.; Slater, C.E.; Pijuan-Galitó, S.; Lis-Slimak, K.; Holley, R.J.; Meade, K.A.; Thompson, A.; Arkill, K.P.; et al. Peptide Gels of Fully-Defined Composition and Mechanics for Probing Cell-Cell and Cell-Matrix Interactions In Vitro. *Matrix Biol.* **2020**, *85–86*, 15–33. [CrossRef]
45. Freeman, F.E.; Kelly, D.J. Tuning Alginate Bioink Stiffness and Composition for Controlled Growth Factor Delivery and to Spatially Direct Msc Fate within Bioprinted Tissues. *Sci. Rep.* **2017**, *7*, 17042. [CrossRef] [PubMed]
46. Lucas, L.; Aravind, A.; Emma, P.; Christophe, M.; Edwin-Joffrey, C. Rheology, Simulation and Data Analysis toward Bioprinting Cell Viability Awareness. *Bioprinting* **2021**, *21*, e00119.
47. Blaeser, A.; Campos, D.F.D.; Puster, U.; Richtering, W.; Stevens, M.M.; Fischer, H. Controlling Shear Stress in 3d Bioprinting Is a Key Factor to Balance Printing Resolution and Stem Cell Integrity. *Adv. Healthc. Mater.* **2016**, *5*, 326–333. [CrossRef]
48. Cidonio, G.; Glinka, M.; Dawson, J.I.; Oreffo, R.O.C. The Cell in the Ink: Improving Biofabrication by Printing Stem Cells for Skeletal Regenerative Medicine. *Biomaterials* **2019**, *209*, 10–24. [CrossRef] [PubMed]
49. Müller, M.; Öztürk, E.; Arlov, Ø.; Gatenholm, P.; Zenobi-Wong, M. Alginate Sulfate–Nanocellulose Bioinks for Cartilage Bioprinting Applications. *Ann. Biomed. Eng.* **2017**, *45*, 210–223. [CrossRef] [PubMed]
50. Longati, P.; Jia, X.; Eimer, J.; Wagman, A.; Witt, M.-R.; Rehnmark, S.; Verbeke, C.; Toftgård, R.; Löhr, M.; Heuchel, R.L. 3D Pancreatic Carcinoma Spheroids Induce a Matrix-Rich, Chemoresistant Phenotype Offering a Better Model for Drug Testing. *BMC Cancer* **2013**, *13*, 95. [CrossRef]
51. Gorchs, L.; Ahmed, S.; Mayer, C.; Knauf, A.; Moro, C.F.; Svensson, M.; Heuchel, R.; Rangelova, E.; Bergman, P.; Kaipe, H. The Vitamin D Analogue Calcipotriol Promotes an Anti-Tumorigenic Phenotype of Human Pancreatic Cafs but Reduces T Cell Mediated Immunity. *Sci. Rep.* **2020**, *10*, 17444. [CrossRef]

polymers

Article

Tissue-Engineered Carotid Artery Interposition Grafts Demonstrate High Primary Patency and Promote Vascular Tissue Regeneration in the Ovine Model

Larisa V. Antonova [1], Evgenia O. Krivkina [1], Viktoriia V. Sevostianova [1,*], Andrey V. Mironov [1], Maria A. Rezvova [1], Amin R. Shabaev [1], Vadim O. Tkachenko [2], Sergey S. Krutitskiy [1], Mariam Yu. Khanova [1], Tatiana Yu. Sergeeva [1], Vera G. Matveeva [1], Tatiana V. Glushkova [1], Anton G. Kutikhin [1], Rinat A. Mukhamadiyarov [1], Nadezhda S. Deeva [1], Tatiana N. Akentieva [1], Maxim Yu. Sinitsky [1], Elena A. Velikanova [1] and Leonid S. Barbarash [1]

1   Research Institute for Complex Issues of Cardiovascular Diseases, 650002 Kemerovo, Russia; antonova.la@mail.ru (L.V.A.); leonora92@mail.ru (E.O.K.); a.mir.80@mail.ru (A.V.M.); rezvovamaria@mail.ru (M.A.R.); shabar@kemcardio.ru (A.R.S.); kss911@mail.ru (S.S.K.); khanovam@gmail.com (M.Y.K.); sergtu@kemcardio.ru (T.Y.S.); matveeva_vg@mail.ru (V.G.M.); bio.tvg@mail.ru (T.V.G.); antonkutikhin@gmail.com (A.G.K.); rem57@rambler.ru (R.A.M.); deevanadusha69@yandex.ru (N.S.D.); t.akentyeva@mail.ru (T.N.A.); max-sinitsky@rambler.ru (M.Y.S.); velikanova_ea@mail.ru (E.A.V.); reception@kemcardio.ru (L.S.B.)
2   Budker Institute of Nuclear Physics of Siberian Branch Russian Academy of Sciences, 630090 Novosibirsk, Russia; vtkachen@mail.ru
*   Correspondence: sevostv@gmail.com; Tel.: +7-9069356076

check for **updates**

Citation: Antonova, L.V.; Krivkina, E.O.; Sevostianova, V.V.; Mironov, A.V.; Rezvova, M.A.; Shabaev, A.R.; Tkachenko, V.O.; Krutitskiy, S.S.; Khanova, M.Y.; Sergeeva, T.Y.; et al. Tissue-Engineered Carotid Artery Interposition Grafts Demonstrate High Primary Patency and Promote Vascular Tissue Regeneration in the Ovine Model. *Polymers* **2021**, *13*, 2637. https://doi.org/10.3390/polym13162637

Academic Editors: João Carlos Silva and Frederico Castelo Ferreira

Received: 19 July 2021
Accepted: 4 August 2021
Published: 8 August 2021

Publisher's Note: MDPI stays neutral with regard to jurisdictional claims in published maps and institutional affiliations.

Abstract: Tissue-engineered vascular graft for the reconstruction of small arteries is still an unmet clinical need, despite the fact that a number of promising prototypes have entered preclinical development. Here we test Poly(3-hydroxybutyrate-co-3-hydroxyvalerate)Poly(ε-caprolactone) 4-mm-diameter vascular grafts equipped with vascular endothelial growth factor (VEGF), basic fibroblast growth factor (bFGF) and stromal cell-derived factor 1α (SDF-1α) and surface coated with heparin and iloprost (PHBV/PCL[VEGF-bFGF-SDF]$^{\text{Hep/Ilo}}$, $n = 8$) in a sheep carotid artery interposition model, using biostable vascular prostheses of expanded poly(tetrafluoroethylene) (ePTFE, $n = 5$) as a control. Primary patency of PHBV/PCL[VEGF-bFGF-SDF]$^{\text{Hep/Ilo}}$ grafts was 62.5% (5/8) at 24 h postimplantation and 50% (4/8) at 18 months postimplantation, while all (5/5) ePTFE conduits were occluded within the 24 h after the surgery. At 18 months postimplantation, PHBV/PCL[VEGF-bFGF-SDF]$^{\text{Hep/Ilo}}$ grafts were completely resorbed and replaced by the vascular tissue. Regenerated arteries displayed a hierarchical three-layer structure similar to the native blood vessels, being fully endothelialised, highly vascularised and populated by vascular smooth muscle cells and macrophages. The most (4/5, 80%) of the regenerated arteries were free of calcifications but suffered from the aneurysmatic dilation. Therefore, biodegradable PHBV/PCL[VEGF-bFGF-SDF]$^{\text{Hep/Ilo}}$ grafts showed better short- and long-term results than bio-stable ePTFE analogues, although these scaffolds must be reinforced for the efficient prevention of aneurysms.

Keywords: tissue-engineered vascular graft; preclinical development; ovine model; carotid artery; primary patency

1. Introduction

Despite immense efforts to reduce morbidity from atherosclerotic vascular disease [1] which have resulted in a steady decrease in the number of coronary artery bypass grafting procedures worldwide, this treatment modality remains common (82 procedures per 100,000 US adults annually) [2,3]. Vascular bypass implies the use of autologous blood vessel conduits (e.g., saphenous vein or internal mammary artery (IMA)) [4,5] while other types of arterial reconstruction involve biostable tubular scaffolds (e.g., Poly(ethylene terephthalate) (PET), expanded poly(tetrafluoroethylene) (ePTFE), or polyurethane prostheses) [6,7]. Yet, the limited availability of autografts (because of prior surgery, extensive atherosclerosis, or anatomical incompatibility) and high rate of thrombotic occlusion and neointimal hyperplasia in small diameter biostable prostheses limit their use in cardiovascular surgery [8,9]. Therefore, biodegradable, tissue-engineered vascular grafts (TEVGs) for the guided regeneration of vascular tissue have become a mainstream approach [8,9].

The fabrication of a commercially available off-the-shelf vascular graft requires excellent biocompatibility of its polymer composition for successful adhesion, migration, and proliferation of peripheral blood-derived cells, as well as to avert calcification [9–11]. Further, the appropriate prototype must possess a good haemocompatibility to prevent acute thrombosis and, ideally, should have a compliance profile similar with native coronary artery or IMA for precluding aneurysms and neointimal hyperplasia [9–11]. Current additive manufacturing technologies permit layer-by-layer incorporation of bioactive factors (e.g., growth factors or chemokines) to the tubular scaffold and their controlled release to guide endothelial/vascular smooth muscle specification and vascular tissue regeneration [11]. Rapid endothelialisation, propagation of contractile and synthetic mesenchymal cells (which can have smooth muscle or fibroblast identity), and production of the extracellular matrix (ECM) are key factors underlying high long-term primary patency of TEVGs [11–13].

Though natural polymers (e.g., collagen, elastin, or Poly(3-hydroxybutyrate-co-3-hydroxyvalerate, PHBV) generally have high biocompatibility, their durability is insufficient to resist the physiological blood pressure [11,12] that requires the addition of synthetic polymers (e.g., poly(ε-caprolactone), PCL) into the blend [14,15]. While endowing the scaffolds with superior mechanical properties, this manoeuvre may negatively affect cell differentiation. Together with turbulent flow conditions and low shear stress, increased rigidity of the grafts results in inadequate cyclic stretch and provokes a contractile-to-synthetic phenotype reprogramming of vascular smooth muscle cells, eventually leading to neointimal hyperplasia [16–18]. Moreover, synthetic vascular smooth muscle cells may further acquire osteochondrogenic specification that might lead to the development of calcifications which cause delamination of the extracellular matrix and are associated with the aneurysmatic dilation of the regenerated blood vessel [18,19]. Another shortcoming of adding synthetic polymers is an excessive infiltration of the graft by immune cells which thereby represent a majority of the cells within the regenerating blood vessel [20–22]. Further, the blending of natural and synthetic polymers does not improve haemocompatibility of the grafts, as rapid and complete endothelialisation of long vascular grafts, which is an important prerequisite to prevent thrombosis, is relatively rare [13,23,24]. To overcome this drawback, antiplatelet or anticoagulant drugs might be conjugated to a scaffold luminal surface [13].

Our group has previously designed an electrospun, small (4 mm) diameter PHBV/PCL vascular graft which employs the principle of guided vascular tissue regeneration through the release of a pro-angiogenic molecule vascular endothelial growth factor (VEGF) incorporated into the inner layer and a potent chemokine stromal cell-derived factor 1α (SDF-1α) incorporated into the outer layer along with another pro-angiogenic protein basic fibroblast growth factor (bFGF) [25–27]. To improve the hemocompatibility of this scaffold, here we attached an anticoagulant heparin (Hep) and a vasodilator/antiplatelet drug iloprost (Ilo) to its luminal surface (PHBV/PCL[VEGF-bFGF-SDF]$^{Hep/Ilo}$). Biostable ePTFE vascular prostheses, which are frequently used in cardiovascular surgery, were selected as a control group [6,7]. For the proper assessment of the long-term primary patency and ultrastructural features, we used a sheep carotid artery interposition model since ovine arterial anatomy, haemodynamic conditions, and coagulation are similar to humans, while the long neck provides easy surgical access [28–30]. In addition, sheep exhibit accelerated rates of vascular calcification that allows evaluating the long-term prosthetic complications as soon as 1–2 years postimplantation [8].

In this study, we found that PHBV/PCL[VEGF-bFGF-SDF]$^{Hep/Ilo}$ grafts demonstrated promising primary patency rate (62.5% and 50% at 24 h and 18 months postimplantation) in contrast to ePTFE conduits all of which were occluded within 24 h after the surgery. The regenerated arteries demonstrated a complete endothelialisation, high vascularisation and hierarchical multilayer structure reminiscent of the native blood vessels. Hence, PHBV/PCL[VEGF-bFGF-SDF]$^{Hep/Ilo}$ grafts might be considered as a potential candidate for the further improvement and additional preclinical testing. However, PHBV/PCL[VEGF-bFGF-SDF]$^{Hep/Ilo}$ grafts suffered from aneurysms, suggesting the need in their mechanical reinforcement.

2. Materials and Methods

2.1. Fabrication of TEVGs

Biodegradable TEVGs (4 mm diameter, ≈400 μm thickness and 4 cm length) were fabricated using emulsion electrospinning (Nanon-01A, MECC, Tokyo, Japan) from PHBV (403105, Sigma-Aldrich, Saint Louis, MO, USA): PCL (440744, Sigma-Aldrich, Saint Louis, MO, USA) (5:10%)/chloroform (366927, Sigma-Aldrich, Saint Louis, MO, USA) solution using the following parameters: 23 kV voltage, 0.5 mL/h feed rate, 2 mm rotating drum diameter, 22G needle, and 150 mm tip-to-collector distance. Abovementioned polymer ratio was determined in our previous studies [25,31,32]. In all these investigations, PHBV/PCL vascular grafts did not show any signs of dissolution as long as 1 year after implantation into rat abdominal aorta. Either VEGF (V7259, Sigma-Aldrich, St. Louis, MO, USA), bFGF (SRP4037, Sigma-Aldrich, St. Louis, MO, USA), or SDF-1α (SRP3276, Sigma-Aldrich, St. Louis, MO, USA) were dissolved in phosphate buffered saline (10010023, Thermo Fisher Scientific, Waltman, MA, USA) to 10 μg/mL concentration and then added to PHBV/PCL/chloroform solution (1:20), with the final concentration of 500 ng/mL. Grafts with the combination of VEGF, bFGF, and SDF-1α were two-layered, with the inner layer fabricated using 27G needle and containing VEGF (500 ng/mL) and the outer layer prepared utilizing 22G needle and containing bFGF and SDF-1α (500 ng/mL each).

2.2. Antithrombotic Modification of TEVG Luminal Surface

To increase the haemocompatibility of PHBV/PCL[VEGF-bFGF-SDF] vascular grafts, we first immobilised a PVP hydrogel at their luminal surface by incubating the prostheses in 5% alcoholic solution of PVP (K90, PanReac AppliChem, Darmstadt, Germany) for 30 min, drying at room temperature (20–24 °C) and sterile conditions for 24 h, and irradiating in the argon atmosphere by a linear particle accelerator (ILU-10, 50 kGy, electron energy 5 mEV, beam power 50 kW, Budker Institute of Nuclear Physics of Siberian Branch Russian Academy of Sciences, Novosibirsk, Russia). Grafts were then incubated in a 0.1 M glycine solution (pH = 2.6, 410225, Sigma-Aldrich, St. Louis, MO, USA) of heparin (125 IE/mL, Moscow Endocrine Plant, Moscow, Russia) and iloprost (0.2 μg/mL, Bayer, Barcelona-, Spain) for 30 min and dried at room temperature (20–24 °C) and sterile conditions for 24 h.

2.3. Evaluation of the Mechanical Properties

To evaluate the mechanical properties of TEVGs before and after Hep/Ilo immobilisation, uniaxial tension test was performed. Grafts were cut in the longitudinal axis using a custom-shaped knife in the Zwick/Roell cutting press (Zwick/Roell, Ulm, Germany). Segments of human internal mammary artery (length = 10 mm), excised during the coronary artery bypass grafting, were used for the control purposes. Tests were performed on the Z-series universal testing machine (Zwick/Roell, Ulm, Germany) with a nominal force of 50 N, limit of permissible error of ±1% and crosshead speed of 50 mm/min. We evaluated ultimate tensile strength, ultimate tensile force, elongation at break and elastic modulus; the latter was measured within the range of physiological pressure (80–120 mmHg). Prior to tensile testing, graft samples were not sterilised.

2.4. Haemocompatibility Testing

Assessment of haemolysis and platelet aggregation upon contact of the blood with a polymer surface was performed according to the ISO 10993.4 standard. For the haemolysis testing, fresh donor blood was collected into the improvacuter tubes (Guangzhou Improve Medical Instruments, Guangzhou, China) containing 3.8% sodium citrate. The citrate-to-blood ratio was 1:9. Polymer (PHBV/PCL[VEGF-bFGF-SDF], PHBV/PCL[VEGF-bFGF-SDF]$^{\text{Hep/Ilo}}$, and ePTFE) samples (25 cm^2, n = 5) were placed into a 10 mL physiological saline (0.9% NaCl solution, Binnofarm, Zelenograd, Russia) and incubated at 37 °C for 2 h to wet TEVGs for ensuring their proper interactions with red blood cells. Then, 200 mL citrated blood was added to each sample followed by repeated incubation at 37 °C for 1 h. Red blood cells were sedimented by centrifugation of the samples at 1446× g (2800 rpm, ELMI CM-6M, ELMI, Riga, Latvia) for 10 min. Optical density of the solution was measured using GENESYS 6 spectrophotometer (Thermo Scientific, Waltham, MA, USA) at 545 nm wavelength. Sample-induced haemolysis was reported as a percentage normalised to the positive control (double distilled water). Physiological saline was used as a negative control, as it is isotonic and therefore does not cause swelling or shrinking of red blood cells.

For the platelet aggregation analysis, we isolated platelet-rich plasma by the centrifugation of the blood at 184× g (1000 rpm, ELMI CM-6M, ELMI, Riga, Latvia) for 10 min. To measure platelet aggregation, we added 250 mL platelet-rich plasma to the polymer (PHBV/PCL[VEGF-bFGF-SDF], PHBV/PCL[VEGF-bFGF-SDF]$^{\text{Hep/Ilo}}$, and ePTFE) samples (0.5 × 0.5 cm^2) for 3 min and then added 25 µL 0.025 M CaCl$_2$ (383147, Sigma-Aldrich, St. Louis, MO, USA) after removing the samples to activate platelet aggregation. Measurement was performed using APACT 4004 platelet aggregometer (LABiTec, Ahrensburg, Germany). Intact platelet-rich and platelet-poor plasma were used as a positive and negative control, respectively. Platelet-poor plasma was obtained by the centrifugation of platelet-rich plasma at 2260× g (3500 rpm, ELMI CM-6M, ELMI, Riga, Latvia) for 20 min.

2.5. Visualisation of TEVG Structure

An assessment of polymer (PHBV/PCL[VEGF-bFGF-SDF], PHBV/PCL[VEGF-bFGF-SDF]$^{\text{Hep/Ilo}}$, and ePTFE) samples (0.5 × 0.5 cm^2) was conducted following gold-palladium sputter coating (15 nm thickness, EM ACE200, Leica Microsystems, Wetzlar, Germany) by scanning electron microscopy (Hitachi S-3400N, Hitachi, Tokyo, Japan) in high vacuum mode at 10 kV voltage.

2.6. Animal Model

The study protocol was approved by the local ethical committee of the Research Institute for Complex Issues of Cardiovascular Diseases (protocol number 20180305, 28.04.2016, Kemerovo, Russia). Animal experiments were performed in accordance with the European Convention for the Protection of Vertebrate Animals (Strasbourg, 1986). For the in vivo experiments, we used female Edilbay sheep of 42–45 kg body weight which were received from the Animal Core Facility of the Research Institute for Complex Issues of Cardiovascular Diseases (Kemerovo, Russia). Sheep were selected for the surgery by Doppler ultrasonography to identify those having carotid artery diameter of 3.8–4.2 mm. Among the tested samples were biodegradable PHBV/PCL[VEGF-bFGF-SDF]$^{\text{Hep/Ilo}}$ TEVGs (n = 8, 7 grafts implanted for 18 months and 1 graft implanted for 6 months to evaluate a biodegradation rate) and biostable ePTFE vascular prostheses (n = 5, all implanted for 6 months, GORE-TEX, ST04010A, W. L. Gore and Associates, Newark, DE, USA).

Anaesthesia was induced with zoletil (Virbac, Carros, France) and maintained by artificial ventilation with sevoflurane (Baxter International, Deerfield, IL, USA) with constant control of heart rate, respiratory rate, and oxygen saturation. Upon achieving access to the carotid artery and intravenous infusion of heparin (5000 IU), we clamped the artery, excised a 4 cm segment, performed end-to-end implantation of a vascular prosthesis using the twisted seam (Prolene 6-0, Ethicon, Somerville, NJ, USA), and conducted the wound closure (Vicryl 2-0, Ethicon, Somerville, NJ, USA). Postoperative care included daily infusions of cefuroxime (Medochemie, Limassol, Cyprus). Graft patency was assessed by Doppler ultrasonography immediately after the surgery and then at the following time points: 24 h, 3 months, 6 months, 12 months, and 18 months postoperation.

2.7. Histological, Immunofluorescent, and Ultrastructural Examination

At 6 or 18 months postoperation, sheep were sacrificed. At the site of the implantation, the graft (or the regenerated artery with the scaffold remnants) was excised and cut into four segments of the equal length.

The first segment was fixed in two changes of 10% neutral phosphate buffered formalin (B06-003, BioVitrum, St. Petersburg, Russia) for 24 h at 4 °C, dehydrated in ascending ethanol series (70, 80, and 95%, 1 h per each) and isopropanol (1 h, 06-002, BioVitrum, St. Petersburg, Russia), impregnated and embedded into paraffin (56 °C, 3 changes, 1 h per each, Paraplast REGULAR, 39601006, Leica), cooled at 4 °C overnight and cut (5 μm sections) on a microtome (Microm HM 325, Thermo Scientific, Waltham, MA, USA). To ensure the proper histological examination, we prepared 12 sections, evenly distributed across the entire excised segment, per slide. Upon the deparaffinisation in three changes of xylene (23400, Electron Microscopy Sciences, Hatfield, PA, USA) and three changes of 95% ethanol, sections were stained with: (1) haematoxylin and eosin (ab245880, Abcam, Cambridge, UK) according to the manufacturer's protocol for the general examination; (2) van Gieson stain (21-020, BioVitrum, St. Petersburg, Russia) as in [31] to distinguish connective and smooth muscle tissue; (3) 2% aqueous alizarin red S (ab146374, Abcam, Cambridge, UK) and DAPI (10 μg/mL, D9542, Sigma-Aldrich, St. Louis, MO, USA) as in [31] for the detection of calcium deposits within the grafts. Visualisation was performed by light or fluorescent microscopy (AxioImager.A1, Carl Zeiss, Oberkochen, Germany).

The second segment was snap-frozen in the optimal cutting temperature medium (Tissue-Tek, 4583, Sakura Finetek, Tokyo, Japan) and cut on a cryostat (Microm HM 525, Thermo Scientific, Waltham, MA, USA) as described above. Sections (5 μm thickness) were then stained with rabbit anti-CD31 (ab28364, Abcam, Cambridge, UK) and mouse anti-α-SMA (ab7817, Abcam, Cambridge, UK); or rabbit anti-vWF (ab6994, Abcam, Cambridge, UK); or rabbit anti-Collagen IV (ab6586, Abcam, Cambridge, UK) and mouse anti-Collagen I (ab23446, Abcam, Cambridge, UK); or rabbit anti-collagen III (NB600-594, Novus Biologicals, Centennial, CO, USA) primary antibodies. Samples were further treated with goat antirabbit highly cross-adsorbed Alexa Fluor 488-conjugated (A11034, Thermo Fisher Scientific, Waltham, MA, USA) and donkey antimouse highly cross-adsorbed Alexa Fluor 555-conjugated (A31570, Thermo Fisher Scientific, Waltham, MA, USA) secondary antibodies. Counterstaining was performed with DAPI (10 μg/mL, 30 min, D9542, Sigma-Aldrich, St. Louis, MO, USA). Visualisation was performed by a confocal microscopy (LSM700, Carl Zeiss, Oberkochen, Germany).

The third segment was fixed in two changes of 10% neutral phosphate buffered formalin for 24 h at 4 °C, postfixed in 1% phosphate buffered osmium tetroxide (OsO4, 19110, Electron Microscopy Sciences, Hatfield, PA, USA) for 24 h, stained in 2% aqueous osmium tetroxide for 48 h, dehydrated in ascending ethanol series (50, 60, 70, 80 and 95%, 15 min per each), stained in 2% alcoholic uranyl acetate (22400-2, Electron Microscopy Sciences, Hatfield, PA, USA) for 5 h, dehydrated in isopropanol for 5 h and acetone (1 h), impregnated with acetone: epoxy resin (Epon, 14120, Electron Microscopy Sciences, Hatfield, PA, USA) mixture (1:1) for 6 h and with epoxy resin for 24 h, and finally embedded into the fresh epoxy resin at 60 °C. Samples were then ground, polished (TegraPol-11, Struers, Copenhagen, Denmark), and counterstained with Reynolds's lead citrate (17810, Electron Microscopy Sciences, Hatfield, PA, USA) for 15 min. After a brief washing in double distilled water, samples were sputter coated (10 nm thickness) with carbon (EM ACE200, Leica, Wetzlar, Germany) and visualised by means of backscattered scanning electron microscopy at 10 or 15 kV voltage (S-3400N, Hitachi, Tokyo, Japan). Elemental analysis was carried out by energy-dispersive X-ray spectroscopy (XFlash 4010, Bruker, Billerica, MA, USA) in a low vacuum mode at 15 kV voltage.

2.8. Transcriptional Profiling

At 18 months postimplantation, the fourth segment of the regenerated carotid arteries replacing the TEVG, and intact contralateral carotid arteries was flushed with TRIzol Reagent (15596018, Thermo Fisher Scientific, Waltham, MA, USA) to collect endothelial RNA. Then, de-endothelialised blood vessels were homogenised in TRIzol Reagent (FastPrep-24 Instrument and Lysing Matrix S, 116925050-CF, MP Biomedicals, Irvine, CA, USA) to extract the remaining RNA of other vascular cell populations. Upon the RNA isolation according to the manufacturer's protocol, reverse transcription (RT) was carried out utilising High-Capacity cDNA Reverse Transcription Kit (4368814, Thermo Fisher Scientific, Waltham, MA, USA). Gene expression was measured by a quantitative polymerase chain reaction (RT-qPCR) using the customised primers (Table 1). Primers were produced using ABI 3900 high-throughput DNA synthesiser (Thermo Fisher Scientific, Waltham, MA, USA) at Evrogen (Moscow, Russia). To perform RT-qPCR, primers (500 nmol/L each), cDNA (20 ng) and PowerUp SYBR Green Master Mix (A25778, Thermo Fisher Scientific, Waltham, MA, USA) were mixed and incubated according to the manufacturer's protocol for Tm $\geq$ 60 °C (fast cycling mode). Technical replicates (n = 3 per each sample of the endothelial lysate or homogenate of the de-endothelialised vascular tissue) were performed in all RT-qPCR experiments. The reaction was considered successful if its efficiency was 90–105% and R2 was $\geq$0.98. Quantification of the mRNA levels (IL1B, IL6, CXCL8, ICAM1, MMP2, KDR, NR2F2, and SNAI2 genes) in the indicated samples was performed by using the $2^{-\Delta\Delta Ct}$ method. Relative transcript levels in the endothelial lysate or de-endothelialised vascular tissue within the regenerated artery replacing the TEVG were expressed as a value relative to the average of two housekeeping genes (GAPDH and B2M) and to the respective RNA fractions of the intact contralateral carotid artery ($2^{-\Delta\Delta Ct}$). Alternatively, relative transcript levels in the endothelial lysate of the regenerated artery were expressed as a value relative to the same housekeeping genes and to the corresponding de-endothelialised vascular tissue. The adjusted values were finally represented as a heat map (green, grey and red colours reflected fold change $\leq$0.50, 0.51–1.99, and $\geq$2.00, respectively).

Table 1. Sequences of customised primers for RT-qPCR.

Gene	Forward Primer Sequence	Reverse Primer Sequence
IL1B	5′-TGCTGAAGGCTCTCCACCTC-3′	5′-ACCCAAGGCCACAGGAATCTT-3′
IL6	5′-TGTCATGGAGTTGCAGAGCAGT-3′	5′-CCAGCATGTCAGTGTGTGTGG-3′
CXCL8	5′-CTTCCAAGCTGGCTGTTGCTC-3′	5′-ATTTGGGGTGGAAAGGTGTGG-3′
ICAM1	5′-GTCACGGGGAACAGATTGTAGC-3′	5′-TGAGTTCTTCACCCACAGGCT-3′
MMP2	5′-ACCCCGCTACGGTTTTCTCG-3′	5′-ATGAGCCAGGAGCCCGTCTT-3′
KDR	5′-ACAGAACCAAGTTAGCCCCATC-3′	5′-TCGCTGGAGTACACAGTGGTG-3′
NR2F2	5′-GCAAGCGGTTTGGGACCTT-3′	5′-GGACAGGTAGGAGTGGCAGTTG-3′
SNAI2	5′-ACCCTGGTTACTGCAAGGACA-3′	5′-GAGCCCTCAGATTGGACCTG-3′
GAPDH	5′-TGGTGAAGGTCGGAGTGAACG-3′	5′-AGGGGTCATTGATGGCAACG-3′
B2M	5′-CCTTCTGTCCCACGCTGAGT-3′	5′-TGGTGCTGCTTAGAGGTCTCG-3′

2.9. Statistical Analysis

Statistical analysis was performed using GraphPad Prism 8 (GraphPad Software, San Diego, CA, USA). Data were presented as the median and interquartile range. Independent groups were compared using Kruskal-Wallis test with Dunn's multiple comparisons test. p values of $\leq$0.05 were regarded as statistically significant.

3. Results

3.1. Hep/Ilo-Treated TEVGs Are Mechanically Competent

To demonstrate an adequate performance within the living organism, TEVGs are required to withstand an initial hydrodynamic pressure burst (i.e., to show a sufficient durability) and to properly adapt to the repeated changes of the blood pressure occurring during the cardiac cycle (i.e., to exhibit a considerable elasticity). We first conducted a tensile testing of PHBV/PCL[VEGF-bFGF-SDF]$^{Hep/Ilo}$ grafts to reveal whether the Hep/Ilo attachment affects their mechanical properties. As a kind of control, we used human internal mammary artery which is often applied for the arterial replacement [5]. Stiffness (reflected by the elastic modulus) of PHBV/PCL[VEGF-bFGF-SDF]$^{Hep/Ilo}$ grafts were 6- and 21-fold higher than in unmodified PHBV/PCL[VEGF-bFGF-SDF] prostheses and IMA, respectively (Table 2). Likewise, chemical immobilisation of heparin and iloprost increased durability but slightly reduced elasticity of TEVGs which still exceeded those of IMA (Table 2). Expectedly, the durability and elasticity of the biostable ePTFE prostheses was significantly higher than in TEVGs and IMA, whereas their stiffness was close to that of the latter conduit (Table 2).

Table 2. Mechanical properties of PHBV/PCL[VEGF-bFGF-SDF] vascular grafts before and after Hep/Ilo coating in comparison with ePTFE prostheses and IMA.

	PHBV/PCL [VEGF-bFGF-SDF] (*n* = 9) Median (IQR)	PHBV/PCL [VEGF-bFGF-SDF]$^{Hep/Ilo}$ (*n* = 9) Median (IQR)	ePTFE (*n* = 9) Median (IQR)	IMA (*n* = 9) Median (IQR)
Ultimate Tensile Strength, MPa	3.05 (2.90; 3.20) [&]	3.94 (3.78–3.99) [&]	22.95 (22.42–23.47) [**]	2.48 (1.36–3.25) [&]
Ultimate Tensile Force, N	2.30 (2.20; 2.50) [#,&]	3.08 (2.94–3.30) [#,&]	21.10 (20.60–21.60) [**]	0.92 (0.59–1.72) [&]
Elongation at Break, %	121.70 (117.1; 129.6) [#,&]	109,17 (92.29–116.06) [#,&]	337.00 (332.00–341.80) [**]	29.72 (23.51–39.62) [&]
Elastic Modulus, MPa	8.60 (8.00; 9.64) [#,&]	49.95 (44.90–54.70) [*,#,&]	1.98 (1.36–2.59)	2.42 (1.87–3.19)
Sample Thickness, mm	0.36 (0.34–0.39)	0.39 (0.39–0.40)	0.46 (0.46–0.46)	0.27 (0.24–0.30)

PHBV/PCL—poly(3-hydroxybutyrate-co-3-hydroxyvalerate), PCL—poly(ε-caprolactone), [VEGF-bFGF-SDF]—combination of vascular endothelial growth factor, basic fibroblast growth factor and stromal cell-derived factor 1α, Hep/Ilo—modification with heparin and iloprost, ePTFE—expanded poly(tetrafluoroethylene), IMA—internal mammary artery, IQR—interquartile range. [*] $p < 0.05$ compared with PHBV/PCL[VEGF-bFGF-SDF] grafts, [#] $p < 0.05$ compared with IMA, [&] $p < 0.05$ compared with ePTFE prostheses, [**] $p < 0.05$ compared with all groups.

3.2. Conjugation of Hep/Ilo with the Luminal Surface of TEVGs Improves Their Haemocompatibility

Haemocompatibility is a key factor in providing short-term patency of the graft [24,33]. Thrombotic occlusion is primarily caused by a platelet aggregation [24,33] which has been measured upon the contact of platelet-rich plasma with unmodified PHBV/PCL[VEGF-bFGF-SDF] grafts or PHBV/PCL[VEGF-bFGF-SDF]$^{Hep/Ilo}$ grafts having an antithrombotic coating. Control PHBV/PCL[VEGF-bFGF-SDF] grafts induced a considerable aggregation of platelets similar to the intact platelet-rich plasma upon the calcium chloride activation (Table 2). Immobilisation of heparin and iloprost at the luminal surface resulted in a 2–3 fold decrease in platelet aggregation (Table 3). Haemolysis did not exceed 0.4% in all experimental groups testifying to the high haemocompatibility of PHBV/PCL[VEGF-bFGF-SDF]$^{Hep/Ilo}$ grafts (Table 3).

Table 3. Haemolysis and platelet aggregation measured upon the contact of the blood or platelet-rich plasma with vascular grafts.

Graft Type	Haemolysis, % Median (IQR)	Maximum Platelet Aggregation, % Median (IQR)
Platelet-rich plasma	-	14.61 (13.63–17.72)
PHBV/PCL[VEGF-bFGF-SDF]	0 (0–0)	17.25 (16.30–17.96)
PHBV/PCL[VEGF-bFGF-SDF][Hep/Ilo]	0.36 (0.36–0.36) **	8.22 (8.13–8.78) *
ePTFE	0.33 (0.21–2.40) **	22.74 (22.45–24.52)

PHBV/PCL—poly(3-hydroxybutyrate-co-3-hydroxyvalerate), PCL—poly(ε-caprolactone), [VEGF-bFGF-SDF]—combination of vascular endothelial growth factor, basic fibroblast growth factor and stromal cell-derived factor 1α, Hep/Ilo—modification with heparin and iloprost, ePTFE—expanded poly(tetrafluoroethylene), IQR—interquartile range. * $p < 0.05$ compared with platelet-rich plasma, PHBV/PCL[VEGF-bFGF-SDF] and ePTFE grafts, ** $p < 0.05$ compared with platelet-rich plasma and PHBV/PCL[VEGF-bFGF-SDF] grafts.

3.3. Immobilisation of Hep/Ilo at the Luminal Surface of TEVGs by Poly(N-vinylpyrrolidone) Preserves Their Structure

Structurally, electrospun PHBV/PCL[VEGF-bFGF-SDF][Hep/Ilo] scaffolds represented a layered microfiber network (Figure 1A,B) which consisted of multiple interconnected pores (Figure 1C,D). Binding of Hep/Ilo to the luminal surface of the PHBV/PCL[VEGF-bFGF-SDF] grafts is achieved via its hydrophilic coating with poly(N-vinylpyrrolidone) (PVP) hydrogel which also occupies the pores (Figure 1C) and prevents protein adsorption [34], potentially being one of the factors behind the reduced induction of platelet aggregation by PHBV/PCL[VEGF-bFGF-SDF][Hep/Ilo] grafts. Experimental washing from PVP remained the pores intact (Figure 1D), confirming retained integrity of the prostheses upon the chemical immobilisation of heparin and iloprost. In contrast to TEVGs, biostable prostheses were composed of wide and thick ePTFE fragments and large pores (Figure 1E,F).

3.4. Hep/Ilo Coating Endows TEVGs with a Considerable Long-Term Primary Patency

We next implanted PHBV/PCL[VEGF-bFGF-SDF][Hep/Ilo] grafts ($n = 8$) into the sheep carotid artery to test their haemo- and bio- compatibility in terms of primary patency which was assessed at 24 h, 3, 6, 12, and 18 months postimplantation by Doppler ultrasonography. Short-term (24 h postimplantation, Figure 2A) and mid-term (3 and 6 months postimplantation, Figure 2B,C) patency of PHBV/PCL[VEGF-bFGF-SDF][Hep/Ilo] grafts was 65% (5 out of 8). At 12 months postimplantation, one of the scaffolds became occluded. Hence, long-term (18 months postimplantation) primary patency of PHBV/PCL[VEGF-bFGF-SDF][Hep/Ilo] grafts was 50% (4/8) (Figure 2D).

At 6 months postimplantation, the diameter of 80% (4/5) of the patent PHBV/PCL[VEGF-bFGF-SDF][Hep/Ilo] grafts increased from 4 to 22 mm, which was indicative of aneurysm formation (Figure 3A,B). The remaining graft without an aneurysm was stenosed and characterised by a low blood flow. A histological examination of the aneurysmatic grafts revealed a neointima (Figure 3C,D) and large amounts of regenerated vascular tissue which was particularly prominent in the medial layer (Figure 3E,F), suggesting an ongoing replacement of the PHBV/PCL scaffold. Resorption of the prosthetic tissue was confirmed by the abundant connective tissue around the polymer fibres (Figure 3D,F). No signs of calcium deposition have been revealed both in the neointima and within the degrading polymer scaffold (Figure 3G,H). The rapid resorption of the scaffold contrasted to both data described in the literature (2–3 years for PCL) [35,36] and our previous results from a rat aortic interposition model [25–27]. The development of the aneurysms indicated the immaturity of the regenerated artery.

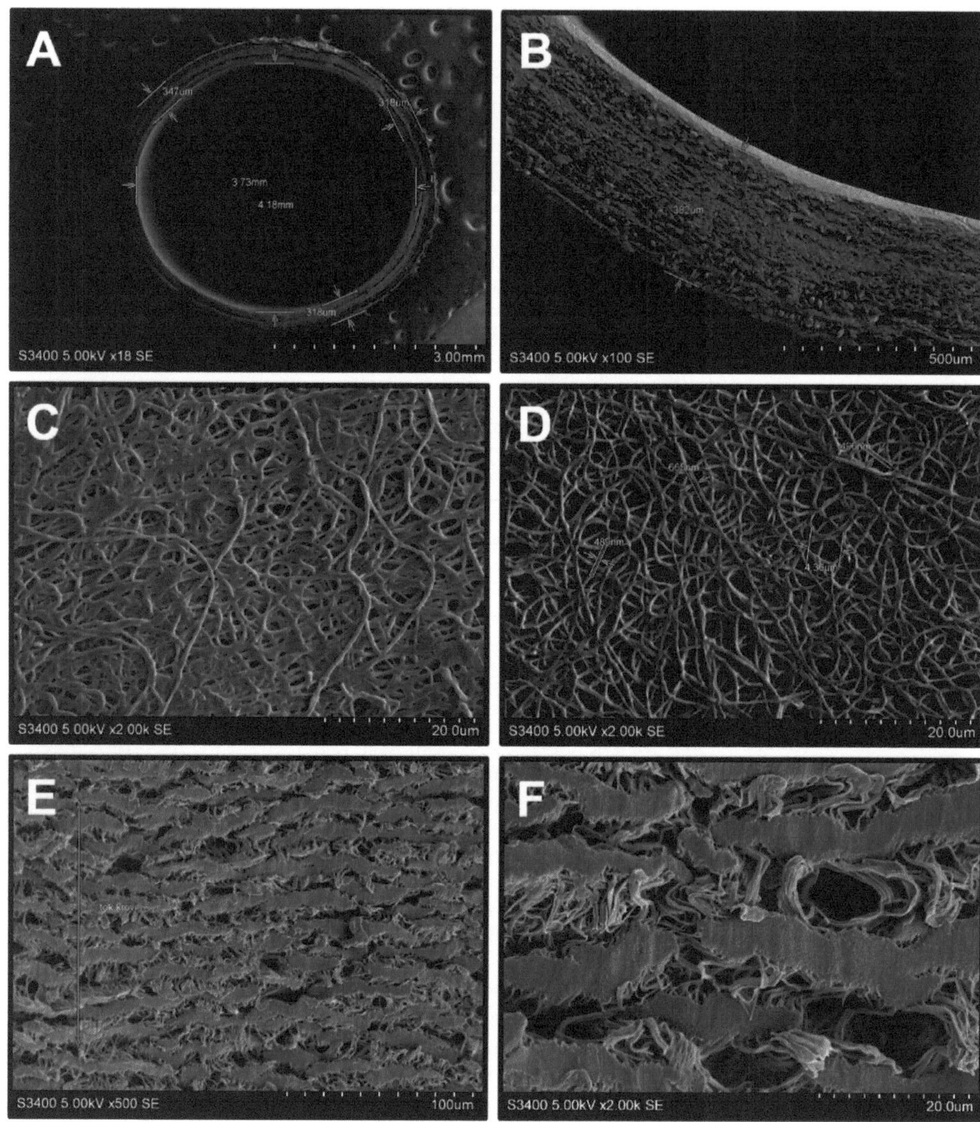

Figure 1. Scanning electron microscopy of biodegradable and synthetic vascular grafts: (**A**) Layer-by-layer hierarchical structure of the PHBV/PCL[VEGF-bFGF-SDF]$^{Hep/Ilo}$ grafts, view from above, ×18 magnification; (**B**) Layer-by-layer hierarchical structure of the PHBV/PCL[VEGF-bFGF-SDF]$^{Hep/Ilo}$ grafts, side view, ×100 magnification; (**C**) Luminal surface of the PHBV/PCL[VEGF-bFGF-SDF]$^{Hep/Ilo}$ grafts before the washing from PVP, ×2000 magnification; (**D**) Luminal surface of the PHBV/PCL[VEGF-bFGF-SDF]$^{Hep/Ilo}$ grafts after the washing from PVP, ×2000 magnification; (**E**) Luminal surface of ePTFE tubular scaffold, ×500 magnification; (**F**) Luminal surface of ePTFE tubular scaffold, ×2000 magnification. Representative images.

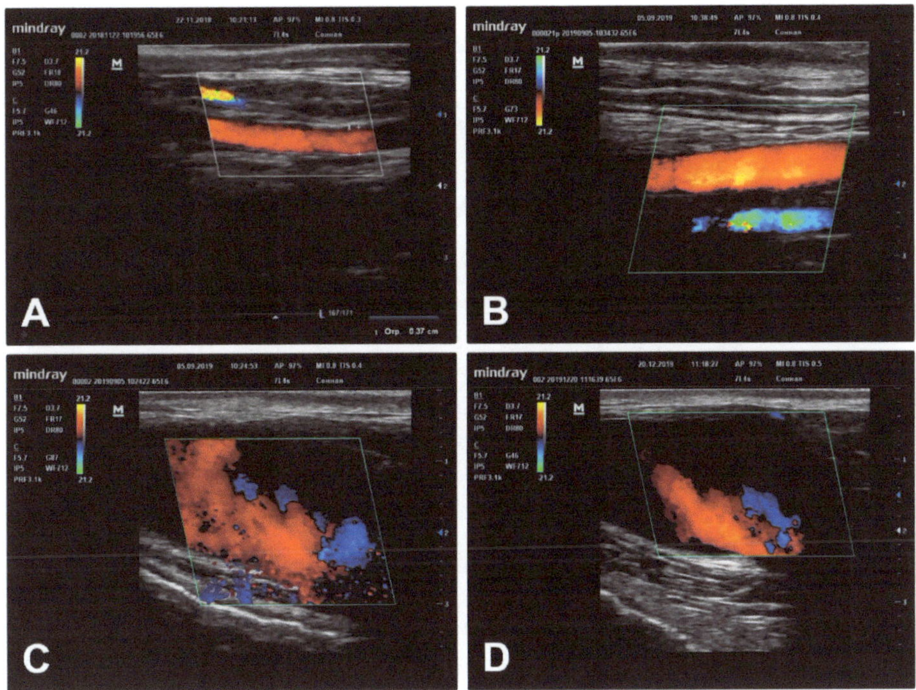

Figure 2. Time-resolved Doppler ultrasound images of the PHBV/PCL[VEGF-bFGF-SDF]$^{Hep/Ilo}$ grafts implanted into the sheep carotid artery. (**A**) 24 h postimplantation; (**B**) 3 months postimplantation; (**C**) 6 months postimplantation; (**D**) 18 months postimplantation. Representative images.

3.5. Cellular and Molecular Composition of PHBV/PCL[VEGF-bFGF-SDF]$^{Hep/Ilo}$ Grafts Is Reminiscent of Native Arteries

To better examine the cell populations within the patent PHBV/PCL[VEGF-bFGF-SDF]$^{Hep/Ilo}$ grafts and structure of the aneurysmatic ECM 6 months postimplantation, we stained excised vascular prostheses for endothelial markers CD31 and von Willebrand factor (vWF), vascular smooth muscle cell (VSMC) marker α-smooth muscle actin (α-SMA), and different types of arterial collagen. The examined grafts were well endothelialised (CD31+ cells at the luminal surface, Figure 4A), contained multiple layers of VSMCs in the medial layer (Figure 4B), and were highly vascularised (multiple vasa vasorum, Figure 4C–F). Similar to the large arteries [37], the microvessels within the graft were positively stained for CD31 and α-SMA (Figure 4C), vWF (Figure 4D) as well as type IV (Figure 4E) and type III collagen (Figure 4F) composing the vascular basement membrane.

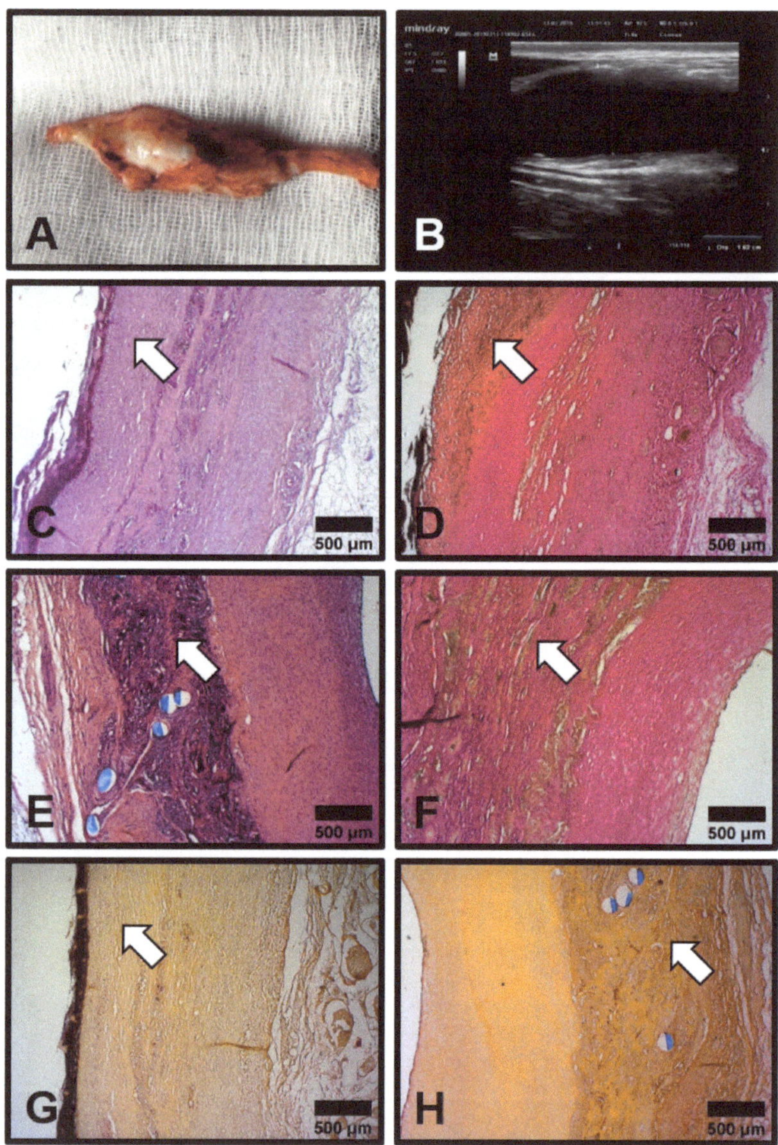

Figure 3. Development of the aneurysms in the patent PHBV/PCL[VEGF-bFGF-SDF]$^{Hep/Ilo}$ grafts 6 months postimplantation. (**A**) Gross examination of the aneurysm which developed through the whole length of the graft; (**B**) Ultrasound examination of the same aneurysm; (**C,D**) Neointima (indicated by white arrows) demarcated from the degrading polymer scaffold by an organised layer of collagen fibres. Haematoxylin and eosin staining (**C**) and van Gieson staining (**D**), ×50 magnification; (**E,F**) Degrading polymer scaffold is partially substituted by collagen bundles (white arrows) formed de novo. Haematoxylin and eosin staining (**E**) and van Gieson staining (**F**), ×50 magnification; (**G,H**) Absence of calcium deposits both within the neointima (**G**, white arrow) and polymer scaffold (**H**, white arrow). Alizarin red S staining, ×50 magnification. Representative images.

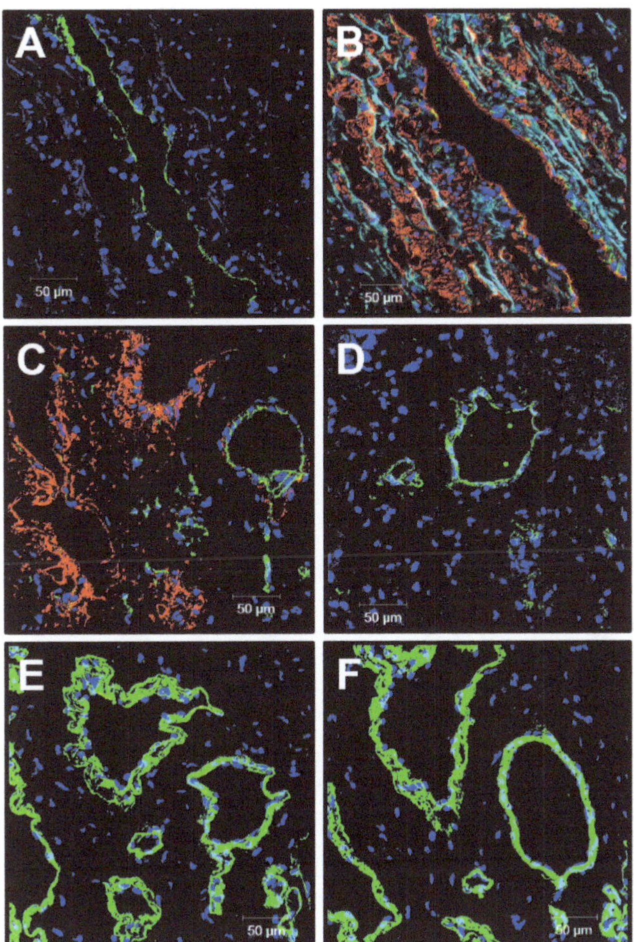

Figure 4. Confocal microscopy examination of the PHBV/PCL[VEGF-bFGF-SDF]$^{Hep/Ilo}$ graft 6 months postimplantation. (**A**) CD31 (a marker of endothelial cells, green colour) staining; (**B**) α-SMA (a marker of VSMCs, red colour) staining; (**C**) Combined CD31 (green colour) and-SMA (red colour) staining of the vasa vasorum within the graft; (**D**) vWF (a marker of endothelial cells and a component of the subendothelial ECM, green colour) staining of the vasa vasorum within the graft; (**E**) Type IV collagen (an ECM protein constituting the vascular basement membrane, green colour) staining; (**F**) Type III collagen (an ECM protein constituting the vascular basement membrane, green colour) staining. 4′,6-diamidino-2-phenylindole (DAPI) counterstaining (nuclei, blue colour). Representative images, ×200 magnification.

At 18 months postimplantation, the PHBV/PCL scaffold was almost resorbed and replaced by a blood vessel which consisted of three layers similar to a contralateral carotid artery. Typically, regenerated arteries were endothelialised (Figure 5A), retained their integrity (Figure 5B), and had ample connective tissue and numerous vasa vasorum in the outer layer resembling the tunica adventitia (Figure 5C). Perivascular adipose tissue was also observed (Figure 5C). However, regenerated arteries were notable for the absence of both elastic fibres and hierarchical orientation of the VSMCs (Figure 5B) that was the most probable cause of the aneurysms.

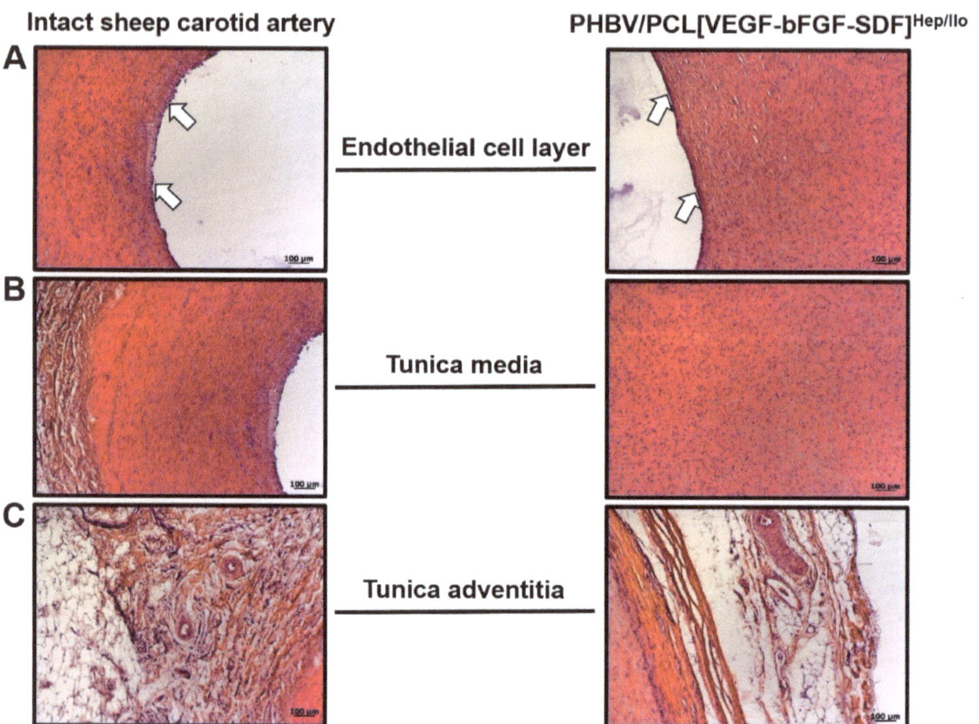

Figure 5. Histological comparison of the intact ovine carotid artery and regenerated artery which replaced PHBV/PCL[VEGF-bFGF-SDF]^Hep/Ilo graft 18 months postimplantation. (**A**) Endothelial cell layer (indicated by white arrows); (**B**) Tunica media; (**C**) Tunica adventitia and perivascular adipose tissue. Haematoxylin and eosin staining, representative images, ×100 magnification.

We then performed an ultrastructural investigation of the regenerated arteries by means of backscattered scanning electron microscopy. Neointima, media, and adventitia were clearly identified. Endothelial cells were either elongated in the direction of flow (Figure 6A) or acquired a polymorphic shape (Figure 6B), possibly owing to the endothelial-to-mesenchymal transition. Neointima consisted of VSMCs confined in the dense ECM (Figure 6C), while the loose medial layer which replaced a resorbable scaffold was populated by macrophages (Figure 6D) and fibroblast-like cells (Figure 6E). Tunica adventitia contained the loose ECM, numerous multinucleated giant cells (Figure 6F), and vasa vasorum (Figure 6G) formed during the polymer resorption.

One of the most hazardous long-term complications of TEVG implantation is calcification, which frequently leads to a delamination of the regenerated vessel and pseudoaneurysm formation [38,39]. Despite the fact that the ovine model is considered to be the "worst case scenario", as sheep are prone to the mineralisation of cardiovascular implants [8], we did not observe calcification loci in PHBV/PCL[VEGF-bFGF-SDF]^Hep/Ilo grafts 6 months postimplantation. At the 18-month time point, one of the regenerated arteries (1/5, 20%) contained a single mineral deposit which had structural and chemical homogeneity (Figures 6H and A1). The calcium to phosphorus ratio of 2.08 was indicative of carbonate-hydroxyapatite (bioapatite) [40,41].

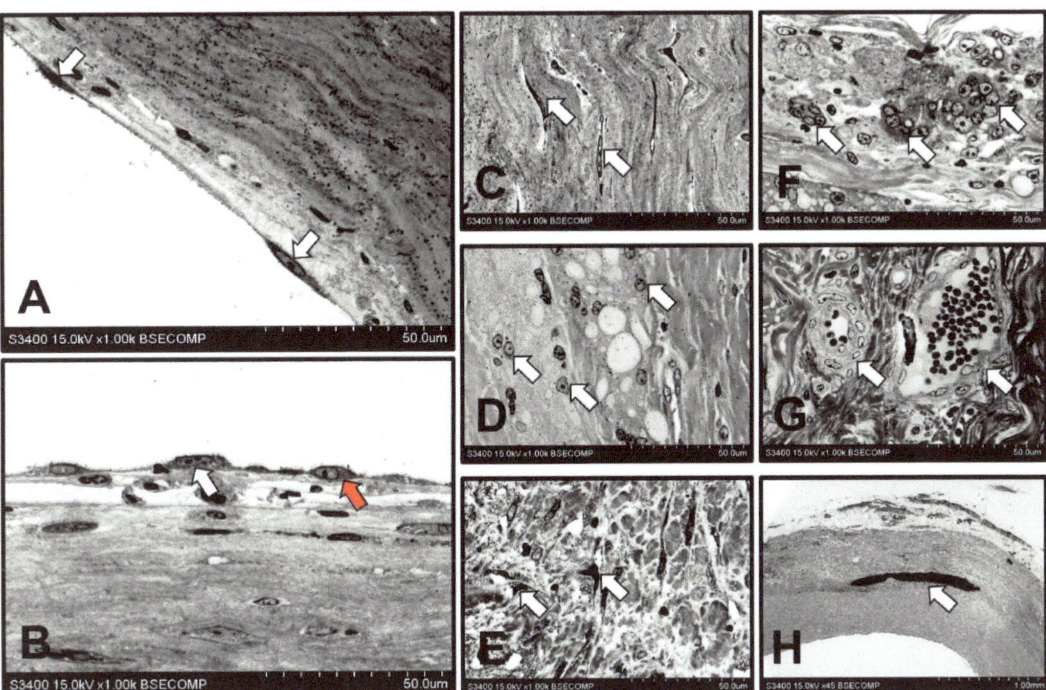

Figure 6. Backscattered scanning electron microscopy examination of the regenerated artery which replaced PHBV/PCL[VEGF-bFGF-SDF]$^{Hep/Ilo}$ graft 18 months postimplantation. (**A**) Monolayer of elongated endothelial cells (indicated by white arrows), ×1000 magnification; (**B**) Combination of elongated endothelial cells (white arrow) and polymorphic endothelial cells (red arrow) suggestive of a transitional phenotype, ×1000 magnification; (**C**) VSMCs in the neointima (white arrows), ×1000 magnification; (**D**) Macrophages in the medial layer (white arrows) which substituted a polymer scaffold, ×1000 magnification; (**E**) Fibroblast-like cells in the medial layer (white arrows), ×1000 magnification; (**F**) Multinucleated giant cells in the tunica adventitia (white arrows), ×1000 magnification; (**G**) Adventitial vasa vasorum (white arrows), ×1000 magnification; (**H**) Calcification on the border of neointimal and medial layers (white arrow), ×45 magnification. Representative images.

Next, we compared the transcriptional profiles of the endothelium and the wall of the regenerated artery in relation to those in contralateral carotid arteries. The excised vessels were flushed with TRIzol to collect the endothelial lysate and further homogenised to extract the RNA from the vascular wall. Reverse transcription-quantitative polymerase chain reaction revealed the abundance of the transcripts associated with inflammation (IL1B, IL6, and CXCL8), ECM remodelling (MMP2) and endothelial-to-mesenchymal transition (SNAI2) in both RNA fractions obtained from the regenerated artery (Figure 7A,B). Endothelial lysate was enriched with the inflammatory transcripts (IL1B, IL6, and ICAM1) and signatures of endothelial reprogramming (a venous transcript NR2F2 and an endothelial-to-mesenchymal transition marker SNAI2) (Figure 7B). These observations suggest that the molecular landscape of TEVGs can differ from the corresponding blood vessels, even with long-term arterial regeneration (18 months postimplantation).

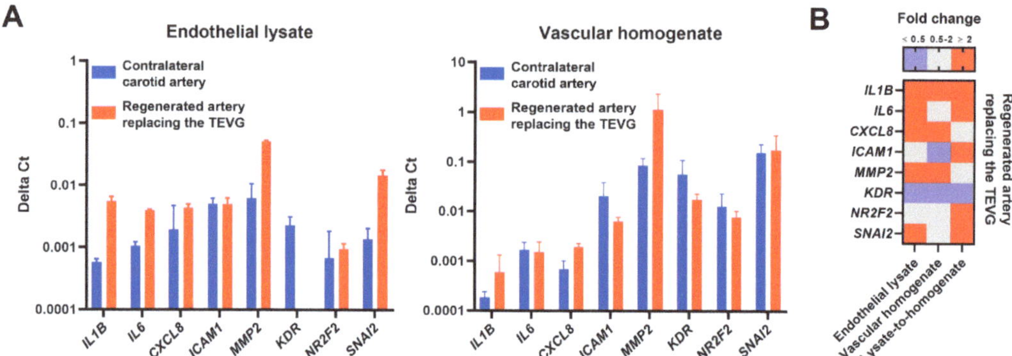

Figure 7. Transcriptional profiling of the regenerated and intact contralateral carotid arteries 18 months postimplantation. (**A**) Delta Ct values for the measured transcripts (IL1B, IL6, CXCL8, ICAM1, MMP2, KDR, NR2F2 and SNAI2) in the endothelial lysate (left) and homogenate of the de-endothelialised blood vessel (right); (**B**) Heat map indicating the differences between the endothelial lysate and vascular homogenate of the regenerated artery in relation to the respective RNA fractions in the contralateral intact artery and between endothelial lysate and homogenate of the de-endothelialised regenerated artery normalised to the mRNA expression in these compartments in the contralateral intact artery. Blue, grey and red colours mean fold change $\leq$0.50, 0.51–1.99, and $\geq$2.00, respectively, compared to the contralateral intact artery or homogenate of the de-endothelialised regenerated artery.

3.6. Biostable ePTFE Vascular Prostheses Undergo Thrombotic Occlusion and Calcification

In contrast to PHBV/PCL[VEGF-bFGF-SDF]$^{\text{Hep/Ilo}}$ grafts, ePTFE vascular prostheses were occluded as early as 24 h postimplantation (Figure 8A,B). At 6 months postimplantation, thrombosis became recanalised (Figure 8C) and biostable grafts were surrounded by a connective tissue capsule (Figure 8D) which was calcified (Figure 8E), suggesting a foreign body reaction. Multiple microcalcifications were also detected within the thrombus (Figure 8F). This histopathological pattern significantly differed from that observed in TEVGs.

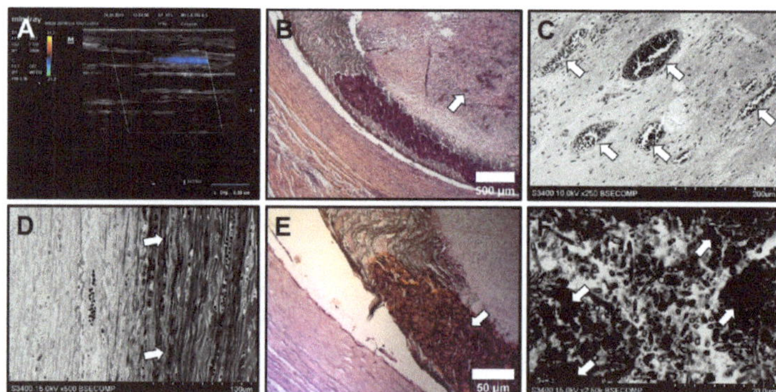

Figure 8. Histological and ultrastructural examination of ePTFE vascular prostheses. (**A**) Detection of thrombosis by Doppler ultrasound examination; (**B**) Thrombotic occlusion (white arrow), haematoxylin and eosin staining, ×50 magnification; (**C**) Recanalised thrombus (recanalising microvessels are indicated by white arrows), backscattered scanning electron microscopy, ×250 magnification; (**D**) Connective tissue capsule (white arrows), backscattered scanning electron microscopy, ×500 magnification. (**E**) Calcification (white arrow) within the connective tissue capsule, alizarin red S staining, ×100 magnification; (**F**) Microcalcifications within the thrombus (white arrows), backscattered scanning electron microscopy, ×250 magnification. Representative images.

Immunofluorescent examination of ePTFE grafts at 6 months postimplantation revealed the absence of the endothelium suggestive of poor biocompatibility (CD31- and largely vWF-negative cells at the luminal surface, Figure 9A,B). VSMCs were identified within the thrombus but not in the medial layer (Figure 9A). Disorganised bundles of type IV collagen were located beneath the luminal surface (Figure 9C) whereas type III and type I collagen were not found (Figure 9C,D).

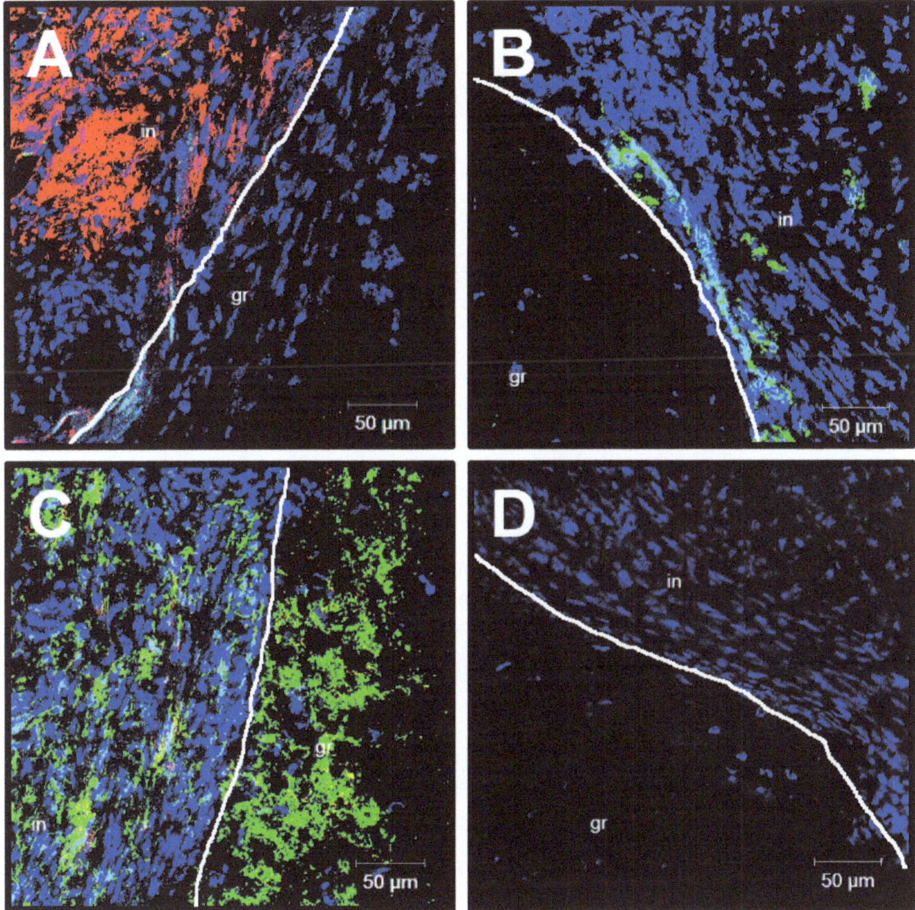

Figure 9. Confocal microscopy examination of ePTFE vascular prostheses 6 months postimplantation. (**A**) CD31 (a marker of endothelial cells, green colour) and α-SMA (a marker of VSMCs, red colour) staining; (**B**) von Willebrand staining (a marker of endothelial cells and a component of the subendothelial ECM, green colour); (**C**) Type IV collagen (an ECM protein constituting the basement membrane, green colour) and type I collagen (an ECM protein characteristic of the adventitia, red colour) staining; (**D**) Type III collagen staining (an ECM protein constituting the basement membrane and tunica media, green colour). DAPI counterstaining (nuclei, blue colour). White line demarcates the graft (gr) from thrombotic masses (in). Representative images, ×200 magnification.

4. Discussion

Despite the fact that numerous rigorous studies and preclinical trials have been performed over the past decade, a commercially available, off-the-shelf TEVG has not been developed to date. The reasons for this include insufficient durability of the prototypes fabricated from biocompatible natural polymers, poor biocompatibility of synthetic polymers, and uncontrollable biodegradation of polymer blends [9–13,42]. Further, small animal models are inappropriate for TEVG testing, as the haemodynamic conditions in the blood vessels and parameters of haemostasis in mice and rats are vastly different from those in humans [28–30]. Yet, the use of large animals such as dogs, swine, or sheep is complicated because of their high cost and difficult handling [28–30]. Among the available animal models, haemostasis and inflammatory response in sheep show the highest similarity to humans, while their carotid artery is sufficiently long for proper mechanical testing and is easily accessible for the surgical intervention and postimplantation visualisation [28–30]. Sheep are also characterised by an accelerated calcification of cardiovascular implants mirroring the clinical scenario in working-age patients [8]. Hence, we chose an ovine carotid artery interposition model for the animal testing of our TEVGs which were previously investigated in rats to verify the positive effects of incorporated growth factors on vascular tissue regeneration [25–27].

Implantation of a TEVG into the circulatory system is often accompanied by a number of significant complications. The first is thrombosis, which occurs within the first 24 h postimplantation and often during surgery in cases of contact activation induced by a negatively charged polymer surface [9,12,33]. Long-term pathological consequences include: (1) neointimal hyperplasia which leads to the restenosis and develops due to the proliferation of mesenchymal cells (i.e., VSMCs and fibroblasts), active synthesis of the ECM, and concurrent vascularisation-driven inflammation; (2) calcification, which provokes delamination and causes mechanical incompetence of the graft, emerging as a result of an uncontrolled degradation of the polymer fibres, incomplete or improper assemble of the ECM and osteogenic differentiation of mesenchymal cells; (3) aneurysms, which may induce graft rupture and occur because of dyscoordinated polymer replacement with the regenerating ECM and uneven distribution of the mechanical load across the graft [12,33].

We previously demonstrated the high primary patency rate of an electrospun scaffold prepared from PHBV/PCL blend enriched with VEGF, bFGF and SDF-1α by using a rat abdominal aorta replacement model [26,27]. The applied electrospinning technique generated nanoscale polymer fibres which formed interconnected microsized pores, thereby increasing the area for cell-ECM interactions [11,43]. Combined with bioactive factors, such construction of the conduit potentiated migration, attachment, and guided differentiation of blood-derived cells, consequently improving the integration of the implant with the host tissues [32]. However, preliminary studies found unacceptably high frequency of thrombosis when this prototype was applied to replace ovine carotid artery. To solve this issue, we immobilised Hep/Ilo on a luminal surface through the PVP linker to endow the graft with antithrombotic properties before the endothelialisation. In this study, we tested the aforementioned TEVG (PHBV/PCL[VEGF-bFGF-SDF]$^{\text{Hep/Ilo}}$) against the biostable ePTFE vascular prosthesis, which is not replaced by de novo ECM and has superior mechanical properties but exhibits poor endothelialisation.

The majority (62.5%) of the PHBV/PCL[VEGF-bFGF-SDF]$^{\text{Hep/Ilo}}$ grafts, but none of the ePTFE prostheses, showed absence of thrombosis, and 50% of TEVGs were patent 18 months postimplantation, providing data of future improvements of this prototype. However, in contrast to ePTFE grafts, almost all regenerated arteries (i.e., vascular tissue which replaced the grafts) suffered from aneurysms. The reason behind this was the unexpectedly [35,36] rapid resorption of the PHBV/PCL scaffold, which lost its integrity 6 months postimplantation and was almost completely resorbed 18 months postimplantation. In rats, all grafts were intact 12 months postimplantation and did not display any signs of aneurysm formation [27,32]. Intriguingly, these results fully correspond to a recent paper which also reported a striking difference in the degradation rates of TEVGs between small and large animal models [44]. Similar to our previous and present research, this study employed rat abdominal aorta and sheep carotid artery interposition models for the implantation of PCL/chitosan vascular graft [44]. In keeping with the abovementioned results, ovine prostheses demonstrated a pronounced degradation concurrently with ECM deposition and absence of calcification at 6 months postimplantation, in contrast to the rat model [44]. Taken together, these results illustrate the importance of TEVG testing in large animal models and underline major drawbacks of small animal models in this regard.

Regenerated arterial tissue was similar to the contralateral artery, also having three layers resembling intima, media, and adventitia. Importantly, it was fully endothelialised, highly vascularised, and populated by VSMCs and macrophages, two major populations which are, along with endothelial cells, responsible for the arterial homeostasis. The regenerated artery was generally resistant to calcification, but did not contain elastic fibres and hierarchical layers of VSMCs, which contributed to its mechanical incompetence and made it unable to restore tissue integrity upon resorption of the PHBV/PCL scaffold. In contrast, all ePTFE grafts were heavily calcified.

Both endothelial lysate and vascular homogenate of the regenerated artery replacing PHBV/PCL[VEGF-bFGF-SDF]$^{Hep/Ilo}$ grafts showed elevated expression of IL1B, CXCL8, and MMP2 genes in comparison with a contralateral carotid artery, suggestive of a cytokine-driven remodelling, in accordance with previous studies [20,45,46]. Unfortunately, studies measuring interleukin-1b, interleukin-6, or interleukin-8 in TEVGs upon their implantation have not been conducted to date. However, matrix metalloproteinase 2 was found to be upregulated 20 [47,48] and even 80 weeks [47] postimplantation of TEVGs in the pulmonary artery [47] or aorta [48] of lambs. Similar results were obtained after the short-term ($\leq$4 weeks) implantation of TEVGs into the inferior vena cava of mice [49]. Intriguingly, we found a pronounced expression of the genes encoding pro-inflammatory molecules (IL1B, IL6, and ICAM1) as well as the gene for the key transcription factor governing endothelial-to-mesenchymal transition (SNAI2) in the endothelial lysate as compared with vascular homogenate upon normalisation to the expression of the respective genes in the contralateral carotid artery. Yet, neither cytokine release nor endothelial-to-mesenchymal transition have been measured in situ after the implantation of TEVGs. Further research in this direction would require transcriptomic and proteomic profiling (e.g., by means of RNA-seq, dot blotting arrays, or mass spectrometry) of endothelial lysate and vascular homogenate at ascending time points following TEVG implantation.

5. Conclusions

The present study demonstrates the suitability of a small-diameter electrospun TEVG fabricated from a biodegradable PHBV/PCL polymer blend, biofunctionalised by the incorporation of bioactive factors (VEGF, bFGF and SDF-1α), and modified with heparin and iloprost to improve antithrombotic properties. Upon implantation into the ovine carotid artery, PHBV/PCL[VEGF-bFGF-SDF]$^{Hep/Ilo}$ grafts exhibited high biocompatibility and calcification resistance along with a moderate haemocompatibility, but were prone to aneurysm formation. Hence, this TEVG prototype needs mechanical reinforcement and further improvement of its antithrombotic coating. Yet, the 50% primary patency at 18 months postimplantation holds promise for the translation of the PHBV/PCL[VEGF-bFGF-SDF]$^{Hep/Ilo}$ vascular graft prototype into future preclinical trials following anti-aneurysmal and antimicrobial modification. Further molecular profiling studies are needed to better understand the arterial regeneration which occurs following the replacement of the PHBV/PCL blend. Relative quantification of endothelial-to-mesenchymal transition and osteogenic transcription factors in comparison with the contralateral ovine carotid artery, as well as interrogation of vascular smooth muscle cell plasticity and macrophage polarisation at ascending time points postimplantation, would be particularly beneficial.

Author Contributions: Conceptualization, L.V.A. and L.S.B.; methodology, L.V.A., E.O.K., V.V.S., A.V.M., M.A.R. and A.G.K.; formal analysis, L.V.A.; investigation, L.V.A., E.O.K., V.V.S., A.V.M., M.A.R., A.R.S., V.O.T., S.S.K., M.Y.K., T.Y.S., V.G.M., T.V.G., A.G.K., R.A.M., N.S.D., T.N.A., M.Y.S. and E.A.V.; resources, L.S.B.; writing—original draft preparation, L.V.A., E.O.K., V.V.S., A.V.M., M.A.R., A.R.S., V.O.T., S.S.K., M.Y.K., T.Y.S., V.G.M., T.V.G., A.G.K., R.A.M., N.S.D., T.N.A., M.Y.S. and E.A.V.; writing—review and editing, L.S.B.; visualization, L.V.A., E.O.K., V.V.S., V.G.M., T.V.G., A.G.K., R.A.M., T.N.A. and E.A.V.; supervision, L.V.A. and L.S.B. All authors have read and agreed to the published version of the manuscript.

Funding: This study was supported by the Complex Program of Basic Research under the Siberian Branch of the Russian Academy of Sciences within the Basic Research Topic of Research Institute for Complex Issues of Cardiovascular Diseases № 0546-2019-0002 "Pathogenetic basis for the development of cardiovascular implants from biocompatible materials using patient-oriented approach, mathematical modelling, tissue engineering, and genomic predictors".

Institutional Review Board Statement: The study was conducted according to the guidelines of the Declaration of Helsinki, and approved by the local Ethics committee of the Research Institute for Complex Issues of Cardiovascular Diseases (protocol number 20180305, 28.04.2016, Kemerovo, Russia).

Informed Consent Statement: Not applicable.

Data Availability Statement: The data presented in this study are available on request from the corresponding author. The data are not publicly available due to the data form part of an ongoing study.

Conflicts of Interest: The authors declare no conflict of interest. The funders had no role in the design of the study; in the collection, analyses, or interpretation of data; in the writing of the manuscript, or in the decision to publish the results.

Appendix A

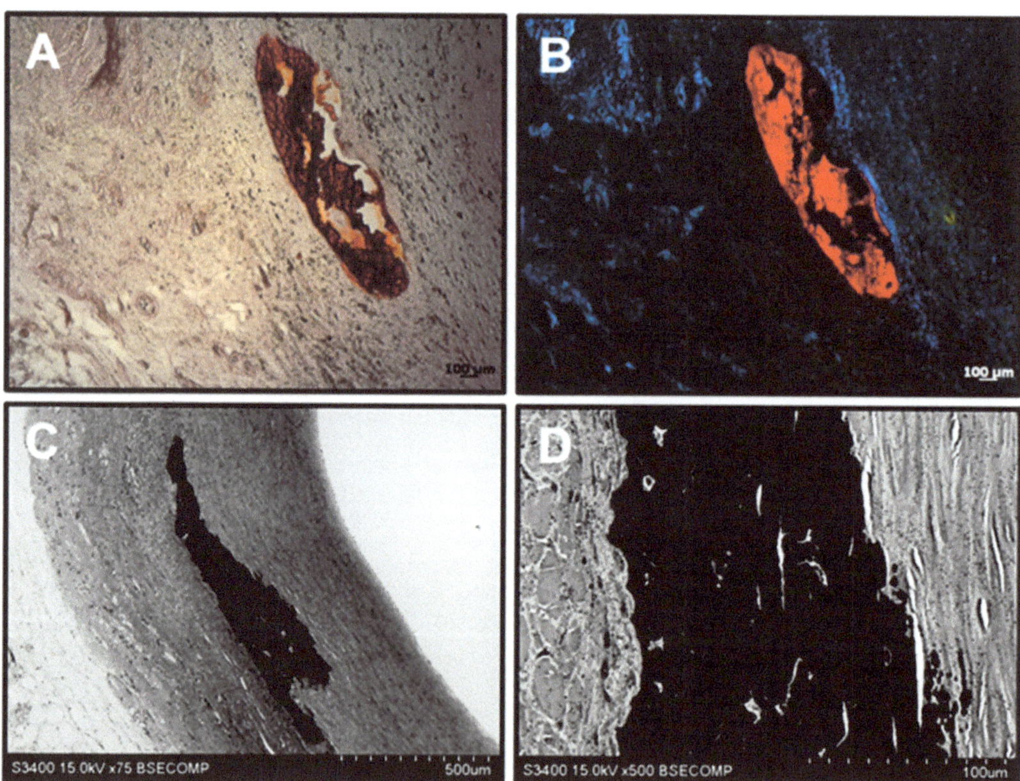

Figure A1. Calcification of a regenerated artery on the border of neointimal and medial layers 18 months postimplantation. (**A**) Alizarin red S staining, light microscopy, ×50 magnification; (**B**) Alizarin red S staining with DAPI counterstaining, ×50 magnification; (**C**) Backscattered scanning electron microscopy, ×75 magnification; (**D**) Backscattered scanning electron microscopy, ×500 magnification.

References

1. Virani, S.S.; Alonso, A.; Benjamin, E.J.; Bittencourt, M.S.; Callaway, C.W.; Carson, A.P.; Chamberlain, A.M.; Chang, A.R.; Cheng, S.; Delling, F.N.; et al. American Heart Association Council on Epidemiology and prevention statistics committee and stroke statistics subcommittee. Heart disease and stroke statistics-2020 Update: A report from the American Heart Association. *Circulation* **2020**, *141*, e139–e596. [CrossRef] [PubMed]
2. Alkhouli, M.; Alqahtani, F.; Kalra, A.; Gafoor, S.; Alhajji, M.; Alreshidan, M.; Holmes, D.R.; Lerman, A. Trends in characteristics and outcomes of patients undergoing coronary revascularization in the United States, 2003–2016. *JAMA Netw. Open* **2020**, *3*, e1921326. [CrossRef]
3. Blumenfeld, O.; Na'amnih, W.; Shapira-Daniels, A.; Lotan, C.; Shohat, T.; Shapira, O.M. Trends in coronary revascularization and ischemic heart disease-related mortality in Israel. *J. Am. Heart Assoc.* **2017**, *6*, e004734. [CrossRef]
4. Caliskan, E.; de Souza, D.R.; Böning, A.; Liakopoulos, O.J.; Choi, Y.H.; Pepper, J.; Gibson, C.M.; Perrault, L.P.; Wolf, R.K.; Kim, K.B.; et al. Saphenous vein grafts in contemporary coronary artery bypass graft surgery. *Nat. Rev. Cardiol.* **2020**, *17*, 155–169. [CrossRef]
5. Saraiva, F.A.; Leite-Moreira, J.P.; Barros, A.S.; Lourenço, A.P.; Benedetto, U.; Leite-Moreira, A.F. Multiple versus single arterial grafting in coronary artery bypass grafting: A meta-analysis of randomized controlled trials and propensity score studies. *Int. J. Cardiol.* **2020**, *320*, 55–63. [CrossRef]
6. Lejay, A.; Vento, V.; Kuntz, S.; Steinmetz, L.; Georg, Y.; Thaveau, F.; Heim, F.; Chakfé, N. Current status on vascular substitutes. *J. Cardiovasc. Surg.* **2020**, *61*, 55–63. [CrossRef]
7. Obiweluozor, F.O.; Emechebe, G.A.; Kim, D.W.; Cho, H.J.; Park, C.H.; Kim, C.S.; Jeong, I.S. Considerations in the development of small-diameter vascular graft as an alternative for bypass and reconstructive surgeries: A review. *Cardiovasc. Eng. Technol.* **2020**, *11*, 495–521. [CrossRef]
8. Matsuzaki, Y.; John, K.; Shoji, T.; Shinoka, T. The evolution of tissue engineered vascular graft technologies: From preclinical trials to advancing patient care. *Appl. Sci.* **2019**, *9*, 1274. [CrossRef]
9. Yuan, H.; Chen, C.; Liu, Y.; Lu, T.; Wu, Z. Strategies in cell-free tissue-engineered vascular grafts. *J. Biomed. Mater. Res. A* **2020**, *108*, 426–445. [CrossRef] [PubMed]
10. Toong, D.W.Y.; Toh, H.W.; Ng, J.C.K.; Wong, P.E.H.; Leo, H.L.; Venkatraman, S.; Tan, L.P.; Ang, H.Y.; Huang, Y. Bioresorbable polymeric scaffold in cardiovascular applications. *Int. J. Mol. Sci.* **2020**, *21*, 3444. [CrossRef]
11. Wang, Z.; Mithieux, S.M.; Weiss, A.S. Fabrication techniques for vascular and vascularized tissue engineering. *Adv. Healthc. Mater.* **2019**, *8*, e1900742. [CrossRef] [PubMed]
12. Carrabba, M.; Madeddu, P. Current strategies for the manufacture of small size tissue engineering vascular grafts. *Front. Bioeng. Biotechnol.* **2018**, *6*, 41. [CrossRef] [PubMed]
13. Radke, D.; Jia, W.; Sharma, D.; Fena, K.; Wang, G.; Goldman, J.; Zhao, F. Tissue engineering at the blood-contacting surface: A review of challenges and strategies in vascular graft development. *Adv. Healthc. Mater.* **2018**, *7*, e1701461. [CrossRef] [PubMed]
14. Del Gaudio, C.; Fioravanzo, L.; Folin, M.; Marchi, F.; Ercolani, E.; Bianco, A. Electrospun tubular scaffolds: On the effectiveness of blending poly(ε-caprolactone) with poly(3-hydroxybutyrate-co-3-hydroxyvalerate). *J. Biomed. Mater. Res. B Appl. Biomater.* **2012**, *100*, 1883–1898. [CrossRef]
15. Bye, F.J.; Bissoli, J.; Black, L.; Bullock, A.J.; Puwanun, S.; Moharamzadeh, K.; Reilly, G.C.; Ryan, A.J.; Neil, S.M. Development of bilayer and trilayer nanofibrous/microfibrous scaffolds for regenerative medicine. *Biomater. Sci.* **2013**, *1*, 942–951. [CrossRef] [PubMed]
16. Pařízek, M.; Novotná, K.; Bačáková, L. The role of smooth muscle cells in vessel wall pathophysiology and reconstruction using bioactive synthetic polymers. *Physiol. Res.* **2011**, *60*, 419–437. [CrossRef]
17. Chan-Park, M.B.; Shen, J.Y.; Cao, Y.; Xiong, Y.; Liu, Y.; Rayatpisheh, S.; Kang, G.C.; Greisler, H.P. Biomimetic control of vascular smooth muscle cell morphology and phenotype for functional tissue-engineered small-diameter blood vessels. *J. Biomed. Mater. Res. A* **2009**, *88*, 1104–1121. [CrossRef]
18. Xie, S.A.; Zhang, T.; Wang, J.; Zhao, F.; Zhang, Y.P.; Yao, W.J.; Hur, S.S.; Yeh, Y.T.; Pang, W.; Zheng, L.S.; et al. Matrix stiffness determines the phenotype of vascular smooth muscle cell in vitro and in vivo: Role of DNA methyltransferase 1. *Biomaterials* **2018**, *155*, 203–216. [CrossRef]
19. Dayekh, K.; Mequanint, K. The effects of progenitor and differentiated cells on ectopic calcification of engineered vascular tissues. *Acta Biomater.* **2020**, *115*, 288–298. [CrossRef]
20. Roh, J.D.; Sawh-Martinez, R.; Brennan, M.P.; Jay, S.M.; Devine, L.; Rao, D.A.; Yi, T.; Mirensky, T.L.; Nalbandian, A.; Udelsman, B.; et al. Tissue-engineered vascular grafts transform into mature blood vessels via an inflammation-mediated process of vascular remodeling. *Proc. Natl. Acad. Sci. USA* **2010**, *107*, 4669–4674. [CrossRef] [PubMed]
21. Lee, K.W.; Gade, P.S.; Dong, L.; Zhang, Z.; Aral, A.M.; Gao, J.; Ding, X.; Stowell, C.E.T.; Nisar, M.U.; Kim, K.; et al. A biodegradable synthetic graft for small arteries matches the performance of autologous vein in rat carotid arteries. *Biomaterials* **2018**, *181*, 67–80. [CrossRef]
22. Stowell, C.E.T.; Li, X.; Matsunaga, M.H.; Cockreham, C.B.; Kelly, K.M.; Cheetham, J.; Tzeng, E.; Wang, Y. Resorbable vascular grafts show rapid cellularization and degradation in the ovine carotid. *J. Tissue Eng. Regen. Med.* **2020**, *14*, 1673–1684. [CrossRef]
23. Cai, Q.; Liao, W.; Xue, F.; Wang, X.; Zhou, W.; Li, Y.; Zeng, W. Selection of different endothelialization modes and different seed cells for tissue-engineered vascular graft. *Bioact. Mater.* **2021**, *6*, 2557–2568. [CrossRef]

24. Sánchez, P.F.; Brey, E.M.; Briceño, J.C. Endothelialization mechanisms in vascular grafts. *J. Tissue Eng. Regen. Med.* **2018**, *12*, 2164–2178. [CrossRef] [PubMed]

25. Antonova, L.V.; Seifalian, A.M.; Kutikhin, A.G.; Sevostyanova, V.V.; Krivkina, E.O.; Mironov, A.V.; Burago, A.Y.; Velikanova, E.A.; Matveeva, V.G.; Glushkova, T.V.; et al. Bioabsorbable bypass grafts biofunctionalised with RGD have enhanced biophysical properties and endothelialisation tested in vivo. *Front. Pharmacol.* **2016**, *7*, 136. [CrossRef] [PubMed]

26. Antonova, L.V.; Sevostyanova, V.V.; Kutikhin, A.G.; Mironov, A.V.; Krivkina, E.O.; Shabaev, A.R.; Matveeva, V.G.; Velikanova, E.A.; Sergeeva, E.A.; Burago, A.Y.; et al. Vascular endothelial growth factor improves physico-mechanical properties and enhances endothelialization of poly(3-hydroxybutyrate-co-3-hydroxyvalerate)/poly(ε-caprolactone) small-diameter vascular grafts in vivo. *Front. Pharmacol.* **2016**, *7*, 230. [CrossRef]

27. Antonova, L.V.; Seifalian, A.M.; Kutikhin, A.G.; Sevostyanova, V.V.; Matveeva, V.G.; Velikanova, E.A.; Mironov, A.V.; Shabaev, A.R.; Glushkova, T.V.; Senokosova, E.A.; et al. Conjugation with RGD peptides and incorporation of vascular endothelial growth factor are equally efficient for biofunctionalization of tissue-engineered vascular grafts. *Int. J. Mol. Sci.* **2016**, *17*, 1920. [CrossRef] [PubMed]

28. Liu, R.H.; Ong, C.S.; Fukunishi, T.; Ong, K.; Hibino, N. Review of vascular graft studies in large animal models. *Tissue Eng. Part B Rev.* **2018**, *24*, 133–143. [CrossRef]

29. Swartz, D.D.; Andreadis, S.T. Animal models for vascular tissue-engineering. *Curr. Opin. Biotechnol.* **2013**, *24*, 916–925. [CrossRef] [PubMed]

30. Thomas, L.V.; Lekshmi, V.; Nair, P.D. Tissue engineered vascular grafts–preclinical aspects. *Int. J. Cardiol.* **2013**, *167*, 1091–1100. [CrossRef] [PubMed]

31. Antonova, L.V.; Silnikov, V.N.; Sevostyanova, V.V.; Yuzhalin, A.E.; Koroleva, L.S.; Velikanova, E.A.; Mironov, A.V.; Godovikova, T.S.; Kutikhin, A.G.; Glushkova, T.V.; et al. Biocompatibility of small-diameter vascular grafts in different modes of RGD modification. *Polymers* **2019**, *11*, 174. [CrossRef]

32. Antonova, L.; Kutikhin, A.; Sevostianova, V.; Velikanova, E.; Matveeva, V.; Glushkova, T.; Mironov, A.; Krivkina, E.; Shabaev, A.; Senokosova, E.; et al. bFGF and SDF-1α improve in vivo performance of VEGF-incorporating small-diameter vascular grafts. *Pharmaceuticals* **2021**, *14*, 302. [CrossRef]

33. Dimitrievska, S.; Niklason, L.E. Historical perspective and future direction of blood vessel developments. *Cold Spring Harb. Perspect. Med.* **2018**, *8*, a025742. [CrossRef] [PubMed]

34. Liu, X.; Xu, Y.; Wu, Z.; Chen, H. Poly(N-vinylpyrrolidone)-modified surfaces for biomedical applications. *Macromol. Biosci.* **2013**, *13*, 147–154. [CrossRef] [PubMed]

35. Sun, H.; Mei, L.; Song, C.; Cui, X.; Wang, P. The in vivo degradation, absorption and excretion of PCL-based implant. *Biomaterials* **2006**, *27*, 1735–1740. [CrossRef]

36. Malikmammadov, E.; Tanir, T.E.; Kiziltay, A.; Hasirci, V.; Hasirci, N. PCL and PCL-based materials in biomedical applications. *J. Biomater. Sci. Polym. Ed.* **2018**, *29*, 863–893. [CrossRef] [PubMed]

37. Farquharson, C.; Robins, S.P. Immunolocalization of collagen types I and III in the arterial wall of the rat. *Histochem. J.* **1989**, *21*, 172–178. [CrossRef] [PubMed]

38. De Valence, S.; Tille, J.C.; Mugnai, D.; Mrowczynski, W.; Gurny, R.; Möller, M.; Walpoth, B.H. Long term performance of polycaprolactone vascular grafts in a rat abdominal aorta replacement model. *Biomaterials* **2012**, *33*, 38–47. [CrossRef]

39. Tara, S.; Kurobe, H.; Rocco, K.A.; Maxfield, M.W.; Best, C.A.; Yi, T.; Naito, Y.; Breuer, C.K.; Shinoka, T. Well-organized neointima of large-pore poly(L-lactic acid) vascular graft coated with poly(L-lactic-co-ε-caprolactone) prevents calcific deposition compared to small-pore electrospun poly(L-lactic acid) graft in a mouse aortic implantation model. *Atherosclerosis* **2014**, *237*, 684–691. [CrossRef]

40. Gibson, I.R.; Bonfield, W. Novel synthesis and characterization of an AB-type carbonate-substituted hydroxyapatite. *J. Biomed. Mater. Res.* **2002**, *59*, 697–708. [CrossRef]

41. Danilchenko, S.N.; Kalinkevich, A.N.; Moskalenko, R.A.; Kuznetsov, V.N.; Kochenko, A.V.; Husak, E.V.; Starikov, V.V.; Liu, F.; Meng, J.; Lü, J. Structural and crystal-chemical characteristics of the apatite deposits from human aortic walls. *Interv. Med. Appl. Sci.* **2018**, *10*, 110–119. [CrossRef]

42. Song, R.; Murphy, M.; Li, C.; Ting, K.; Soo, C.; Zheng, Z. Current development of biodegradable polymeric materials for biomedical applications. *Drug Des. Dev. Ther.* **2018**, *12*, 3117–3145. [CrossRef] [PubMed]

43. Ren, X.; Feng, Y.; Guo, J.; Wang, H.; Li, Q.; Yang, J.; Hao, X.; Lv, J.; Ma, N.; Li, W. Surface modification and endothelialization of biomaterials as potential scaffolds for vascular tissue engineering applications. *Chem. Soc. Rev.* **2015**, *44*, 5680–5742. [CrossRef]

44. Fukunishi, T.; Ong, C.S.; Yesantharao, P.; Best, C.A.; Yi, T.; Zhang, H.; Mattson, G.; Boktor, J.; Nelson, K.; Shinoka, T.; et al. Different degradation rates of nanofiber vascular grafts in small and large animal models. *J. Tissue Eng. Regen. Med.* **2020**, *14*, 203–214. [CrossRef]

45. Tara, S.; Kurobe, H.; Maxfield, M.W.; Rocco, K.A.; Yi, T.; Naito, Y.; Breuer, C.K.; Shinoka, T. Evaluation of remodeling process in small-diameter cell-free tissue-engineered arterial graft. *J. Vasc. Surg.* **2015**, *62*, 734–743. [CrossRef] [PubMed]

46. Wolf, F.; Paefgen, V.; Winz, O.; Mertens, M.; Koch, S.; Gross-Weege, N.; Morgenroth, A.; Rix, A.; Schnoeringm, H.; Chalabi, K.; et al. MR and PET-CT monitoring of tissue-engineered vascular grafts in the ovine carotid artery. *Biomaterials* **2019**, *216*, 119228. [CrossRef]

47. Cummings, I.; George, S.; Kelm, J.; Schmidt, D.; Emmert, M.Y.; Weber, B.; Zünd, G.; Hoerstrup, S.P. Tissue-engineered vascular graft remodeling in a growing lamb model: Expression of matrix metalloproteinases. *Eur. J. Cardio Thorac. Surg.* **2012**, *41*, 167–172. [CrossRef] [PubMed]
48. Shum-Tim, D.; Stock, U.; Hrkach, J.; Shinoka, T.; Lien, J.; Moses, M.A.; Stamp, A.; Taylor, G.; Moran, A.M.; Landis, W.; et al. Tissue engineering of autologous aorta using a new biodegradable polymer. *Ann. Thorac. Surg.* **1999**, *68*, 2298–2304. [CrossRef]
49. Naito, Y.; Williams-Fritze, M.; Duncan, D.R.; Church, S.N.; Hibino, N.; Madri, J.A.; Humphrey, J.D.; Shinoka, T.; Breuer, C.K. Characterization of the natural history of extracellular matrix production in tissue-engineered vascular grafts during neovessel formation. *Cells Tissues Organs* **2012**, *195*, 60–72. [CrossRef] [PubMed]

Article

Electrospun Biodegradable α-Amino Acid-Substituted Poly(organophosphazene) Fiber Mats for Stem Cell Differentiation towards Vascular Smooth Muscle Cells

Meng Wang, Shigang Lin and Kibret Mequanint *

Department of Chemical and Biochemical Engineering, The University of Western Ontario, London, ON N6A 5B9, Canada; mwang529@uwo.ca (M.W.); slin45@uwo.ca (S.L.)
* Correspondence: kmequani@uwo.ca

Abstract: Mesenchymal stem cells, derived from human-induced pluripotent stem cells (iPSC), are valuable for generating smooth muscle cells (SMCs) for vascular tissue engineering applications. In this study, we synthesized biodegradable α-amino acid-substituted poly(organophosphazene) polymers and electrospun nano-fibrous scaffolds (~200 nm diameter) to evaluate their suitability as a matrix for differentiation of iPSC-derived mesenchymal stem cells (iMSC) into mature contractile SMCs. Both the polymer synthesis approach and the electrospinning parameters were optimized. Three types of cells, namely iMSC, bone marrow derived mesenchymal stem cells (BM-MSC), and primary human coronary artery SMC, attached and spread on the materials. Although L-ascorbic acid (AA) and transforming growth factor-beta 1 (TGF-β1) were able to differentiate iMSC along the smooth muscle lineage, we showed that the electrospun fibrous mats provided material cues for the enhanced differentiation of iMSCs. Differentiation of iMSC to SMC was characterized by increased transcriptional levels of early to late-stage smooth muscle marker proteins on electrospun fibrous mats. Our findings provide a feasible strategy for engineering functional vascular tissues.

Keywords: biodegradable poly(organophosphazenes); human mesenchymal stem cells; vascular smooth muscle cells; vascular tissue engineering; electrospinning

Citation: Wang, M.; Lin, S.; Mequanint, K. Electrospun Biodegradable α-Amino Acid-Substituted Poly(organophosphazene) Fiber Mats for Stem Cell Differentiation towards Vascular Smooth Muscle Cells. *Polymers* **2022**, *14*, 1555. https://doi.org/10.3390/polym14081555

Academic Editors: João Carlos Silva and Frederico Castelo Ferreira

Received: 15 March 2022
Accepted: 6 April 2022
Published: 11 April 2022

Publisher's Note: MDPI stays neutral with regard to jurisdictional claims in published maps and institutional affiliations.

1. Introduction

Biodegradable poly(organophosphazenes), substituted with α-amino acids and hereafter abbreviated as PαAPz, are important biomaterials for drug delivery and tissue engineering applications [1–4]. Unlike biodegradable polyesters (e.g., poly(lactide), poly(glycolide), and their copolymers), PαAPz releases non-acidic and buffering degradation products comprised mainly of phosphate, ammonia, and the corresponding side groups [5,6]. Several previous studies investigated PαAPz primarily, for the regeneration of musculoskeletal tissues, and demonstrated their ability to support osteoblast adhesion and proliferation with minimal inflammatory responses and enhanced bone growth [1–3,7–9]. To introduce porosity to bone tissue engineering PαAPz scaffolds, either dynamic solvent sintering of pre-formed microspheres [10,11] or electrospinning of PαAPz blended with nanohydroxyapatite [12] have been studied. In one study [13], where a PαAPz was electrospun for bone tissue engineering, the fiber diameter was large (~55 μm) making it challenging for osteoblasts to see the true 3D topography of the fibers. In addition to bone cell culture, some studies investigated electrospun PαAPz for their potential use of an endothelial monolayer cell culture [14,15]. In general, electrospun scaffolds are preferred in comparison to other forms of porous scaffolds to better mimic the fibrous extracellular matrix [16,17].

While the above-mentioned studies demonstrated the versatility of PαAPz, primarily in bone regeneration, their utility in vascular tissue engineering has not been studied. Vascular smooth muscle cells (VSMCs) are the major cellular components to engineer vascular tissues. Since VSMCs are particularly sensitive to acidic degradation products, we believe that PαAPz that degrade into non-acidic by-products could be excellent biomaterials for smooth muscle tissue engineering. Considering that harvesting primary VSMCs from patients is not feasible due to their inaccessible anatomical location (e.g., coronary artery), mesenchymal stem cells (MSCs) are ideal cell sources for vascular tissue engineering. MSCs are multipotent cells commonly characterized by their ability to differentiate into cell types of mesodermal origin [18–23]. MSCs can be isolated from many adult tissue sources in a non-invasive manner [24,25]; however, their self-renewal and differentiation potential is dependent on the cell source of isolation [26]. It may also be affected by donor aging and environmental stresses from in vitro cultivation [27]. To circumvent these challenges, the use of iPSCs for generating MSCs is beneficial to preserve their high differentiation potential [28,29].

MSCs can be differentiated into VSMCs to engineer vascular tissues for morphogenesis and functional studies. VSMCs express smooth muscle phenotypic marker genes and proteins, such as α-SMA and calponin, which are early and mid-stage differentiation markers, as well as myosin heavy chain (MHC), smoothelin (SMTN), and smoothelin-B, which represent late-stage differentiation stages. Expression of early and mid-stage markers only indicates progression towards a smooth muscle lineage and are also expressed in other cells; however, the late-stage marker proteins MHC and SMTN [30,31] are exclusively expressed in mature contractile VSMCs. In view of this, biomaterials designed for vascular tissue engineering must promote the differentiation of MSCs to VSMCs by providing mechanical and topological cues [32]. Although MSC differentiation to vascular smooth muscle cells on electrospun fibers have recently been reported, the fibers were prepared from a non-degradable polyacrylonitrile [33] or else the fibers were used only to deliver a biochemical factor [34].

The objectives of this study were twofold. First, we aim to synthesize electrospinnable PαAPz by modifying the synthesis procedure. This is important since most PαAPz are generally difficult for producing electrospun fibrous mats without being blended with other polymers, and only a few previous studies attempted doing this. Secondly, we aim to use the fibrous mats for the differentiation of iPSC-derived MSCs towards vascular smooth muscle cells.

2. Experimental

2.1. Materials

Hexachlorocyclotriphosphazene (HCCP) was obtained from Sigma Aldrich (Milwaukee, WI, USA) and stored in the desiccator until used. Anhydrous tetrahydrofuran (THF) and glass distilled hexanes-190 were acquired from Caledon Labs (Georgetown, ON, Canada). Triethylamine (NEt$_3$), chloroform (CF), and dimethyl sulfoxide (DMSO) were purchased from Sigma Aldrich (Milwaukee, WI, USA). L-alanine ethyl ester hydrochloride (H-Ala-Oet · HCl) and L-phenylalanine ethyl ester hydrochloride (H-Phe-Oet · HCl) were from Alfa Aesar (Ward Hill, MA, USA) and stored in the fridge. Krytox Performance Lubricant GPL207 high-temperature grease was purchased from DuPont (Wilmington, DE, USA). Unless specified otherwise, all chemicals and solvents were used as received.

2.2. Synthesis of PαAPz

Polydichlorophosphazene (PDCP) was prepared using thermal ring-opening polymerization (TROP) with three different approaches, namely: grease sealing, flame sealing, and HCCP recrystallization, followed by flame sealing. In the grease sealing approach, approximately 1 g HCCP was added into a dry glass reactor fitted with a stopper, and argon was purged to displace any residual air on the walls of the reactor. For the flame sealing and recrystallization (FS-R) approach, HCCP was first recrystallized by vacuum sublimation at 100 °C and 20 mmHg before use. Following recrystallization, approximately 2 g of HCCP was added into a glass ampoule, connected to a vacuum line, and flame-sealed using a propane torch. Alternatively, the vacuum step could be replaced by filling the ampoules with dry argon, followed by flame sealing. The sealed ampoules were placed in the oven at 230 °C for 72 h. The samples were recovered from the ampoules, dissolved in THF, and purified by precipitating in hexanes three times to remove the unreacted HCCP and crosslinked PDCP. PDCP recovery and purification were similar for grease sealing and flame sealing.

2.3. Macromolecular Substitution

The macromolecular substitution reaction was carried out by a one-step approach at room temperature, as described before [35]. Briefly, after flame drying, a 50 mL flask of either H-Ala-OEt · HCl (1.70 g, 11.1 mmol, 2.6 equiv.) or H-Phe-OEt · HCl (2.16 g, 11.1 mmol, 2.6 equiv.) was added into the reaction flask, and then, the flask was sealed with a rubber septum. Next, 10 mL of THF and 4 mL of triethylamine were injected into the flask through the septum. Then, PDCP (0.50 g, 4.3 mmol of N = P-Cl$_2$ units) obtained from the TROP was dissolved in 10 mL of anhydrous THF and injected into the reaction flask through the septum. The mixture was stirred for 72 h, filtered to remove the insoluble salt, and the product was dissolved in anhydrous THF and purified by repeated precipitation in hexanes. The PαAPz polymers were then dried in a vacuum and stored at 4 °C.

2.4. PDCP and PαAPz Characterization with ^{31}P-NMR and ^{1}H-NMR

Nuclear Magnetic Resonance (NMR) spectroscopy was performed on a Varian INOVA 400 MHz spectrometer (^{1}H 400.1 MHz, and ^{31}P{1H 161.8 MHz, Varian Canada Inc., Mississauga, ON, Canada). Chemical shifts are reported in parts per million (ppm). All chemicals are dissolved in chloroform (CDCl$_3$) with a concentration of ~40 mg/mL. Chemical shifts were relative to chloroform at δ = 7.27 ppm.

2.5. Preparation of PαAPz Thin Films

PαAPz was dissolved in THF to form a 1 wt.% solution. Then 67.8 μL of the solution was dipped onto a glass coverslip with a diameter of 12 mm. After drying, all of the film was sterilized under the UV light for 30 min and pre-treated with Hank's Balanced Salt Solution (HBSS) overnight.

2.6. Electrospinning of PαAPz

The electrospinning equipment consists of 0.5 mL glass syringes, blunt-tip 22-gauge stainless steel needles, high voltage DC power supply (7–20 kV, ES30P, Gamma High Voltage, Ormond Beach, FL, USA), syringe pump (KD101, KD Scientific, Holliston, MA, USA), and rotating collector. The PαAPz were electrospun at various concentrations and spinning parameters. The morphology of the PαAPz fibrous mat was evaluated using SEM (S-2600N Hitachi, Tokyo, Japan). Fibrous samples were sputter-coated with gold/palladium (K550X, sputter coater, Emitech Ltd., Ashford, UK) and scanned at a working distance of 9 mm and a constant accelerating voltage of 5 kV. Analysis of the SEM was performed with ImageJ software (NIH, Bethesda, MD, USA).

2.7. Cell Culture Studies on PαAPz and Smooth Muscle Cell Differentiation

Mesenchymal stem cells derived from induced pluripotent stem cells (iPSCs), hereinafter named iMSCs (kindly donated by Dr. Dale Laird, Western University, Canada), and human bone marrow-derived mesenchymal stem cells (BMMSCs, Lonza, Walkersville, MD, USA; PT-2501) were used. iMSCs were grown on gelatin (Sigma-Aldrich Canada Co., Oakville, ON, Canada) coated dishes in mesenchymal stem cell expansion media (MSCEM, Cedarlane Labs, Burlington, ON, Canada; HMSC.E.MEDIA-450) supplemented with 10% fetal bovine serum (FBS), 1% l-glutamine, and 1% penicillin/streptomycin (all from Fisher Scientific, Whitby, ON, Canada). Media were changed every other day. Human coronary artery smooth muscle cells (HCASMCs, Lonza, Cohasset, MN, USA. CC-2583) were maintained in Smooth Muscle Cell Growth Medium 2 BulletKit (SmGM-2, Lonza, Cohasset, MN, USA). Cells between passage numbers 4–11 were used for experiments. Cells were cultured in humidified incubators at 37 °C and 5% CO$_2$. Differentiation of stem cells to smooth muscle cell lineage was induced using L-ascorbic acid (L-AA, Sigma-Aldrich Canada Co., Oakville, ON, Canada) and transforming growth factor (TGF-β1, R&D Systems, Minneapolis, MN, USA) in high glucose DMEM, modified based on reported procedures [36]. Briefly, iMSCs were plated either on gelatin-coated dishes for 2D study (2×10^4 cells/cm^2 in 6-cm Corning cell culture dish) or on gelatin-coated electrospun fibrous mats for 3D study (20×10^4 cells per mat with a size of 2 cm $\times$ 2 cm) and pre-cultured in MSCEM to reach 70% of confluence, followed by changing culture media to differentiation condition (DC), composed of DMEM supplemented with 1% FBS plus 82.5 μg/mL [37] L-AA or 2 ng/mL TGF-β1. The induced smooth muscle-like cells were characterized by the detection of smooth muscle marker genes and proteins using qRT-PCR, Western blot analysis, as well as immunofluorescence staining. All fibrous mats were coated with 0.1% gelatin at 37 °C for 1 h before cell seeding. Non-induced iMSCs were grown in regular MSCEM as the undifferentiated growth control (GC).

2.8. Quantitative Real-Time qPCR and Western Blot Analysis

Total RNA was extracted from cells using TRIzol™ Reagent (Life Technologies Inc., Burlington, ON, Canada) following the manufacturer's instructions. One μg of total RNA was reverse transcribed into complementary DNA (cDNA) using the Promega™ Random Hexamers protocol (Fisher Scientific, Whitby, ON, Canada). qRT-PCR was conducted in four repeats using a CFX96™ Real-Time System (C1000 Touch Thermal Cycler; Bio-Rad, Mississauga, ON, Canada), and human genes of interest were determined with SsoAdvanced Universal SYBR® Green Supermix (Bio-Rad, Mississauga, ON, Canada), according to the recommended procedures. The sequences of primers are presented in Table 1. The results were analyzed with the comparative threshold cycle method and normalized with human 18S as an endogenous reference.

Table 1. qRT-PCR primer sequences for human-specific mRNA amplification.

Gene	Forward Primer ($5'{\rightarrow}3'$)	Reverse Primer ($3'{\rightarrow}5'$)
Human *Acta2*	CAA GTG ATC ACC ATC GGA AAT G	GAC TCC ATC CCG ATG AAG GA
Human *Cnn1*	TGA AGC CCC ACG ACA TTT TT	GGG TGG ACT GCA CCT GTG TA
Human *Myh11*	GTC CAG GAG ATG AGG CAG AAA C	GTC TGC GTT CTC TTT CTC CAG C
Human SMTN	CAG GAC AAC AAG GAG AAC TGG	CAG TCA ATT CCT CCA CAT CGT
Human 18S	GCG GTT CTA TTT TGT TGG TTT	CTC CGA CTT TCG TTC TTG ATT

Western blotting was performed to evaluate the expression levels of specific smooth muscle marker proteins. Cell lysates were separated via SDS-PAGE, transferred to nitrocellulose membrane, and immunoblotted using the following primary antibodies: SM-MHC (rabbit, 1:1000; Alfa Aesar (Haverhill, MA, USA), BT-562), SMTN-B (rabbit, 1:1000; Santa Cruz (Dallas, TX, USA), sc-28562), and β-Tubulin (mouse, 1:250; ThermoFisher Scientific (Waltham, MA, USA), MA5-11732). Primary antibody labeling was detected using HRP-conjugated goat anti-rabbit or anti-mouse secondary antibodies, and the ECL detection system.

2.9. Immunofluorescence Microscopy

Cells were fixed in 10% formalin for 10 min at ambient temperature, followed by three washes in PBS. All samples were permeabilized with 0.1% Triton X-100 for 10 min, and then, they were blocked in 2% BSA, with 22.52 mg/mL of glycine in PBS, for 30 min at ambient temperature. Samples were labeled with the following antibodies overnight at 4 °C: α-SMA (mouse, 1:100; Santa Cruz (Dallas, TX, USA), sc-32251), SM-MHC (rabbit, 1:100; Alfa Aesar (Haverhill, MA, USA), BT-562), SMTN-B (rabbit, 1:100; Santa Cruz (Dallas, TX, USA), sc-28562), and SMTN (rabbit, 1:100; Santa Cruz (Dallas, TX, USA), sc-166292). Primary antibody binding was detected using Alexa Fluor® 555 goat anti-rabbit IgG or Alexa Fluor® 488 goat anti-mouse IgG as secondary antibodies (1:300) addition with Alexa Fluor® 568-

or Alexa Fluor® 488 phalloidin (1:100 dilution, all from Life Technologies, Burlington, ON, Canada), in some cases. The cells were counterstained with 4′6-diamidino-2-phenylindole (DAPI, 300 nM in PBS, Life Technologies, Burlington, ON, Canada) for labeling nuclei. Coverslips were mounted on microscope slides with PermaFluor™ Mounting Medium (Fisher Scientific™, Whitby, ON, Canada) and sealed with clear nail enamel. Images were taken with a Zeiss LSM 800 confocal microscope (Zeiss, Toronto, ON, Canada) equipped with $10\times/25\times$ (water) lenses and analyzed using ZEN 3.1 (blue edition) software.

2.10. Statistical Analysis

Data are presented as mean $\pm$ SD unless otherwise indicated. At least three independent experiments were analyzed by one-way ANOVA with Student's two-tailed independent sample *t* test. All analyses were done using GraphPad 5.0 statistical software (GraphPad Software) and significance was assigned for $p < 0.05$

3. Results and Discussion

3.1. PαAPz Synthesis and Characterization

The synthesis of PαAPz is a two-step procedure. In the first step, the thermal ring-opening polymerization of HCCP at high temperature (ca. 230 °C) produces PDCP. In the second step, the chlorine atoms attached to the central phosphorous are replaced with side chains via a macromolecular substitution reaction at room temperature (Figure 1A). The thermal ring-opening reaction requires a strict anhydrous environment and specialized equipment. In this study, we attempted to simplify the complexity involved in the thermal ring-opening reaction while, at the same time, reducing the undesirable hydrolysis and crosslinking. Since the primary objective in the ring-opening reaction is to avoid moisture, we attempted six different approaches (Table 2). In the first approach (GS method), we used high-temperature fluorinated grease as a potential moisture barrier when sealing the reaction content using a glass stopper [35]. The main component of this grease is perfluoroalkyl ether, which was used for food packaging and proved to have the ability to provide moisture resistance [38]. With the GS method, the product crosslinked after 25 h reaction, so the reaction was stopped at this time. In a modification of the GS procedure, we purged the glass reaction vial with argon and sealed it with the same grease and glass stopper, with the expectation that the argon may displace any residual moisture/air present around the walls of the glass vial due to its high density (GS-Ar method). Indeed, the conversion of HCCP to PDCP increased from 54% to 86% when argon was used to displace air, although it was crosslinked after 54 h of reaction. For the remaining four thermal ring-opening reaction methods, flame sealing of the glass ampoule was used. Thus, flame sealing (FS), flame sealing using argon (FS-Ar), flame sealing under vacuum (FS-vacuum), and recrystallization of HCCP followed by flame sealing under vacuum (R-FS-vacuum) were explored. Both FS-vacuum and R-FS-vacuum methods produced PDCP that was not crosslinked during 92 h of reaction at 230 °C. In addition, the conversion of HCCP to PDCP increased to 95%; thus, FS-vacuum and R-FS-vacuum methods were taken as appropriate methods to produce linear PDCP for the subsequent macromolecular substitution with α-amino acid esters. The thermal ring-opening polymerization time of HCCP to produce PDCP is up to 120 h, and the goal is to get very high conversions without crosslinking. Since the reaction occurs in the melt, the occurrence of crosslinking is easily detected by visually inspecting if the melt still flows when the ampoule is held upside down.

^{31}P-NMR spectra for HCCP, linear PDCP, PαAPz, and ^{1}H-NMR spectrum of poly [bis (ethyl alanato) phosphazene] (PαAPz-A) are shown in Figure 1B,C. Consistent with the literature, HCCP shows a single peak at 21.09 ppm, and PDCP, which is the product from the thermal ring-opening reaction after purification by precipitation, showed a singlet at approximately −17 ppm. In Figure 1B, representative ^{31}P-NMR spectra of PDCP from grease sealing and flame sealing methods before purification are also shown (B2, B3).

Table 2. Comparison of different methods to synthesize PDCP.

Method	Reaction Time at 230 °C	Integration of +20 ppm Peak	Integration of −17 ppm Peak	% Conversion
GS	25 h	1	1.19	54
GS-Ar	54 h	1	6.51	86
FS	58 h	1	5.70	85
FS-Ar	93 h	1	5.32	84
FS-Vacuum	92 h	1	18.91	95
R-FS-Vacuum	92 h	1	15.38	94

GS: grease sealing. GS-Ar: grease sealing with Ar. FS: flame sealing. FS-Ar: flame sealing with Ar. FS-Vacuum: flame sealing under vacuum. R-FS-Ar: recrystallization followed by flame sealing under vacuum.

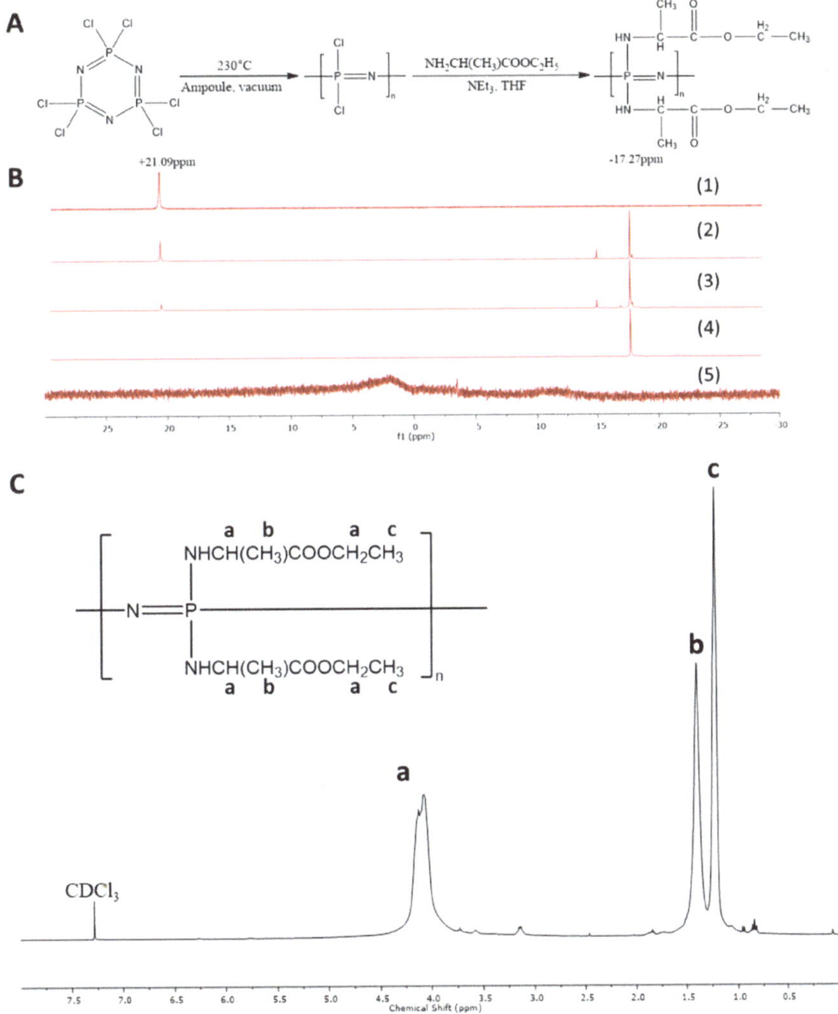

Figure 1. (**A**) Synthesis scheme of PαAPz-A. (**B**) ^{31}P NMR of (1) HCCP, (2) Crude PDCP from GS method, (3) Crude PDCP from FS, and (4) purified PDCP, (5) PαAPz-A. (**C**) ^{1}H NMR of PαAPz-A.

In the ^{31}P-NMR spectrum of PαAPz, after the macromolecular substitution with alanine ester, the peak at −17.27 ppm, representing the phosphorous atom on PDCP, disappeared and there is only a small peak at +1.41 ppm (Figure 1B5) which lies between the reported values of +1.2 to +1.9 for PαAPz [39]. The complete disappearance of the peak at −17.27 ppm means that most, if not all, Cl atoms in the PDCP have been replaced. In the ^{1}H-NMR, all the expected peaks for the ethyl ester of alanine are shown, which confirmed the successful substitution. As reported before [35], the hydrogen connected to the amino group was not seen in the ^{1}H-NMR spectrum. In addition, the ^{1}H-NMR demonstrated that the unreacted amino acid had been fully removed during the precipitation process. In the present study, we used L-alanine and L-phenylalanine as the preferred amino acids for the following two reasons. (i) From our previous study [35], we observed that both of these amino acids produced PαAPz that have comparable molecular weights that could be suitable for electrospinning. (ii) Both of these amino acids are neutral and non-polar, yet the glass transition temperatures (T_g) of the corresponding PαAPz are different. PαAPz-A has a T_g that is in the range of −20 °C to −10 °C making it flexible at a cell culture temperature of 37 °C, whereas PαAPz-F has a T_g above 37 °C [35,40] that could display some brittleness during cell culture handling and subsequent staining.

3.2. Electrospinning of PαAPz

Although PαAPz have been used for drug delivery studies in solid and film forms, electrospinning allows for the formation of fibers that can be utilized for tissue engineering. However, standalone fiber formation from PαAPz is often challenging except for a few reports [15,41]. Therefore, we have studied a wide range of electrospinning parameters and solvent combinations to produce fibrous mats (Table 3). The PαAPz-A and PαAPz-F prepared in the current study were electrospinnable from a solution of THF or CF, but that depended on the method of the PDCP preparation. Although PαAPz-A and PαAPz-F were successfully synthesized from PDCP that was prepared by all the methods listed in Table 2, only PDCP from FS-Vacuum and R-FS-Vacuum methods resulted in PαAPz that could be consistently electrospun. PDCP prepared by grease sealing resulted in a corresponding PαAPz that led to electrosprayed particles instead of fibers attributed to hydrolysis of PDCP during its synthesis and subsequent lower molecular weight of PαAPz. It is known that lower molecular weight polymers often produce electrosprayed particles rather than continuous fibers [42–44] due to the lower solution viscosity that makes the Taylor cone unstable during electrospinning [45]. Figure 2 shows the SEM images of the electrospun PαAPz fiber mats that were synthesized from PDCP and, in turn, were prepared from the FS method, with both low magnification (large scan area) and high magnification (small scan area). The fibers were free from defects and had diameters below 0.7 μm with averages between 0.1 μm and 0.3 μm.

Table 3. Studied electrospinning parameters for PαAPz.

Sample Abbreviations	Solvent Ratio	Distance (cm)	Voltage (kV)	Concentration (wt%)	Flow Rate (mL/h)
PDCP from FS-vacuum method					
PαAPz-A	CF:DMSO (3:1)	12–15(*12)	15–20(*20)	7.5–12.5(*10)	0.2–0.6(*0.2)
PαAPz-F	THF:CF (9:1)	9–15(*12)	12–20(*12)	10–15(*10)	0.2–0.6(*0.2)
PDCP from R-FS-vacuum method					
PαAPz-A	CF:DMSO (3:1)	12–15(*12)	15–20(*15)	7.5–12.5(*10)	0.2–0.6(*0.2)
PαAPz-A	THF:CF (9:1)	12–15(*12)	15–20(*15)	7.5–12.5(*10)	0.2–0.6(*0.2)
PαAPz-F	THF:CF (9:1)	9–15(*12)	12–20(*12)	10–15(*10)	0.2–0.6(*0.2)

* Optimal electrospinning parameters. CF: Chloroform, THF: tetrahydrofuran, DMSO: dimethylsulfoxide. A—Alanine and F—Phenylalanine.

For the PαAPz-A polymers, the CF:DMSO solvent combination was used to electrospin them. When the CF alone was used, the needle tip was clogged due to the rapid evaporation of the solvent, while the fibers collected also had defects due to beading. Thus, 25% DMSO was added to address both challenges. As an alternative, it was also possible to electrospin PαAPz-A polymers from THF:CF (9:1) ratio using the indicated parameters of Table 3. It is noteworthy that PαAPz-F polymers were only electrospun in the THF:CF (9:1) ratio. The fiber size distribution appeared to be influenced by the PDCP synthesis approach rather than the solvent combination since the recrystallization of HCCP, followed by flame sealing, produced PαAPz that have slightly higher solution viscosity, at the same concentration, compared with those PαAPz that were prepared from PDCP without HCCP recrystallization.

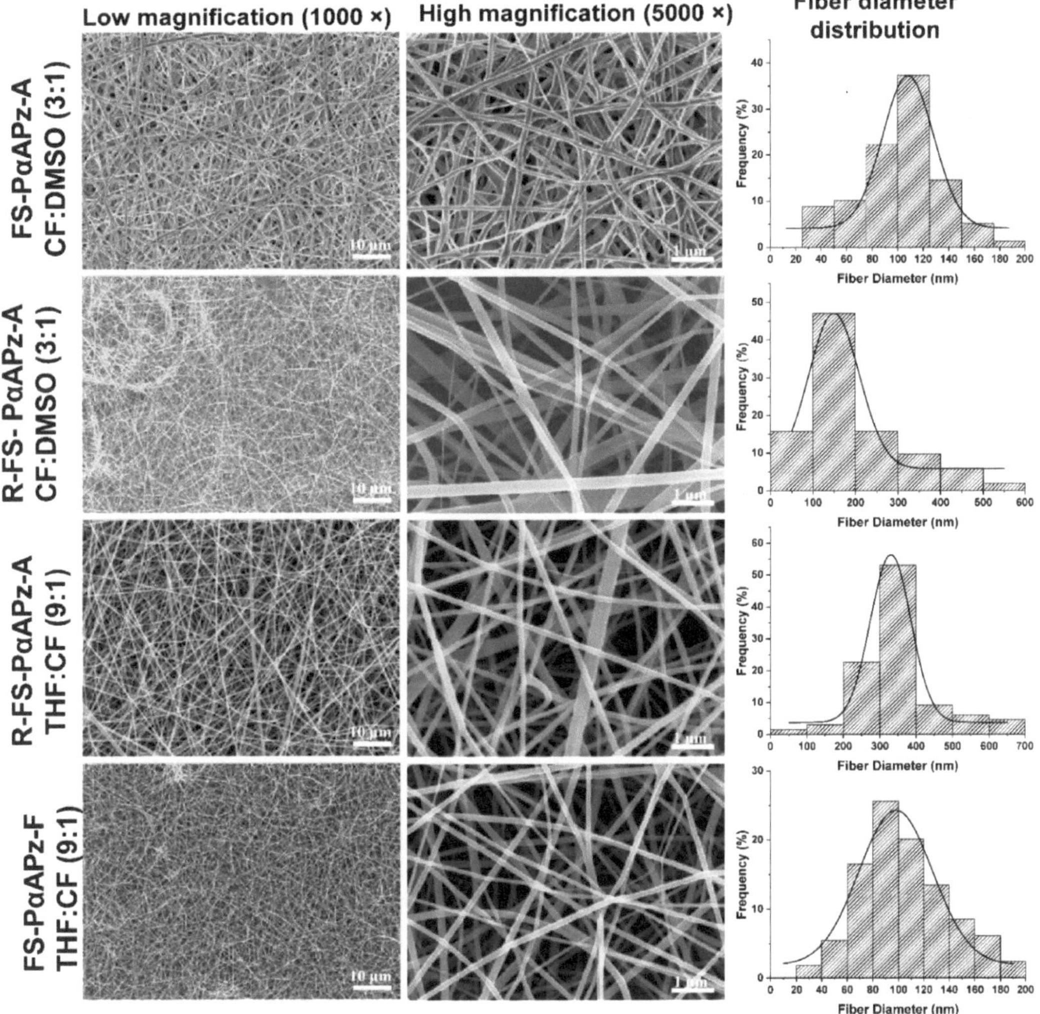

Figure 2. SEM images of electrospun PαAPz derived from L-phenylalanine and L-alanine, as well as the corresponding histogram, showing the fiber diameter distribution. For each SEM image, the PDCP synthesis method (FS—flame sealing; R-FS: recrystallization followed by flame sealing) and the optimum solvent ratio are shown. Other electrospinning parameters are listed in Table 3.

3.3. Short-Term In Vitro Degradation Study

PαAPz are biodegradable polymers due to the presence of α-amino acid residues. While this is a desired property for tissue engineering, rapid degradation (within 3–5 days) of the electrospun fibers is a disadvantage, since infiltration of seeded cells into the porous mats will be minimal. Cell infiltration into electrospun fibers is generally challenging, and early degradation exacerbates it as the fibers fuse and thus limit infiltration. To gain an insight into the early degradation of the fiber mats, samples were incubated in PBS at 37 °C for up to 5 days. As presented in Figure 3, the fiber morphologies of PαAPz-A did not change during the 5 days incubation; however, progressive fiber thickening, flattening, and eventual fusion was observed for the PαAPz-F. After five days of incubation, the fibrous mats lost their porosity considerably, presumably due to the propensity of L-phenylalanine to hydrolysis compared with alanine. In addition, PαAPz-F mats displayed some brittleness, making them difficult for cell culture handling and subsequent staining. All cell culture experiments were therefore conducted on PαAPz-A.

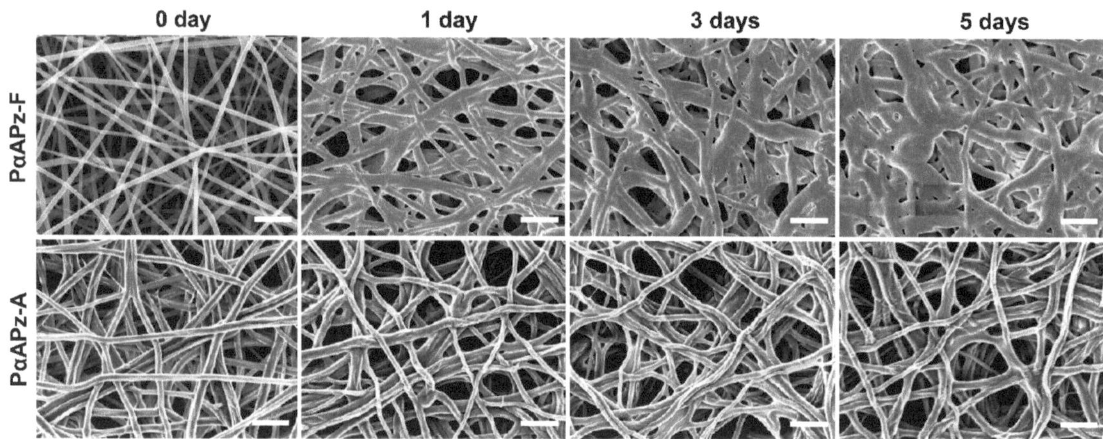

Figure 3. SEM images showing the short-term in vitro degradation study of electrospun mats fabricated from PαAPz, derived from the amino acids L-phenylalanine (PαAPz-F) and L-alanine (PαAPz-A). PαAPz-A fibers lost their porous morphology faster than PαAPz-A fibers. Scale bar = 1 μm and is applicable to all images.

3.4. Cell Adhesion and Morphology on PαAPz-A Films

To investigate in vitro cell adhesion and morphology, three distinct cell types were seeded on PαAPz-A films over a three-day period. Since primary human vascular cells are not readily available, we evaluated the potential of obtaining them by differentiating bone marrow-derived mesenchymal stem cells (BMMSC) and mesenchymal stems cells (iMSC) from induced pluripotent stem cells. Human coronary artery smooth muscle cells (HCASMC) were used as control. Confocal microscopy images shown in Figure 4A demonstrated that all three types of cells attached and spread on the surface of PαAPz-A films after three days of culture and were morphologically indistinguishable from cells cultured on the glass coverslips. All three types of cells showed abundant F-actin expression; however, BMMSCs had a better attachment, as judged by the dense cell layer observed. As we have initially seeded the same number of cells, and the culture time is only three days, the observed high cell density in BMMSC is related to high retention rather than cell growth. Although cell retention is high for BMMSC, their application as a cell source is not as robust compared with iMSC since they require invasive procedures to obtain them, and they have replicative senescence during expansion [46]. Compared to primary HCASMC, iMSC's retention on the PαAPz-A film was high, suggesting that these cells were viable. Thus, iMSC attachment and growth were further evaluated on the electrospun fibrous mats by F-actin fluorescence staining after culturing iMSCs on PαAPz-A fibrous mats for 2 days (Figure 4(B1)) and 4 days (Figure 4(B2)). As seen from these confocal images, iMSCs were well-spread on the fibrous mats, characterized by well-defined F-actin stress fibers with a typical small cell body that is of long and thin morphology, which is characteristic of stem cells, suggesting that the scaffold microenvironment was favorable for cell attachment and growth.

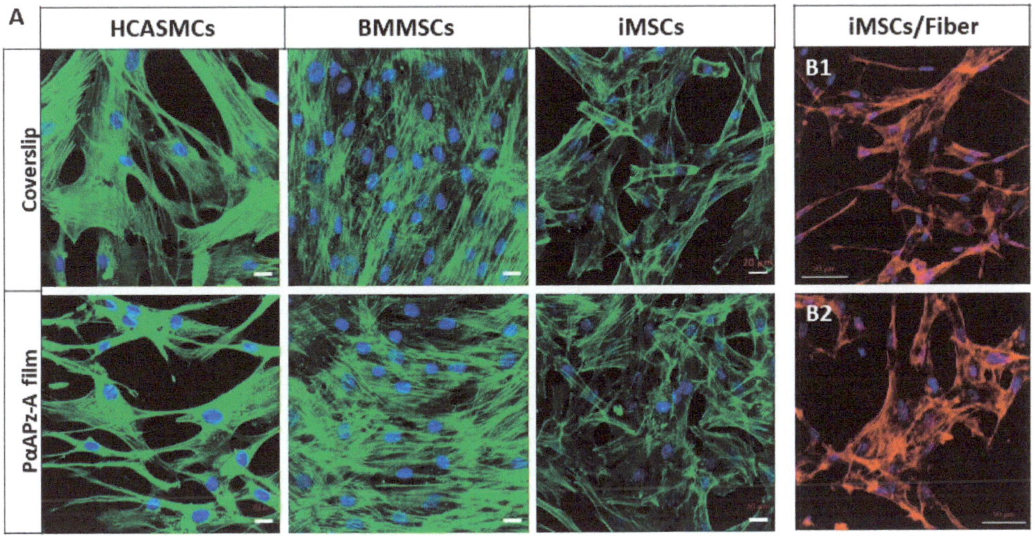

Figure 4. (**A**) Fluorescence images of different cell types on PαAPz-A films and on coverslips. After 3 days of culture, F-actin is labeled by Phalloidin in green, and nuclei are labeled by DAPI in blue. Scale bar = 20 μm. (**B1,B2**) Fluorescence images of iMSCs after 2 days (**B1**) and 4 days (**B2**) of culture on PαAPz-A electrospun fibrous mats. Red channel is F-actin, and nuclei are labeled by DAPI in blue. Scale bar = 50 μm.

3.5. Differentiation Potential of iMSCs towards Smooth Muscle Phenotype

After establishing that PαAPz-A fibrous mats were able to support attachment and spreading of iMSCs, we next investigated if these cells have the differentiation potential towards the vascular smooth muscle phenotype. We chose iMSCs because BMMSCs have a relative ease to undergo osteogenic and adipogenic differentiation compared to vascular differentiation, which is more complex [47–49]. Although direct differentiation of partially induced pluripotent stem cell to a VSMC has been reported [50], those obtained from a fully induced pluripotent cells via a mesenchymal precursor cells are the most established method [51–53]. To determine the differentiation capacity of iMSCs into smooth muscle phenotype, the expression of smooth muscle-specific marker genes and proteins was analyzed under 2D conventional tissue culture conditions, and results are shown in Figure 5. iMSCs were plated on gelatin-coated culture dishes in stem cell culture media. Once the cells reach 70% of confluency, the culture media were changed to the following for seven days: (i) differentiation culture (DC) media; (ii) DC + 82.5 μg/mL L-Ascorbic acid (L-AA); (iii) DC + 2 ng/mL TGF-β1. A regular growth condition (GC) was used as a control. As shown in Figure 5A, DC media was able to significantly upregulate the expression of smooth muscle marker gene *Acta2* ($p < 0.05$), *Cnn1* ($p < 0.05$), and *SMTN* ($p < 0.001$) but not the late-stage marker *Myh11* ($p > 0.05$), which suggested that iMSCs possess an intrinsic differentiation capacity towards the smooth muscle phenotype. Upon the addition of AA, the expression levels of *Cnn1* and *SMTN* were further increased.

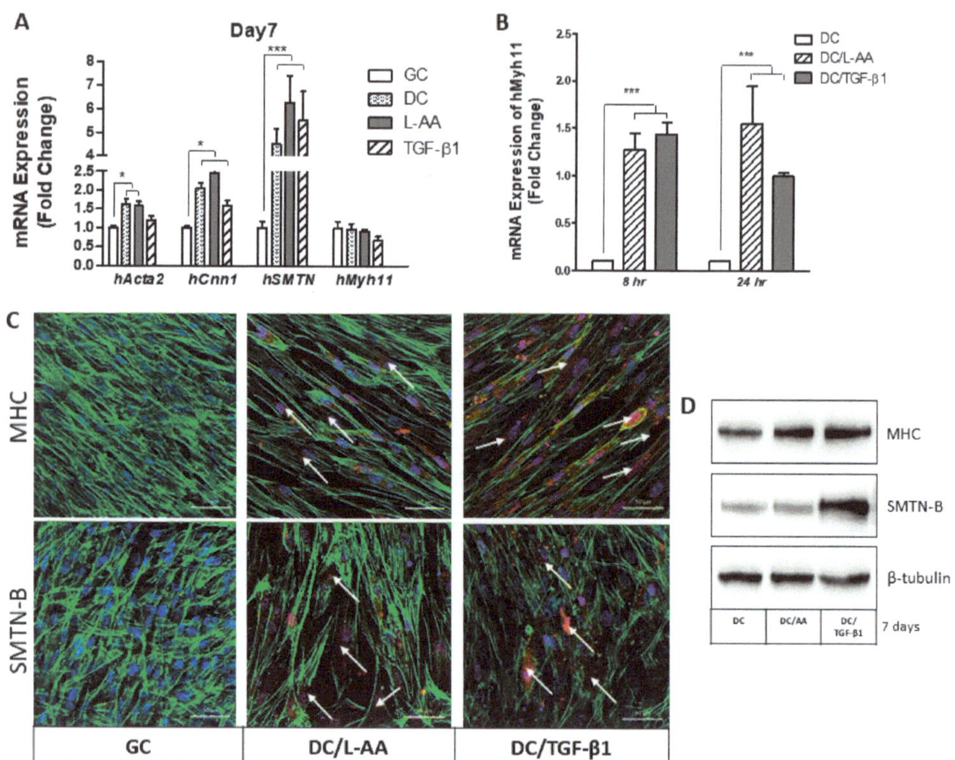

Figure 5. Differentiation potential of iMSCs towards vascular smooth muscle lineage in conventional 2-D cultures. (**A**) iMSC mRNA expression of *hActa2*, *hCnn1*, *hSMTN*, and *hMyh11* was analyzed using Quantitative real-time polymerase chain reaction (qRT-PCR) after cell treatment with L-AA and TGF-β1 for 7 days. GC stands for regular growth condition of iMSCs, and DC stands for differentiation inducible condition by exchanging culture media to DMEM with 1% FBS. (**B**) mRNA expression of *hMyh11* by iMSCs after 8 h and 24 h of cultivation under DC. (**C**) iMSCs were cultured on coverslips, under differential conditions, for 7 days, MHC and SMTN-B were detectable with either AA or TGF-β1 pre-treatment. Green channel is F-actin, and Red is MHC/SMTN-B. Nuclei are labeled by DAPI in Blue. Scale bar = 50 μm. (**D**) Western blot analysis revealed that vascular smooth muscle contractile marker SMTN-B was up-regulated by pre-treatment with TGF-β1, whereas MHC was slightly up-regulated by either L-AA or TGF-β1 pre-treatment. * $p < 0.05$; *** $p < 0.001$.

TGF-β1 is one of the major cytokines that can induce VSMC differentiation of stem/progenitor cells, alone or in combination with other factors. Interestingly, the addition of TGF-β1 into DC media did not provide a synergistic effect on mRNA expression. Moreover, the late-stage marker *Myh11* was less responsive to differentiation inducers after 7 days of culture, which suggested that the *Myh11* marker gene may have been expressed earlier than 7 days. To verify the expression pattern of *Myh11*, iMSCs were differentiated for 8 h and 24 h, using L-AA and TGF-β1, in DC media. As shown in Figure 5B, *Myh11* was notably upregulated by L-AA but was less responsive to TGF-β1. We subsequently performed immunofluorescence staining to detect proteins of MHC and SMTN, two key smooth muscle markers, and the representative confocal images are presented in Figure 5C. Both marker proteins were extensively expressed in DC-treated cells with the addition of L-AA and TGF-β1. Moreover, the expressions of two SMC-specific contractile proteins were detectable, after seven days of culture, using Western blot analysis (Figure 5D). Although MHC protein remains unchanged by either L-AA or TGF-β1 treatment, protein expression of SMTN-B, a specific differentiation marker exclusively expressed in vascular smooth muscle tissues, was upregulated with the addition of TGF-β1.

3.6. Differentiation of iMSCs towards Smooth Muscle Phenotype on Electrospun PαAPz-A Fibrous Mats

The preceding cell differentiation results were done on conventional cell culture dishes to rule out the effect of the fiber mat topography. To determine if the microenvironment provided by electrospun PαAPz-A fibrous mats is favorable for vascular differentiation of iMSC, we performed immunofluorescence staining of key vascular smooth muscle markers (α-SMA, MHC, SMTN, and SMTN-B) that represent different stages of smooth muscle differentiation. Stem cell culture on synthetic materials is challenging, as it may transform them into subpopulations with reduced multipotency and fibrotic character; therefore, the microenvironment provided by PαAPz-A fibrous mats with the right physical cues is indispensable for the highly efficient proliferation and differentiation of iMSCs. After seven days of differentiation on PαAPz-A fibrous mats, protein markers α-SMA (green, Figure 6(A1)) and MHC (green, Figure 6(B1)) were observed in L-AA-treated iMSCs, and the fluorescence intensity of early-stage marker protein α-SMA was further enhanced in the presence of TGF-β1 (Figure 6(A2)); however, we did not observe a further increase in the late-stage protein marker MHC (green, Figure 6(B2)) with the addition of TGF-β1. After 12 days of iMSC differentiation, a more contractile phenotype of smooth muscle cells was observed, as verified by immunofluorescence staining of two contractile-specific protein markers, SMTN (Figure 6(C1)) and MHC (Figure 6(C2)), respectively. In addition, SMTN-B is a long isoform of smoothelin that is specifically detectable in vascular smooth muscle tissues and used to discriminate VSMCs from myofibroblasts [54,55]. Consistent with the observation of Western blot analysis in Figure 5D, the co-localization of SMTN-B and F-actin (Figure 6(D1,D2)) confirmed the iMSC differentiation into the smooth muscle phenotype on PαAPz-A fibrous mats. Overall, the immunostaining of SMTN and MHC for cells cultured on the PαAPz-A strongly indicates robust expression compared to conventional cultures shown in Figure 5. Furthermore, differentiated cells were aligned and elongated on the fibrous mats, suggesting directional orientation, presumably due to cues provided by the fiber topography.

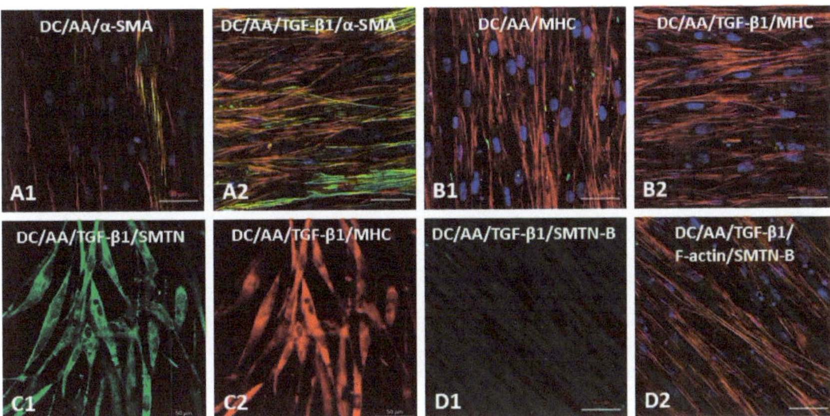

Figure 6. Characteristic analysis of smooth muscle markers by iMSCs seeded on PαAPz-A fibrous mats addition with L-AA or L-AA/TGF-β1 in DC culture media. After 7 days of pre-differentiation, α-SMA (**A1,A2**) and MHC (**B1,B2**) were determined by immunofluorescence staining. (**C1,C2**) Representative confocal images of SMTN (green) and MHC (red) by iMSCs after 12 days of cultivation. (**D1,D2**) Co-localization of SMTN-B (green) and F-actin (red) by iMSCs after 12 days of cultivation. Scale bar = 50 μm.

PαAPz polymers have been previously studied, almost exclusively, as orthopedic biomaterials, including tendon and ligament [2,3], but their use in vascular smooth muscle differentiation of stem cell tissues has not been reported before. Regarding other vascular cells, to the best of our knowledge, only two previous studies [14,15] reported endothelial cell interaction with PαAPz fibers. Furthermore, standalone electrospinning of PαAPz is challenging, and it is often modified with other polymers such as polycaprolactone [8,35,56,57]. In this study, we showed for the first time that iMSCs could differentiate to vascular smooth muscle cells on standalone electrospun PαAPz-A mats. The presence of elongated and spindle-shaped cells, as well as abundant expression of specific proteins related the vascular smooth muscle cell lineage, demonstrated differentiation. These studies provide strong evidence on the utility of electrospun PαAPz-A as a conducive microenvironment for stem cell differentiation.

4. Conclusions

In this study, we first synthesized two PαAPz polymers from L-alanine and L-phenyl alanine. Both PαAPz were electrospinnable and produced beads free fibers with average diameters between 0.1 μm and 0.3 μm. The electrospun fibers from L-phenyl alanine-based PαAPz started to degrade faster than the corresponding PαAPz from L-alanine and became brittle. The PαAPz fibrous mats from L-alanine were able to support adhesion and spreading of iMSC, BM-MSC, and primary human coronary artery SMC. More importantly, electrospun fibers from PαAPz from L-alanine were able to promote differentiation of iMSCs towards SMC lineage. Taken together, our data suggest the utility of PαAPz for differentiating stem cells for tissue engineering applications.

Author Contributions: Conceptualization, K.M.; Formal analysis, M.W. and S.L.; Funding acquisition, K.M.; Investigation, M.W. and S.L.; Methodology, M.W., S.L. and K.M.; Supervision, K.M.; Writing—original draft, M.W. and S.L.; Writing—review & editing, K.M. All authors have read and agreed to the published version of the manuscript.

Funding: This work was supported by the Natural Science and Engineering Research (NSERC) of Canada (Grant # RGPN-2018-06310).

Institutional Review Board Statement: Not applicable.

Informed Consent Statement: Not applicable.

Data Availability Statement: The data presented in this study are available on request from the corresponding author.

Conflicts of Interest: The authors declare no conflict of interest.

References

1. Ogueri, K.S.; Allcock, H.R.; Laurencin, C.T. Polyphosphazene polymers: The next generation of biomaterials for regenerative engineering and therapeutic drug delivery. *J. Vac. Sci. Technol. B. Nanotechnol. Microelectron.* **2020**, *38*, 030801. [CrossRef] [PubMed]
2. Ogueri, K.S.; Ogueri, K.S.; Ude, C.C.; Allcock, H.R.; Laurencin, C.T. Biomedical applications of polyphosphazenes. *Med. Devices Sens.* **2020**, *3*, 10113. [CrossRef] [PubMed]
3. Ogueri, K.S.; Escobar Ivirico, J.L.; Nair, L.S.; Allcock, H.R.; Laurencin, C.T. Biodegradable polyphosphazene-based blends for regenerative engineering. *Regen. Eng. Transl. Med.* **2017**, *3*, 15–31. [CrossRef] [PubMed]
4. Shi, J.; Yu, L.; Ding, J. PEG-based thermosensitive and biodegradable hydrogels. *Acta Biomater.* **2021**, *128*, 42–59. [CrossRef] [PubMed]
5. Decollibus, D.P.; Marin, A.; Andrianov, A.K. Effect of environmental factors on hydrolytic degradation of water-soluble polyphosphazene polyelectrolyte in aqueous solutions. *Biomacromolecules* **2010**, *11*, 2033–2038. [CrossRef] [PubMed]
6. Andrianov, A.K.; Marin, A. Degradation of polyaminophosphazenes: Effects of hydrolytic environment and polymer processing. *Biomacromolecules* **2006**, *7*, 1581–1586. [CrossRef] [PubMed]
7. Ogueri, K.S.; Jafari, T.; Escobar Ivirico, J.L.; Laurencin, C.T. Polymeric biomaterials for scaffold-based bone regenerative engineering. *Regen. Eng. Transl. Med.* **2019**, *5*, 128–154. [CrossRef] [PubMed]
8. Peach, M.S.; Ramos, D.M.; James, R.; Morozowich, N.L.; Mazzocca, A.D.; Doty, S.B.; Allcock, H.R.; Kumbar, S.G.; Laurencin, C.T. Engineered stem cell niche matrices for rotator cuff tendon regenerative engineering. *PLoS ONE* **2017**, *12*, e0174789. [CrossRef]

9. Baillargeon, A.L.; Mequanint, K. Biodegradable polyphosphazene biomaterials for tissue engineering and delivery of therapeutics. *Biomed. Res. Int.* **2014**, *2014*, 761373. [CrossRef]

10. Brown, J.L.; Nair, L.S.; Laurencin, C.T. Solvent/non-solvent sintering: A novel route to create porous microsphere scaffolds for tissue regeneration. *J. Biomed. Mater. Res. B* **2008**, *86*, 396–406. [CrossRef]

11. Nukavarapu, S.P.; Kumbar, S.G.; Brown, J.L.; Krogman, N.R.; Weikel, A.L.; Hindenlang, M.D.; Nair, L.S.; Allcock, H.R.; Laurencin, C.T. Polyphosphazene/nano-hydroxyapatite composite microsphere scaffolds for bone tissue engineering. *Biomacromolecules* **2008**, *9*, 1818–1825. [CrossRef] [PubMed]

12. Bhattacharyya, S.; Kumbar, S.G.; Khan, Y.M.; Nair, L.S.; Singh, A.; Krogman, N.R.; Brown, P.W.; Allcock, H.R.; Laurencin, C.T. Biodegradable polyphosphazene-nanohydroxyapatite composite nanofibers: Scaffolds for bone tissue engineering. *J. Biomed. Nanotechnol.* **2009**, *5*, 69–75. [CrossRef] [PubMed]

13. Conconi, M.T.; Lora, S.; Menti, A.M.; Carampin, P.; Parnigotto, P.P. In vitro evaluation of poly[bis(ethyl alanato)phosphazene] as a scaffold for bone tissue engineering. *Tissue Eng.* **2006**, *12*, 811–819. [CrossRef] [PubMed]

14. Conconi, M.T.; Lora, S.; Baiguera, S.; Boscolo, E.; Folin, M.; Scienza, R.; Rebuffat, P.; Parnigotto, P.P.; Nussdorfer, G.G. In vitro culture of rat neuromicrovascular endothelial cells on polymeric scaffolds. *J. Biomed. Mater. Res. A* **2004**, *71*, 669–674. [CrossRef] [PubMed]

15. Carampin, P.; Conconi, M.T.; Lora, S.; Menti, A.M.; Baiguera, S.; Bellini, S.; Grandi, C.; Parnigotto, P.P. Electrospun polyphosphazene nanofibers for in vitro rat endothelial cells proliferation. *J. Biomed. Mater. Res. A* **2007**, *80*, 661–668. [CrossRef]

16. Srinath, D.; Lin, S.; Knight, D.K.; Rizkalla, A.S.; Mequanint, K. Fibrous biodegradable l-alanine-based scaffolds for vascular tissue engineering. *J. Tissue Eng. Regen. Med.* **2014**, *8*, 578–588. [CrossRef]

17. Allo, B.A.; Lin, S.; Mequanint, K.; Rizkalla, A.S. Role of bioactive 3D hybrid fibrous scaffolds on mechanical behavior and spatiotemporal osteoblast gene expression. *ACS Appl. Mater. Interfaces* **2013**, *5*, 7574–7583. [CrossRef]

18. Afra, S.; Matin, M.M. Potential of mesenchymal stem cells for bioengineered blood vessels in comparison with other eligible cell sources. *Cell Tissue Res.* **2020**, *380*, 1–13. [CrossRef]

19. Gong, Z.; Niklason, L.E. Small-diameter human vessel wall engineered from bone marrow-derived mesenchymal stem cells (hMSCs). *FASEB J.* **2008**, *22*, 1635–1648. [CrossRef]

20. Gu, W.; Hong, X.; Le Bras, A.; Nowak, W.N.; Issa Bhaloo, S.; Deng, J.; Xie, Y.; Hu, Y.; Ruan, X.Z.; Xu, Q. Smooth muscle cells differentiated from mesenchymal stem cells are regulated by microRNAs and suitable for vascular tissue grafts. *J. Biol. Chem.* **2018**, *293*, 8089–8102. [CrossRef]

21. Liu, C.; Tsai, A.L.; Li, P.C.; Huang, C.W.; Wu, C.C. Endothelial differentiation of bone marrow mesenchyme stem cells applicable to hypoxia and increased migration through Akt and NFκB signals. *Stem Cell Res. Ther.* **2017**, *8*, 29. [CrossRef] [PubMed]

22. Pittenger, M.F.; Martin, B.J. Mesenchymal stem cells and their potential as cardiac therapeutics. *Circ. Res.* **2004**, *95*, 9–20. [CrossRef] [PubMed]

23. Zhou, R.; Zhu, L.; Fu, S.; Qian, Y.; Wang, D.; Wang, C. Small diameter blood vessels bioengineered from human adipose-derived stem cells. *Sci. Rep.* **2016**, *6*, 35422. [CrossRef] [PubMed]

24. Beeravolu, N.; McKee, C.; Alamri, A.; Mikhael, S.; Brown, C.; Perez-Cruet, M.; Chaudhry, G.R. Isolation and characterization of mesenchymal stromal cells from human umbilical cord and fetal placenta. *J. Vis. Exp.* **2017**, *122*, e55224. [CrossRef]

25. Bieback, K.; Kern, S.; Kocaömer, A.; Ferlik, K.; Bugert, P. Comparing mesenchymal stromal cells from different human tissues: Bone marrow, adipose tissue and umbilical cord blood. *Biomed. Mater. Eng.* **2008**, *18*, S71–S76.

26. Sensebé, L.; Gadelorge, M.; Fleury-Cappellesso, S. Production of mesenchymal stromal/stem cells according to good manufacturing practices: A review. *Stem Cell Res. Ther.* **2013**, *4*, 66. [CrossRef]

27. Wagner, W.; Horn, P.; Castoldi, M.; Diehlmann, A.; Bork, S.; Saffrich, R.; Benes, V.; Blake, J.; Pfister, S.; Eckstein, V.; et al. Replicative senescence of mesenchymal stem cells: A continuous and organized process. *PLoS ONE* **2008**, *3*, e2213. [CrossRef]

28. Kang, L.; Wang, J.; Zhang, Y.; Kou, Z.; Gao, S. iPS cells can support full-term development of tetraploid blastocyst-complemented embryos. *Cell Stem Cell* **2009**, *5*, 135–138. [CrossRef]

29. Park, I.H.; Zhao, R.; West, J.A.; Yabuuchi, A.; Huo, H.; Ince, T.A.; Lerou, P.H.; Lensch, M.W.; Daley, G.Q. Reprogramming of human somatic cells to pluripotency with defined factors. *Nature* **2008**, *451*, 141–146. [CrossRef]

30. Hsia, L.T.; Ashley, N.; Ouaret, D.; Wang, L.M.; Wilding, J.; Bodmer, W.F. Myofibroblasts are distinguished from activated skin fibroblasts by the expression of AOC3 and other associated markers. *Proc. Natl. Acad. Sci. USA* **2016**, *113*, E2162–E2171. [CrossRef]

31. Von Kleeck, R.; Castagnino, P.; Roberts, E.; Talwar, S.; Ferrari, G.; Assoian, R.K. Decreased vascular smooth muscle contractility in Hutchinson-Gilford Progeria Syndrome linked to defective smooth muscle myosin heavy chain expression. *Sci. Rep.* **2021**, *11*, 10625. [CrossRef] [PubMed]

32. Rothdiener, M.; Hegemann, M.; Uynuk-Ool, T.; Walters, B.; Papugy, P.; Nguyen, P.; Claus, V.; Seeger, T.; Stoeckle, U.; Boehme, K.A.; et al. Stretching human mesenchymal stromal cells on stiffness-customized collagen type I generates a smooth muscle marker profile without growth factor addition. *Sci. Rep.* **2016**, *6*, 35840. [CrossRef] [PubMed]

33. Fakhrieh, M.; Darvish, M.; Ardeshirylajimi, A.; Taheri, M.; Omrani, M.D. Improved bladder smooth muscle cell differentiation of the mesenchymal stem cells when grown on electrospun polyacrylonitrile/polyethylene oxide nanofibrous scaffold. *J. Cell Biochem.* **2019**, *120*, 15814–15822. [CrossRef] [PubMed]

34. Ezhilarasu, H.; Sadiq, A.; Ratheesh, G.; Sridhar, S.; Ramakrishna, S.; Ab Rahim, M.H.; Yusoff, M.M.; Jose, R.; Reddy, V.J. Functionalized core/shell nanofibers for the differentiation of mesenchymal stem cells for vascular tissue engineering. *Nanomedicine* **2019**, *14*, 201–214. [CrossRef]

35. Baillargeon, A.L.; Penev, K.I.; Mequanint, K. One-pot substitution approach for the syntheses of nonfunctional and functional poly [(amino acid ester) phosphazene] Biomaterials. *Macromol. Mater. Eng.* **2017**, *302*, 1600318. [CrossRef]

36. Luo, J.; Qin, L.; Kural, M.H.; Schwan, J.; Li, X.; Bartulos, O.; Cong, X.Q.; Ren, Y.; Gui, L.; Li, G.; et al. Vascular smooth muscle cells derived from inbred swine induced pluripotent stem cells for vascular tissue engineering. *Biomaterials* **2017**, *147*, 116–132. [CrossRef]

37. Atchison, L.; Zhang, H.; Cao, K.; Truskey, G.A. A Tissue engineered blood vessel model of Hutchinson-Gilford Progeria Syndrome using human iPSC-derived smooth muscle cells. *Sci. Rep.* **2017**, *7*, 8168. [CrossRef]

38. Glenn, G.; Shogren, R.; Jin, X.; Orts, W.; Hart-Cooper, W.; Olson, L. Per- and polyfluoroalkyl substances and their alternatives in paper food packaging. *Compr. Rev. Food Sci. F* **2021**, *20*, 2596–2625. [CrossRef]

39. Lemmouchi, Y.; Schacht, E.; Dejardin, S. Biodegradable poly[(amino acid ester)phosphazenes] for biomedical applications. *J. Bioact. Compat. Pol.* **1998**, *13*, 4–18. [CrossRef]

40. Nichol, J.L.; Morozowich, N.L.; Allcock, H.R. Biodegradable alanine and phenylalanine alkyl ester polyphosphazenes as potential ligament and tendon tissue scaffolds. *Polym. Chem.* **2013**, *4*, 600–606. [CrossRef]

41. Bhattacharyya, S.; Nair, L.S.; Singh, A.; Krogman, N.R.; Greish, Y.E.; Brown, P.W.; Allcock, H.R.; Laurencin, C.I. Electrospinning of poly[bis(ethyl alanato) phosphazene] nanofibers. *J. Biomed. Nanotechnol.* **2006**, *2*, 36–45. [CrossRef]

42. Härdelin, L.; Perzon, E.; Hagström, B.; Walkenström, P.; Gatenholm, P. Influence of molecular weight and rheological behavior on electrospinning cellulose nanofibers from ionic liquids. *J. Appl. Polym. Sci.* **2013**, *130*, 2303–2310. [CrossRef]

43. Gómez-Mascaraque, L.G.; Sanchez, G.; López-Rubio, A. Impact of molecular weight on the formation of electrosprayed chitosan microcapsules as delivery vehicles for bioactive compounds. *Carbohydr. Polym.* **2016**, *150*, 121–130. [CrossRef] [PubMed]

44. Colmenares-Roldán, G.J.; Quintero-Martínez, Y.; Agudelo-Gómez, L.M.; Rodríguez-Vinasco, L.F.; Hoyos-Palacio, L.M. Influence of the molecular weight of polymer, solvents and operational condition in the electrospinning of polycaprolactone. *Rev. Fac. Ing. Univ. Antioq.* **2017**, *84*, 35–45. [CrossRef]

45. Nezarati, R.M.; Eifert, M.B.; Cosgriff-Hernandez, E. Effects of humidity and solution viscosity on electrospun fiber morphology. *Tissue Eng. Part C Methods* **2013**, *19*, 810–819. [CrossRef] [PubMed]

46. Bertolo, A.; Mehr, M.; Janner-Jametti, T.; Graumann, U.; Aebli, N.; Baur, M.; Ferguson, S.J.; Stoyanov, J.V. An in vitro expansion score for tissue-engineering applications with human bone marrow-derived mesenchymal stem cells. *J. Tissue Eng. Regen. Med.* **2016**, *10*, 149–161. [CrossRef]

47. Wang, C.; Meng, H.; Wang, X.; Zhao, C.; Peng, J.; Wang, Y. Differentiation of bone marrow mesenchymal stem cells in osteoblasts and adipocytes and its role in treatment of osteoporosis. *Med. Sci. Monit.* **2016**, *22*, 226–233. [CrossRef]

48. Xu, L.; Liu, Y.; Sun, Y.; Wang, B.; Xiong, Y.; Lin, W.; Wei, Q.; Wang, H.; He, W.; Li, G. Tissue source determines the differentiation potentials of mesenchymal stem cells: A comparative study of human mesenchymal stem cells from bone marrow and adipose tissue. *Stem Cell Res. Ther.* **2017**, *8*, 275. [CrossRef]

49. Meng, X.; Chen, M.; Su, W.; Tao, X.; Sun, M.; Zou, X.; Ying, R.; Wei, W.; Wang, B. The differentiation of mesenchymal stem cells to vascular cells regulated by the HMGB1/RAGE axis: Its application in cell therapy for transplant arteriosclerosis. *Stem Cell Res. Ther.* **2018**, *9*, 85. [CrossRef]

50. Maguire, E.M.; Xiao, Q.; Xu, Q. Differentiation and application of induced pluripotent stem cell-derived vascular smooth muscle cells. *Arter. Thromb. Vasc. Biol.* **2017**, *37*, 2026–2037. [CrossRef]

51. Bajpai, V.K.; Mistriotis, P.; Loh, Y.H.; Daley, G.Q.; Andreadis, S.T. Functional vascular smooth muscle cells derived from human induced pluripotent stem cells via mesenchymal stem cell intermediates. *Cardiovasc. Res.* **2012**, *96*, 391–400. [CrossRef] [PubMed]

52. Sundaram, S.; One, J.; Siewert, J.; Teodosescu, S.; Zhao, L.; Dimitrievska, S.; Qian, H.; Huang, A.H.; Niklason, L. Tissue-engineered vascular grafts created from human induced pluripotent stem cells. *Stem Cells Transl. Med.* **2014**, *3*, 1535–1543. [CrossRef] [PubMed]

53. Eoh, J.H.; Shen, N.; Burke, J.A.; Hinderer, S.; Xia, Z.; Schenke-Layland, K.; Gerecht, S. Enhanced elastin synthesis and maturation in human vascular smooth muscle tissue derived from induced-pluripotent stem cells. *Acta Biomater.* **2017**, *52*, 49–59. [CrossRef] [PubMed]

54. Hinz, B.; Phan, S.H.; Thannickal, V.J.; Galli, A.; Bochaton-Piallat, M.L.; Gabbiani, G. The myofibroblast: One function, multiple origins. *Am. J. Pathol.* **2007**, *170*, 1807–1816. [CrossRef]

55. Yoshida, T.; Owens, G.K. Molecular determinants of vascular smooth muscle cell diversity. *Circ. Res.* **2005**, *96*, 280–291. [CrossRef]

56. Peach, M.S.; Kumbar, S.G.; James, R.; Toti, U.S.; Balasubramaniam, D.; Deng, M.; Ulery, B.; Mazzocca, A.D.; McCarthy, M.B.; Morozowich, N.L.; et al. Design and optimization of polyphosphazene functionalized fiber matrices for soft tissue regeneration. *J. Biomed. Nanotechnol.* **2012**, *8*, 107–124. [CrossRef]

57. Peach, M.S.; James, R.; Toti, U.S.; Deng, M.; Morozowich, N.L.; Allcock, H.R.; Laurencin, C.T.; Kumbar, S.G. Polyphosphazene functionalized polyester fiber matrices for tendon tissue engineering: In vitro evaluation with human mesenchymal stem cells. *Biomed. Mater.* **2012**, *7*, 045016. [CrossRef]

Article

The Collagen Origin Influences the Degradation Kinetics of Guided Bone Regeneration Membranes

Marta Vallecillo-Rivas [1,2], Manuel Toledano-Osorio [1,2,*], Cristina Vallecillo [1,2], Manuel Toledano [1] and Raquel Osorio [1]

1   Faculty of Dentistry, Colegio Máximo de Cartuja s/n, University of Granada, 18071 Granada, Spain; mvallecillo@correo.ugr.es (M.V.-R.); cvallecillorivas@hotmail.com (C.V.); toledano@ugr.es (M.T.); rosorio@ugr.es (R.O.)
2   Medicina Clínica y Salud Pública PhD Programme, 18071 Granada, Spain
*   Correspondence: mtoledano@correo.ugr.es; Tel.: +34-958-243-789

Abstract: Collagen membranes are currently the most widely used membranes for guided bone regeneration; however, their rapid degradation kinetics means that the barrier function may not remain for enough time to permit tissue regeneration to happen. The origin of collagen may have an important effect on the resistance to degradation. The aim of this study was to investigate the biodegradation pattern of five collagen membranes from different origins: Biocollagen, Heart, Evolution X-fine, CopiOs and Parasorb Resodont. Membranes samples were submitted to different degradation tests: (1) hydrolytic degradation in phosphate buffer saline solution, (2) bacterial collagenase from *Clostridium histolyticum* solution, and (3) enzyme resistance using a 0.25% porcine trypsin solution. Immersion periods from 1 up to 50 days were performed. At each time point, thickness and weight measurements were performed with a digital caliper and an analytic microbalance, respectively. ANOVA and Student–Newman–Keuls tests were used for comparisons ($p < 0.05$). Differences between time-points within the same membranes and solutions were assessed by pair-wise comparisons ($p < 0.001$). The Evolution X-fine collagen membrane from porcine pericardium attained the highest resistance to all of the degradation tests. Biocollagen and Parasorb Resodont, both from equine origin, experienced the greatest degradation when immersed in PBS, trypsin and *C. histolyticum* during challenge tests. The bacterial collagenase solution was shown to be the most aggressive testing method.

Keywords: bone regeneration; collagen membrane; hydrolytic degradation; collagen origin; bacterial collagenase; trypsin digestion; degradation testing; degradation kinetics

Citation: Vallecillo-Rivas, M.; Toledano-Osorio, M.; Vallecillo, C.; Toledano, M.; Osorio, R. The Collagen Origin Influences the Degradation Kinetics of Guided Bone Regeneration Membranes. *Polymers* **2021**, *13*, 3007. https://doi.org/10.3390/polym13173007

Academic Editors: João Carlos Silva and Frederico Castelo Ferreira

Received: 7 August 2021
Accepted: 3 September 2021
Published: 5 September 2021

Publisher's Note: MDPI stays neutral with regard to jurisdictional claims in published maps and institutional affiliations.

1. Introduction

Nowadays, both in oral implantology and periodontology, the management and treatment of alveolar deficiencies are major clinical issues. Guided bone regeneration (GBR), a technique based on the use of membranes creating space to be filled with new bone, was originally hypothesized more than three decades ago [1,2]. This concept implies the exclusion of those cells types that proliferate faster than bone cells, such as those destined to form epithelial and connective tissue [3,4]. At the same time, space is maintained to allow cells capable of forming bone to grow [3]. These selective repopulation procedures are applied in regenerative periodontal clinical situations and to treat various intraoral bone defects, not limited to implant purposes [5].

For a material to be used as a membrane for GBR, some fundamental requirements must be fulfilled: biocompatibility, cell-occlusiveness, space maintenance, tissue integration and easy manipulation [1,6–9]. Volumetric stability of the membrane over time is also pivotal, in order to remain for long enough for regeneration to take place [10]. A membrane should be capable of being degraded or resorbed at the same rate that bone growth occurs [7]. It has been established that for periodontal regeneration, 4–6 weeks are necessary; however, for GBR procedures, a longer period of more than 6 months is recommended [1,11]. Membranes for GBR can be broadly divided into two main groups: resorbable and non-resorbable. Non-resorbable membranes remain stable over time without undergoing degradation processes. On the contrary, these membranes require a second surgery for their removal [8,12]. Due to this, resorbable membranes are used more frequently for GBR, alleviating the discomfort of the patient caused by a second surgery, avoiding tissue damage and the risk of additional morbidity [7]. The main disadvantage of these membranes is the unpredictable reabsorption time and the degree of degradation [13]. It has been described that increased degradation of collagen membranes is mainly caused by the enzymatic activity of infiltrating macrophages and polymorphonuclear leukocytes [11]. Among the different resorbable membranes available, collagen membranes are the most widely studied and investigated [6].

Although the clinical application of collagen membranes in bone regeneration is well established, only a limited number of studies have investigated their resorption patterns. These studies have stated that the degradation of collagen membranes can be produced between 4 days and 6 weeks after surgical placement [14]. Collagen membranes have been shown to exhibit excellent biocompatibility, but rapid resorption. When tested in a rabbit calvarium model, different degradation patterns have been described depending on the origin of the collagen membranes [15]. They may come from the tendon, skin, pericardium, or other regions of purified bovine, equine or porcine collagen [16]. Membranes made from porcine collagen have an open and porous collagen network, but also dense. The inherent open pores of native porcine skin facilitate the migration of blood vessels into the defect area, allowing for rapid vascularization of the underlying wound bed, while the density of the membrane maintains a barrier against ingrowth of soft tissue [16]. The collagen fibers in tendons are predominantly oriented longitudinally and, to a lesser extent, also transversely and horizontally. Furthermore, tendon collagen fibers are crimped, in contrast to skin fibers, which are even fibers [17]. The pericardium has to withstand the forces of the heart muscle; it has an exceptionally dense collagen-like structure, which gives it rigidity and multidirectional tear resistance [17]. In addition, equine tendon collagen retains a partial lateral packing, thanks to the higher content of lysine and hydroxylysine compared to other mammal tendon collagens. This feature explains why collagen and collagen-based devices from horse tendons may be intrinsically more resistant to degradation [18].

Differences in collagen tissue structure lead to different degradation responses and clinical outcomes [15,19]. It has also been suggested that other characteristics may influence the degradation behavior of a membrane, such as porosity, thickness and weight. The thickness and weight will also affect the mechanical properties of the membrane. Porosity allows nutrients infiltration into the defect, which promotes bone growth, although excessively large pores can make the membranes less effective as a barrier against soft tissue cells [11,20]. Variations in composition, structure and properties will definitely influence the degree of degradation. The reabsorption rate of commercially available collagen membranes can vary broadly. For this reason, it is important to choose a membrane that maintains its structural integrity and mechanical properties for long enough to permit the proliferation and maturation of the desired cells within the wound [5]. Performing in vitro studies will allow for detecting the influence of the composition of the membrane and their properties in the duration of the degradation process. Hence, in the present study, the aim was to investigate the biodegradation pattern of five collagen membranes from different origin, in a period of 1 to 50 d. A qualitative microstructural evaluation and a quantitative analysis of collagen membrane degradation were performed. The null hypotheses that were tested were that: (i) the five membranes for guided bone regeneration do not present the same degradation pattern over time; and (ii) the five membranes do not resist the different degradation processes in a similar way (hydrolytic, bacterial collagenase and enzyme resistance).

2. Materials and Methods

2.1. Membranes Description

Five experimental guided bone regeneration (GBR) collagen membranes were tested. Membranes are commercially available and CE-certified for oral applications and all have heterologous origin. The 0.2 mm-thickness resorbable membranes investigated were: (1) Biocollagen (Bioteck by Bioteck S.p.A, Torino, Italy); (2) Heart (Bioteck by Bioteck S.p.A, Torino, Italy); (3) Evolution X-fine (Osteobiol by Tecnoss, Torino, Italy); (4) CopiOs (Zimmer Biomet Dental, FL, USA); and (5) Parasorb Resodont (Resorba by Resorba Medical GmbH, Nürnberg, Germany). According to the manufacturers, all of the resorbable membranes used in the present study must be previously hydrated. Biocollagen is derived from equine type I collagen obtained from the Achilles tendon. Heart is an equine pericardium membrane, where molecular links among its constituents (collagen and elastin) are preserved. Evolution X-fine is a resorbable pericardium porcine membrane obtained from heterologous mesenchymal tissue; its structure is made of dense collagen fibers. CopiOs is a membrane made from the bovine pericardium based on type I and III collagen. Finally, Parasorb Resodont is a resorbable membrane made of natural cross-linked equine collagen fibers without chemical additives. The main characteristics of the membranes used in this biodegradation study are listed in Table 1.

Table 1. Commercial name, collagen type, collagen origin and clinical characteristics (cross-link, estimated duration of the barrier effect) of the tested collagen membranes. The information was obtained from the membrane providers.

Membrane Comercial Name	Collagen Type	Origin	Cross-Link	Estimated Barrier Effect Duration
Biocollagen	Collagen type I	Equine tendon	No	4–6 weeks
Heart	NR	Equine pericardium	No	12–16 weeks
Evolution X-fine	Collagen fibers	Porcine pericardium	No	8 weeks
CopiOs	Collagen type I	Bovine pericardium	No	16–24 weeks
Parasorb Resodont	Collagen fibers	Equine	Natural	NR

NR: not reported.

2.2. Degradation Assays

Membranes samples were cut to a size of 10×10 mm^2. Three specimens of each membrane type were employed for each degradation test ($n = 3$). Three different degradation tests were performed [1,11].

(1) Hydrolytic degradation test: membranes were immersed in phosphate buffer saline solution (PBS) at 37 °C [21].

(2) Enzyme resistance test: membranes were submerged in a 0.25% porcine trypsin solution (Sigma-Aldrich, St. Louis, MO, USA) and were incubated at 37 °C [6].

(3) Bacterial collagenase resistance test: a collagenase solution from *Clostridium histolyticum* bacteria Type V (Sigma-Aldrich, St Louis, MO, USA) was used to submerge the membranes. It is a mixture of several different enzymes containing; collagenase, nonspecific proteases, clostripain, neutral protease, and aminopeptidase activities. These enzymes act together to cause a breakdown of tissues. The specific activity is ≥ 125 CDU/mg solid. A collagenase concentration of 2 IU/mL in 50 mM Tris HCl (pH 7.4), containing 10 mM CaCl$_2$, was used [22,23].

After each 48 h, degradation solutions were removed carefully through suction and renewed [24].

Membranes were retrieved from the media after the following time periods: 1 h, 6 h, 24 h, 48 h, 7 d, 14 d, 28 d and 50 d. After each period of immersion, membranes were completely dried in a vacuum chamber. Weight (W) was then measured with an analytic balance (A&D-Instruments, Frankfurt, Germany) with an accuracy of 0.0001 g; the complete device was mounted on an antivibratory table. Thickness (Th) was also measured at random positions by means of a digital caliper (Mitutoyo 293-561, Tokyo, Japan) [1,11].

2.3. Statistical Analysis

The normal distribution of data was established by the Kolmogórov–Smirnov test ($p > 0.05$). Therefore, parametric tests were employed. Multiple ANOVA models were used to assess the influence of the independent variables (degradation solution, type of matrix and immersion time) on the dependent variables (weight and thickness). Analyses of interactions were also carried out. ANOVA and Student–Newman–Keuls post hoc comparisons were performed to determine differences between materials and degradation solutions. To permit these comparisons, the variables weight and thickness were converted into the percentage of variation with respect to the initial measurement following the equation:

$$\text{Percentage of loss} = [(X_0 - X_t)/X_0] \times 100, \tag{1}$$

where X_0 is the initial weight or thickness of the specimen, and X_t is the specimen's weight or thickness at each time-point (t).

Pairwise comparisons were performed to ascertain for differences between immersion time-points within the same membrane and solution experimental group. Significance was always considered at $p < 0.05$, except for pairwise comparisons, where a Bonferroni's correction was applied and $p < 0.001$ was set. Statistical analysis was performed using SPSS 25.0 (SPSS Inc., Chicago, IL, USA) software package.

2.4. Light Microscopy Analysis

At the end of the storage period of 7 d, specimens were observed under a stereomicroscope Olympus SZ-60 (Olympus, Tokyo, Japan) for microstructural analysis. Images were taken at $60\times$ and $120\times$ magnifications.

3. Results

The thickness loss values of the membranes, expressed in percentage, across the different degradation tests and times, are represented in Figure 1. The thickness values in mm of the five tested membranes (Biocollagen, Heart, Evolution X-fine, CopiOs and Parasorb Resodont) into the different solutions (PBS, trypsin and *C. histolyticum* collagenase) over the different immersion time periods (1 h, 6 h, 24 h, 48 h, 7 d, 14 d, 28 d and 50 d) are shown in Table 2. Comparisons between membranes within the same solution are expressed in Figure 1, while the loss of thickness after each immersion time-point compared to the initial value of each of the membranes type is shown in Table 2.

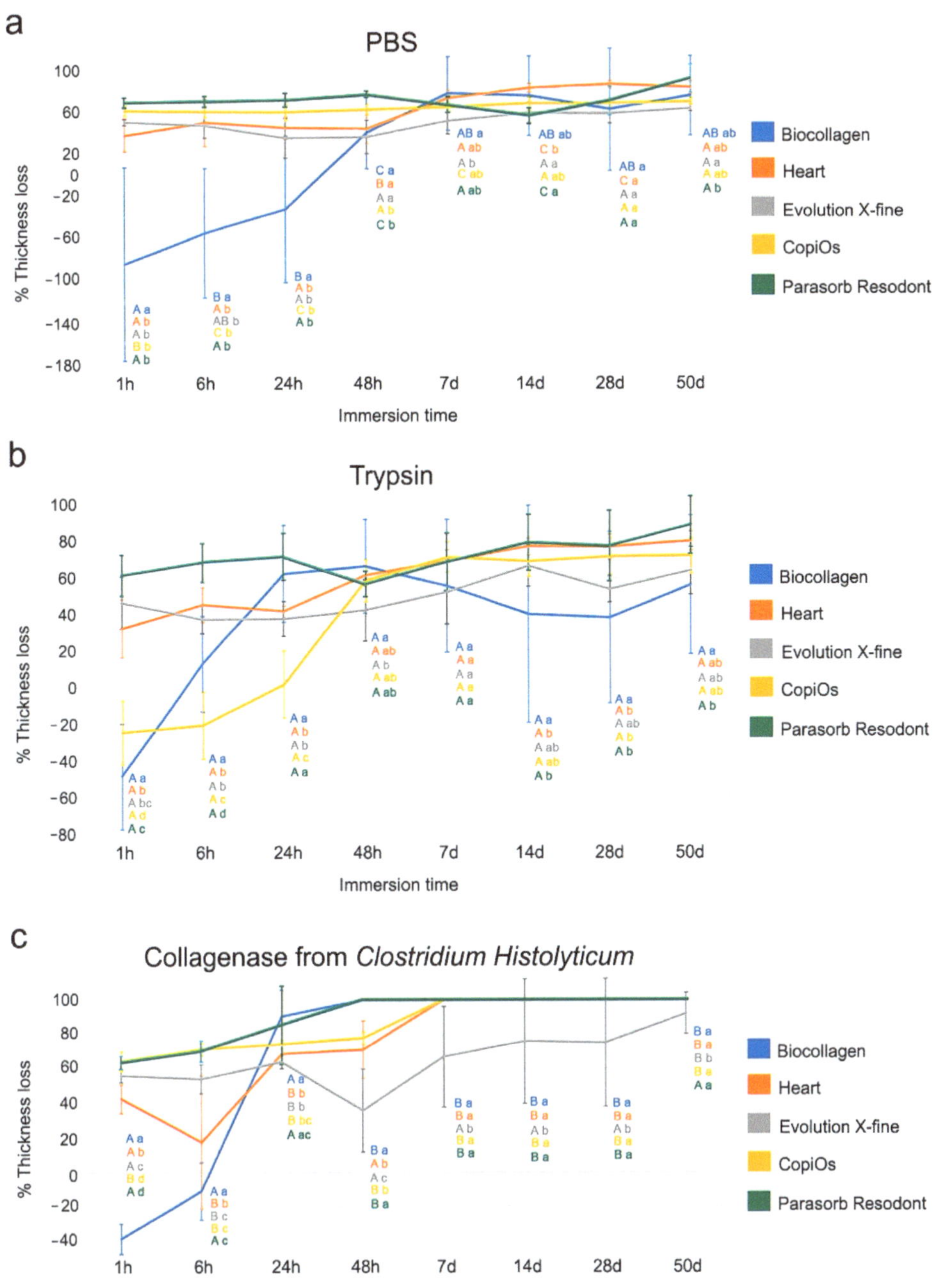

Figure 1. Percentage of thickness loss of the five tested membranes after being submitted to (**a**) PBS, (**b**) trypsin and (**c**) *C. histolyticum* collagenase over the different time-points. Values are expressed as mean and standard deviation. Significant differences among membranes are shown with low-case letters and capital letters indicate differences between the different degradation tests within the same membrane. Student-Newman-Keuls tests were performed for multiple comparisons ($p < 0.05$).

Table 2. (**a**) Thickness values (mm) of the five membranes (Biocollagen, Heart, Evolution X-fine, CopiOs and Parasorb Resodont) after being submitted to: PBS, trypsin and *C. histolyticum*, from 1 h to 50 d. Values are presented as means and standard deviations. (**b**) *p* values resulted from pairwise comparisons between membranes' thicknesses before immersion (initial thickness or thickness at t0: Th0) and after the different immersion periods. Statistical significance was set at $p < 0.001$.

(a)

	Biocollagen			Heart			Evolution X-Fine			CopiOs			Parasorb Resodont		
	Tryp.	CH	PBS	Tryp.	CH	PBS	Tryp.	CH	PBS	Tryp.	CH	PBS	Tryp.	CH	PBS
t0	0.17 (0.01)	0.17 (0.01)	0.17 (0.01)	0.25 (0.07)	0.3 (0.17)	0.53 (0.11)	0.17 (0.06)	0.18 (0.05)	0.2 (0.05)	0.32 (0.09)	0.4 (0.06)	0.26 (0.07)	0.17 (0.01)	0.15 (0.01)	0.18 (0.01)
1 h	0.25 (0.06)	0.23 (0.02)	0.31 (0.16)	0.17 (0.08)	0.16 (0.08)	0.33 (0.09)	0.09 (0.04)	0.08 (0.02)	0.1 (0.03)	0.39 (0.11)	0.14 (0.03)	0.1 (0.01)	0.07 (0.02)	0.05 (0.01)	0.05 (0.01)
6 h	0.15 (0.05)	0.19 (0.04)	0.26 (0.11)	0.13 (0.02)	0.2 (0.07)	0.26 (0.11)	0.11 (0.04)	0.09 (0.03)	0.11 (0.04)	0.37 (0.06)	0.12 (0.02)	0.1 (0.01)	0.05 (0.02)	0.04 (0.01)	0.05 (0.01)
24 h	0.07 (0.05)	0.02 (0.03)	0.23 (0.12)	0.14 (0.04)	0.09 (0.03)	0.29 (0.04)	0.11 (0.05)	0.07 (0.02)	0.14 (0.06)	0.31 (0.11)	0.11 (0.02)	0.1 (0.01)	0.05 (0.02)	0.02 (0.03)	0.05 (0.02)
48 h	0.06 (0.04)	0 (0)	0.1 (0.06)	0.09 (0.03)	0.07 (0.02)	0.29 (0.06)	0.1 (0.05)	0.11 (0.02)	0.13 (0.06)	0.12 (0.02)	0.09 (0.02)	0.1 (0.01)	0.08 (0.02)	0 (0)	0.04 (0.01)
7 d	0.08 (0.06)	0 (0)	0.04 (0.06)	0.08 (0.03)	0 (0)	0.14 (0.03)	0.08 (0.04)	0.07 (0.06)	0,1 (0,04)	0.09 (0.02)	0 (0)	0.09 (0.02)	0.05 (0.03)	0 (0)	0.06 (0.01)
14 d	0.1 (0.1)	0 (0)	0.04 (0.06)	0.05 (0.01)	0 (0)	0.09 (0.02)	0.06 (0.03)	0.06 (0.08)	0.08 (0.03)	0.09 (0.03)	0 (0)	0.08 (0.01)	0.03 (0.03)	0 (0)	0.08 (0.01)
28 d	0.11 (0.08)	0 (0)	0.06 (0.09)	0.06 (0.02)	0 (0)	0.08 (0.01)	0.07 (0.02)	0.06 (0.09)	0.09 (0.04)	0.09 (0.04)	0 (0)	0.08 (0.02)	0.04 (0.03)	0 (0)	0.05 (0.02)
50 d	0.07 (0.07)	0 (0)	0.04 (0.06)	0.05 (0.01)	0 (0)	0.09 (0.02)	0.07 (0.05)	0.02 (0.03)	0.07 (0.02)	0.08 (0.02)	0 (0)	0.08 (0.02)	0.02 (0.03)	0 (0)	0.01 (0.02)

(b)

	Tryp.	CH	PBS	Tryp.	CH	PBS	Tryp.	CH	PBS	Tryp.	CH	PBS	Tryp.	CH	PBS
0–1 h	**0.001**	**<0.001**	0.025	**<0.001**	0.002	**<0.001**	**<0.001**	**<0.001**	**<0.001**	0.002	**<0.001**	**<0.001**	**<0.001**	**<0.001**	**<0.001**
0–6 h	0.184	0.0719	0.027	**<0.001**	0.114	**<0.001**	**<0.001**	**<0.001**	**<0.001**	0.007	**<0.001**	**<0.001**	**<0.001**	**<0.001**	**<0.001**
0–24 h	**<0.001**	**<0.001**	0.191	**<0.001**	0.002	**<0.001**	**<0.001**	**<0.001**	**<0.001**	0.864	**<0.001**	**<0.001**	**<0.001**	**<0.001**	**<0.001**
0–48 h	**<0.001**	**<0.001**	0.009	**<0.001**	0.003	**<0.001**	**<0.001**	0.005	**<0.001**	**<0.001**	**<0.001**	**<0.001**	**<0.001**	**<0.001**	**<0.001**
0–7 d	**0.001**	**<0.001**	**<0.001**	**<0.001**	0.001	**<0.001**	**<0.001**	**<0.001**	**<0.001**	**<0.001**	**<0.001**	**<0.001**	**<0.001**	**<0.001**	**<0.001**
0–14 d	0.081	**<0.001**	**<0.001**	**<0.001**	0.001	**<0.001**	**<0.001**	**<0.001**	**<0.001**	**<0.001**	**<0.001**	**<0.001**	**<0.001**	**<0.001**	**<0.001**
0–28 d	0.036	**<0.001**	**0.011**	**<0.001**	0.001	**<0.001**	**<0.001**	**<0.001**	**<0.001**	**<0.001**	**<0.001**	**<0.001**	**<0.001**	**<0.001**	**<0.001**
0–50 d	**0.001**	**<0.001**	**<0.001**	**<0.001**	0.001	**<0.001**	**<0.001**	**<0.001**	**<0.001**	**<0.001**	**<0.001**	**<0.001**	**<0.001**	**<0.001**	**<0.001**

Tryp: trypsine; CH: *Clostridium histolyticum* collagenase; PBS: phosphate buffered saline.

3.1. Thickness Evaluation after PBS Degradation Assay

The comparison of the thickness loss, expressed as a percentage, suffered by membranes submitted to the PBS degradation assay is shown in Figure 1a. At 1 h, Biocollagen's thickness increased by more than 80% due to water absorption during first immersion, while the rest of membranes experienced a marked loss of thickness of over 30% (Figure 1a). After 24 h of storage, all membranes suffered a reduction in Th percentage, with Biocollagen being the membrane whose weight loss values were lower. The trend was as follows: Parasorb Resodont > CopiOs > Heart > Evolution X-fine > Biocollagen. For Parasorb Resodont, CopiOs, Heart and Evolution X-fine, the trend was maintained, and the thickness loss values were similar at 24 and 48 h. However, Biocollagen from 24 to 48 h suffered a marked reduction in thickness, with a loss of over 70%. At the 7 d time-point, Biocollagen attained the highest and Evolution X-fine the lowest percentage thickness values; the trend was as follows: Biocollagen > Heart > Parasorb Resodont > CopiOs > Evolution X-fine (Figure 1a). All membranes ended 50 d of storage with more than 50% thickness loss. The final trend was: Parasorb Resodont > Heart > Biocollagen > CopiOs > Evolution X-fine (Figure 1a).

In general terms, the five membranes had an increasing loss of thickness over different time-points when comparing to the initial, except for Biocollagen, which at 1 h did not experience any loss of thickness (Table 2). None of the membranes were completely degraded in this medium. However, all membranes ended 50 d of storage with lower thickness values than the initial value (Table 2).

3.2. Thickness Evaluation after Trypsin Degradation Assay

The percentage of thickness loss suffered by the different membranes studied when they were immersed in trypsin is shown in Figure 1b. At the 1 h time-point, Biocollagen and CopiOs did not suffer loss of percentage thickness values, experiencing a gain of over 40% and 20%, respectively. At the same time-point, Heart, Evolution X-fine and Parasorb Resodont showed a loss of Th values of over 30% (Figure 1b). From 1 to 48 h, Biocollagen and CopiOs suffered a marked loss of thickness. The rest of the membranes suffered a gradual loss of Th, with values very similar to those obtained in the first moment of immersion. After 7 d of storage, Parasorb Resodont, Heart and CopiOs performed similarly, with the highest loss of percentage thickness values, while Biocollagen reached intermediate performance and Evolution X-fine attained the lowest loss of percentage thickness values (Figure 1b). At 14 and 28 d, the trend was as follows: Parasorb Resodont $\geq$ Heart > CopiOs > Evolution X-fine > Biocollagen. After 50 d of immersion, all membranes suffered a reduction in percentage thickness of above 50%. The trend was as follows: Parasorb Resodont > Heart > CopiOs > Evolution X-fine > Parasorb Resodont (Figure 1b).

In general terms, all membranes suffered changes in their thickness throughout the different time-points (Table 2). All membranes endured 50 d of storage, suffering a statistically significant loss of thickness (Figure 1b).

3.3. Thickness Evaluation after C. histolyticum Collagenase Degradation Assay

The comparison of the thickness loss, expressed as a percentage, suffered by membranes submitted to collagenase degradation assay is shown in Figure 1c. At the 1 h time-point, CopiOs attained the highest loss of percentage thickness values, close to 70%, this being similar to the loss experienced by Parasorb Resodont. Heart and Evolution X-fine experienced a thickness loss between 40 and 60%. Finally, Biocollagen did not experience a loss of thickness; instead, its thickness increased by almost 40%. The trend was as follows: CopiOs $\geq$ Parasorb Resodont > Evolution X-fine > Heart > Biocollagen (Figure 1c). After 48 h of storage, Parasorb Resodont and Biocollagen completely degraded. After 7 d of immersion, Heart and CopiOs totally degraded, with a 100% reduction in Th values (Figure 1c). Evolution X-fine was the only membrane which overcame 50 d of storage with high loss of percentage thickness values of nearly 90%.

In general terms, this degradation assay was the most dangerous to the membranes studied (Table 2). All membranes suffered a marked reduction in their thickness values over time when compared to the initial time-point (Table 2). Parasorb Resodont, Biocollagen, Heart and CopiOs could not endure more than 7 d of storage, while Evolution X-fine reached 50 d of immersion with low thickness values.

The weight (W) loss values, expressed in percentage, over distinct degradation tests and times, are shown in Figure 2. The weight (W) values in grams of the five membranes studied (Biocollagen, Heart, Evolution X-fine, CopiOs and Parasorb Resodont) subjected to different solutions (PBS, trypsin and *C. histolyticum* collagenase) during various submersion time periods are presented in Table 3. Comparisons between membranes within the same solution are shown in Figure 2, while the weight after each immersion time-point and comparisons with initial weight values are displayed in Table 3.

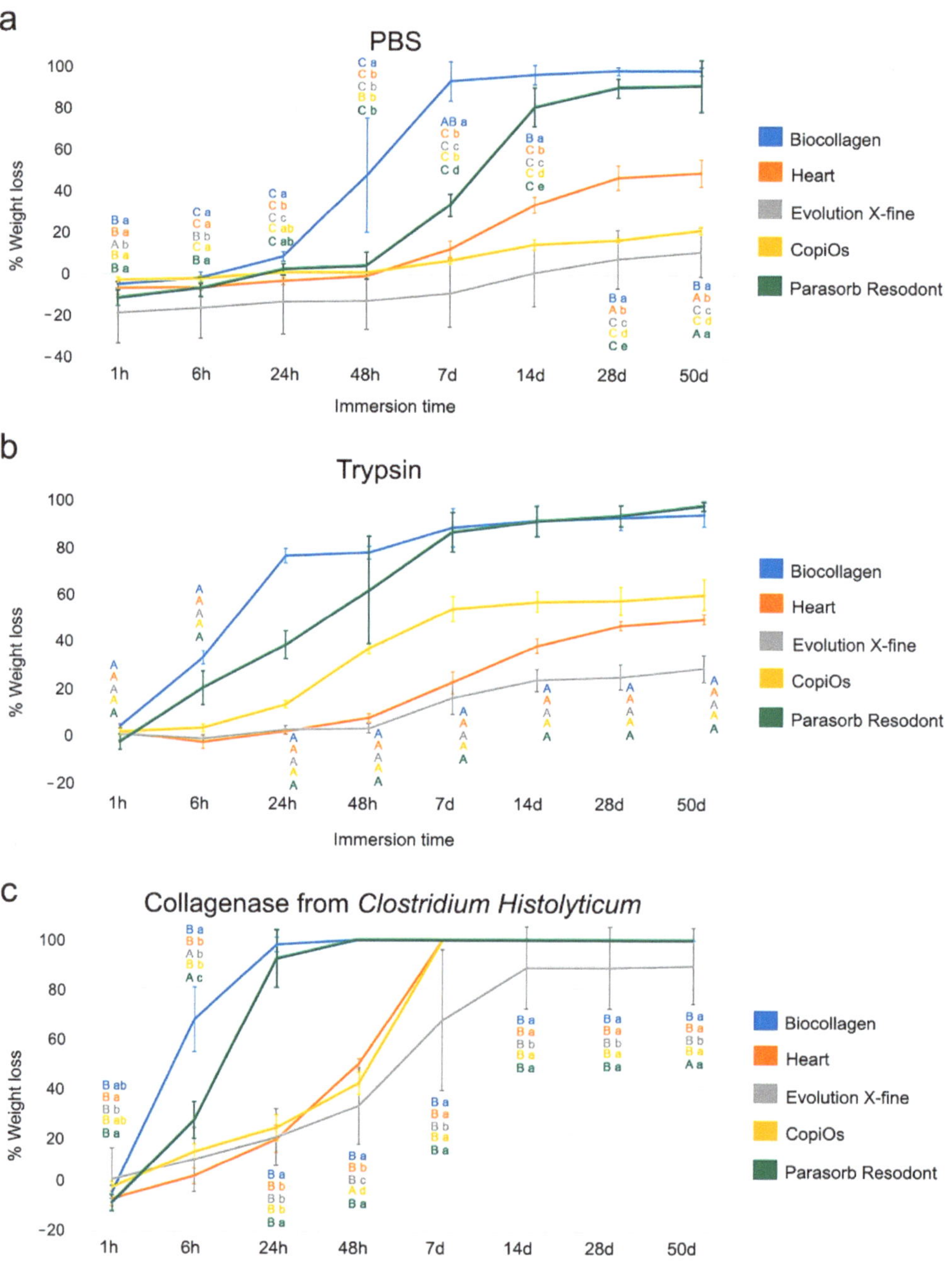

Figure 2. Percentage of weight loss of the five membranes studied submitted to: (**a**) PBS, (**b**) trypsin and (**c**) C. histolyticum collagenase over different time-points. Values expressed as mean and standard deviation. Significant differences among membranes are shown with lowercase letters and with capital letters for differences between degradation tests within the same membrane. Student-Newman-Keuls tests were performed for multiple comparisons ($p < 0.05$).

Table 3. (**a**) Weight values (g) of the five membranes (Biocollagen, Heart, Evolution X-fine, CopiOs and Parasorb Resodont) after being submitted to PBS, trypsin and *C. histolyticum* for 1 h to 50 d. Values are presented as means and standard deviations. (**b**) Obtained *p* values after pairwise comparisons between membrane weights before immersion (initial weight or weight at t0: W0) and after the different immersion periods. Significance was considered at $p < 0.001$.

(a)

	Biocollagen			Heart			Evolution X-Fine			CopiOs			Parasorb Resodont		
	Tryp.	CH	PBS	Tryp.	CH	PBS	Tryp.	CH	PBS	Tryp.	CH	PBS	Tryp.	CH	PBS
t0	12.27 (0.33)	11.95 (0.37)	11.63 (0.59)	8.91 (2.43)	8.54 (1.04)	10.83 (0.99)	7.21 (1.8)	8.34 (2.31)	7.9 (1.39)	12.95 (2.17)	13.94 (0.59)	11.55 (1.41)	3.77 (0.3)	3.96 (0.25)	3.82 (0.37)
1 h	11.66 (0.28)	12.43 (0.28)	12.17 (0.52)	8.75 (2.55)	9.1 (0.99)	11.55 (0.95)	7.13 (1.93)	8.01 (1.72)	9.32 (1.79)	12.63 (1.98)	14.22 (0.83)	11.87 (1.47)	3.83 (0.27)	4.29 (0.32)	4.24 (0.33)
6 h	8.10 (0.19)	3.94 (1.64)	11.79 (0.45)	9.07 (2.4)	8.33 (1.18)	11.5 (1.03)	7.21 (1.74)	7.41 (1.99)	9.11 (1.73)	12.41 (1.91)	12.18 (0.21)	11.77 (1.31)	2.98 (0.41)	2.95 (0.41)	4.07 (0.48)
24 h	2.78 (0.31)	0.25 (0.37)	10.59 (0.48)	8.68 (2.43)	7.05 (1.26)	11.16 (0.99)	6.94 (1.72)	6.75 (2.13)	8.86 (1.7)	11.13 (1.94)	10.78 (0.59)	11.41 (1.35)	2.29 (0.36)	0.31 (0.47)	3.71 (0.38)
48 h	2.6 (0.25)	0 (0)	5.87 (2.97)	8.16 (2.32)	4.36 (0.46)	10.88 (0.97)	6.9 (1.72)	5.91 (2.52)	8.82 (1.57)	8.02 (1.36)	8.23 (0.46)	11.42 (1.4)	1.41 (0.91)	0 (0)	3.66 (0.57)
7 d	1.29 (0.98)	0 (0)	0.73 (1.1)	6.9 (2.23)	0 (0)	9.5 (1.18)	6.06 (1.9)	3.21 (2.8)	8.55 (1.72)	5.76 (0.28)	0 (0)	10.75 (1.49)	0.49 (0.34)	0 (0)	2.52 (0.29)
14 d	0.92 (0.77)	0 (0)	0.36 (0.54)	5.45 (1.63)	0 (0)	7.17 (0.98)	5.49 (1.67)	1.08 (1.62)	7.78 (1.68)	5.38 (0.31)	0 (0)	9.88 (1.49)	0.32 (0.26)	0 (0)	0.71 (0.37)
28 d	0.76 (0.62)	0 (0)	0.15 (0.22)	4.69 (1.44)	0 (0)	5.75 (1.01)	5.39 (1.66)	1.08 (1.62)	7.27 (1.61)	5.29 (0.15)	0 (0)	9.59 (1.25)	0.22 (0.17)	0 (0)	0.36 (0.17)
50 d	0.61 (0.59)	0 (0)	0.15 (0.23)	4.43 (1.34)	0 (0)	5.48 (0.97)	5.14 (1.65)	1 (1.5)	6.97 (1.34)	4.93 (0.18)	0 (0)	9.03 (1.04)	0.05 (0.08)	0 (0)	0.32 (0.49)
(b)															
0–1 h	<0.001	0.002	<0.001	0.099	<0.001	<0.001	0.189	0.393	0.06	0.04	0.011	<0.001	0.228	<0.001	<0.001
0–6 h	<0.001	<0.001	0.146	0.047	0.062	<0.001	1	0.072	0.013	0.001	<0.001	0.002	<0.001	<0.001	0.003
0–24 h	<0.001	<0.001	<0.001	<0.001	<0.001	<0.001	<0.001	0.005	0.047	<0.001	<0.001	0.045	<0.001	<0.001	0.008
0–48 h	<0.001	<0.001	0.001	<0.001	<0.001	0.352	<0.001	<0.001	0.035	<0.001	<0.001	<0.001	<0.001	<0.001	0.054
0–7 d	<0.001	<0.001	<0.001	<0.001	<0.001	<0.001	<0.001	<0.001	0.161	<0.001	<0.001	<0.001	<0.001	<0.001	<0.001
0–14 d	<0.001	<0.001	<0.001	<0.001	<0.001	<0.001	<0.001	<0.001	0.788	<0.001	<0.001	<0.001	<0.001	<0.001	<0.001
0–28 d	<0.001	<0.001	<0.001	<0.001	<0.001	<0.001	<0.001	<0.001	0.12	<0.001	<0.001	<0.001	<0.001	<0.001	<0.001
0–50 d	<0.001	<0.001	<0.001	<0.001	<0.001	<0.001	<0.001	<0.001	0.024	<0.001	<0.001	<0.001	<0.001	<0.001	<0.001

Tryp: Trypsine; CH: *Clostridium histolyticum* collagenase; PBS: phosphate buffered saline.

3.4. Weight Evaluation after PBS Degradation Assay

The weight loss percentage suffered by membranes submitted to PBS degradation assay is shown in Figure 2a. From the first immersion period up to the 6 h time-point, membranes hardly suffered from weight loss. The highest loss of weight corresponded to CopiOs at the 1 and 6 h time-points, this being similar to Biocollagen after 6 h of storage. From 24 to 48 h, Biocollagen suffered a marked loss of weight, while the rest of the membranes maintained their weight values. At the 48 h time-point, Evolution X-fine attained the lowest and Biocollagen the highest loss of percentage weight values. The trend was as follows: Biocollagen > Parasorb Resodont > CopiOs > Heart > Evolution X-fine (Figure 2a). After 7 d of immersion, this trend changed as follows: Biocollagen > Parasorb Resodont > Heart > CopiOs > Evolution X-fine, and it was maintained until 50 d of storage (Figure 2a). So, at the end of the study, Biocollagen and Parasorb Resodont experienced a loss of weight of above 90%, Heart above 40%, CopiOs above 20% and Evolution X-fine experienced an almost 10% weight loss.

In general terms, all membranes suffered significant weight loss after completing the different study periods, except for Evolution X-fine, which never attained significance

(Table 3). After 50 d of storage, Biocollagen reached the highest loss of percentage weight values, and Evolution X-fine the lowest (Figure 2a).

3.5. Weight Evaluation after Trypsin Degradation Assay

The comparison of the weight loss, expressed as a percentage, suffered by membranes submitted to collagenase degradation assay is shown in Figure 2b. After 1 h of immersion, Biocollagen attained the highest loss of percentage weight values, but these values were below 10%. At this time-point, the trend was as follows: Biocollagen > CopiOs > Evolution X-fine ≥ Heart > Parasorb Resodont (Figure 2b). At the 48 h and 7 d time-points, the highest loss of weight corresponded to Biocollagen, followed by Parasorb Resodont. CopiOs and Heart performed intermediately, and Evolution X-fine showed the lowest loss of percentage weight values. After 28 d of storage, the trend was as follows: Parasorb Resodont ≥ Biocollagen > CopiOs > Heart > Evolution X-fine. This trend was maintained until the final stage of study (Figure 2a). At 50 d of immersion, Parasorb Resodont and Biocollagen attained the highest loss of weight values, these being over 90%. CopiOs and Heart performed intermediately, with nearly 60% and 50% weight loss. Finally, Evolution X-fine exceeded 50 d of storage, with weight loss close to 30%.

In general terms, all membranes showed significant weight loss over time in the trypsin degradation assay after completing the different study periods (Table 3).

3.6. Weight Evaluation after C. histolyticum Collagenase Degradation Assay

The percentage of weight loss suffered by membranes when immersed in *C. histolyticum* is shown in Figure 1c. At the 6 h time-point, all membranes experienced a loss of weight, this value being higher for Biocollagen, followed by Parasorb Resodont, CopiOs, Evolution X-fine and, finally, Heart. After 24 h of storage, the highest loss of percentage weight values corresponded to Biocollagen and the lowest to Heart. The trend found at this time-point was as follows: Biocollagen > Parasorb Resodont > CopiOs > Evolution X-fine > Heart (Figure 2c). At the 48 h time-point, Biocollagen and Parasorb Resodont were completely degraded. After 7 d of storage, CopiOs and Heart were fully degraded. Evolution X-fine, with a loss of percentage weight of close to 90%, overcame 50 d of immersion (Figure 2c).

In general terms, the bacterial collagenase solution was found to be the most aggressive test, causing the complete degradation of all membranes before 7 d. All membranes experienced changes in their weight over time, when compared to the initial values (Table 3).

3.7. Matrices Morphological Analysis

Figures 3 and 4 show light micrographs taken from the different surfaces of membranes tested before being submitted to immersion in the degradation solutions. GBR collagen membranes have two different surfaces: one smooth and the other rough. Compact collagen fibers with cellular occlusive properties constituted a smooth surface (Figure 3). Evolution X-fine, CopiOs and Parasorb Resodont presented a profile of dense distribution of filaments with fewer detectable pores (Figure 3e–j). Surfaces with randomly distributed collagen fibers in Biocollagen (Figure 3a,b) and with a trabecular structure in Heart (Figure 3c,d) were observed, respectively. The second layer, presented in Figure 4, consisted of a rough collagen structure designed to be positioned in direct contact with the bone surface or the bone defect. Parasorb Resodont showed a semi-compact collagen surface where the pores and the collagen fibers were hardly visible (Figure 4i,j). Biocollagen had collagen fibers that were absorbed throughout the surface of the membrane formed by lattices and pores aiming to allow cell adhesion (Figure 4a,b). Heart, Evolution X-fine and CopiOs presented collagen fibers arranged in layers that randomly run parallel to the surface. Light microscopy clearly showed regions with larger pores (more than 100 μm) located in areas where collagen fibrils were absent.

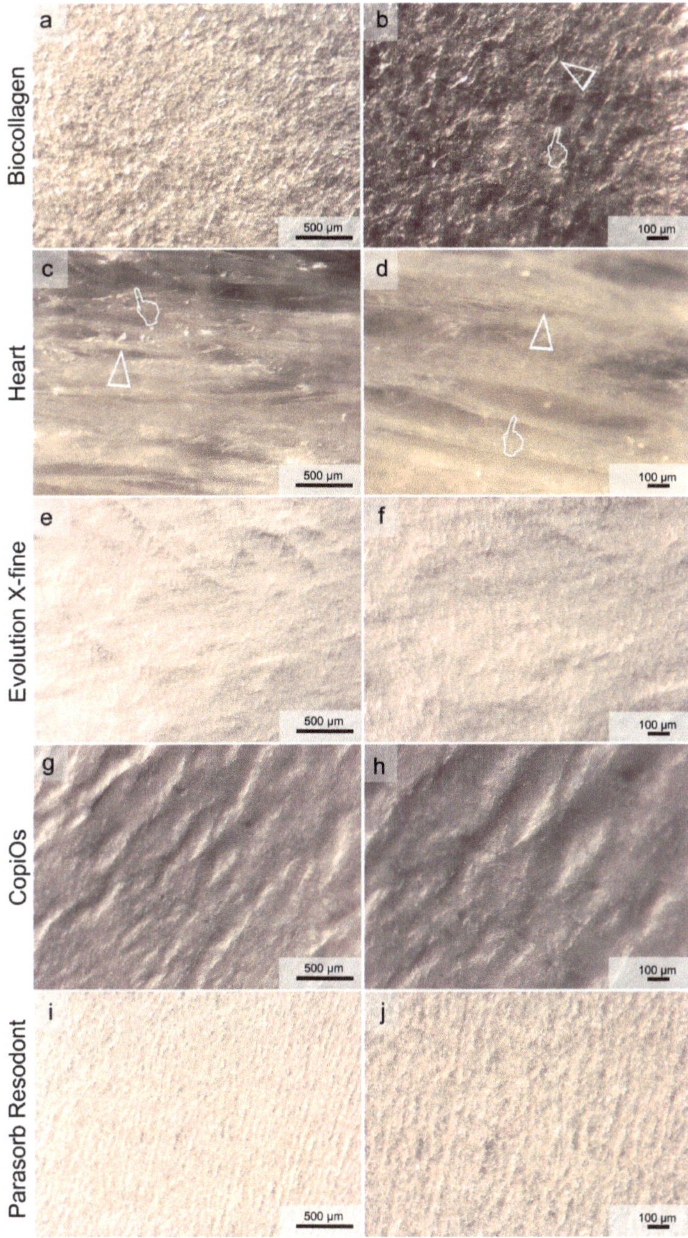

Figure 3. Light micrographs taken of the smooth surface of the membranes before immersion. (**a**) Biocollagen surface at 60× magnification. (**b**) Biocollagen at 120× magnification. (**c**) Heart surface at 60× magnification. (**d**) Heart at 120× magnification. (**e**) Evolution X-fine surface at 60× magnification. (**f**) Evolution X-fine surface at 120× magnification. (**g**) CopiOs surface at 60× magnification. (**h**) CopiOs surface at 120× magnification. (**i**) Parasorb Resodont surface at 60× magnification. (**j**) Parasorb Resodont surface at 120× magnification. In (**a,c,e,g,i**), the scale bar is 500 µm, and it is 100 µm in (**b,d,f,h,j**). Pointers indicate pores of different size across the membrane's surface. Arrow heads indicate collagen fibers.

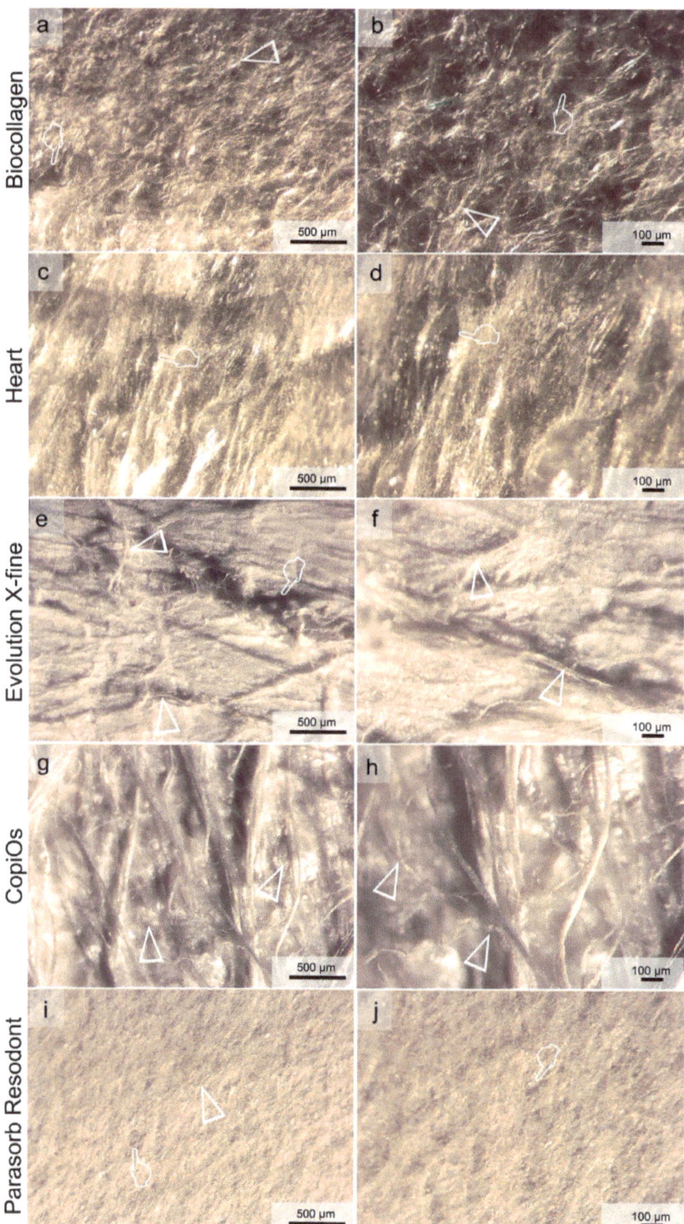

Figure 4. Light micrographs taken of the rough surface of the membranes before immersion. (**a**) Biocollagen surface at 60× magnification. (**b**) Biocollagen at 120× magnification. (**c**) Heart surface at 60× magnification. (**d**) Heart surface at 120× magnification. (**e**) Evolution X-fine surface at 60× magnification. (**f**) Evolution X-fine surface at 120× magnification. (**g**) CopiOs surface at 60× magnification. (**h**) CopiOs surface at 120× magnification. (**i**) Parasorb Resodont surface at 60× magnification. (**j**) Parasorb Resodont surface at 120× magnification. In (**a,c,e,g,i**), the scale bar is 500 μm, it is and 100 μm in (**b,d,f,h,j**). Pointers indicate pores of different size across the membrane's surface. Arrow heads indicate collagen fibers. Arrows point to mineral deposits on the membrane's surface.

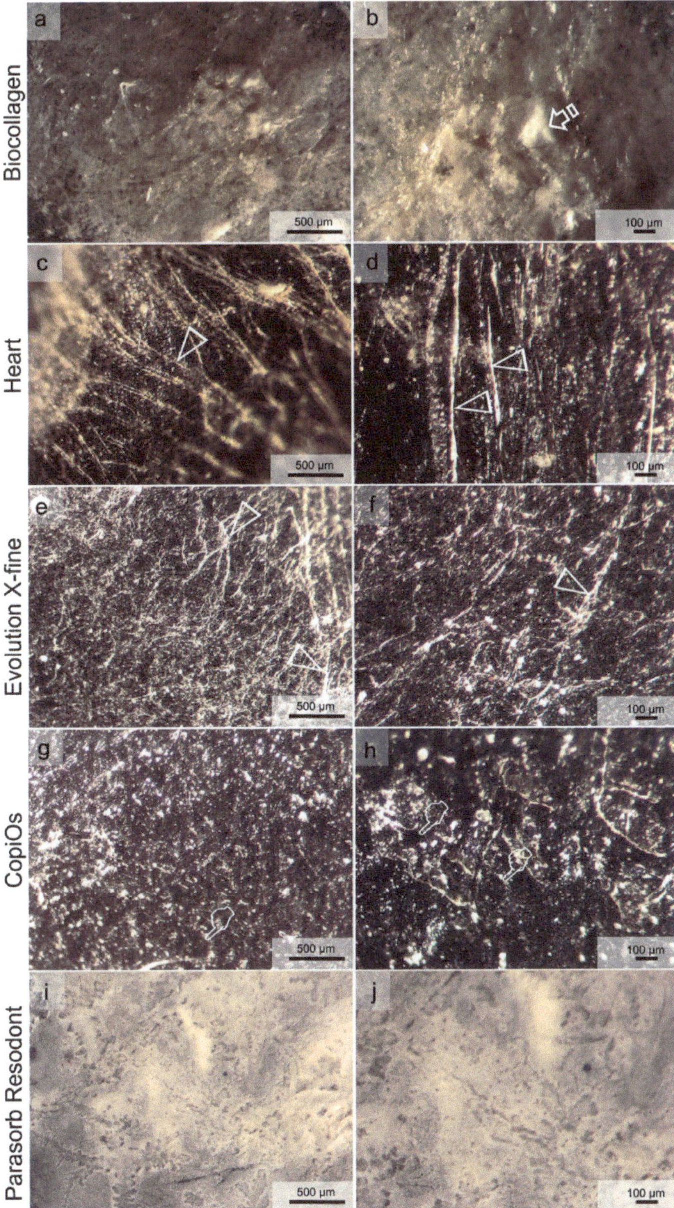

Figure 5. Light micrographs taken of membranes after immersion in the PBS degradation test. (**a**) Biocollagen surface at $60\times$ magnification. (**b**) Biocollagen at $120\times$ magnification. (**c**) Heart surface at $60\times$ magnification. (**d**) Heart surface at $120\times$ magnification. (**e**) Evolution X-fine surface at $60\times$ magnification. (**f**) Evolution X-fine surface at $120\times$ magnification. (**g**) CopiOs surface at $60\times$ magnification. (**h**) CopiOs surface at $120\times$ magnification. (**i**) Parasorb Resodont surface at $60\times$ magnification. (**j**) Parasorb Resodont surface at $120\times$ magnification. The scale bar is 500 μm in (**a,c,e,g,i**), and 100 μm in (**b,d,f,h,j**). Pointers indicate pores of different size across the membrane's surface. Arrow heads indicate collagen fibers. Arrows point to mineral deposits on the membrane's surface.

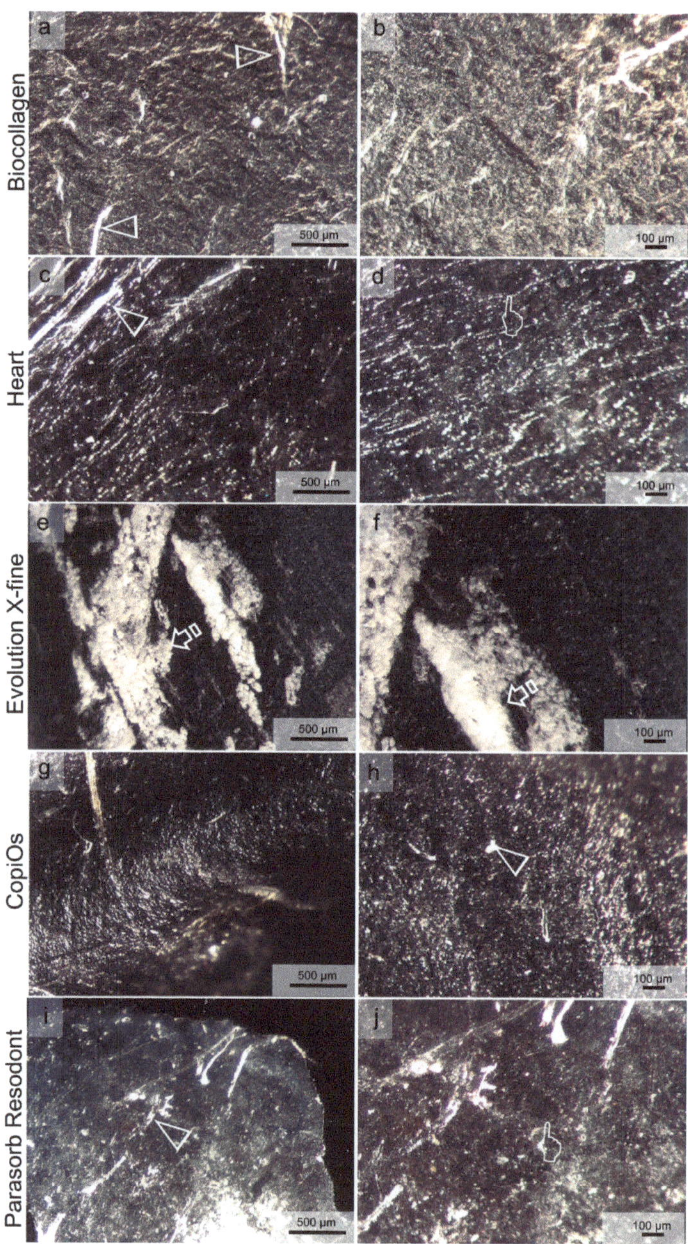

Figure 6. Light micrographs taken of membranes after immersion in the Trypsin degradation test. (**a**) Biocollagen surface at 60× magnification. (**b**) Biocollagen at 120× magnification. (**c**) Heart surface at 60× magnification. (**d**) Heart surface at 120× magnification. (**e**) Evolution X-fine surface at 60× magnification. (**f**) Evolution X-fine surface at 120× magnification. (**g**) CopiOs surface at 60× magnification. (**h**) CopiOs surface at 120× magnification. (**i**) Parasorb Resodont surface at 60× magnification. (**j**) Parasorb Resodont surface at 120× magnification. The scale bar is 500 μm in (**a,c,e,g,i**), and 100 μm in (**b,d,f,h,j**). Pointers indicate pores of different sizes across the membrane's surface. Arrow heads indicate collagen fibers.

Figures 5 and 6 display the appearance of the membranes after 7 days of immersion in PBS and Trypsin. The effect of the membrane degradation prevented us from distinguishing rough from smooth membrane surfaces in both degradation tests. After immersion in PBS, Biocollagen showed a dense surface without appreciation of collagen fiber distribution, but some mineral deposits were deposited on its surface (Figure 5a,b). Multiple bundles of collagen of different sizes were aligned in parallel and were separated from each other by irregular spaces in Heart, Evolution X-fine and CopiOs. Parasorb Resodont showed a regular surface where pores and collagen fibers were hardly visible (Figure 5i,j). After the trypsin test, Biocollagen demonstrated less uniformity regarding collagen fiber orientation, with a random distribution of collagen fibers (Figure 6a,b). Heart presented multiple bundles of collagen distributed in layers (Figure 6c,d). The fiber network was difficult to appreciate in the surface of Evolution X-fine, CopiOs and Parasorb Resodont. A large mineral condensation was found on Evolution X-fine's surface (Figure 6e,f).

4. Discussion

The present study examined the effect of different immersion media on the biodegradation pattern of five commercially available membranes over time. The presented data demonstrated that the membranes suffered a continuous degradation process throughout the multiple storage periods and did not resist to degradation assays equally. Evolution X-fine was the most resistant membrane to the degradation testing, assessed by weight and thickness (Figures 1 and 2).

Biocollagen, Heart, Evolution X-fine, CopiOs and Parasorb Resodont are resorbable membranes mainly composed of collagen. The donor area and the origin of collagen vary among membranes; this may explain the differences between degradation behavior [15]. Measured by weight, Biocollagen obtained the highest percentage of degradation in the initial periods (Figure 2). However, observing the results presented for thickness, Biocollagen did not suffer a loss of thickness until 6 or 24 h of storage. Furthermore, the final thickness loss of Biocollagen was less than in the other membranes (Figure 1). Parasorb Resodont degraded by almost 90%, reaching the highest percentages of thickness loss after 50 d of immersion in PBS and trypsin, and it was completely degraded after 7 days in *C. histolyticum* collagenase solution. Signs of degradation showed that the loss of the membranes' structural integrity occurred before even 6 h. Our results agree with previous findings of Toledano et al. [11], who demonstrated that collagen membranes submitted to degradation tests attained a significant reduction in thickness and weight before 4 h of immersion. Zhao et al. [25] reported a disintegration of membrane material following 3 weeks of implantation in rats' subcutaneous pouches. In addition, Calciolari et al. [14] clinically applied collagen membranes for GBR and found a significant reduction in membrane thickness from 7 to 30 days of healing. One possible reason for this performance is that both Biocollagen and Parasorb Resodont are made of equine collagen. Gallo et al. [18] stated that collagen from horses is the most resistant to degradation. In contrast, in the present research, it has been observed that among all of the membranes evaluated, those of equine origin demonstrated faster degradation. In addition, Heart, an equine collagen membrane, performed intermediate in loss of weight and thickness values, being the third membrane with the highest degradation percentages. In accordance with our findings, Toledano et al. [11] submitted three membranes, two from equine origin and one from porcine origin to different degradation media. Derma Fine, a porcine collagen-based membrane, demonstrated the greatest resistance to all degradation challenges under in vitro conditions [11]. Differences in membranes degradation may be attributed to variations in membrane composition [15,19]. Thus, the first null hypothesis has to be rejected.

Guided bone regeneration (GBR) is based on the concept of cell exclusion and space maintenance to allow sufficient time for new bone formation [1]. For regeneration to occur, the membrane is expected to remain intact up to 4–6 months [18]. Evolution X-fine performed as the most resistant membrane to the degradation tests followed by CopiOs. In addition to their origin, porcine and bovine, respectively, these membranes presented the least porous surface when they were evaluated with optical microscopy before immersion (Figures 3 and 4). Morphological changes have been observed on collagen matrixes after degradation. These are the disarrangement of the collagen net and pore enlargement through the widening of the interfibrillar spaces. These findings are consistent with previously described alterations that were found after water degradation of human collagen [26]. Previous studies correlating the membrane structure with degradation reported that thicker collagen fibers, higher collagen density and a more compact structure may be produce slower biodegradation [27,28].

When comparing membranes surface, Evolution X-fine surface was smoother and more homogenous to any other membrane type regardless of the side observed. A less porous membrane characteristic seems to avoid excessive humidity and the consequent loss of the physical properties of the membrane [29]. It is generally accepted that the absorption ability of barrier membranes greatly varies depending on their origin and composition [30]. Therefore, the macroscopic structure and origin of these membranes explains their greater resistance to degradation. Caballé-Serrano et al. [19] found that porcine origin barrier membranes and porous membranes showed a high wettability. In addition, Toledano et al. [11] also observed that a specific distribution of the collagen fibers and the presence of pores influenced the degradation behavior. Evolution X-fine, made from collagen fibers from porcine pericardium, attained a reduction of less than 50% in thickness and 20% in weight, for the media PBS and trypsin (Figures 1 and 2). Measuring the weight and thickness of the membrane after different immersion times makes it possible to quantitatively evaluate how a membrane behaves in a medium. Thickness features could lead to a remarkable interpretation, since the barrier effect depends on the ability to maintain a minimum thickness for the full period, enabling the desired cell exclusion. Several studies have investigated whether the addition of a second membrane layer prevents rapid reabsorption and maintains the essential function of the membrane [1,15]. The obtained results show that collagen degrades at the same rate; however, having a double thickness results in a significantly greater amount of residual collagen [1,15]. This makes possible to extend the functional time of GBR membranes. Consequently, in addition to the origin and macroscopic structure, the initial thickness of the membrane is also an aspect to be considered.

The main disadvantages of collagen-based membranes are their lack of rigidity and fast degradation kinetics by endogenous collagenases, so that the barrier function may not remain long enough for tissue regeneration [31]. From 0 to 1 h of immersion, weight and thickness values increased for Biocollagen. This could be explained by the major water absorption that likely occurs during the first immersion. Biocollagen comes from equine tendons, where collagen fibrils are coupled via proteoglycans (PGs) linkages [32]. PGs are one type of non-collagenous protein, and primarily contain a core protein and glycosaminoglycans (GAGs). GAGs are highly polar and negatively charged, thus having a strong tendency towards attracting water molecules into the matrix. On the other hand, water also functions as a plasticizer, thus diminishing mechanical properties and facilitating the degradation process [33].

In the present study, *C. histolyticum* was the most aggressive medium for membrane degradation, as four out of five collagen membranes could not endure more than 7 d of storage (Figures 1c and 2c). Evolution X-fine was the only membrane that reached 50 d of immersion, but with a great rate of resorption in *C. histolyticum* (Tables 2 and 3). The rapid degradation of collagen membranes in this medium has been previously verified by other authors, Vallecillo et al. [34] observed early degradation when three collagen matrices were submitted to the *C. histolyticum* assay. Fibro-Gide, a cross-linked collagen matrix, was uniquely able to withstand 7 d of storage in a bacterial collagenase test. The authors attributed the greater resistance of this matrix compared to the others studied to the cross-linking of collagen. Cross-linked methods have been proposed in order to decelerate collagen-based membrane resorption [10,25,26]. Toledano et al. [11] showed that samples subjected to *C. histolyticum* challenge lost more thickness and weight than those in trypsin and PBS. Degradation in bacterial collagenase in vitro is an adequate test to assess and predict membrane stability in the presence of wound dehiscence [18]. Sela et al. [17] faced collagen membranes for GBR to different enzymes produced by periodontopathic bacteria. They suggested that bacterial collagenase is more aggressive in degrading collagen membranes than the proposed enzyme cocktail [17]. Compared to the enzyme and hydrolytic degradation assays, bacterial collagenase has been shown to be more effective in degrading collagen; all membranes suffered an increased loss of thickness and weight, regardless of their origin [11]. When the samples were immersed in PBS and Trypsin, loss was experienced, but they all endured 50 d of immersion. However, in *C. histolyticum*, membranes were not able to resist degradation beyond 7 d, except for Evolution X-fine. For this reason, the second null hypothesis has also rejected.

The present study demonstrates the in vitro degradability behavior and morphological analysis of five commercially available collagen-based membranes from different heterologous origins. This in vitro study allows for detecting the influence of the initial composition and origin of the collagen membrane in the biodegradation pattern, regardless of the cellular reaction of the recipient organism. The present results may be useful to surgeons to better select a GBR membrane according to the procedure to be performed and the risk of wound dehiscence. Evolution X-fine is made from porcine pericardium collagen and has shown the highest resistance to the tested degradation challenges, even in collagenase from *C. histolyticum*. This superior resistance added to a more homogeneous surface prolongs the barrier effect of the membrane, even in case of being exposed to a more aggressive environment. Therefore, Evolution X-fine should be considered as a potential candidate for use in those GBR procedures prone to cross-infection and in several challenging pathologies in which bacteria may alter the positive clinical outcome, including periodontal diseases [35].

Three degradation tests were performed, which represent more aggressive conditions for collagen membranes than just submerged healing. This allows the membranes to be evaluated in the worst scenario, which may occur clinically in the presence of wound dehiscence. In clinical practice, it is common to observe membrane exposure, especially during the early stages of healing [14]. As the membrane is exposed to the oral environment, saliva and bacteria will trigger more rapid degradation of the biomaterial, compromising the success of bone regeneration [14].

Nevertheless, the results of this study should be considered with caution due to its limitations. The correlation between membranes' topography, collagen fibrils' nanostructure and mechanical properties was not studied. Further research will be conducted in order to ascertain whether these changes in collagen nanostructure may also be related to a decrease in the mechanical properties of collagen membranes through the ongoing degradation process. Additionally, the impact of embedded cells on membrane degradation is missing. Despite the limitations of the study, our in vitro findings are intended to give a first estimation of the membranes behavior, but it should also be considered that in vivo, a more complex set of different cell types, body fluids and possible microbial contamination may influence the membrane degradation process, with the oral cavity environment being an scenario that is difficult to mimic [27,28]. Therefore, it might be challenging to directly transfer these data to the clinical practice. Authors encourage researchers to continue evaluating how to retard membranes resorption. Membranes doped with antibacterial agents could also be prepared and studied aiming to counteract the rapid degradation in the presence of bacterial enzymes.

5. Conclusions

Evolution X-fine collagen membrane from porcine pericardium origin has been proven to be the least affected when immersed in PBS, trypsin and *C. histolyticum*. Biocollagen and Parasorb Resodont, both of equine origin, experienced the greatest degradation, measured in thickness and weight, under in vitro conditions. The bacterial collagenase solution acted as the most aggressive test against membranes, producing complete degradation of four membranes before 7 days of immersion. The loss of structural integrity and pores is presented after storage processes, demonstrating the occlusive effect of GBR membranes to exclude the epithelium and connective tissue.

Author Contributions: Conceptualization, M.V.-R., M.T. and R.O.; Formal analysis, M.T., M.T.-O., C.V., M.V.-R. and R.O.; Funding acquisition, M.T. and R.O.; Investigation, M.T., M.T.-O., M.V.-R., C.V. and R.O.; Methodology, M.T.-O., C.V., M.V.-R. and R.O.; Project administration, M.T. and R.O.; Supervision, M.T., R.O. and M.T.-O.; Validation, M.T., M.T.-O. and R.O.; Visualization; M.T., M.T.-O. and R.O.; Writing—original draft, C.V., M.T.-O., M.V.-R., M.T. and R.O.; Writing—review and editing, M.V.-R., M.T.-O., C.V., M.T. and R.O. All authors have read and agreed to the published version of the manuscript.

Funding: This work was supported by: (1) the Ministry of Economy and Competitiveness and European Regional Development Fund [Project PID2020-114694RB-100 MINECO/AEI/FEDER/UE], (2) University of Granada/Regional Government of Andalusia Research Fund from Spain and European Regional Development Fund (A-BIO-157-UGR-18/FEDER).

Institutional Review Board Statement: Not applicable.

Informed Consent Statement: Not applicable.

Data Availability Statement: The data presented in this study are available on request from the corresponding author.

Conflicts of Interest: The authors report no conflicts of interest related to this in vitro study.

References

1. Kozlovsky, A.; Aboodi, G.; Moses, O.; Tal, H.; Artzi, Z.; Weinreb, M.; Nemcovsky, C.E. Bio-Degradation of a Resorbable Collagen Membrane (Bio-Gide) Applied in a Double-Layer Technique in Rats. *Clin. Oral Implants Res.* **2009**, *20*, 1116–1123. [CrossRef] [PubMed]
2. Dahlin, C.; Linde, A.; Gottlow, J.; Nyman, S. Healing of Bone Defects by Guided Tissue Regeneration. *Plast Reconstr. Surg.* **1988**, *81*, 672–676. [CrossRef]
3. Omar, O.; Elgali, I.; Dahlin, C.; Thomsen, P. Barrier Membranes: More than the Barrier Effect? *J. Clin. Periodontol.* **2019**, *46* Suppl. 21, 103–123. [CrossRef]
4. Wessing, B.; Lettner, S.; Zechner, W. Guided Bone Regeneration with Collagen Membranes and Particulate Graft Materials: A Systematic Review and Meta-Analysis. *Int. J. Oral Maxillofac. Implants* **2018**, *33*, 87–100. [CrossRef]
5. Sbricoli, L.; Guazzo, R.; Annunziata, M.; Gobbato, L.; Bressan, E.; Nastri, L. Selection of Collagen Membranes for Bone Regeneration: A Literature Review. *Materials* **2020**, *13*, 786. [CrossRef] [PubMed]
6. An, Y.-Z.; Kim, Y.-K.; Lim, S.-M.; Heo, Y.-K.; Kwon, M.-K.; Cha, J.-K.; Lee, J.-S.; Jung, U.-W.; Choi, S.-H. Physiochemical Properties and Resorption Progress of Porcine Skin-Derived Collagen Membranes: In Vitro and in Vivo Analysis. *Dent. Mater. J.* **2018**, *37*, 332–340. [CrossRef]
7. Zhang, Y.; Zhang, X.; Shi, B.; Miron, R. Membranes for Guided Tissue and Bone Regeneration. *Ann. Oral Maxillofac. Surg.* **2013**, *1*. [CrossRef]
8. Elgali, I.; Turri, A.; Xia, W.; Norlindh, B.; Johansson, A.; Dahlin, C.; Thomsen, P.; Omar, O. Guided Bone Regeneration Using Resorbable Membrane and Different Bone Substitutes: Early Histological and Molecular Events. *Acta Biomater.* **2016**, *29*, 409–423. [CrossRef]
9. Toledano-Osorio, M.; Manzano-Moreno, F.J.; Ruiz, C.; Toledano, M.; Osorio, R. Testing Active Membranes for Bone Regeneration: A Review. *J. Dent.* **2021**, *105*, 103580. [CrossRef] [PubMed]
10. Thoma, D.S.; Zeltner, M.; Hilbe, M.; Hämmerle, C.H.F.; Hüsler, J.; Jung, R.E. Randomized Controlled Clinical Study Evaluating Effectiveness and Safety of a Volume-Stable Collagen Matrix Compared to Autogenous Connective Tissue Grafts for Soft Tissue Augmentation at Implant Sites. *J. Clin. Periodontol.* **2016**, *43*, 874–885. [CrossRef] [PubMed]
11. Toledano, M.; Asady, S.; Toledano-Osorio, M.; García-Godoy, F.; Serrera-Figallo, M.-A.; Benítez-García, J.A.; Osorio, R. Differential Biodegradation Kinetics of Collagen Membranes for Bone Regeneration. *Polymers* **2020**, *12*, 1290. [CrossRef]
12. Tal, H.; Kozlovsky, A.; Artzi, Z.; Nemcovsky, C.E.; Moses, O. Long-Term Bio-Degradation of Cross-Linked and Non-Cross-Linked Collagen Barriers in Human Guided Bone Regeneration. *Clin. Oral Implants Res.* **2008**, *19*, 295–302. [CrossRef] [PubMed]
13. Soldatos, N.K.; Stylianou, P.; Koidou, V.P.; Angelov, N.; Yukna, R.; Romanos, G.E. Limitations and Options Using Resorbable versus Nonresorbable Membranes for Successful Guided Bone Regeneration. *Quintessence Int.* **2017**, *48*, 131–147. [CrossRef]
14. Calciolari, E.; Ravanetti, F.; Strange, A.; Mardas, N.; Bozec, L.; Cacchioli, A.; Kostomitsopoulos, N.; Donos, N. Degradation Pattern of a Porcine Collagen Membrane in an in Vivo Model of Guided Bone Regeneration. *J. Periodontal Res.* **2018**, *53*, 430–439. [CrossRef]

15. Von Arx, T.; Broggini, N.; Jensen, S.S.; Bornstein, M.M.; Schenk, R.K.; Buser, D. Membrane Durability and Tissue Response of Different Bioresorbable Barrier Membranes: A Histologic Study in the Rabbit Calvarium. *Int. J. Oral Maxillofac. Implants* **2005**, *20*, 843–853.
16. Rothamel, D.; Schwarz, F.; Fienitz, T.; Smeets, R.; Dreiseidler, T.; Ritter, L.; Happe, A.; Zöller, J. Biocompatibility and Biodegradation of a Native Porcine Pericardium Membrane: Results of in Vitro and in Vivo Examinations. *Int. J. Oral Maxillofac. Implants* **2012**, *27*, 146–154. [PubMed]
17. Meyer, M. Processing of Collagen Based Biomaterials and the Resulting Materials Properties. *Biomed. Eng. Online* **2019**, *18*, 24. [CrossRef]
18. Gallo, N.; Natali, M.L.; Sannino, A.; Salvatore, L. An Overview of the Use of Equine Collagen as Emerging Material for Biomedical Applications. *J. Funct. Biomater.* **2020**, *11*, 79. [CrossRef] [PubMed]
19. Caballé-Serrano, J.; Munar-Frau, A.; Delgado, L.; Pérez, R.; Hernández-Alfaro, F. Physicochemical Characterization of Barrier Membranes for Bone Regeneration. *J. Mech. Behav. Biomed. Mater.* **2019**, *97*, 13–20. [CrossRef]
20. Zhang, H.Y.; Jiang, H.B.; Ryu, J.-H.; Kang, H.; Kim, K.-M.; Kwon, J.-S. Comparing Properties of Variable Pore-Sized 3D-Printed PLA Membrane with Conventional PLA Membrane for Guided Bone/Tissue Regeneration. *Materials* **2019**, *12*, 1718. [CrossRef]
21. Sunandhakumari, V.J.; Vidhyadharan, A.K.; Alim, A.; Kumar, D.; Ravindran, J.; Krishna, A.; Prasad, M. Fabrication and In Vitro Characterization of Bioactive Glass/Nano Hydroxyapatite Reinforced Electrospun Poly(ε-Caprolactone) Composite Membranes for Guided Tissue Regeneration. *Bioengineering* **2018**, *5*, 54. [CrossRef] [PubMed]
22. Sela, M.N.; Kohavi, D.; Krausz, E.; Steinberg, D.; Rosen, G. Enzymatic Degradation of Collagen-Guided Tissue Regeneration Membranes by Periodontal Bacteria. *Clin. Oral Implants Res.* **2003**, *14*, 263–268. [CrossRef] [PubMed]
23. Bozkurt, A.; Apel, C.; Sellhaus, B.; van Neerven, S.; Wessing, B.; Hilgers, R.-D.; Pallua, N. Differences in Degradation Behavior of Two Non-Cross-Linked Collagen Barrier Membranes: An in Vitro and in Vivo Study. *Clin. Oral Implants Res.* **2014**, *25*, 1403–1411. [CrossRef] [PubMed]
24. Wong, C.; Yoganarasimha, S.; Carrico, C.; Madurantakam, P. Incorporation of Fibrin Matrix into Electrospun Membranes for Periodontal Wound Healing. *Bioengineering* **2019**, *6*, 57. [CrossRef]
25. Zhao, S.; Pinholt, E.M.; Madsen, J.E.; Donath, K. Histological Evaluation of Different Biodegradable and Non-Biodegradable Membranes Implanted Subcutaneously in Rats. *J. Craniomaxillofac. Surg.* **2000**, *28*, 116–122. [CrossRef] [PubMed]
26. Hashimoto, M.; Tay, F.C.M.; Ohno, H.; Sano, H.; Kaga, M.; Yiu, C.; Kumagai, H.; Kudou, Y.; Kubota, M.; Oguchi, H. SEM and TEM Analysis of Water Degradation of Human Dentinal Collagen. *J. Biomed. Mater. Res.* **2003**, *66*, 287–298. [CrossRef]
27. Cavalu, S.; Roiu, G.; Pop, O.; Heredea, D.A.P.; Costea, T.O.; Costea, C.F. Nano-Scale Modifications of Amniotic Membrane Induced by UV and Antibiotic Treatment: Histological, AFM and FTIR Spectroscopy Evidence. *Materials* **2021**, *14*, 863. [CrossRef]
28. Ratiu, C.; Brocks, M.; Costea, T.; Moldovan, L.; Cavalu, S. PRGF-Modified Collagen Membranes for Guided Bone Regeneration: Spectroscopic, Microscopic and Nano-Mechanical Investigations. *Appl. Sci.* **2019**, *9*, 1035. [CrossRef]
29. Silva, M.J.; Gonçalves, C.P.; Galvão, K.M.; D'Alpino, P.H.P.; Nascimento, F.D. Synthesis and Characterizations of a Collagen-Rich Biomembrane with Potential for Tissue-Guided Regeneration. *Eur. J. Dent.* **2019**, *13*, 295–302. [CrossRef]
30. Caballé-Serrano, J.; Abdeslam-Mohamed, Y.; Munar-Frau, A.; Fujioka-Kobayashi, M.; Hernández-Alfaro, F.; Miron, R. Adsorption and Release Kinetics of Growth Factors on Barrier Membranes for Guided Tissue/Bone Regeneration: A Systematic Review. *Arch. Oral Biol.* **2019**, *100*, 57–68. [CrossRef]
31. Bornert, F.; Herber, V.; Sandgren, R.; Witek, L.; Coelho, P.G.; Pippenger, B.E.; Shahdad, S. Comparative Barrier Membrane Degradation over Time: Pericardium versus Dermal Membranes. *Clin. Exp. Dent. Res.* **2021**. [CrossRef]
32. Franchi, M.; Trirè, A.; Quaranta, M.; Orsini, E.; Ottani, V. Collagen Structure of Tendon Relates to Function. *Sci. World J.* **2007**, *7*, 404–420. [CrossRef] [PubMed]
33. Wang, X.; Xu, H.; Huang, Y.; Gu, S.; Jiang, J.X. Coupling Effect of Water and Proteoglycans on the In Situ Toughness of Bone. *J. Bone Miner. Res.* **2016**, *31*, 1026–1029. [CrossRef] [PubMed]
34. Vallecillo, C.; Toledano-Osorio, M.; Vallecillo-Rivas, M.; Toledano, M.; Osorio, R. In Vitro Biodegradation Pattern of Collagen Matrices for Soft Tissue Augmentation. *Polymers* **2021**, *13*, 2633. [CrossRef] [PubMed]
35. Toledano-Osorio, M.; Manzano-Moreno, F.J.; Toledano, M.; Medina-Castillo, A.L.; Costela-Ruiz, V.J.; Ruiz, C.; Osorio, R. Doxycycline-Doped Polymeric Membranes Induced Growth, Differentiation and Expression of Antigenic Phenotype Markers of Osteoblasts. *Polymers* **2021**, *13*, 1063. [CrossRef]

Article

The Effect of Adding Modified Chitosan on the Strength Properties of Bacterial Cellulose for Clinical Applications

Anna Lipovka [1,*,†], Alexey Kharchenko [2,†], Andrey Dubovoy [1,3,†], Maxim Filipenko [4,†], Vyacheslav Stupak [2,†], Alexander Mayorov [5,†], Vladislav Fomenko [6,†], Pavel Geydt [7,†] and Daniil Parshin [1,†]

1   Lavrentyev Institute of Hydrodynamics of the Siberian Branch of the Russian Academy of Sciences, 630090 Novosibirsk, Russia; dubovoy@neuronsk.ru (A.D.); danilo.skiman@gmail.com (D.P.)
2   Novosibirsk Research Institute of Traumatology and Orthopaedics n.a. Ya.L. Tsivyan, 630090 Novosibirsk, Russia; alexdok2000@gmail.com (A.K.); stupak@niito.ru (V.S.)
3   Federal Neurosurgical Center, 630048 Novosibirsk, Russia
4   Institute of Chemical Biology and Fundamental Medicine of the Siberian Branch of the Russian Academy of Sciences, 630090 Novosibirsk, Russia; mlfilipenko@gmail.com
5   Institute of Laser Physics of the Russian Academy of Sciences, 630090 Novosibirsk, Russia; aleksander.mayorov@gmail.com
6   N.N. Vorozhtsov Novosibirsk Institute of Organic Chemistry of the Russian Academy of Sciences, 630090 Novosibirsk, Russia; vladislav@ngs.ru
7   Novosibirsk State University, 630090 Novosibirsk, Russia; p.geydt@nsu.ru
*   Correspondence: lipovkaai@yandex.ru
†   These authors contributed equally to this work.

Citation: Lipovka, A.; Kharchenko, A.; Dubovoy, A.; Filipenko, M.; Stupak, V.; Mayorov, A.; Fomenko, V.; Geydt, P.; Parshin, D. The Effect of Adding Modified Chitosan on the Strength Properties of Bacterial Cellulose for Clinical Applications. *Polymers* **2021**, *13*, 1995. https://doi.org/10.3390/polym13121995

Academic Editors: João Carlos Silva and Frederico Castelo Ferreira

Received: 30 April 2021
Accepted: 13 June 2021
Published: 18 June 2021

Publisher's Note: MDPI stays neutral with regard to jurisdictional claims in published maps and institutional affiliations.

Abstract: Currently, several materials for the closure of the dura mater (DM) defects are known. However, the long-term results of their usage reveal a number of disadvantages. The use of antibiotics and chitosan is one of the major trends in solving the problems associated with infectious after-operational complications. This work compares the mechanical properties of samples of bacterial nanocellulose (BNC) impregnated with Novochizol™ and vancomycin with native BNC and preserved and native human DM. An assessment of the possibility of controling the mechanical properties of these materials by changing their thickness has been performed by statistical analysis methods. A total of 80 specimens of comparable samples were investigated. During the analysis, the results obtained, the factor of Novochizol™ addition has provided a statistically significant impact on the strength properties (Fisher Criteria p-value 0.00509 for stress and 0.00112 for deformation). Moreover, a stronger relationship between the thickness of the samples and their ultimate load was shown: $R^2 = 0.236$ for BNC + Novochizol™ + vancomycin, compared to $R^2 = 0.0405$ for native BNC. Using factor analysis, it was possible to show a significant effect of modified chitosan (Novochizol™) on the ultimate stress (p-value = 0.005).

Keywords: tissue biomechanics; dura matter; bacterial nanocellulose; factor analysis; dura substitutes; chitosan gel; Novochizol™; vancomycin

1. Introduction

The substance of the brain and spinal cord is covered by three layers of connective tissue, collectively called the meninges: the soft (pia mater), the arachnoid membrane, and the dura (dura mater). The meninges contain blood vessels and are surrounded by cerebrospinal fluid (CSF). The outer layer of three layers of membrane, i.e., the dura mater, is an irregularly shaped, thick, whitish sheath of dense fibrous tissue with a large number of elastic fibers.

In the case of injuries, the oncological process of the brain and spinal cord, and during various neurosurgical interventions, the dura mater is dissected to provide access to the nerve structures located underneath. After the operation, the skull is sealed by suturing the dura mater. Often, for a number of reasons, this cannot be achieved, so during the postoperative period, liquorrhea (leakage of cerebrospinal fluid) develops in the area of the surgery, which leads to communication of the cranial cavity with the external environment. Such a postoperative complication is the most dangerous, as it often leads to the development of severe purulent complications from the central nervous system. With the development of such complications, intensive conservative therapy is needed in the form of prescribing antibacterial drugs, and repeated surgical interventions aimed at sealing the cranial cavity and spinal canal with the additional imposition of repeated sealed sutures on the dura mater are also required in some cases. Effective closure of the dura mater defect helps to minimize liquorrhea and promotes normal wound healing [1–3].

In addition, in violation of the dura mater integrity, fistulas and pseudomeningocele are formed (Figure 1). The latter is defined as a pathological extradural accumulation of cerebrospinal fluid in soft tissues communicating through a defect in the dura mater with the arachnoid space of the brain [4,5]. Figure 1 shows a cavity in the soft tissues of the cervical spine filled with cerebrospinal fluid (pseudomeningocele), which connects to the cerebrospinal fluid space of the spinal cord after removal of the extramedullary tumor and non-hermetic suturing of the dura mater. Plastic surgery on the dura mater is also necessary in cases of restoration of the lost part of the intrinsic dura mater associated with its invasion by a tumor, developed cerebral edema, elimination of cerebrospinal fluid fistulas, increase in the subdural space in Arnold–Chiari malformation, and myelomeningocele surgery [6].

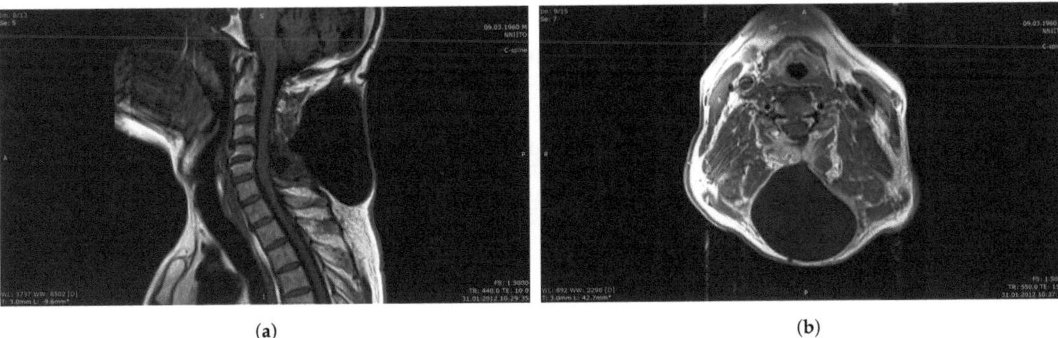

(a) (b)

Figure 1. MRI of the cervical spine after removal of the spinal cord tumor and leaky suturing of the dura mater with the pseudomeningocele formation. (**a**): Sagittal plane, (**b**): Horizontal plane.

From the available statistics of complications based on the clinical data, it can be concluded that it is necessary to repair dura mater defects using various types of implants since it reduces the number of postoperative complications [7–9]. For this, materials of various origins are used [10]:

1. Autograft (fascia lata, fascia of the temporal muscle, pericranium et al. [11–13])
2. Xenograft (collagen implants: DuraGen, Lyoplant® et al. [7,14])
3. Synthetic: absorbable (PGA, copolymer of L-lactic acid and epsilon-caprolactone, or copolymer of lactide and polydioxanone) and non-absorbable (ePTF, Polyesterurethane-Neuro Patch, polypropylene G-patch [15–17])
4. Biopolymers (Chitosan, bacterial cellulose and et al. [18,19])

Sealing of the skull by dura mater repair using the body's own tissues (autografts) is the most effective way [20]. However, the use of the patient's autotissue for these purposes is limited by the laboriousness of the technique (splitting the dura mater according to the technique of N.N. Burdenko). In addition, the use of a portion of the hip fascia is an additional traumatic factor for the patient, fraught with lengthening the duration of the surgery and possible complications at the donor site [2]. The use of allografts (preserved cadaveric dura mater) is currently prohibited due to the possibility of transmission of infections and viruses, as well as difficulties in procurement and storage of material [21].

In the modern neurosurgery practice, there are a number of materials approved for use in the clinic for the closure of dura mater defects: DuraGen, DuraGen Plus, DuraGen Suturable collagen implants, Lyoplant®. All of them are pure collagen implants that are produced from lyophilized cattle pericardium. Collagen implants are the gold standard in neurosurgery for dura mater repair. However, all of these implants are imperfect and have their disadvantages.

Lyodura is a material from cadaveric dura mater produced by the German manufacturer B. Braun Melsungen AG and was the main source of the prion disease outbreak (Creutzfeld–Jakob disease). Laboratory studies have shown that standard methods of decontamination and sterilization may not be sufficient to completely eliminate prion contamination of surgical instruments after surgical treatment [22,23]. It is possible that collagen implants can transmit prions.

Synthetic implants often carry a high risk of infectious complications [24–26]. Synthetic grafts are often rigid and can cause inflammatory and foreign body reactions. These reactions can create inflammation of the surrounding tissue and brain, excessive production of fibrin during graft encapsulation, meningitis, graft rejection, scarring, and delayed bleeding, which often require reoperation [27].

Long-term results of using implants are not always completely satisfactory both for the patient and for the attending physician. Consequently, there is a need to find and develop materials (DM substitutes) that will be free from the drawbacks of existing implants and will have improved properties.

The general problems of using dura mater implants are as follows:

1. Allogeneic tissues: xenograft tissue can cause its rejection [28].
2. Dura mater and surrounding tissue adhesion: after the restoration, the dura mater implants have a different degree of adhesion, mainly associated with the inflammatory response, physical and chemical properties of the material. The lower the content of protein and fat in the material, the lower the degree of adhesion [29].
3. Development of aseptic inflammation. In addition, the use of allogeneic and xenogenic materials can lead to the spread of pathogens among humans and animals, prions and viruses [30].
4. Bleeding: incipient granulation tissue that regenerates and covers the graft material can cause bleeding. There may be a gap between the material and the elaborated network of neocapillaries covering the matrix. The capillaries are fragile, bleeding can occur when the implant is mixed, and a subdural hematoma can form in the cranial cavity [31].
5. Development of liquorrhea as a result of the lack of reliable sealing of the dura mater defect [32].
6. The occurrence of epileptic seizures, as a result of the development of meningeal adhesions [33].

The ideal material for plastics of the dura mater has not yet been created, but the requirements for the design of the material to be used to close the dura mater defect are that it should not induce an immunological or inflammatory response, should not be neurotoxic, carcinogenic, should provide a hermetic closure, and retain its shape after use and stay durable. An ideal DM substitute should not pose a risk of transmission of viral and prion infections. The implant should be capable of being stored for a long time, retain its valuable properties, and have a low cost.

In our opinion, a promising direction is the use of implants made of bacterial nanocellulose, since this material meets the above-mentioned requirements.

In neurosurgery, it is important that new materials for dura mater restoration not only perform a barrier function but also have necessary mechanical strength. This is because increased intracranial pressure can rupture a new dura mater implant after decompression craniotomy.

In the work of Dutta P.K et al. [34], the effect of the selected bacterial strain on the growth of the BNC polymer was investigated. The authors have shown a correlation between the ultimate strength and the concentration of the polymer in the hydrogel, while the ultimate deformation did not correlate with the concentration of the polymer. The choice of a specific strain of bacteria for the production of BNC material, as can be seen from their work, is also important.

In the article by Kizmazoglu C. et al. [35], it was shown that mechanical properties of the DM substitutes vary significantly and often greatly differ from the properties of DM itself. Nevertheless, the studied DM substitutes have shown their effectiveness. Moreover, the tensile strength of these substitutes was measured to be exceeding the DM strength for some materials, while it was lower than DM for others. After comparing the results from similar studies, it can be noted that the difference in strength in DM and its substitute can be significant, up to 10 times.

Such differences can be explained not only by different approaches to measuring deformation but also by different approaches to material fixing. For example, in the works [35,36], fixing was used in clamps with a uniform contact surface. On the one hand, this simplifies the manufacture of such clamps, but on the other hand, it requires significant pressure on the sample to achieve a larger contact surface and, therefore, a greater frictional force to hold the sample. This, as seen in [36], dramatically affects the boundary deformation of the sample. In our approach, a developed technique of fixing the sample was used, which increased the area of the contact surface of the clamps and the sample due to the irregularities of the clamps themselves (due to the presence of special grooves and protrusions).

Chitosan has important properties such as biocompatibility, biodegradability, hydrophilicity, non-toxicity, high bioavailability, favorable water permeability, the ability to form films, gels, and nanoparticles. Chitosan nanoparticles (Novochizol™) were included in composite material for plastics of DM defects. Novochizol™ has a positive charge, which allows it to interact well with various types of molecules. It is believed that this positive charge is responsible for the antimicrobial activity of chitosan through interaction with negatively charged cell membranes of microorganisms [37]. By adding an antibiotic to composite material, we enhance the antibacterial effect of the composite material. If one has a native bacterial pulp antibiotic without chitosan, then the excretion of the antibiotic occurs within one hour. Novochizol™ forming a film gives a slow release of glycopeptide vancomycin. In contrast to chitosan, which is a linear polymer, Novochizol™ has a globular, near-spherical shape, owing to intramolecular cross-linking. Such a molecular design confers several advantages to Novochizol™ over chitosan, including the functioning as an active ingredients carrier (Table 1). It should also be noted that due to the fact that Novochizol™ is based on nanospheres, the diffusion rate is much higher than for ordinary linear chitosan, which is especially important when these materials are impregnated.

Table 1. Characteristics of Novochizol™ of importance in plant treatment, in comparison with chitosan.

Characteristic	Chitosan	Novochizol™
Solubility (pH < 6)	Yes	Yes
Solubility (pH > 6)	No	Yes (dispersion)
Viscosity	High	Low
Biodegradability	Fast	Slow
Chemical stability	Low	High
Frost resistance	No	Yes
Physical states	Modifiable only through chemical reactions with other compounds	Modifiable by changing the degree on intramolecular cross linking

These properties of chitosan (see Figure 2) are ideal for use in a system of hydrophilic antibiotics with slow release, which is extremely important for the prevention of infectious complications during the closure of dura mater defects. In addition, vancomycin is used to enhance the antibacterial properties of chitosan. However, when native BNC is impregnated without Novochizol™, vancomycin is rapidly washed out [38].

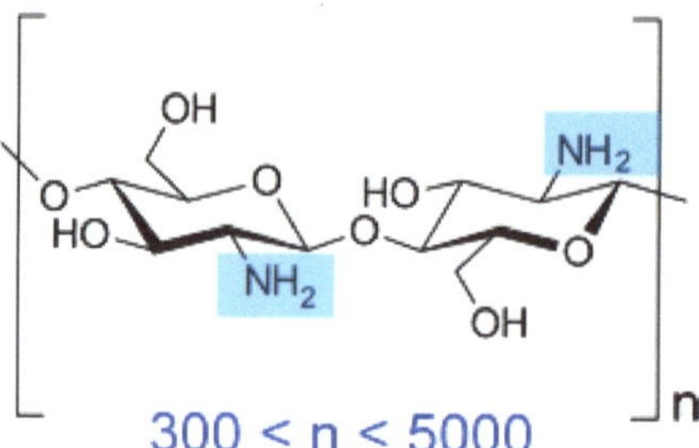

Figure 2. Spatial structure of the chitosan molecule—schematically.

Despite the fact that the mechanical properties of native cellulose have been studied for a long time in the literature, and the results of the effect of nanoadditives on the mechanical properties of DM substitutes have been reported before (Tutopatch®), the controlling of properties by material thickness has not been published previously. Remarkably, it has paramount importance for clinical patient-specific applications.

The purpose of this work is to compare the mechanical properties of samples of BNC impregnated with Novochizol™ and vancomycin with native BNC, cadaveric DM (preserved with formalin) and native human DM to assess the possibility of controlling the mechanical properties of the material by changing its thickness and also to evaluate how the additives affect the mechanical properties of the composite.

2. Materials and Methods

2.1. Ethical Protocol and Transportation

According to the established protocol, fresh dura mater samples were obtained during microsurgical treatment of patients with cerebral vascular pathologies. With unexpressed cerebral edema, the excess part of the dura mater was excised for research. Our clinical approach to tissue harvesting assumes the maximum safety of tissue extraction for the patient; therefore, only a small fragment of the dura mater was taken, which in any case would have been excised. This part of the study was carried out jointly with the Federal Neurosurgical Center of Novosibirsk. Harvested tissues were preserved with 0.9% saline at +2–5 °C during transportation and storage until the experiment (12–48 h, which is a standard period in the literature.).

Cadaveric preserved DM tissues were selected in the laboratory for the preparation and preservation of tissues at the Novosibirsk Research Institute of Traumatology and Orthopedics n.a. Tsivyan, according to the local ethical protocol.

2.2. Methods for the Synthesis Native BNC

Native BNC was obtained by standard cultivation in the medium of the bacterial strain Komagataeibacter Xylinus JCM 7644 at the Institute of Chemical Biology and Fundamental Medicine SB RAS.

2.3. Methods for the Synthesis of Modified BNC

The BNC + N + V samples were prepared under special conditions. The native BNC, grown at the Institute of Chemical Biology and Fundamental Medicine of the SB RAS, was transferred to the Vorozhtsov Institute of Organic Chemistry SB RAS for treating with Novochizol™ (Registered International trademark Novochizol No. 1540749, and in U.S. Patent and Trademark Office No. 6297647). Chitosan nanospheres—Novochizol™ were provided by NOVOCHIZOL SA (Monthey, Switzerland, www.novochizol.ch, accessed on 23 March 2021). The degree of deacetylation was no less than 90%, and the mass is 500 kDa.

Novochizol™ aqueous solutions were obtained by dissolving succinic acid (500 mg per 100 mL sterile water), gradually adding Novochizol™ (1000 mg per 100 mL succinic acid solution) under sonication, and sonicating the mixture for one hour, using model UZTA-0.4/22-OM sonicator (U-sonic, Biysk, Russia) at maximum power. Sterile water was added to compensate for evaporation caused by the prolonged sonication. The solution was filter-sterilized using 0.45 µm apyrogenic acetate cellulose filters (Minisart®, Sartorius Stedim Biotech Göttingen, Germany). Thereafter, 1000 mg of vancomycin hydrochloride was added to the resulting solution, and the solution was subjected to the same ultrasonic treatment for 5 min. The solution was filter-sterilized again using 0.45 µm apyrogenic acetate cellulose filters (Minisart®, Sartorius Stedim Biotech Göttingen, Germany), then stored as a 1% stock at +4 °C and used within one week.

Samples of BC were individually immersed in a Novochizol™ solution in a plastic tube (50 mL) at a ratio of sample volume to Novochizol™ of 1:10. The samples in Novochizol™ were then treated with an ultrasonic bath at +37 °C for 10 min. After that, the samples impregnated with Novochizol™ were neutralized with 1M aqueous ammonia to pH7, washed with sterile water and stored at +4 °C until implantation. The fundamental differences of Novochizol™ from chitosan are indicated in Supplementary Materials (Figure S1).

2.4. Elemental Analysis of BNC Samples

Elemental analysis of unimplanted samples was performed to identify possible damage as a result of their coating and sonication. Elemental analysis was carried out on an instrument "Carlo Erba Strumentazione, Elemental Analyzer-Mod. 1106, Milano, Italy".

2.5. Atomic Force Microscopy

Atomic Force Microscopy (AFM) measurements were done in Semicontact Tapping mode on the Solver Next SPM station (NT-MDT, Russia) in standard atmospheric room conditions (RH 30%, temperature 25 °C). Semisoft AFM probes NSG01 (NT-MDT, Russia) with resonant frequency 138 kHz and nominal tip radius 10 nm were used for the studies with the optimized setpoint force. All recordings were done with a scan rate of 0.5 Hz and resolution 512×512 points. Further data treatment involved only removal of inclination without additional data filtration.

2.6. Thickness Measurement and Material Cutting

The thickness of the investigated samples was preliminarily measured in the laboratory of the Institute of Laser Physics SB RAS. The thickness of 4 groups of materials was measured by a highly sensitive LVDT sensor, rigidly connected on one axis with a precision wheel 2 mm wide. The wheel evenly moves along the tissue flap from left to right and from top to bottom, transmitting vertical vibrations to the LVDT sensor, the signal from which is sent to a personal computer, where the developed software directly displays the thickness of the biological material. Before the cutting, a "thickness map" of the tissue flap with an accuracy of ±10 µm is created, where the selected thickness range corresponds to the color specified by the operator (Figure 3). To conduct a reliable experiment, the same material thickness in the middle part of the "dog-bone" shape was chosen.

For cutting out samples from the sheet, a laboratory installation for cutting out biological tissue "Melaz Cardio" (LLC "Lasarus") was used. The cutting was carried out in a continuous mode using a CO_2 laser, the power of which can reach 40 W, which was regulated depending on the material. Cutting speed was 600 mm/min, while power was 33%.

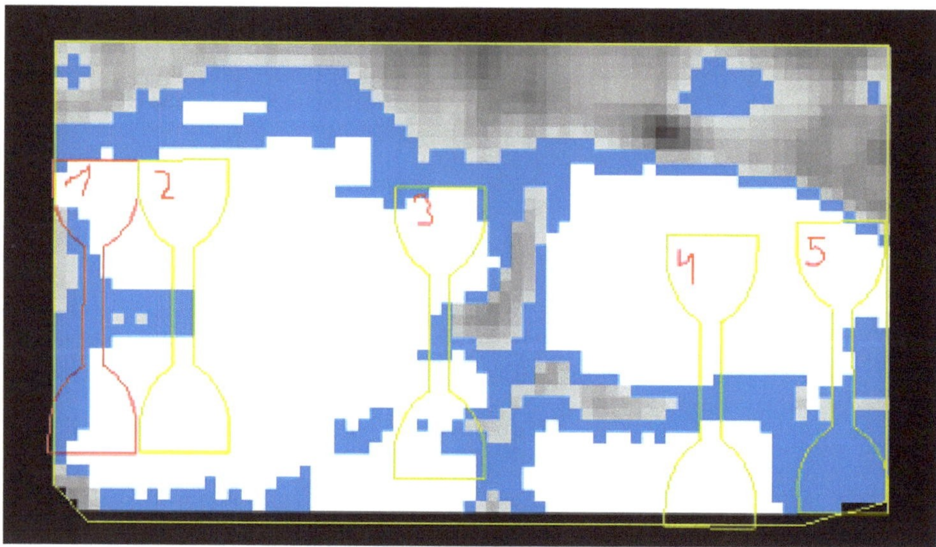

Figure 3. The image was obtained on a laser thickness gauge with markings for subsequent cutting of tissue with a laser. The white zone corresponds to a thickness of 1.0 mm, the blue zone—0.7 mm, the gray zone—0.3 mm.

2.7. Mechanical Test Protocol

All mechanical tests were carried out at the Lavrentyev Institute of Hydrodynamics SB RAS on a universal tensile testing machine INSTRON 5944 with a thermostatic biobath. During the test, each sample was subjected to cyclic loading with a starting displacement of 0.25 mm at the first cycle, with a displacement step of 0.25 mm and a crosshead speed of 2 mm/min. After delivery to the laboratory, the thickness of the sample was measured (for DM). Then the sample was fixed in an Instron 5944 tensile machine, and a series of experiments were performed (Figure 4b,c). A dog-bone shape was cut from the cadaveric material with the same parameters as for BNC and the BNC + N + V composite material. For fresh dura mater samples obtained during neurosurgical interventions, a rectangular sample was used for testing, mainly due to the small size and irregular shape of the original samples, as well as to simplify data processing. A videoextensometer (lens from Fujifilm) was used to measure local strain in the specimens, when their size allowed. Small bright plastic marks (1.5 mm in diameter) were attached to the sample with the water-proof glue for the extensometer to capture gauge length (Figure 4) during the loading. The focal length of the video extensometer objective is 16 mm.

The video extensometer used to measure local deformation allows minimizing the boundary effects, which are prone to happen in the case of measuring only the crosshead strain. The known parasitic effects caused by the clamps, such as slippage of the sample, were considered. If the clamps were too loose, then the part of material under them could extend, as well as the part outside the clamps, or even slip out [39]. Uneven deformation of the material (demonstrated, e.g., in [36] between the clamps also motivated the acquisition of deformation data in two ways: by considering edge effects and minimizing them. In addition, damage to the material could be introduced while fastening the sample. This established the zones of already partly ruptured material, which negatively affected the accuracy of the obtained experimental data. All of that emphasizes the need to use other means of measurement apart from the crosshead strain.

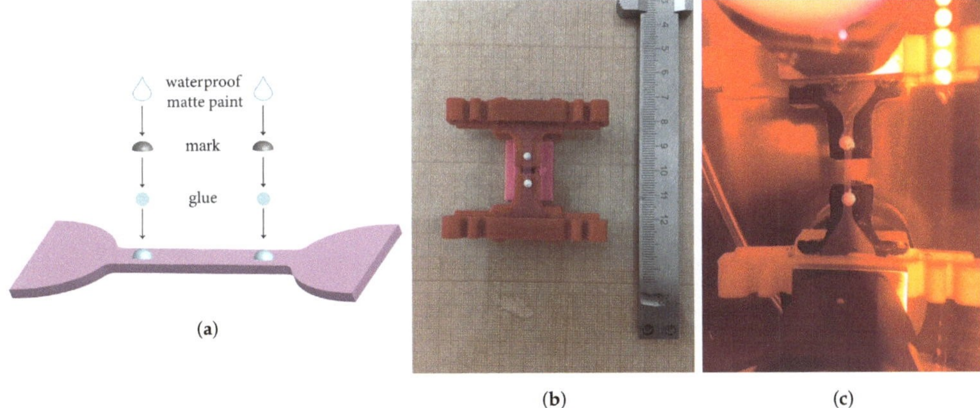

Figure 4. (**a**)—The technique of attaching marks for extensometer, (**b**)—preparation of the specimens, (**c**)—specimen in the tensile machine.

While carrying out the experiment, such a well-known phenomenon for biological tissues as preconditioning [40] was taken into account. This technique was applied for the initial stages (1st–5th stages of elongation, depending on the sample), and during the next stages, the influence of this condition was not noticed. During this study, it was established that was no need to perform more than two preconditioning cycles. For each stage of the experiment, the specimen's initial elongation was the same, i.e., the machine's clamps returned to the original program-defined position after the completion of each stage of the loading. During the experiment, the sample was positioned in the sodium chloride solution heated to the human body temperature. In each experiment, the sample lost its elasticity. For each sample, its loading was performed until it became separated into two disconnected segments (or a visible discontinuity of the sample appeared).

2.8. Statistical Analysis

To analyze the results, the following software were used: the MS Office 2016 (MS Excel with custom extensions) package licensed by LIH SB RAS and free software environment R. Truncated samples relative to the original ones (10% of the highest and lowest results were discarded) were formed, which is determined by outliers in the data that are observed not only in this work but also in the literature [41].

3. Results
3.1. Results of AFM of BNC

The analysis of AFM topography images provided observation of separate and bundled cellulose nanofibrils with the diameter in the range of 30–50 nm (see Figure 5a), which were evenly spread around the surface (as seen on Figure 5b). These observations were supported by Scanning Electron Microscopy data (not shown) and verified by the splitter-fiber structure of the sample.

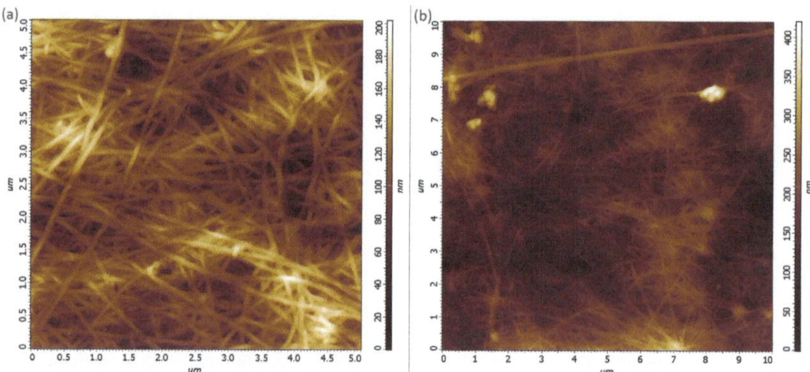

Figure 5. Topography of the BNC sample obtained by Atomic Force Microscopy for 5×5 µm area (**a**) and 10×10 µm area (**b**).

3.2. Elemental Analysis of Novochizol™

The following element ratio was established for Novochizol™ (see Table 2).

Table 2. Results of elemental analysis of BNC, BNC + N, BNC + N + V .

Sample Type	C,%	H,%	N,%
BNC	39.73	7.11	7.14
BNC + N	42.31	7.37	4.85
BNC + N + V	46.4	6.64	4.86

That elemental analysis of BNC + N sample shows that Novochizol™ and cellulose are in approximately equal weight proportions.

3.3. Characteristics of a Sample of Tested Specimens

A total of 43 experiments were carried out on uniaxial mechanical loading of BNC specimens and 22 BNC + N + V specimens. In addition, tissues of the dura mater (4 samples) of healthy patients and samples of the cadaveric dura mater (13 samples) were examined using the same technique. After processing the test data, it was decided to present the results of the measurements made with the data taken from the traverse displacement of the stretching machine (Figure 6a) and the data of the video extensometer (Figure 6b). It is visible that the data on the ultimate deformation of the samples can differ significantly.

To analyze the mechanical properties of such materials, it is important to analyze not only their ultimate stress indices but also the elastic modulus at small deformations, since exactly the small deformations of the DM are the most physiological type of deformations of such a material. Thus, the atypical behavior of the material in the region of small deformations with its adequate limiting mechanical characteristics may indicate an unsatisfactory quality of the material.

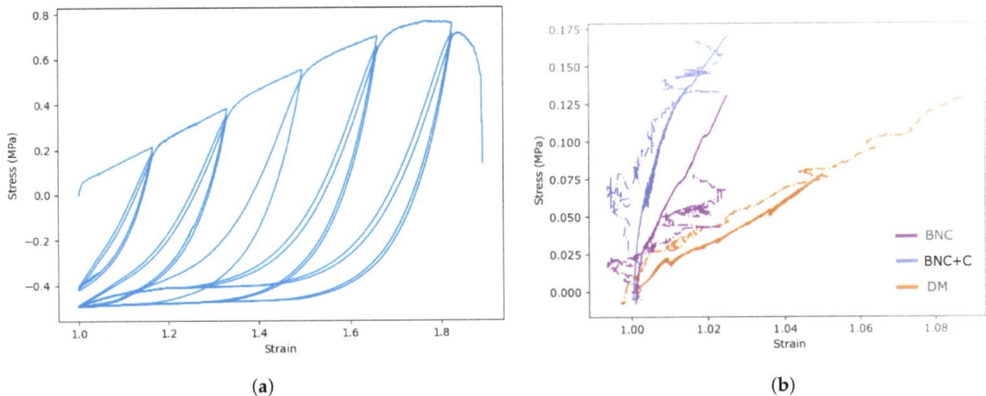

Figure 6. (**a**)—The course of the tensile test, (**b**)—first stage of the testing for different materials. The continuous line denotes the crosshead strain, and the dotted line denotes the video extensometer strain.

3.4. Statistical Analysis of Mechanical Test Results

In the course of the study, it was shown that the ultimate stress of the new composite is similar and slightly exceeds the ultimate stress of the material grown according to the standard technique: 0.75 vs. 0.58 MPa (+29.31%) (Table 3).

Table 3. Mean values of the material parameters (measurement error 0.6%).

ID	Ultimate Strain	Ultimate Stress, MPa	Young Modulus (Small Deformations)
DM (cadaveric)	3.08	2.58	1.15
DM (fresh)	2.99	2.29	1.27
BNC	1.09	0.58	31.6
BNC + N + V	1.12	0.75	34.37

The ultimate deformation of the cadaveric dura mater is in accordance with Figure 7a 1.001, the ultimate deformation of BNC is 1.002, and the ultimate deformation of the BNC + N + V composite is 1.125. Thus, the ultimate deformation of the new BNC + N + V composite is 12.4% higher than the cadaveric dura mater.

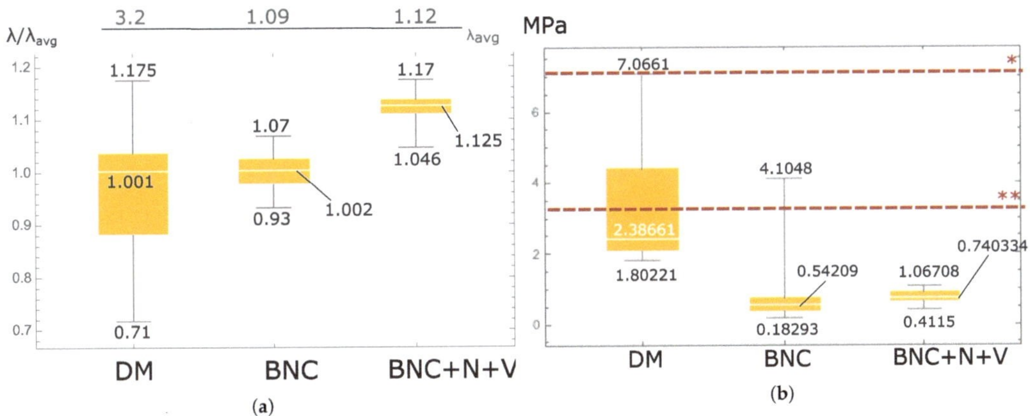

Figure 7. Boxplot for values of ultimate strain (**a**) and stress (**b**). The lines (*) and (**) correspond to data of the same value from [35] of DM and Tutopatch® material. Reprinted from [42].

In the course of the study, in accordance with Figure 7b, it was shown that the ultimate stress of the new composite (BNC + N + V) slightly exceeds the limiting stress of the native BNC material grown according to the standard method: 0.74 and 0.54 MPa. The ultimate stress of the new composite (BNC + N + V) is 36.54% higher than the limiting stress of the native BNC. The limiting stress of preserved cadaveric DM (2.387 MPa) is 336.6% higher than that of native BNC. The results of the ultimate stress were also compared with the results of the DM and Tutopatch® presented in [35]. Tutopatch® (Tutogen Medical GmbH, Neunkirchen am Brand, Germany), which is produced from bovine pericardium, is xenogeneic and exposed to the Tutoplast process [43]). The Tutoplast process is chemical sterilization, which increases the strength of bovine pericardium to enzymatic breakdown and decreases its antigenicity.

Statistical analysis of the dependency between the thickness of the samples and their ultimate values of stress and strain, as well as the Young's modulus at small deformations, showed that no significant differences between the two types of material (BNC vs. BNC + N + V) can be observed from their Young's modulus. Regarding the relationship between ultimate stress and deformation, the relationship for the standard BNC cultivation method is more linear ($R = 0.306$), at the same time, this only indicates that this relationship is less linear for BNC + N + V material (Figure 8). Analyzing the relationship between the thickness of the material and its ultimate stress, it can be seen (Figure 9) that this relationship is more linear for the new material ($R^2 = 0.235362$ for BNC + N + V, compared to $R^2 = 0.040461$ for native BNC), and in this case, this indicates more predictable strength characteristics of the material during its growth. Therefore, with the thickness–stress relationship being linear, it is enough to carry out the measurements of the material thickness at the same definite time intervals rather than building a complex polynomial or recursive model for measuring the thickness of the grown material, which is an undoubted technological advantage of such a material.

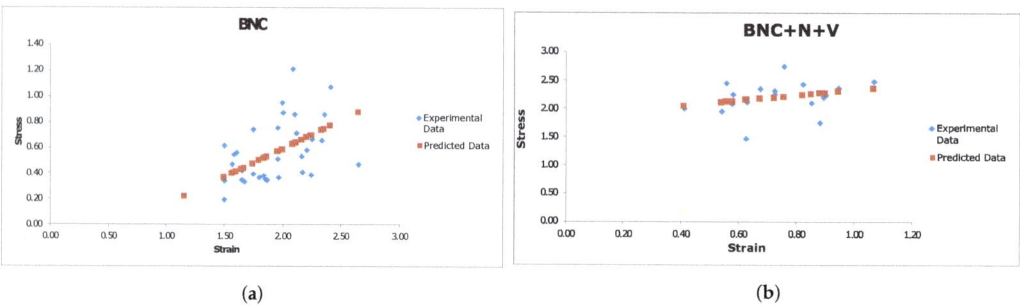

(a)　　　　　　　　　　　　　　　　　(b)

Figure 8. Linear regression for BNC (**a**) and BNC + N + V (**b**) samples for stress-strain relationship. Reprinted from [42].

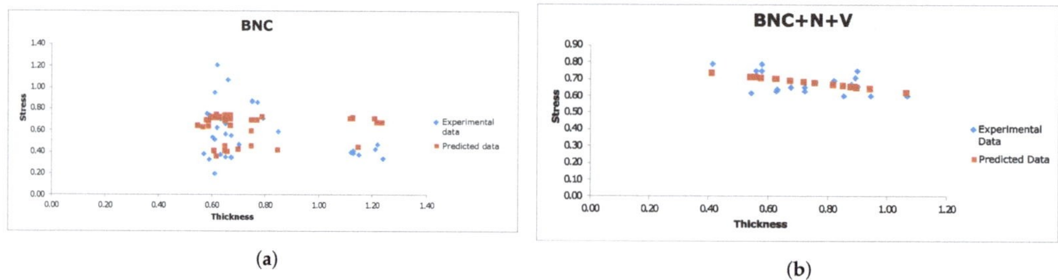

(a)　　　　　　　　　　　　　　　　　(b)

Figure 9. Linear regression for BNC (**a**) and BNC + N + V (**b**) samples for the stress–thickness relationship. Each blue dot represents one sample. The red dots represent the result predicted by the linear regression. Reprinted from [42].

An analysis of variance was also performed for the values of ultimate stress and deformation with respect to such a factor as the presence of Novochizol™ in the material (Figure 10). The results showed a significant effect on both parameters (*p*-value 0.005 for stress and 0.001 for deformation). That is, the impregnation of the material with Novochizol™ statistically significantly increases both the strength and elasticity of the material.

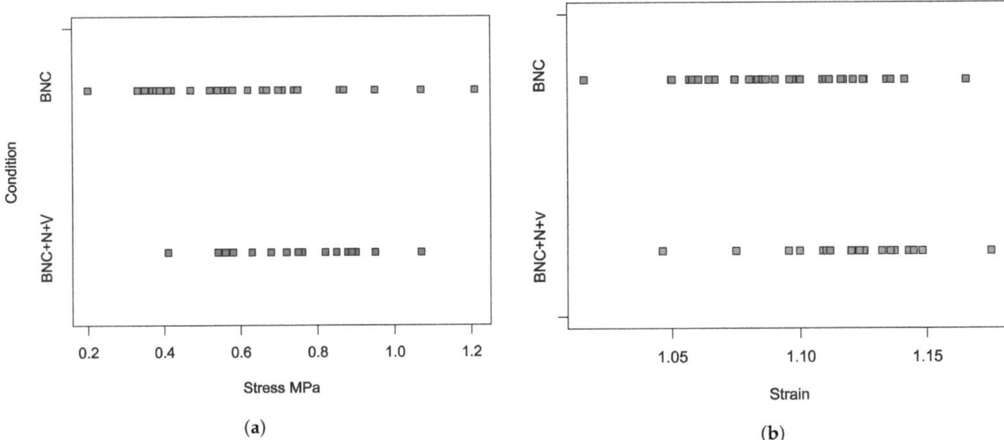

Figure 10. An illustration for analysis of variance. (**a**): the values of the ultimate stress for material with and without chitosan, (**b**): for the ultimate deformation. (Each point represents one sample).

4. Discussion

Bacterial Nanocellulose (BNC), synthesized by the strain bacterium Komagataeibacter xylinus JCM 7644, consists of a biogenic structure of nanofibers formed by self-assembly. BNC has a higher water retention capacity, excellent biocompatibility, a high degree of crystallinity and, therefore, a high tensile strength and fine mesh compared to pure natural biodegradable polymers such as collagen, chitin and gelatin [44–47].

BNC possesses high mechanical properties, which are required in most cases when the material is used as a base in the tissue engineering. With a reticular structure and very small pore sizes, protofibrils of bacterial nanocellulose intertwine to form a large surface area. The fibrous structure of BNC consists of a three-dimensional network of nanofibrils connected by intrafibrillar hydrogen bonds, which allows it to maintain a constantly wet state of the hydrogel, as well as the high strength of BNC [47,48]. In the last decade, BNC has been widely used to create biocompatible prostheses in human tissues [19,47]. For materials used as dura mater transplants, especially those with high hydrophilicity, very important properties are the deposition of drugs and a decrease in toxicity by reducing leaching and, as a consequence, local action and achieving prolonged drug release.

For this, a variety of polymer sustained-release drugs with different physical properties have been developed. Such formulations have been shown to be effective in increasing the release time when relatively hydrophobic and water-insoluble drugs are used. However, there is still a need for new compositions and methods that could reduce the diffusion of the drug and eliminate the "explosion" effect for drugs that are highly soluble in water.

Currently, the world literature is actively highlighting the prospect of using a natural polymer of chitosan [49–52]. It is noteworthy that the entire 62nd issue of Advanced Drug Delivery Reviews (2010) was devoted to this material. Chitosan, which is a mucopolysaccharide, resembles the structure of the polymer lining the intima of blood vessels. It is not surprising that it has complete biocompatibility with human tissues. Its low toxicity, ability to enhance regenerative processes during wound healing, and biodegradability of chitosan materials are of particular interest for medical use [18,53,54].

Biodegradation of Novochizol™ will lead to an increase in the duration of the release of antibacterial and antimycotic drugs, which prolongs their effect in the tissues adjacent to the composite material. Due to the use of the BNC + N + V composite, the number of bacterial and fungal complications during neurosurgical interventions will decrease.

In this regard, it should be noted that, given the degradation over time of BNC with medicinal coatings (additives), namely, the release of the medicinal agent, the change in mechanical properties is of great interest. In Reference [55], structural changes in the polymer are described as a function of changes in the concentration of the drug additive.

In other studies, only the mechanical properties of BNC are considered, from the basic uniaxial test protocol [56] to the advanced measurement techniques that allow characterizing the material in sufficient detail (to determine not only Young's modulus but also Poisson's ratio by performing a compression test) [57].

With all the advantages of BNC as a transport and separation barrier for the human brain, the main problem during its growth is the unevenness of tissue thickness, which, during decompression craniectomy and with the emerging cerebral edema, going beyond the trepanation window, can lead to the occurrence of hernias and ruptures in the BNC film after its implantation. Therefore, the strength properties of materials for meringoplasty are very important.

This question poses an additional challenge, since, as our results show, regulating the thickness of bacterial cellulose allows one to regulate its strength properties. However, such a dependence, which is also shown in the Results section, is clearly non-linear, which makes it difficult to grow this composite.

In terms of mechanical properties, plastic materials used to replace DM defects should be as close as possible to the properties of natural DM. The mechanical properties of the material (implants) include ultimate stresses and deformation, which are described in [35,36], as well as strength characteristics in the area of small deformations (Young's modulus, dependence of deformation on stress), which was studied.

The most common method for studying the strength properties of DM is the study of the mechanical properties of cadaveric specimens. This is usually due to ethical standards (inability to take a DM sample from a healthy person within the framework of ethical protocols) and established clinical protocols. However, the use of formalin significantly changes the strength properties of the bacterial cellulose material [58]. In turn, obtaining fresh samples of dura mater is difficult since, in the general case, excision of dura mater is not provided for by the protocols and can be performed in exceptional cases.

The mechanical properties of BNC were studied both to recognize the strength properties of this material for the purposes of implantation instead of DM [35] and for other applications in the sense of implantation, for example, in [59]. Undoubtedly, the test method is determined by the application of the target material. In this case, as a DM substitute, the implanted BNC is experiencing stretching stresses; therefore, it seems natural to use an uniaxial mechanical test [60,61]. In this paper, the most advanced method of uniaxial testing is applied: the use of the most sensitive dynamometer (10N Load Cell), the use of a thermostatic bath and the use of a video extensometer to fix the truth, and calculating relative deformations. In addition, we used dog-bone-shaped samples for the most correct localization of the sample break zone during a mechanical experiment [62]. To provide the possibility of using the results of mechanical tests by other researchers, we provide access to the results of the experiments in the Supplementary Materials (Table S1).

The limitations of this work include a small number of dura mater samples from healthy patients. Although, it is worth noting that this indicator is rather an overview at this stage of the study and does not affect the conclusions drawn. In addition, during neurosurgical intervention, dura mater samples usually cannot be removed, unlike cerebral aneurysm tissues [63], which can be excised, and sometimes must be removed, e.g., if there is a mass effect on the patient's brain. In addition, recently, the technique of multi-axis experiments has been developed, despite the fact that the advantage of multi-axis tests [64] is hampered by the complexity of the interpretation of such tests [65]. It is impossible to disregard the usefulness of such tests; therefore, similar tests should be performed in the future.

5. Conclusions

In this work, a comparison of the strength properties of two cohorts of a bacterial nanocellulose polymer (native and impregnated with Novochizol™+vancomycin) was conducted and showed the effect of the additives on the strength characteristics (the additive plays the role of a reinforcement factor). The strength characteristics of the BNC + N + V composite polymer have been compared with the DM material (fresh and cadaveric), and similar characteristics of the ultimate stress for the BNC + N + V and DM materials were shown, which points to the adequacy of the considered BNC options. The results obtained shed light on the change in the strength characteristics upon addition of Novochizol™ and on the possibility of controlling the strength properties of the polymer using the polymer thickness. In the future, it is planned to carry out punch tests on a hard surface. Particularly, the anticipated study of the curved surface, analogous to the material near the skull, seems promising for the development of real applications.

Supplementary Materials: The following are available online at https://www.mdpi.com/article/10.3390/polym13121995/s1, Figure S1: Novochizol structure, Table S1: Results of the mechanical testing.

Author Contributions: Clinical application of BNC materials, manuscript preparation of the Introduction section, A.K. and V.S.; mechanical testing, manuscript preparation of Results and Discussion, processing of the data, A.L.; providing the DM samples, A.D.; the cultivation of the samples, M.F.; preparation of the samples, A.M. and V.F.; mechanical testing, statistical processing of research results, A.L. and D.P.; AFM and analysis, P.G. All authors have read and agreed to the published version of the manuscript.

Funding: A.L. and D.P. thank the Government of the Russian Federation (grant number 14.W03.31.0002) for providing the equipment and support for measuring the strength properties of cadaveric DM and BNC and analyzing measurement results; A.D. thanks the Russian Foundation for Basic Research (grant number 19-48-540010) for providing and mechanical testing fresh DM samples.

Acknowledgments: Authors would like to acknowledge the Multi-Access Chemical Research Center SB RAS for spectral and analytical measurements. Authors thanks CKP "VTAN" at NSU and CKP "Nanostructures" at ISP SB RAS for access to the AFM and SEM equipment.

Conflicts of Interest: V.V. Fomenko is the inventor of the patent covering Novochizol™ technology ("New method of synthesis of chitosan derivatives and uses thereof", application number EP20156732.8) and uses and owns stocks in Novochizol SA.

Sample Availability: Raw data of the mechanical tests are available as the Mendeley storage item connected with the research.

Abbreviations

The following abbreviations are used in this manuscript:

DM	Dura Mater
BNC	Bacterial Nanocellulose
BNC + N	composite Bacterial Nanocellulose + Novochizol™
BNC + N + V	composite Bacterial Nanocellulose + Novochizol™ + vancomycin
LVDT sensor	Linear Variable Differential Transformer sensor
CSF	Cerebrospinal fluid
MRI	Magnetic resonance imaging
AFM	Atomic force microscopy
SEM	Scanning electron microscope

References

1. Fontana, R.; Talamonti, G.; D'Angelo, V.; Arena, O.; Monte, V.; Collice, M. Spontaneous haematoma as unusual complication of silastic dural substitute. Report of 2 cases. *Acta Neurochir.* **1992**, *115*, 64–66. [CrossRef] [PubMed]
2. Van Calenbergh, F.; Quintens, E.; Sciot, R.; Van Loon, J.; Goffin, J.; Plets, C. The use of Vicryl Collagen® as a dura substitute: A clinical review of 78 surgical cases. *Acta Neurochir.* **1997**, *139*, 120–123. [CrossRef] [PubMed]
3. Allen, K.; Isaacson, B.; Kutz, J.; Purcell, P.; Roland, P. The Association of Meningitis with Postoperative Cerebrospinal Fluid Fistula. *J. Neurol. Surg. Part Skull Base* **2012**, *73*, 401–404. [CrossRef]
4. Couture, D.; Branch, C.L. Spinal pseudomeningoceles and cerebrospinal fluid fistulas. *Neurosurg. Focus* **2003**, *15*, 1–5. [CrossRef] [PubMed]
5. Knight, S. Pseudomeningocele. In *Encyclopedia of Clinical Neuropsychology*; Kreutzer, J.S., DeLuca, J., Caplan, B., Eds.; Springer International Publishing: Cham, Switzerland, 2018; pp. 2869–2874. [CrossRef]
6. Esposito, F.; Cappabianca, P.; Fusco, M.; Cavallo, L.M.; Bani, G.G.; Biroli, F.; Sparano, A.; de Divitiis, O.; Signorelli, A. Collagen-only biomatrix as a novel dural substitute: Examination of the efficacy, safety and outcome: Clinical experience on a series of 208 patients. *Clin. Neurol. Neurosurg.* **2008**, *110*, 343–351. [CrossRef]
7. Sun, H.; Wang, H.; Diao, Y.; Tu, Y.; Li, X.; Zhao, W.; Ren, J.; Zhang, S. Large retrospective study of artificial dura substitute in patients with traumatic brain injury undergo decompressive craniectomy. *Brain Behav.* **2018**, *8*, e00907. [CrossRef] [PubMed]
8. Bi, X.; Liu, B.; Mao, Z.; Wang, C.; Dunne, N.; Fan, Y.; Li, X. Applications of materials for dural reconstruction in pre-clinical and clinical studies: Advantages and drawbacks, efficacy, and selections. *Mater. Sci. Eng. C* **2020**, *117*, 111326. [CrossRef]
9. Callovini, G.M.; Bolognini, A.; Callovini, T.; Giordano, M.; Gazzeri, R. Treatment of CSF leakage and infections of dural substitute in decompressive craniectomy using fascia lata implants and related anatomopathological findings. *Br. J. Neurosurg.* **2021**, *35*, 18–21. [CrossRef]
10. Knopp, U.; Christmann, F.; Reusche, E.; Sepehrnia, A. A new collagen biomatrix of equine origin versus a cadaveric dura graft for the repair of dural defects–a comparative animal experimental study. *Acta Neurochir.* **2005**, *147*, 877–887. [CrossRef]
11. Warren, W.L.; Medary, M.B.; Dureza, C.D.; Bellotte, J.B.; Flannagan, P.P.; Oh, M.Y.; Fukushima, T. Dural repair using acellular human dermis: Experience with 200 cases: Technique assessment. *Neurosurgery* **2000**, *46*, 1391–1396. [CrossRef]
12. Eichberg, D.G.; Ali, S.C.; Buttrick, S.S.; Komotar, R.J. The use of dehydrated amniotic membrane allograft for augmentation of dural closure in craniotomies and endoscopic endonasal transphenoidal surgeries. *Br. J. Neurosurg.* **2018**, *32*, 516–520. [CrossRef] [PubMed]
13. Morales-Avalos, R.; Soto-Domínguez, A.; García-Juárez, J.; Saucedo-Cardenas, O.; Bonilla-Galvan, J.R.; Cardenas-Serna, M.; Guzmán-López, S.; Elizondo-Omaña, R.E. Characterization and morphological comparison of human dura mater, temporalis fascia, and pericranium for the correct selection of an autograft in duraplasty procedures. *Surg. Radiol. Anat.* **2017**, *39*, 29–38. [CrossRef] [PubMed]
14. Zerris, V.A.; James, K.S.; Roberts, J.B.; Bell, E.; Heilman, C.B. Repair of the dura mater with processed collagen devices. *J. Biomed. Mater. Res. Part B* **2007**, *83*, 580–588. [CrossRef] [PubMed]
15. Schmalz, P.; Griessenauer, C.; Ogilvy, C.S.; Thomas, A.J. Use of an absorbable synthetic polymer dural substitute for repair of dural defects: A technical note. *Cureus* **2018**, *10*, e2127. [CrossRef]
16. MacEwan, M.R.; Kovacs, T.; Osbun, J.; Ray, W.Z. Comparative analysis of a fully-synthetic nanofabricated dura substitute and bovine collagen dura substitute in a large animal model of dural repair. *Interdiscip. Neurosurg.* **2018**, *13*, 145–150. [CrossRef]
17. Pandit, V.A.; Sharma, R.K.; Bhaskar, S.; Kindra, A.S.; Choudhary, A.; Gupta, L. A Randomised Interventional Study to Compare Autologous and Nonautologous Dural Substitutes Among Traumatic Brain Injury Patients. *Indian J. Neurotrauma* **2021**, *18*, 26–31. [CrossRef]
18. Pogorielov, M.; Kravtsova, A.; Reilly, G.; Deineka, V.; Tetteh, G.; Kalinkevich, O.; Pogorielova, O.; Moskalenko, R.; Tkach, G. Experimental evaluation of new chitin–chitosan graft for duraplasty. *J. Mater. Sci. Mater. Med.* **2017**, *28*, 34. [CrossRef]
19. Rosen, C.L.; Steinberg, G.K.; DeMonte, F.; Delashaw, J.B., Jr.; Lewis, S.B.; Shaffrey, M.E.; Aziz, K.; Hantel, J.; Marciano, F.F. Results of the prospective, randomized, multicenter clinical trial evaluating a biosynthesized cellulose graft for repair of dural defects. *Neurosurgery* **2011**, *69*, 1093–1104. [CrossRef]
20. Laun, A.; Tonn, J.; Jerusalem, C. Comparative study of lyophilized human dura mater and lyophilized bovine pericardium as dural substitutes in neurosurgery. *Acta Neurochir.* **1990**, *107*, 16–21. [CrossRef]
21. Yamada, K.; Miyamoto, S.; Takayama, M.; Nagata, I.; Hashimoto, N.; Ikada, Y.; Kikuchi, H. Clinical application of a new bioabsorbable artificial dura mater. *J. Neurosurg.* **2002**, *96*, 731–735. [CrossRef]
22. Bonda, D.J.; Manjila, S.; Mehndiratta, P.; Khan, F.; Miller, B.R.; Onwuzulike, K.; Puoti, G.; Cohen, M.L.; Schonberger, L.B.; Cali, I. Human prion diseases: Surgical lessons learned from iatrogenic prion transmission. *Neurosurg. Focus* **2016**, *41*, E10. [CrossRef]
23. Brooke, F.J.; Boyd, A.; Klug, G.M.; Masters, C.L.; Collins, S.J. Lyodura use and the risk of iatrogenic Creutzfeldt–Jakob disease in Australia. *Med. J. Aust.* **2004**, *180*, 177–181. [CrossRef] [PubMed]
24. Barbolt, T.A.; Odin, M.; Léger, M.; Kangas, L.; Holste, J.; Liu, S.H. Biocompatibility evaluation of dura mater substitutes in an animal model. *Neurol. Res.* **2001**, *23*, 813–820. [CrossRef]
25. Yoshioka, N. Cranial reconstruction following the removal of an infected synthetic dura mater substitute. *Plast. Reconstr. Surg. Glob. Open* **2014**, *2*. [CrossRef] [PubMed]

26. Malliti, M.; Page, P.; Gury, C.; Chomette, E.; Nataf, F.; Roux, F.X. Comparison of deep wound infection rates using a synthetic dural substitute (neuro-patch) or pericranium graft for dural closure: A clinical review of 1 year. *Neurosurgery* **2004**, *54*, 599–604. [CrossRef]

27. Parlato, C.; Granata, R.; Moraci, A.; Accardo, M. Dural reconstruction in meningioma surgery. In *Monleon D. Meningiomas-Management and Surgery*; InTechOpen: London, UK, 2012; pp. 103–124.

28. Platt, J.L. Knocking out xenograft rejection. *Nat. Biotechnol.* **2002**, *20*, 231–232. [CrossRef]

29. Shintani, T. Experimental studies on prevention of adhesion of tissue surrounding dura mater and morphological changes in the spinal cord following laminectomy (author's transl). *Nihon Seikeigeka Gakkai Zasshi* **1980**, *54*, 1477–1495.

30. Rohde, S. Inflammatory diseases of the meninges. In *Inflammatory Diseases of the Brain*; Springer: New York, NY, USA, 2009; pp. 169–183.

31. Takai, K.; Taniguchi, M. Intracranial hypotension with coma: Microsurgical repair of a spinal ventral dural tear and drainage of subdural hematoma with intracranial pressure monitoring. *World Neurosurg.* **2018**, *118*, 269–273. [CrossRef]

32. Danilova, D.; Gorbunova, L.; Tsybusov, S.; Uspensky, I.; Kravets, L.Y. Materials for Plastic Surgery of the Dura Mater: History and Current State of the Problem. *Mod. Technol. Med.* **2018**, *10*, 194–201. [CrossRef]

33. Zaidi, H.A.; Pendleton, C.; Cohen-Gadol, A.A.; Quinones-Hinojosa, A. Harvey Cushing's repair of a dural defect after a traumatic brain injury: Novel use of a fat graft. *World Neurosurg.* **2011**, *75*, 696–699. [CrossRef]

34. Chen, S.Q.; Lopez-Sanchez, P.; Wang, D.; Mikkelsen, D.; Gidley, M.J. Mechanical properties of bacterial cellulose synthesised by diverse strains of the genus Komagataeibacter. *Food Hydrocoll.* **2018**, *81*, 87–95. [CrossRef]

35. Kizmazoglu, C.; Aydin, H.E.; Kaya, I.; Atar, M.; Husemoglu, B.; Kalemci, O.; Sozer, G.; Havitcioglu, H. Comparison of biomechanical properties of dura mater substitutes and cranial human dura mater: An in vitro study. *J. Korean Neurosurg. Soc.* **2019**, *62*, 635–642. [CrossRef] [PubMed]

36. Zwirner, J.; Scholze, M.; Waddell, J.N.; Ondruschka, B.; Hammer, N. Mechanical Properties of Human Dura Mater in Tension–An Analysis at an Age Range of 2 to 94 Years. *Sci. Rep.* **2019**, *9*, 16655. [CrossRef] [PubMed]

37. Shariatinia, Z. Pharmaceutical applications of chitosan. *Adv. Colloid Interface Sci.* **2019**, *263*, 131–194. [CrossRef]

38. Lazarini, S.C.; de Aquino, R.; Amaral, A.C.; Corbi, F.C.; Corbi, P.P.; Barud, H.S.; Lustri, W.R. Characterization of bilayer bacterial cellulose membranes with different fiber densities: A promising system for controlled release of the antibiotic ceftriaxone. *Cellulose* **2016**, *23*, 737–748. [CrossRef]

39. Saville, B. *Physical Testing of Textiles*; Elsevier: Amsterdam, The Netherlands, 1999.

40. Ogden, R. Large Deformation Isotropic Elasticity: On the Correlation of Theory and Experiment for Compressible Rubberlike Solids. *Proc. R. Soc. Lond. A Math. Phys. Sci.* **1972**, *326*, 565. [CrossRef]

41. Schisterman, E.F.; Cole, S.R.; Ye, A.; Platt, R.W. Accuracy loss due to selection bias in cohort studies with left truncation. *Paediatr. Perinat. Epidemiol.* **2013**, *27*, 491–502. [CrossRef] [PubMed]

42. Parshin, D.; Lipovka, A.; Kharchenko, A. Investigation of the mechanical properties of a composite material of chitosan-vancomycin-nanocellulose nanoparticles of bacterial origin to close dura mater defects. In Proceedings of the 11th World Biomaterial Congress, Online, 11–15 December 2020.

43. Rasch, A.; Naujokat, H.; Wang, F.; Seekamp, A.; Fuchs, S.; Klüter, T. Evaluation of bone allograft processing methods: Impact on decellularization efficacy, biocompatibility and mesenchymal stem cell functionality. *PLoS ONE* **2019**, *14*, e0218404. [CrossRef]

44. Czaja, W.; Krystynowicz, A.; Bielecki, S.; Brown, R.M., Jr. Microbial cellulose—The natural power to heal wounds. *Biomaterials* **2006**, *27*, 145–151. [CrossRef]

45. Czaja, W.K.; Young, D.J.; Kawecki, M.; Brown, R.M. The future prospects of microbial cellulose in biomedical applications. *Biomacromolecules* **2007**, *8*, 1–12. [CrossRef]

46. Klemm, D.; Heublein, B.; Fink, H.P.; Bohn, A. Cellulose: Fascinating biopolymer and sustainable raw material. *Angew. Chem. Int. Ed.* **2005**, *44*, 3358–3393. [CrossRef] [PubMed]

47. Kharchenko, A.V.; Stupak, V.V. Bacterial nanocellulose as a plastic material for closure of defects of the dura mater: Literature review. *J. Spine Surg.* **2019**, *16*, 62–73. [CrossRef]

48. Czaja, W.; Romanovicz, D.; Malcolm Brown, R. Structural investigations of microbial cellulose produced in stationary and agitated culture. *Cellulose* **2004**, *11*, 403–411. [CrossRef]

49. Sergeevichev, D.; Fomenko, V.; Strelnikov, A.; Dokuchaeva, A.; Vasilieva, M.; Chepeleva, E.; Rusakova, Y.; Artemenko, S.; Romanov, A.; Salakhutdinov, N.; et al. Botulinum Toxin-Chitosan Nanoparticles Prevent Arrhythmia in Experimental Rat Models. *Mar. Drugs* **2020**, *18*, 410. [CrossRef]

50. Sergeevichev, D.S.; Krasilnikova, A.A.; Strelnikov, A.G.; Fomenko, V.V.; Salakhutdinov, N.F.; Romanov, A.B.; Karaskov, A.M.; Pokushalov, E.A.; Steinberg, J.S. Globular chitosan prolongs the effective duration time and decreases the acute toxicity of botulinum neurotoxin after intramuscular injection in rats. *Toxicon* **2018**, *143*, 90–95. [CrossRef] [PubMed]

51. Romanov, A.; Strelnikov, A.; Sergeevichev, D.; Salakhutdinov, N.; Fomenko, V.; Shabanov, V.; Losik, D.; Mikheenko, I.; Karaskov, A.; Pokushalov, E. P1711The influence of the new pharmaceutical composition containing botulinum toxin on different pharmacological models of heart rhythm disorders. *Eur. Heart J.* **2017**, *38*. [CrossRef]

52. Krasilnikova, A.A.; Sergeevichev, D.S.; Fomenko, V.V.; Korobeynikov, A.A.; Vasilyeva, M.B.; Yunoshev, A.S.; Karaskov, A.M.; Pokushalov, E.A. Globular chitosan treatment of bovine jugular veins: Evidence of anticalcification efficacy in the subcutaneous rat model. *Cardiovasc. Pathol.* **2018**, *32*, 1–7. [CrossRef]

53. Jayakumar, R.; Prabaharan, M.; Kumar, P.S.; Nair, S.; Tamura, H. Biomaterials based on chitin and chitosan in wound dressing applications. *Biotechnol. Adv.* **2011**, *29*, 322–337. [CrossRef]

54. Dutta, P.K.; Dutta, J.; Tripathi, V. Chitin and chitosan: Chemistry, properties and applications. *J. Sci. Ind. Res.* **2004**, *63*, 20–31.

55. Ifuku, S.; Nogi, M.; Abe, K.; Handa, K.; Nakatsubo, F.; Yano, H. Surface modification of bacterial cellulose nanofibers for property enhancement of optically transparent composites: Dependence on acetyl-group DS. *Biomacromolecules* **2007**, *8*, 1973–1978. [CrossRef]

56. Scionti, G. Mechanical properties of bacterial cellulose implants. *Beilstein J. Nanotechnol.* **2010**, *4*, 325–329.

57. Nakayama, A.; Kakugo, A.; Gong, J.P.; Osada, Y.; Takai, M.; Erata, T.; Kawano, S. High mechanical strength double-network hydrogel with bacterial cellulose. *Adv. Funct. Mater.* **2004**, *14*, 1124–1128. [CrossRef]

58. Dapson, R. Macromolecular changes caused by formalin fixation and antigen retrieval. *Biotech. Histochem.* **2007**, *82*, 133–140. [CrossRef] [PubMed]

59. Nimeskern, L.; Martínez Ávila, H.; Sundberg, J.; Gatenholm, P.; Müller, R.; Stok, K.S. Mechanical evaluation of bacterial nanocellulose as an implant material for ear cartilage replacement. *J. Mech. Behav. Biomed. Mater.* **2013**, *22*, 12–21. [CrossRef]

60. Arianita, A.; Amalia, B.; Pudjiastuti, W.; Melanie, S.; Fauzia, V.; Imawan, C. Effect of glutaraldehyde to the mechanical properties of chitosan/nanocellulose. *J. Phys. Conf. Ser.* **2019**, *1317*, 012045. [CrossRef]

61. Van Noort, R.; Martin, T.; Black, M.; Barker, A.; Montero, C. The mechanical properties of human dura mater and the effects of storage media. *Clin. Phys. Physiol. Meas.* **1981**, *2*, 197–203. [CrossRef]

62. Sang, C.; Maiti, S.; Fortunato, R.N.; Kofler, J.; Robertson, A.M. A uniaxial testing approach for consistent failure in vascular tissues. *J. Biomech. Eng.* **2018**, *140*, 0610101–06101010. [CrossRef]

63. Parshin, D.; Lipovka, A.; Yunoshev, A.; Ovsyannikov, K.; Dubovoy, A.; Chupakhin, A. On the optimal choice of a hyperelastic model of ruptured and unruptured cerebral aneurysm. *Sci. Rep.* **2019**, *9*, 15865. [CrossRef]

64. Laurence, D.W.; Homburg, H.; Yan, F.; Tang, Q.; Fung, K.M.; Bohnstedt, B.N.; Holzapfel, G.A.; Lee, C.H. A pilot study on biaxial mechanical, collagen microstructural, and morphological characterizations of a resected human intracranial aneurysm tissue. *Sci. Rep.* **2021**, *11*, 3525. [CrossRef]

65. Pensalfini, M.; Meneghello, S.; Lintas, V.; Bircher, K.; Ehret, A.E.; Mazza, E. The suture retention test, revisited and revised. *J. Mech. Behav. Biomed. Mater.* **2018**, *77*, 711–717. [CrossRef]

 *polymers*

Article

Triborheological Study under Physiological Conditions of PVA Hydrogel/HA Lubricant as Synthetic System for Soft Tissue Replacement

Laura C. Duque-Ossa, Gustavo Ruiz-Pulido and Dora I. Medina *

Tecnologico de Monterrey, School of Engineering and Science, Atizapan de Zaragoza, Estado de Mexico 52926, Mexico; A01747663@itesm.mx (L.C.D.-O.); A01166117@itesm.mx (G.R.-P.)
* Correspondence: dora.medina@tec.mx

Abstract: In soft tissue replacement, hydrophilic, flexible, and biocompatible materials are used to reduce wear and coefficient of friction. This study aims to develop and evaluate a solid/liquid triborheological system, polyvinyl alcohol (PVA)/hyaluronic acid (HA), to mimic conditions in human synovial joints. Hydrogel specimens prepared via the freeze–thawing technique from a 10% (w/v) PVA aqueous solution were cut into disc shapes (5 ± 0.5 mm thickness). Compression tests of PVA hydrogels presented a Young's modulus of 2.26 ± 0.52 MPa. Friction tests were performed on a Discovery Hybrid Rheometer DHR-3 under physiological conditions using 4 mg/mL HA solution as lubricant at 37 °C. Contact force was applied between 1 and 20 N, highlighting a coefficient of friction change of 0.11 to 0.31 between lubricated and dry states at 3 N load (angular velocity: 40 rad/s). Thermal behavior was evaluated by differential scanning calorimetry (DSC) in the range of 25–250 °C (5 °C/min rate), showing an endothermic behavior with a melting temperature (Tm) around 231.15 °C. Scanning Electron Microscopy (SEM) tests showed a microporous network that enhanced water content absorption to 82.99 ± 1.5%. Hydrogel achieved solid/liquid lubrication, exhibiting a trapped lubricant pool that supported loads, keeping low coefficient of friction during lubricated tests. In dry tests, interstitial water evaporates continuously without countering sliding movement friction.

Keywords: triborheology; polyvinyl alcohol hydrogel; hyaluronic acid lubricant; wear; coefficient of friction

 check for
updates

Citation: Duque-Ossa, L.C.; Ruiz-Pulido, G.; Medina, D.I. Triborheological Study under Physiological Conditions of PVA Hydrogel/HA Lubricant as Synthetic System for Soft Tissue Replacement. *Polymers* **2021**, *13*, 746. https://doi.org/10.3390/polym13050746

Academic Editors: João Carlos Silva and Frederico Castelo Ferreira

Received: 20 January 2021
Accepted: 13 February 2021
Published: 28 February 2021

Publisher's Note: MDPI stays neutral with regard to jurisdictional claims in published maps and institutional affiliations.

1. Impact Statement

Musculoskeletal diseases represent one of the main causes of labor and academic absence in the world, so they represent a public health problem that causes economic, time, and production losses [1]. Osteoarthritis is the most common degenerative disease associated with synovial joints, cartilage and bone wear, bone volume loss and deterioration, osteophyte formation, and inflammation [2,3]. Around the world, it is present in more than 25% of the adult population [2]. However, most treatments are ineffective over the long term, and thus, cartilage replacement surgery is commonly used for treating the disease. The procedure involves the use of synthetic elements which emulate soft tissues and synovial joint behavior. In order to determine components that can be used for treatments, this study proposes the use of biocompatible triborheological systems under physiological conditions, operating with synthetic materials with non-isotropic surfaces in terms of roughness.

2. Introduction

The human body suffers degradation and loss of motor skills over time, affecting quality of life in relation to the mental and physical health of individuals in areas of functional independence [4,5]. The joint cavity is composed by cartilaginous tissue, synovial fluid, subchondral bone, and sometimes menisci [3]. The synovial fluid (SF) is a biological lubricant composed by mucopolysaccharides, proteins, and lipids, which form a protective lubricant film for articular joints by reducing friction and wear [6–8]. The articular cartilage (AC) creates a fibrous connection between 2 or more bones [9,10]. It is integrated by chondrocytes and an extracellular matrix (ECM), which give strength and help to generate good mechanical properties against stress loads [11,12].

When an articulation begins its failure stage, it means that its AC has lost its properties, decreasing its volume and deteriorating [10,13,14]. This leads to the appearance of diseases such as osteoarthritis (OA) [15,16]. OA is a degenerative disease associated with wear and tear of joints such as the knee, hip, foot, hand, spine, and elbow, among others (Figure 1), presenting loss of AC, osteophyte formation, inflammation, alteration of SF composition, subchondral bone sclerosis, and varying degrees of synovitis [2,3,17,18]. Worldwide, it is present in more than 25% of the population over 18 years of age and is responsible for an economic cost of around 89.1 billion USD per year worldwide. [2,17]. OA pathology is derived not only from age, but from lifestyle, weight, overuse, injury, and genetic and gender conditions [2,19,20]. For example, cervical facet joint OA affects older adults, increasing its incidence with age [21,22]. Its prevalence is around 54–67% of spinal pain symptoms [23,24]. For hip OA, the population over 85 years old presents a risk of around 25% of suffering from the disease. [25]. In the case of knees, it is estimated that 6% of adults worldwide have signs of OA [26].

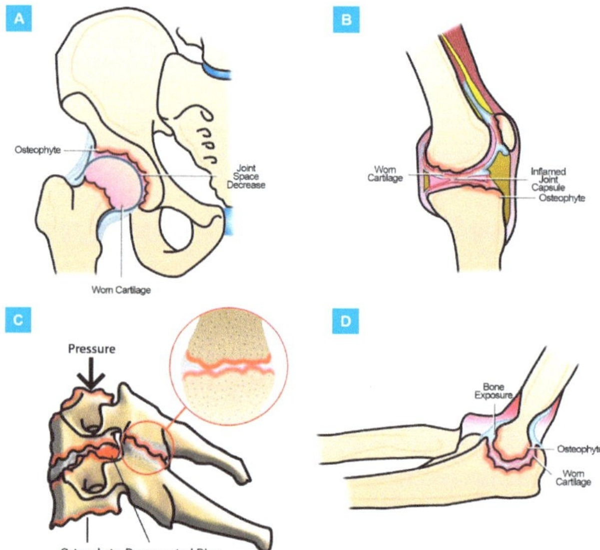

Figure 1. Damage explanation in the joints: (**A**) Hip osteoarthritis (OA); (**B**) knee OA; (**C**) facet OA; (**D**) elbow OA.

Medical and engineering researchers have tried to reduce the progress of OA and correct its effects with pharmacological interventions, physical therapy, cell therapy, and tissue engineering through surgical methods [3,27,28]. AC as an avascular tissue does not regenerate easily, so surgical methods are most commonly used for restoring or replacing cartilage surfaces [10,29]. Tissue engineering has generated advances by repairing cartilage and developing implants for surgery, considering human body mechanics and its response to external agents [30,31]. Cell culture techniques, clinical trials, scaffolds, and synthetic materials are also used [30,32].

The use of biomaterials has focused on polymers, mainly hydrogels [10]. They have hydrophilic structures (high water content), softness, flexibility, and biocompatibility [33–35].

Synthetic hydrogels can mimic AC behavior in terms of lubrication and mechanical performance. They can even improve upon AC's toughness and strength, providing extraordinary mechanical properties [36–38]. Within the polymer group, polyvinyl alcohol (PVA) is the most relevant in current studies [39–41]. PVA has been used to improve and to regenerate different tissues and organs in the human body, such as arterial phantoms, corneal implants, and cartilage tissue replacements [41]. Specifically, for AC substitution, several sliding studies have demonstrated its performance under diverse temperatures, charges, and velocity conditions. Thus, PVA is considered an excellent choice in tissue engineering for friction reduction and low wear results [42–44].

For tests, the use of lubricants to minimize damage in artificial cartilage is imperative. Hyaluronic acid (HA), a component of synovial fluid, is used as an alternative to increase joint lubrication due to its viscoelastic properties. [45–47]. This viscous fluid provides protection against articular wear [48,49]. Today, HA is commonly used in tribological tests that involve the study of PVA hydrogels [50].

Due to the above, the goal of this research was to evaluate the potential of PVA hydrogels for soft tissue replacement, such as AC, using HA as lubricant, under physiological conditions of sample temperature (37 °C) and using a saline solution as a solvent for sample preparation to recreate fluid conditions in the human body. Additionally, pressures applied during the tests were equivalent to the impact upon joints on a daily basis.

3. Materials and Methods

3.1. Preparation of PVA Hydrogels

The specimens were produced from a 10% PVA aqueous solution (Sigma Aldrich, St. Louis, MO, USA, 13×10^4 Daltons, 99% hydrolyzed, 10% w/w) in distilled water by the freeze–thawing technique (FT) [51]. The mixture was stirred continuously at 80 °C for 3 h using a magnetic stirrer to achieve complete homogenization. The solution was placed in glass petri dishes and four FT cycles were performed. The freezing stage was carried out for 12 h at −20 °C, while the thawing stage was carried out at room temperature for 12 h.

For triborheological and mechanical tests, FT hydrogel samples were cut in a disc shape with 5 ± 0.5 mm thickness and 45 mm diameter.

HA lubricant solution was prepared using 4 mg of HA powder (Lifecore Biomedical, Chaska, MN, USA, 8×10^5 Daltons) per mL of saline solution (0.15 M NaCl) under stirring conditions of 150 rpm for 6 h.

3.2. Equilibrium Water Content Measurements

For equilibrium water content measurement (EWCM), the procedure proposed by Sardinha et al. in 2013 was followed. PVA hydrogels samples were cut into 25×25 mm squares with 5 ± 0.5 mm thickness, weighted, and submerged in distilled water at ambient temperature until they reached a stable weight. Samples were removed from water and weighted again. Weight difference between the hydrated hydrogel (Wh) and the dehydrated hydrogel (Wd) gives the EWCM in percentage, using Equation (1) [44].

$$EWCM\ (\%) = \frac{Wh - Wd}{Wh} \times 100 \tag{1}$$

3.3. Mechanical Tests

Compression tests were carried out, emulating joint cavity conditions in a Shidmadzu UH universal testing machine (Shimadzu Corporation, Kyoto, Japan): humidity, preheating at 37 °C, compression speed of 1 mm/min with a maximum load of 4500 N. Considering Equations (2)–(4), and values obtained from stress vs. strain curves, the parameters were calculated as described by Gupta et al. [52,53]:

$$\sigma = \frac{F}{A} \tag{2}$$

$$\varepsilon = \frac{\delta}{L_0} = \frac{L_0 - L_f}{L_0} \tag{3}$$

$$E = \frac{\sigma}{\varepsilon} \tag{4}$$

where σ is the stress, F is the applied load, A is the area where the load is applied, ε is the strain, δ is the calibrated length, L_0 is the initial length, L_f is the final length, and E is the modulus of elasticity or Young's modulus [52].

3.4. Rheological and Friction Tests

Tests were performed with a Discovery Hybrid Rheometer DHR-3 (TA Instruments, New Castle, DE, USA, maximum charge 50 N) using cone-and-plate (0.9969°) configuration for rheological characterization (Figure 2a) and plate-and-plate configuration for the triborheological part (Figure 2b).

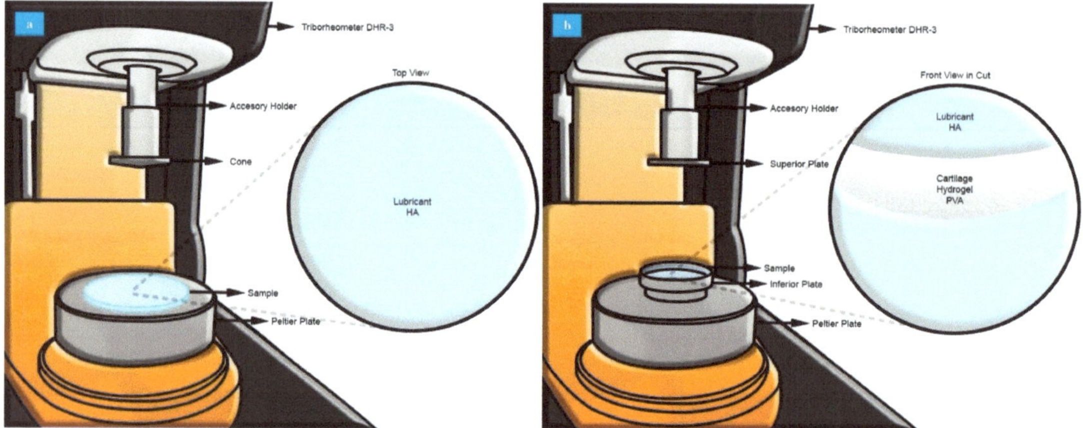

Figure 2. Characterization in a Discovery Hybrid Rheometer DHR-3. (**a**) Rheological study using cone-and-plate stainless steel configuration with an angle of 0.09969°, a diameter of 60 mm, and a truncation gap of 23 um. Samples used were 4 mg/mL hyaluronic acid (HA) solution. (**b**) Triborheological study using plate-and-plate stainless steel configuration with a diameter of 40 mm. Samples used were 10% polyvinyl alcohol (PVA) hydrogel with 5.0 ± 0.5 mm thickness and 45 mm diameter, hydrated with 4 mg/mL HA solution.

All the experiments were performed at 37 °C to mimic the physiological conditions of the human body. HA in 4 mg/mL concentration in saline solution 0.10 M NaCl was used as lubricant fluid [54]. Samples were submerged at all times in lubricant to ensure their complete hydration. Velocity was fixed between 0.1 and 100 rad/s, and contact force was applied in a range of 1–20 N. The coefficient of friction changes were continuously monitored during tests.

Rheology tests were carried out in 3 steps: conditioning sample (37 °C during 10 s), frequency sweep (0.01–100 Hz, 10% strain), and flow sweep (shear rate 0.001–100 s^{-1}). Triborheological tests were conducted in 2 steps: conditioning options active (37 °C for all charges: 1 N, 3 N, 5 N, 10 N, 15 N, 20 N) and flow sweep (angular velocity 0.1–100 rad/s).

3.5. Surface Analysis

The morphology of the PVA hydrogel and identification of wear mechanism traces was observed by JEOLJSM-6360 scanning electron microscopy (SEM, JEOL USA, Peabody, MA, United States) at 20 kV. Samples were managed using high vacuum at level 3 of the beam diameter, working with secondary electrons at magnifications between 50 and 1000×. Samples were cut into squares of 10 × 10 mm (thickness of 5 ± 0.5 mm) and gold-sputtered to enhance conductivity. Samples were analyzed in surface distribution, porosity, and wear tracks before and after triborheological tests.

3.6. Thermal Behavior

The thermal properties of dried hydrogels were measured by differential scanning calorimetry (DSC, Q10 TA Instruments, United States). Dried hydrogels (2–5 mg) were placed in an aluminum pan and heated at a rate of 5 °C/min from 25–250 °C in a nitrogen atmosphere. The melting enthalpy, ΔH_m, was determined by integrating the area under the melting peak over the range 208–242 °C. The crystallinity (Xc) was calculated by

$$X_c = \frac{\Delta H_m}{\Delta H_c} \times 100\% \tag{5}$$

where ΔH_c = 138.6 J/g, which represents the heat required for melting a 100% crystalline PVA sample [55].

4. Results and Discussion

4.1. Surface Morphology and Water Content

Naturally, AC has a high water content to mitigate the stress generated from contact forces applied during the movement of synovial joints. Hydrogels must present similar swelling properties to correctly mimic the natural AC performance in terms of super water-absorbing capability, elasticity, compressive mechanical properties, wear reduction, and low coefficient of friction during sliding [44,56,57]. For that reason, the water content absorption of PVA hydrogels was analyzed by measuring the equilibrium water content in percentage (EWCM %). Results showed an average EWCM of 82.99 ± 1.5% in accordance with Table 1.

Table 1. Water content measurements of dry and hydrated samples at different weights.

Test	W_h (g)	W_s (g)	EWCM (%)
1	1.48	0.23	84.46
2	1.72	0.26	84.88
3	1.73	0.26	84.97
4	1.91	0.31	83.77
5	2.45	0.42	82.86
6	2.54	0.44	82.68
7	2.70	0.49	81.85
8	3.11	0.59	81.03
9	3.32	0.50	84.94
10	3.38	0.64	81.07
11	3.45	0.67	80.58
12	3.62	0.56	84.53
13	3.72	0.63	83.06
14	4.32	0.79	81.50
15	4.73	0.82	82.66
Average			82.99
Standard Deviation			1.50

The percentage of water stored in the hydrogel is not dependent on the initial weight of the sample. Even thickness variation does not affect the swelling range, which is maintained around the same value. The swelling behavior of hydrogels is a consequence of the diffusion of water molecules inside the polymer network [58], which depends mainly on the sample's porosity and porous size, which were similar in all the samples as all were prepared under the same conditions [57]. Therefore, the obtained swelling values for the material are adequate for its purpose, because the water content of a natural CA is around 80% [59,60].

Unworn surfaces of PVA hydrogel were visualized in SEM equipment for surface distribution and porosity evaluations. According to Figure 3a, the surface was smooth and largely homogeneous despite the presence of inclusions adhered from the environment (dust particles) that adhered during the process of dehydration. Moreover, SEM micrographs of PVA hydrogels showed the porous network that results from the spaces generated by the ice crystals formed during the FT process [56]. Those pores allowed water absorption, reducing hydrogel wear (Figure 3b). Similar results have been reported by Sardinha et al. and Li et al. in studies of PVA hydrogels under similar conditions [43,44].

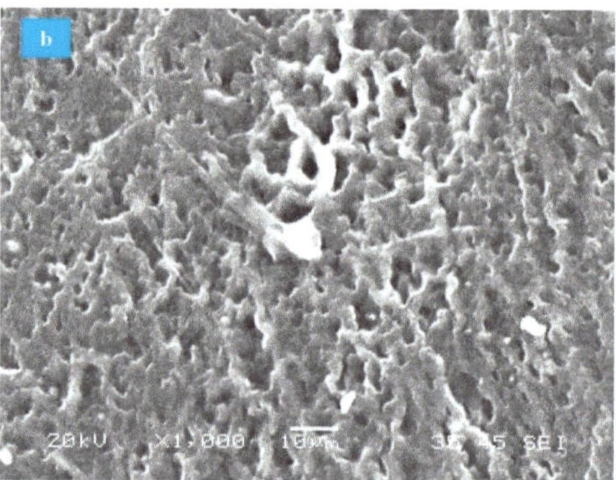

Figure 3. SEM of PVA hydrogels. (**a**) Surface distribution, (**b**) porosity (High vacuum, 50–1000×, secondary electrons).

4.2. Mechanical Tests

The stress–strain curves obtained from 7 tests of compression after 4 FT cycles are reported in Figure 4a. All samples exhibited a non-linear behavior and were plotted from the data obtained by the instrument. Using a linear regression, the average Young's modulus obtained was 2.26 ± 0.52 MPa. The displacement observed was caused by the standard deviation present in the initial thickness of the samples.

In accordance with the curves, the behavior of the hydrogel samples accurately fits an exponential relationship. For example, test 1 exhibited a correlation coefficient R^2 higher than 0.9, indicating a positive proportionality of the effort against deformation, which explains the viscoelastic characteristics of the material (flexibility and rapid response to external loads) [61].

Figure 4b shows that when the deformation rises from 33.7% to 70%, Young's modulus increments from 0.05 MPa to 2.7 MPa, which represents an increase of 51 times. This mechanical behavior is expected, and comparable to natural AC. When the cartilage is subjected to a considerable high load activity (such as running or jumping), the compression module increases to resist high stress, reducing deformation and preventing cartilage damage [61].

Measurement of Young's modulus exhibited similar results to previous studies that involved mechanical characteristics of natural and synthetic cartilages. The viscoelastic behavior of the samples during the tests is shown in Figure 4c, through images that reveal the deformation (sample 5) at certain times.

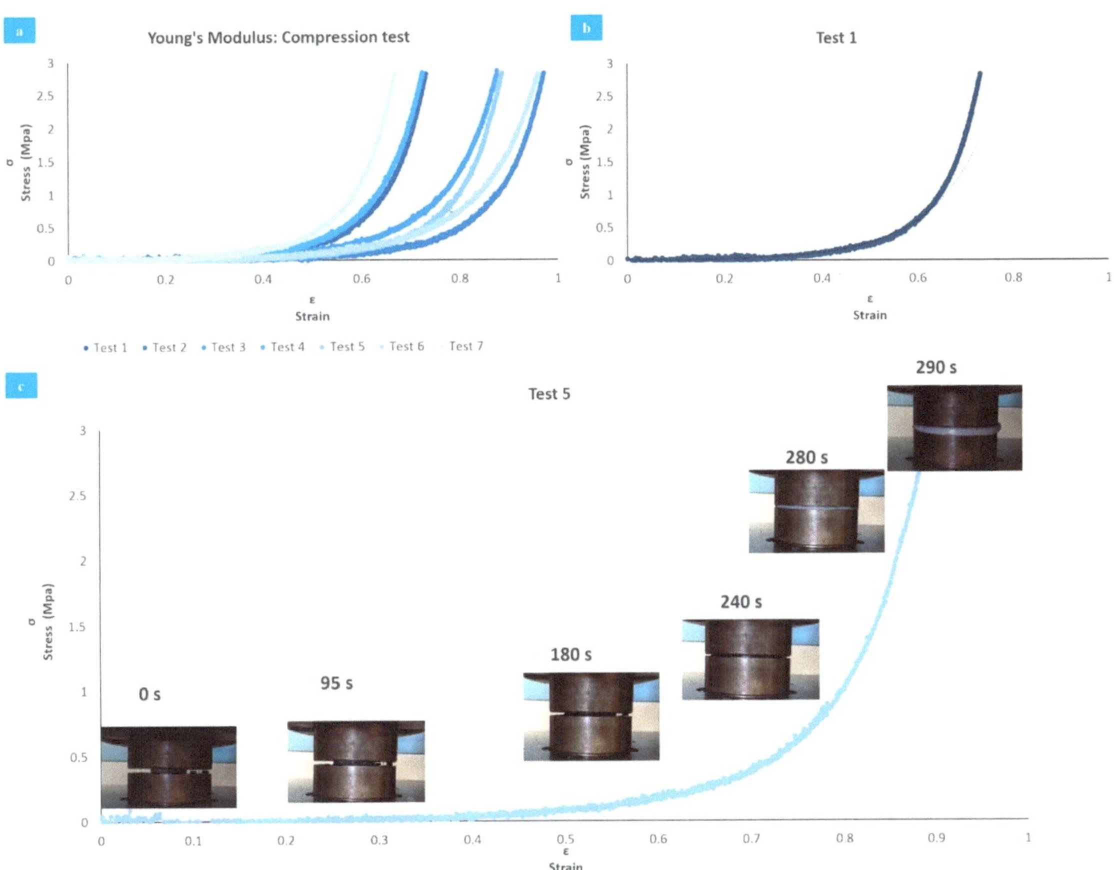

Figure 4. Compression tests (Up to 4500 N, compression rate 1 mm/min). (**a**) Linear regression of all tests; (**b**) exponential behavior of compression modulus in test 1; (**c**) test 5 deformation over time.

4.3. Rheological and Friction Tests

Rheological test data are displayed in Figure 5. HA presented a non-Newtonian behavior, where the viscosity decreases (Figure 5a) as the shear stress increases (Figure 5b), in a clear example of shear thinning [62].

The evaluation of complex viscosity (η^*), elastic (G'), and viscous (G'') modulus (Figure 5c) showed that at low frequencies, the sample exhibited an elastic behavior ($G' > G''$), while at high frequencies, it presented a viscous behavior ($G'' > G'$). This behavior is typical of substances in solution, known as "weak gels", and is contrary to the results of pure HA, which initially behaves viscously ($G'' > G'$) [54,63–65]. Additionally, the number of FT cycles should increase the elastic modulus, which is related to the gel strength and porosity, because the cycles stimulate the formation of crystalline junction zones among PVA chains and the appearance of hydrogen bonds resulting in PVA densification. As more couplings and bonds are formed, the sol–gel transition occurs and the sample no longer flows as a polymer solution, and its viscoelastic properties exhibit a solid-like behavior as its macromolecules lose their identity and become part of a large three-dimensional network, expanding throughout the entire sample volume [58].

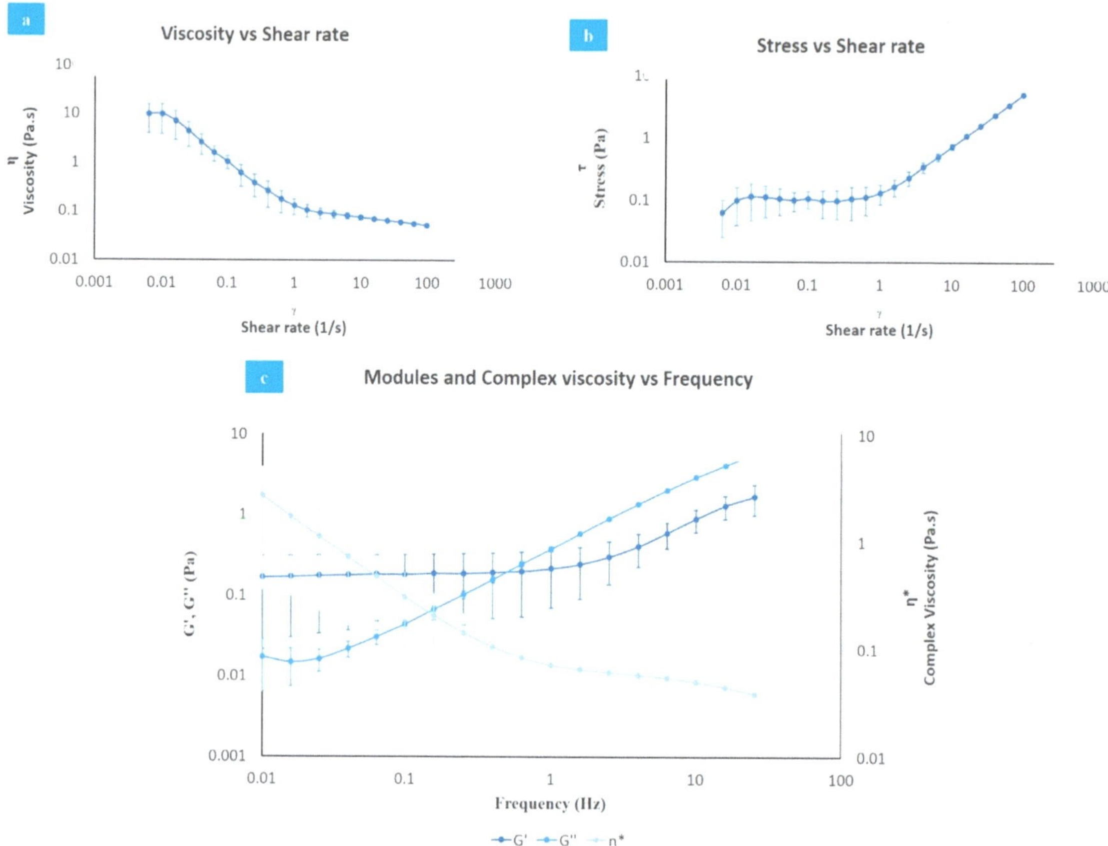

Figure 5. Rheological test. (**a**) Viscosity vs. shear rate; (**b**) shear stress vs. shear rate; (**c**) modules and complex viscosity vs. frequency (humidity, 37 °C, 0.01–100 Hz, 0.001–100 s^{-1}).

The crossover frequency from a viscous to an elastic state demonstrates the viscoelastic behavior expected from a lubricant. The changing behavior of the modulus indicates an increased deformation of the cartilage at high frequencies, enhancing its resistant properties [66].

Data obtained of the coefficient of friction of PVA synthetic cartilage against stainless steel plates under different factors are shown in Figure 6. Figure 6a,b indicates that coefficient of friction is affected by angular velocity, increasing between 0 and 4 rad/s, followed by a quick decrease among 4–10 rad/s. Finally, it rises slowly again after 10 rad/s. The speed generates the same effect on the sample, regardless of whether it is lubricated or not. Specifically, coefficient of friction increased from 0.19 to 0.3 when the velocity incremented from 0 to 4 rad/s and decreased from 0.3 to 0.14 when velocity moved from 4–10 rad/s under 5 N load.

As seen in Figure 6c, lubrication causes a decrease in the coefficient of friction. For a 3 N load and an angular velocity of 40 rad/s, it goes from 0.11 to 0.31 when switching from HA lubrication to a dry state. In a dry state, the large amount of water accumulated by the samples contributes to counteracting the sliding movement, while in a lubricated state, the transition of the lubricant from elastic to a viscous state minimizes the friction generated at high angular velocity values.

Pressure also generates an effect upon coefficient of friction. For 0.46 PSI with HA lubricant, the coefficient of friction was 0.14 at an angular velocity of 10 rad/s. For the same conditions, the coefficient of friction increased to 0.24 under a 1.82 PSI load as Figure 6d shows. Meanwhile, in dry samples, the coefficient of friction increased exponentially after 0.91 PSI.

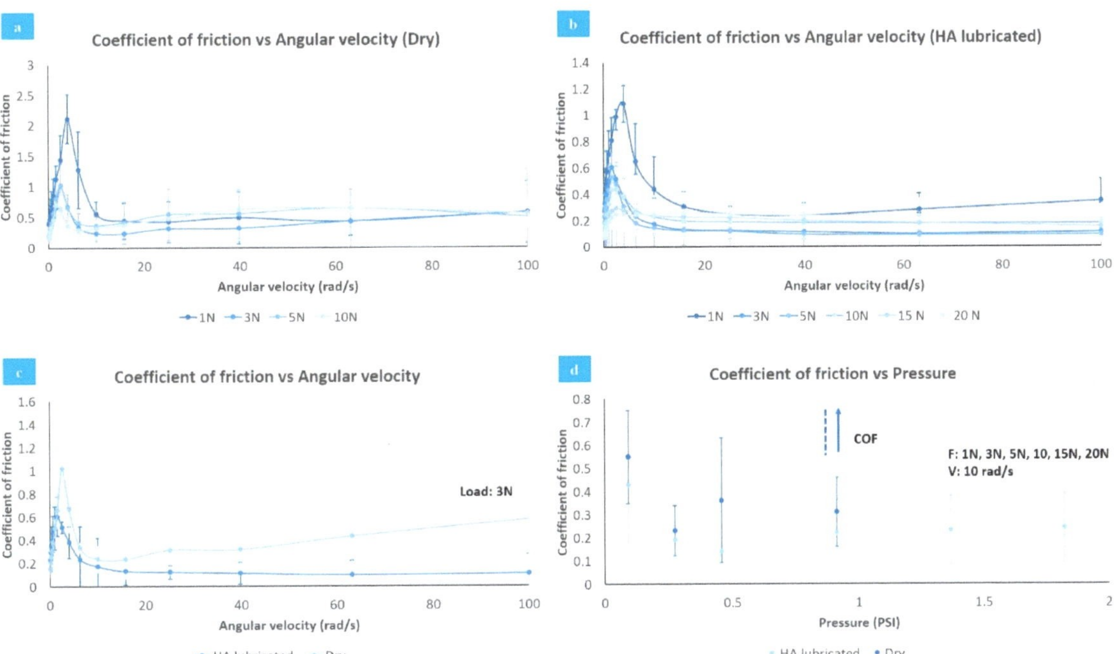

Figure 6. Coefficient of friction vs. angular velocity. (**a**) Dry; (**b**) HA lubricated; (**c**) coefficient of friction vs. angular velocity: Lubrication comparative; (**d**) coefficient of friction vs/pressure (0.1–100 rad/s, 1–20 N).

Under an applied load, the coefficient of friction decreases, followed by an increase as the water or HA (depending on the conditions of the experiment) is squeezed out from the PVA hydrogel through its porous network. As the interstitial fluid expands from the porous structure to the conjunction and suffers instantaneous pressurization supporting a fraction of the applied load, it produces a coefficient of friction reduction.

This behavior could be explained by comparing 3 parts of the lubrication mechanisms of natural articular cartilage. Boundary lubrication is an important lubrication mechanism in natural synovial joints, forming a molecular film [45]. This film takes up the boundary lubricants (HA, lubricin, among others), providing a defense against contact [67]. PVA hydrogels do not present a molecular layer that would recreate boundary lubrication similar to synovial joints, but the initial administration of lubricant prior to each test mimics this function to an extent. HA, at the beginning of the test, forms a small film that avoids hard contact between PVA hydrogel and stainless-steel plate, diminishing surface wear. This film is provided by a lubricant pool (trapped lubricant) that gets out from the hydrogel, creating an interface between the steel and the hydrogel that reduces the effective contact area between surfaces, and consequently the coefficient of friction (Figure 7a) [68]. Moreover, the liquid layer could be continuously fed by the trapped and external lubricant through hydrogel swelling to maintain its pressurization and to form new layers during sliding [69]. Interstitial fluid rehydration is limited by hydrogel swelling capacity.

After a while, during contact between both surfaces, only a small portion of the rough parts (nano asperities) of the stainless-steel plate interact directly, producing friction. This is a consequence of the presence of mixed lubrication, where the HA supports a fraction of the load. However, continuous increase in the load causes the displacement of the lubricant from the hydrogel surface, enlarging the contact area, which leads to coefficient of friction increase [70]. This behavior is known as hydrodynamic lubrication. In this case, high pressurization of the interstitial fluids occurs under applied normal loads as the fluid supports most of the loads transmitted through articular surfaces [71]. As such, friction force in the contact surface is reduced considerably, ensuring that coefficient of friction maintains low values. By contrast, under dry sliding conditions, the liquid inside the porous structure expands as lubricant during the initial period. Afterward, the superficial liquid evaporates without replacing, which produces an increase in the coefficient of friction and surface wear. In natural articular cartilage, mixed and hydrodynamic behavior are explained by solid/liquid theory that considers the collagen–proteoglycan network as a solid phase interacting with interstitial liquid and ions as a fluidic phase [72]. In Figure 7b, all mechanisms applied to the solid/liquid system form by PVA hydrogel and HA as a lubricant can be observed.

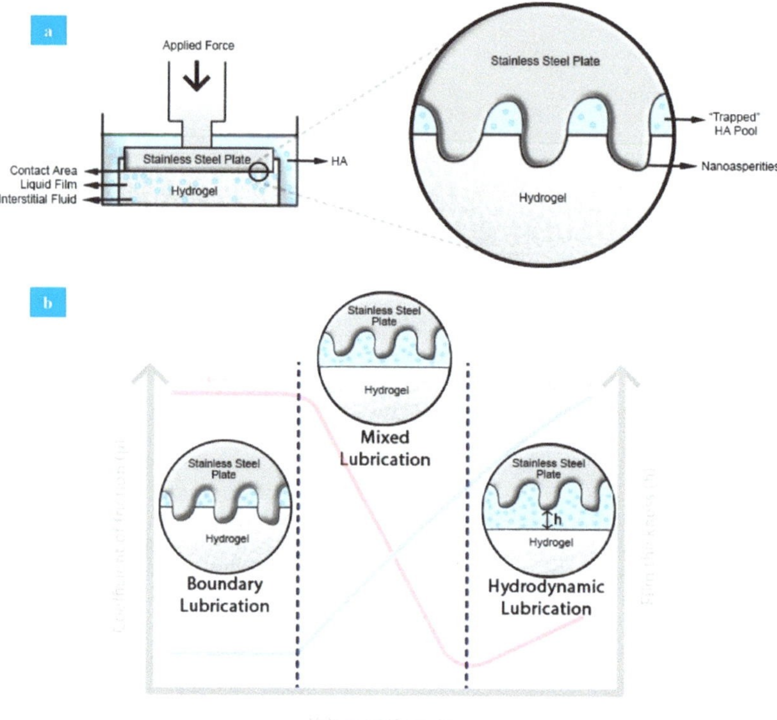

Figure 7. (**a**) Lubrication mechanism in PVA hydrogel under applied force in solid–solid contact. (**b**) Lubrication mechanisms applied to a PVA hydrogel/HA lubricant system.

In general, all results reported are comparable to previous multifactor studies. Pan et al. and Shi et al. showed that an increment in the coefficient of friction is related to the increasing load, as well as with the effect of lubrication or dryness of the sample [72,73]. Likewise, values reported for coefficient of friction in natural AC and synthetic PVA AC are similar. According to Kobayashi et al., natural AC has a coefficient of around 0.1 and synthetic PVA cartilage made by FT cycles between 0.7 and 0.8 [74]. Thus, PVA hydrogel shows a higher coefficient of friction than natural articular cartilage.

4.4. Thermal Behavior

The DSC melting curves of dried PVA hydrogels of different weights are shown in Figure 8. Peaks reported at around 232.31 °C represent the approximate melting temperature, Tm, of the samples. The values obtained are related to sample crystallinity, as has been expressed in previous studies [55]. The melting enthalpy and crystallinity percentages of PVA hydrogel samples were calculated by analyzing the area under the curve from the melting peaks using Universal Analysis 2000 (TA Instruments) software, as shown in Table 2.

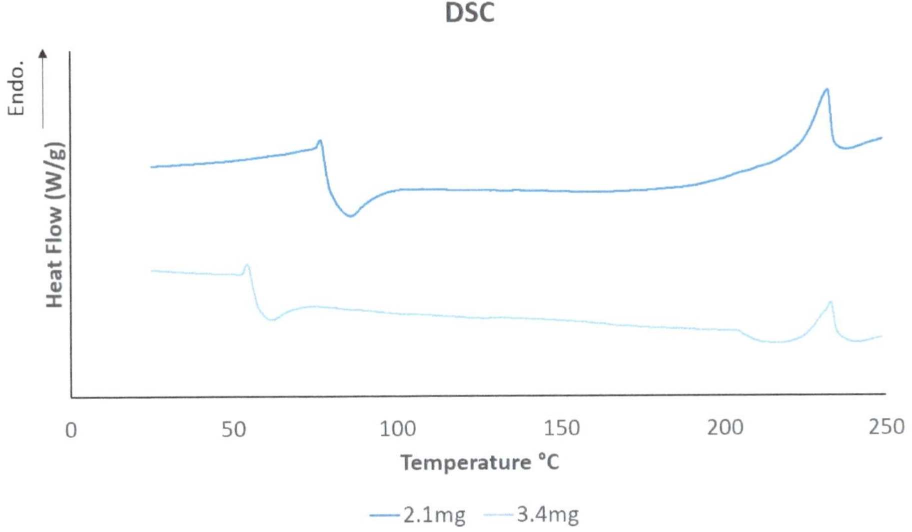

Figure 8. Differential scanning calorimetry (DSC) thermal diagram of PVA samples at different weights.

Table 2. DSC results for melting temperature and crystallinity. Measurements were made on a Q10 TA Instruments. Samples were heated at a rate of 5 °C/min from 25–250 °C in a nitrogen atmosphere.

Sample Weight (mg)	T_m (°C)	ΔH_m (J/g)	X_c (%)
2.1	231.83	13.92	10.04
3.4	232.79	9.78	7.06

In general, PVA hydrogels prepared by the FT technique exhibited similar crystallinity degrees to values reported in literature [75]. Crystallinity can be affected by factors like FT cycles, concentration of aqueous solution, and molecular weight of PVA [55,76].

5. Conclusions

The mechanical and triborheological properties of PVA synthetic cartilage/HA lubricant were investigated, and they demonstrated attributes for the generation of potential prototypes for AC replacement by testing, under physiological conditions, the effects of swelling, load, angular velocity, and lubrication. Surface morphologies of the samples were analyzed with SEM and mechanical properties were determined by a universal machine. PVA showed an average EWCM of 82.99 ± 1.5%, which is comparable to natural AC's EWCM of 80%. Additionally, the compression Young's modulus was higher than 2, providing good resistance to stress by reducing deformation and preventing cartilage damage. In rheology studies, the crossover frequency influenced the behavior of the lubricant, changing its nature from elastic state to a more viscous form and indicating a good performance.

It was also confirmed, through triborheological tests, that the coefficient of friction exhibited values between 0.1 and 0.4, which are comparable with those of the coefficient of natural AC that is around 0.1. Therefore, 10% PVA hydrogel lubricated with 4 mg/mL HA could represent a great mechanical alternative for future advances in replacing damaged AC in human joints under moderate load.

Author Contributions: Conceptualization, L.C.D.-O. and D.I.M.; methodology, L.C.D.-O. and D.I.M.; validation, L.C.D.-O. and G.R.-P.; formal analysis, L.C.D.-O., G.R.-P. and D.I.M.; investigation, L.C.D.-O., G.R.-P. and D.I.M.; data curation, L.C.D.-O.; writing—original draft preparation, L.C.D.-O. and G.R.-P.; writing—review and editing, D.I.M.; supervision, D.I.M.; project administration, D.I.M.; funding acquisition, D.I.M. All authors have read and agreed to the published version of the manuscript.

Funding: This research received no external funding.

Institutional Review Board Statement: Not applicable.

Informed Consent Statement: Not applicable.

Data Availability Statement: The data presented in this study are available on request from the corresponding author.

Acknowledgments: Research supported by Tecnologico de Monterrey and Mexican National Council for Science and Technology (CONACYT). The authors would like to make a most cordial acknowledgment to the research group of Dora Medina, Carlos Rojo, Ranieri Durante, and Jose Antonio Garcia for their support during the research.

Conflicts of Interest: The authors declare no conflict of interest.

References

1. OMS. Prevención de Trastornos Musculoesqueléticos en el Lugar de Trabajo: Serie Protección de la Salud de los Trabajadores n 5; 2004. Available online: https://www.who.int/occupational_health/publications/muscdisorders/es/ (accessed on 1 February 2021).
2. Chen, D.; Shen, J.; Zhao, W.; Wang, T.; Han, L.; Hamilton, J.L.; Im, H.-J. Osteoarthritis: Toward a comprehensive understanding of pathological mechanism. *Bone Res.* **2017**, *5*, 16044. [CrossRef]
3. Clouet, J.; Vinatier, C.; Merceron, C.; Potvaucel, M.; Maugars, Y.; Weiss, P.; Grimandi, G.; Ghicheux, J. From osteoarthritis treatments to future regenerative therapies for cartilage. *Drug Descov. Today* **2009**, *14*, 913–925. [CrossRef]
4. Mahir, L.; Belhaj, K.; Zahi, S.; Zanmasso, H.; Lmidmani, F.; el Fatimi, A. Impact of knee osteoarthritis on the quality of life. *Ann. Phys. Rehabil. Med.* **2016**, *59*, 155–159. [CrossRef]
5. Kumar, H.N.H.; Nagaraj, K.; Luthra, K.; Gupta, P.; Sapar, P.; Gupta, S.; Tyagi, A. Health-related quality of life among osteoarthritis patients attending primary care clinics of mangalore city. *Int. J. Med. Public Health* **2015**, *5*, 55–58. [CrossRef]
6. Stevenson, M.; Jaggard, H.A.; Akhbari, U.; Vaghela, P.; Gupte, C.; Cann, P. The role of denatured synovial fluid proteins in the lubrication of artificial joints. *Biotribology* **2019**, *17*, 49–63. [CrossRef]
7. Al-Shakarchi, I.; Coakley, G. Synovial fluid tests. *Medicine* **2018**, *46*, 166–169. [CrossRef]
8. Horibata, S.; Yarimitsu, S.; Fujie, H. Effect of synovial fluid pressurization on the biphasic lubrication property of articular cartilage. *Biotribology* **2019**, *19*, 100098. [CrossRef]
9. Fox, A.J.S.; Bedi, A.; Rodeo, S.A. The basic science of articular cartilage: Structure, composition, and function. *Sports Health* **2009**, *1*, 461–468. [CrossRef]
10. Liao, I.-C.; Moutos, F.T.; Estes, B.T.; Zhao, X.; Guilak, F. Composite three-dimensional woven scaffolds with interpenetrating network hydrogels to create functional synthetic articular cartilage. *Adv. Funct. Mater.* **2013**, *23*, 5833–5839. [CrossRef]
11. Vinatier, C.; Mrugala, D.; Jorgensen, C.; Ghicheux, J.; Noel, D. Cartilage engineering: A crucial combination of cells, biomaterials and biofactors. *Trends Biotechnol.* **2009**, *27*, 307–314. [CrossRef]
12. Fritz, J.; Janssen, P.; Gaissmaier, C.; Schewe, B.; Weise, K. Articular cartilage defects in the knee basics, therapies and results. *Injury* **2008**, *39*, 50–57. [CrossRef] [PubMed]
13. Grenier, S.; Bhargava, M.M.; Torzilli, P.A. An in vitro model for the pathological degradation of articular cartilage in osteoarthritis. *J. Biomech.* **2014**, *47*, 645–652. [CrossRef] [PubMed]
14. Henao-Murillo, L.; Ito, K.; van Donkelaar, C.C. Collagen damage location in articular cartilage differs if damage is caused by excessive loading magnitude or rate. *Ann. Biomed. Eng.* **2018**, *46*, 605–615. [CrossRef]
15. Pap, T.; Korb-pap, A. Cartilage damage in osteoarthritis and rheumatoid arthritis—Two unequal siblings. *Nat. Rev. Rheumatol.* **2015**, *11*, 606–615. [CrossRef] [PubMed]

16. Stender, M.E.; Carpenter, R.D.; Regueiro, R.A.; Ferguson, V.L. An evolutionary model of osteoarthritis including articular cartilage damage, and bone remodeling in a computational study. *J. Biomech.* **2016**, *49*, 3502–3508. [CrossRef] [PubMed]
17. Carlson, A.; Rawle, R.; Wallace, C.; Brooks, E.; Adams, E.; Greenwood, M.; Olmer, M.; Lotz, M.; Bothner, B.; June, R. Characterization of synovial fluid metabolomic phenotypes of cartilage morphological changes associated with osteoarthritis. *Osteoarthr. Cartil.* **2019**, *27*, 1174–1184. [CrossRef]
18. Loeser, R.F.; Goldring, S.R.; Scanzello, C.R.; Goldring, M.B. Osteoarthritis: A disease of the joint as an organ. *Arthritis Rheum.* **2012**, *64*, 1697–1707. [CrossRef]
19. Zhang, Y.; Jordan, J.M. Epidemiology of osteoarthritis. *Clin. Geriatr. Med.* **2010**, *26*, 355–369. [CrossRef]
20. Loeser, R.F. Aging and osteoarthritis. *Curr. Opin. Rheumatol.* **2011**, *23*, 492–496. [CrossRef] [PubMed]
21. Park, W.M.; Kim, K.; Kim, Y.H. Changes in range of motion, intradiscal pressure, and facet joint force after intervertebral disc and facet joint degeneration in the cervical spine. *J. Mech. Sci. Technol.* **2015**, *29*, 3031–3038. [CrossRef]
22. Gellhorn, A.C.; Katz, J.N.; Suri, P. Osteoarthritis of the spine: The facet joints. *Nat. Rev. Rheumatol.* **2013**, *9*, 216–224. [CrossRef] [PubMed]
23. del Seguro Social, I.M. *Abordaje Diagnóstico del Dolor de Cuello en la Población Adulta en el Primer Nivel de Atención*; IMSS: Mexico City, Mexico, 2013.
24. Manchikanti, L.; Boswell, M.V.; Singh, V.; Pampati, V.; Damron, K.S.; Beyer, C.D. Prevalence of facet joint pain in chronic spinal pain of cervical, thoracic, and lumbar regions. *BMC Musculoskelet. Disord.* **2004**, *5*, 15. [CrossRef]
25. Murphy, N.J.; Eyles, J.P.; Hunter, D.J. Hip osteoarthritis: Etiopathogenesis and implications for management. *Adv. Ther.* **2016**, *33*, 1921–1946. [CrossRef]
26. Wood, A.M.; Brock, T.M.; Heil, K.; Holmes, R.; Weusten, A. A review on the management of hip and knee osteoarthritis. *Int. J. Chronic Dis.* **2013**, *2013*. [CrossRef]
27. Vincent, T.L.; Watt, F.E. Osteoarthritis. *Medicine* **2017**, *46*, 187–195. [CrossRef]
28. Conaghan, P.G.; Dickson, J.; Grant, R.L. Care and management of osteoarthritis in adults: Summary of nice guidance. *BMJ* **2008**, *336*, 502–503. [CrossRef] [PubMed]
29. Moffat, K.L.; Goon, K.; Moutos, F.T.; Estes, B.T.; Oswald, S.J.; Zhao, X.; Guilak, F. Composite cellularized structures created from an interpenetrating polymer network hydrogel reinforced by a 3d woven scaffold. *Macromol. Biosci.* **2018**, *18*, 1800140. [CrossRef]
30. Fisher, M.B.; Mauck, R.L. Tissue engineering and regenerative medicine: Recent innovations and the transition to translation. *Tissue Eng. Part B Rev.* **2013**, *19*, 1–13. [CrossRef] [PubMed]
31. Coburna, J.M.; Gibsonb, M.; Monagleb, S.; Pattersonb, Z.; Elisseeff, J.H. Bioinspired nanofibers support chondrogenesis for articular cartilage repair. *Proc. Natl. Acad. Sci. USA* **2012**, *109*, 10012–10017. [CrossRef]
32. Lysaght, M.J.; Jaklenec, A.; Deweerd, E. Great expectations: Private sector activity in tissue engineering, regenerative medicine, and stem cell therapeutics. *Tissue Eng. Part A* **2008**, *14*, 305–315. [CrossRef]
33. Ahmed, E.M. Hydrogel: Preparation, characterization, and applications: A review. *J. Adv. Res.* **2015**, *6*, 105–121. [CrossRef]
34. Calo, E.; Khutoryanskiy, V.V. Biomedical applications of hydrogels: A review of patents and commercial products. *Eur. Polym. J.* **2015**, *65*, 252–267. [CrossRef]
35. Kuoa, C.K.; Lia, W.-J.; Maucka, R.L.; Tuan, R.S. Cartilage tissue engineering: Its potential and uses. *Curr. Opin. Rheumatol.* **2006**, *18*, 64–73. [CrossRef] [PubMed]
36. Zhao, X. Designing toughness and strength for soft materials. *Proc. Natl. Acad. Sci. USA* **2017**, *114*, 8138–8140. [CrossRef] [PubMed]
37. Lin, S.; Yuk, H.; Zhang, T.; Parada, G.A.; Koo, H.; Yu, C.; Zhao, X. Stretchable hydrogel electronics and devices. *Adv. Mater.* **2016**, *28*, 4497–4505. [CrossRef] [PubMed]
38. Liu, X.; Tang, T.-C.; Tham, E.; Yuk, H.; Lin, S.; Lu, T.K.; Zhao, X. Stretchable living materials and devices with hydrogel—Elastomer hybrids hosting programmed cells. *Proc. Natl. Acad. Sci. USA* **2017**, *114*, 2200–2205. [CrossRef] [PubMed]
39. Osada, Y. Polymer gels as artificial soft tissue. *Polym. Sci. Ser. C* **2017**, *59*, 3–10. [CrossRef]
40. Murakami, T.; Yarimitsu, S.; Nakashima, K.; Sakai, N.; Yamaguchi, T.; Sawae, Y.; Suzuki, A. Biphasic and boundary lubrication mechanisms in artificial hydrogel cartilage: A review. *Proc. IMechE Part H J. Eng. Med.* **2015**, *229*, 864–878. [CrossRef]
41. Jiang, S.; Liu, S.; Feng, W. Pva hydrogel properties for biomedical application. *J. Mech. Behav. Biomed. Mater.* **2011**, *4*, 1228–1233. [CrossRef]
42. Li, F.; Su, Y.; Wang, J.; Wu, G.; Wang, C. Influence of dynamic load on friction behavior of human articular cartilage, stainless steel and polyvinyl alcohol hydrogel as artificial cartilage. *J. Mater. Sci. Mater. Med.* **2010**, *21*, 147–154. [CrossRef]
43. Li, F.; Zhang, G.; Wang, A.; Guo, F. The effects of surface mechanical deformation and bovine serum albumin on the tribological properties of polyvinyl alcohol hydrogel as an artificial cartilage. *Adv. Mater. Sci. Eng.* **2017**. [CrossRef]
44. Sardinha, V.; Lima, L.; Belangero, W.; Zavaglia, C.; Bavaresco, V.; Gomes, J. Tribological characterization of polyvinyl alcohol hydrogel as substitute of articular cartilage. *Wear* **2013**, *301*, 218–225. [CrossRef]
45. McNary, M.; Athanasiou, K.; Reddi, A. Engineering lubrication in articular cartilage. *Tissue Eng. Part B Rev.* **2012**, *18*, 88–100. [CrossRef] [PubMed]
46. Seror, J.; Zhu, L.; Goldberg, R.; Day, A.J.; Klein, J. Supramolecular synergy in the boundary lubrication of synovial joints. *Nat. Commun.* **2015**, *6*, 6497. [CrossRef]

47. Corvelli, M.; Che, B.; Saeui, C.; Singha, A.; Elisseeff, J. Biodynamic performance of hyaluronic acid versus synovial fluid of the knee in osteoarthritis. *Methods* **2015**, *84*, 90–98. [CrossRef]
48. Singh, A.; Corvelli, M.; Unterman, A.S.; Wepasnick, K.A.; McDonnell, P.; Elisseeff, J.H. Enhanced lubrication on tissue and biomaterial surfaces through peptide-mediated binding of hyaluronic acid. *Nat. Mater.* **2014**, *13*, 988–995. [CrossRef] [PubMed]
49. Swann, D.A.; Radin, E.L.; Nazimiec, M.; Weisser, P.A.; Curran, N.; Lewinnek, G. Role of hyaluronic acid in joint lubrication. *Ann. Rheum. Dis.* **1974**, *33*, 318–326. [CrossRef] [PubMed]
50. Li, F.; Wang, A.; Wang, C. Analysis of friction between articular cartilage and polyvinyl alcohol hydrogel artificial cartilage. *Eng. Nano Eng. Approaches Med. Devices* **2016**, *27*, 87–95. [CrossRef]
51. Wan, W.; Bannerman, A.D.; Yang, L.; Mak, H. Poly (vinyl alcohol) cryogels for biomedical applications. *Adv. Polym. Sci.* **2014**, *263*, 283–321. [CrossRef]
52. Hibbeler, R.C. *Mecca de Materiales*; Educaci, P., Ed.; Pearson: Mexico City, Mexico, 2011; ISBN 139786073205597.
53. Gupta, S.; Sinha, S.; Sinha, A. Composition dependent mechanical response of transparent poly (vinyl alcohol) hydrogels. *Colloids Surf. B Biointerfaces* **2010**, *78*, 115–119. [CrossRef]
54. Dodero, A.; Williams, R.; Gagliardi, S.; Vicini, S.; Alloisio, M.; Castellano, M. A micro-rheological and rheological study of biopolymers solutions: Hyaluronic acid. *Carbohydr. Polym.* **2019**, *203*, 249–355. [CrossRef]
55. Jiang, H.; Zuo, Y.; Zhang, L.; Li, J.; Zhang, A.; Li, Y.; Yang, X. Property-based design: Optimization and characterizationof polyvinyl alcohol (pva) hydrogel and pva-matrix composite for artificial cornea. *J. Mater. Sci. Mater. Med.* **2014**, *25*, 941–952. [CrossRef]
56. Huange, M.; Cai, D.; Liu, Y.; Sun, J.; Wang, J. Investigation of a-pva/s-pva hydrogels prepared by freezing-thawing method. *Fibers Polym.* **2012**, *13*, 955–962. [CrossRef]
57. Marrella, A.; Lagazzo, A.; Dellacasa, E.; Pasquini, C.; Finocchio, E. 3d porous gelatin/pva hydrogel as meniscus substitute using alginate microparticles as porogens. *Polymers* **2018**, *10*, 380. [CrossRef]
58. Bercea, M.; Morariu, S.; Teodorescu, M. Rheological investigation of poly (vinyl alcohol)/poly(n-vinyl pyrrolidone) mixture in aqueous solution solution and hydrogel state. *J. Polym. Res.* **2016**, *23*, 142–151. [CrossRef]
59. Moore, D. Articular Cartilage. 2018. Available online: https://www.orthobullets.com/basic-science/9017/articular-cartilage (accessed on 1 February 2021).
60. Wang, M.; Peng, Z. Wear in human knees. *Biosurface Biotribol.* **2015**, *1*, 98–112. [CrossRef]
61. Li, W.; Wang, D.; Yang, W.; Song, Y. Compressive mechanical properties and microstructure of pva-ha hydrogels for cartilage repair. *RSC Adv.* **2016**, *6*, 20166–20172. [CrossRef]
62. Mackley, M. Part IIb Chemical Engineering Tripos. Rheology and Processing. UK. 2011. Available online: http://www.malcolmmackley.com/wp-content/uploads/2012/11/Rheology-lectures-2011-section1_.pdf (accessed on 1 February 2021).
63. Ambrosio, L.; Borzacchiello, A.; Netti, P.A.; Nicolais, L. Rheological study on hyaluronic acid and its derivative solutions. *J. Macromol. Sci. Part A Pure Appl. Chem.* **2007**, *36*, 991–1000. [CrossRef]
64. Falcone, S.J.; Palmeri, D.M.; Berg, R.A. Rheological and cohesive properties of hyaluronic acid. *J. Biomed. Mater. Res. Part A* **2006**, *76*, 721–728. [CrossRef]
65. Braithwaite, G.J.C.; Daley, M.J.; Toledo-Velasquez, D. Rheological and molecular weight comparisons of approved hyaluronic acid products—Preliminary standards for establishing class III medical device equivalence. *J. Biomater. Sci.* **2016**, *27*, 235–246. [CrossRef]
66. Pan, Y.; Xiong, D.; Gao, F. Viscoelastic behavior of nano-hydroxyapatite reinforced poly (vinyl alcohol) gel biocomposite as an articular cartilage. *J. Mater. Sci. Mater. Med.* **2007**, *19*, 1963–1969. [CrossRef]
67. Chan, M.; Neu, C.; Duraine, G.; Komvopoulos, K.; Reddi, A. Atomic force microscope investigation of the boundary-lubricant layer in articular cartilage. *Osteoarthr. Cartil.* **2010**, *18*, 956. [CrossRef] [PubMed]
68. Kumar, P.; Oka, M.; Ikeuchi, K.; Shimizu, K.; Yamamuro, T.; Okumura, H.; Kotoura, Y. Low wear rate of uhmwpe against zirconia ceramic (y-psz) incomparison to alumina ceramic and sus 316l alloy. *J. Biomed. Mater. Res. Part A* **1991**, *25*, 813–828. [CrossRef] [PubMed]
69. Caligaris, M.; Ateshian, G. Effect of sustained internal fluid pressurization under migrating contact area, and boundary lubrication by synovial fluid, on cartilage friction. *Osteoarthr. Cartil.* **2008**, *16*, 1220–1227. [CrossRef]
70. Shi, Y.; Xiong, D.S.; Li, J.L. The friction properties of polyvinyl alcohol/graphene oxide hydrogels as cartilage replacement. *Key Eng. Mater.* **2017**, *739*, 152–156. [CrossRef]
71. Park, R.K.; Nicoll, S.; Ateshian, G. Cartilage interstitial fluid load support in unconfined compression. *J. Biomech.* **2003**, *36*, 1785–1796. [CrossRef]
72. Shi, Y.; Xiong, D. Microstructure and friction properties of pva/pvp hydrogels for articular cartilage repair as function of polymerization degree and polymer concentration. *Wear* **2013**, *305*, 280–285. [CrossRef]
73. Pan, Y.; Xiong, D.; Chen, X. Friction characteristics of poly (vinyl alcohol) hydrogel as an articular cartilage biomaterial. *Key Eng. Mater.* **2007**, *330*, 1297–1300. [CrossRef]
74. Kobayashi, M.; Hyu, H.S. Development and evaluation of polyvinyl alcohol-hydrogels as an artificial articular cartilage for orthopedic implants. *Materials* **2010**, *3*, 2753–2771. [CrossRef]

75. Ricciardi, R.; Auriemma, F.; Gaillet, C.; de Rosa, C.; Laupretre, F. Investigation of the crystallinity of freeze/thaw poly (vinyl alcohol) hydrogels by different techniques. *Macromolecules* **2004**, *37*, 9510–9516. [CrossRef]
76. Peppas, A.N.; Hassan, M.C. Structure and morphology of freeze/thawed pva hydrogels. *Macromolecules* **2000**, *33*, 2472–2479. [CrossRef]

Article

Engineering of Optimized Hydrogel Formulations for Cartilage Repair

Yao Fu [†], Bram Zoetebier [*,†], Sanne Both, Pieter J. Dijkstra and Marcel Karperien [*]

Developmental BioEngineering, TechMed Centre, Faculty of Science and Technology, University of Twente, P.O. Box 217, 7500 AE Enschede, The Netherlands; fuyao1024@gmail.com (Y.F.); s.k.both@utwente.nl (S.B.); p.j.dijkstra@utwente.nl (P.J.D.)
* Correspondence: B.Zoetebier@utwente.nl (B.Z.); Marcel.Karperien@utwente.nl (M.K.)
† These two authors contributed equally.

Abstract: The ideal scaffold for cartilage regeneration is expected to provide adequate mechanical strength, controlled degradability, adhesion, and integration with the surrounding native tissue. As it does this, it mimics natural ECMs functions, which allow for nutrient diffusion and promote cell survival and differentiation. Injectable hydrogels based on tyramine (TA)-functionalized hyaluronic acid (HA) and dextran (Dex) are a promising approach for cartilage regeneration. The properties of the hydrogels used in this study were adjusted by varying polymer concentrations and ratios. To investigate the changes in properties and their effects on cellular behavior and cartilage matrix formation, different ratios of HA- and dextran-based hybrid hydrogels at both 5 and 10% w/v were prepared using a designed mold to control generation. The results indicated that the incorporation of chondrocytes in the hydrogels decreased their mechanical properties. However, rheological and compression analysis indicated that 5% w/v hydrogels laden with cells exhibit a significant increase in mechanical properties after 21 days when the constructs are cultured in a chondrogenic differentiation medium. Moreover, compared to the 10% w/v hydrogels, the 5% w/v hybrid hydrogels increased the deposition of the cartilage matrix, especially in constructs with a higher Dex–TA content. These results indicated that 5% w/v hybrid hydrogels with 25% HA–TA and 75% Dex–TA have a high potential as injectable scaffolds for cartilage tissue regeneration.

Keywords: hydrogel scaffolds; cartilage regeneration; mechanical properties; stem cells; matrix formation

Citation: Fu, Y.; Zoetebier, B.; Both, S.; Dijkstra, P.J.; Karperien, M. Engineering of Optimized Hydrogel Formulations for Cartilage Repair. *Polymers* **2021**, *13*, 1526. https://doi.org/10.3390/polym13091526

Academic Editor: João Carlos Silva

Received: 1 April 2021
Accepted: 1 May 2021
Published: 10 May 2021

Publisher's Note: MDPI stays neutral with regard to jurisdictional claims in published maps and institutional affiliations.

1. Introduction

Articular cartilage is a firm, smooth, viscoelastic padding at the ends of bones that ensure smooth, frictionless, and pain-free joint movement [1,2]. Articular cartilage tissue is highly hydrated and consists of approximately 70% water and 30% extracellular matrix (ECM) [3]. The structure and function of articular cartilage depends on the molecular composition of its ECM, mainly collagens and proteoglycans [4]. Cartilage has a limited capacity for self-repair due to its avascular nature and the low mitotic activity of chondrocytes [5]. Without proper treatment, damaged articular cartilage will deform, causing chronic pain and joint disability.

Over the past decades, several pharmacological and regenerative therapies have been developed [6]. An ideal scaffold for cartilage regeneration is expected to provide adequate mechanical strength, controlled degradability, adhesion, and integration with the surrounding native tissue. As it does this, it mimics natural ECMs functions, which allow for nutrient diffusion and the promotion of cell survival and differentiation [7,8]. It is anticipated that the development of such an effective biomaterial would significantly enhance the potential to develop effective therapies for tissue regeneration and function improvement. One of these promising regenerative therapies is the use of in situ-forming (injectable) hydrogels: a three-dimensional (3D) scaffold that mimics the hydrated environment of articular cartilage and facilitates cell proliferation, differentiation and matrix production by using encapsulated cells [9]. Injectable hydrogels enable a perfect match with irregular cartilage defects and a proper alignment with the surrounding tissues [9,10]. Meanwhile, from the clinical point of view, implantation surgery can be avoided and replaced by a simple, minimally invasive injection [9]. Moreover, bioactive molecules or cultured cells can simply be mixed into the hydrogel precursors prior to injection [11]. Therefore, they are promising materials that can function as scaffolds for chondrocyte culture and cartilage regeneration.

The development of injectable hydrogels as scaffolds for cartilage tissue engineering must meet certain essential conditions: biocompatibility, biodegradability, biofunctionality, and suitable mechanical strength. In our earlier studies, we developed an injectable hybrid hydrogel composed of a hyaluronic acid (HA) backbone with tyramine conjugated dextran (Dex–TA) sidechains [12]. The hydrogel gelates in situ via a biocompatible, enzymatic crosslinking reaction that forms covalent TA–TA bonds and has been shown to support the survival and growth of incorporated chondrocytes and mesenchymal stem cells as well as the deposition of a new matrix in vitro [13,14]. Using a similar mechanism, we also showed efficient gel formation after mixing HA–TA and Dex–TA conjugates [15]. The advantage of these hybrid, injectable hydrogels is that multiple functionalities can be included in one gel system to fine-tune physical properties, proteolytic degradation, and extracellular matrix production. Moreover, these hydrogels can be tailored for stiffness and degradation rate by varying polymer concentrations and ratios. In this study, we determined the optimal concentration and ratio for cell growth and matrix formation in HA–TA and Dex–TA hybrid hydrogels. To obtain these results, hydrogels at different conjugate concentrations and ratios were laden with bovine chondrocytes (bCHs), and the cartilaginous specific matrix formed in the cell/gel constructs over time was analyzed. Furthermore, physical properties like storage moduli and morphology of the hydrogels were examined.

2. Materials and Methods

2.1. Materials

Dextran (40 kDa, pharmaceutical grade) was purchased from Pharmacosmos, Holbæk. Denmark. Sodium hyaluronate (27 kDa, pharmaceutical grade) was purchased from Contipro Pharma, Dolní Dobrouč, Czech Republic. Tyramine (99%), DMF (anhydrous, 99.8%), LiCl (99.0%), p-nitrophenyl chloroformate (PNC, 96%), pyridine (anhydrous, 99.8%), DMSO-d6 (99.9%), NaCl ($\geq$99.0%), D_2O (99.9 atom % D), horseradish peroxidase (HRP, 325 units/mg solid) and hydrogen peroxide (30%) were purchased from Sigma-Aldrich, St. Louis, MO, USA. Tyramine HCl salt (99%) was obtained from Acros Organics, Fair Lawn, NJ, USA. 4-(4,6-Dimethoxy-1,3,5-triazin-2-yl)-4-methylmorpholinium chloride (DMTMM, 97%) was purchased from Fluorochem Ltd., Hadfield, UK. Ethanol ($\geq$99.9%) and diethyl ether ($\geq$99.7%) were purchased from Merck, Kenilworth, NJ, USA. Milli-Q water was used from the Milli-Q Advantage A10 system equipped with a 0.22 µm Millipak®-40 Express filter.

2.2. Synthesis of Dextran-Tyramine and Hyaluronic Acid-Tyramine

Dextran–tyramine was synthesized by the activation of dextran with PNC and subsequent aminolysis with tyramine adapted from Ramirez et al. [16]. Hyaluronic acid—tyramine was prepared by amidation of the HA carboxyl groups with tyramine by using a procedure adapted from Rydergren [17] and D'Este et al. [18]. Detailed description of polymers synthesis can be found in the Supplementary Methods. Synthesis and characterization of Dex–TA and HA–TA polymers are described in Supplementary Figure S1. The Dex–TA used in this study had a substitution degree of 10%; i.e., 10% of the monosaccharides of dextran was modified. HA–TA had a substitution degree of 10%; i.e., 10% of the carboxylic acid groups of hyaluronic acid was modified.

2.3. Cell Culture and Expansion

Bovine chondrocytes (bCHs) were isolated from cartilage knee biopsies of full thickness from six-month old female calves according to the previously reported protocol [19]. bCHs were expanded in a chondrocyte proliferation medium (Dulbecco's modified Eagle's medium (DMEM; Gibco, Billings, MT, USA) and supplemented by 10% fetal bovine serum (FBS; Gibco), 0.2 mM ascorbic acid 2-phosphate (Sigma), 0.4 mM proline (Sigma), 1x nonessential amino acids (Gibco), 100 U/mL penicillin, and 100 µg/mL streptomycin (Invitrogen, Carlsbad, CA, USA)). The medium was refreshed twice a week, and cells were used for experiments at passage 3.

2.4. Hydrogel Formation

To prepare identical hydrogel samples, we designed a mold (Supplementary Figure S2). In brief, the hydrogels were prepared in a PTFE mold to produce six identical hydrogels 8 mm wide and 1.5 mm high. After dissolving the tyramine-conjugated polymers in sterile phosphate buffered saline (PBS), the polymer solution with horseradish peroxidase (HRP, 40 units/mL; Sigma-Aldrich, St. Louis, MO, USA) was incubated overnight at 4 °C on a rollerbank. The mixture was then combined with bCHs in a concentration of 10 million cells/mL. Cell-free controls were also prepared. Freshly prepared hydrogen peroxide (H_2O_2) was added to the mixture and immediately transferred to the mold using a 1 mL pipette after a brief vortex. The final gel concentrations were a 10% w/v or 5% w/v polymer, 10 million/mL bCHs, 4 U/mL HRP and 0.03% H_2O_2 (for the 10% w/v polymer) or 0.015% H_2O_2 (for the 5% w/v polymer). HA–TA and Dex–TA were combined in 5 ratios (100:0, 75:25, 50:50, 25:75, and 0:100), and were represented by groups A, B, C, D, and E respectively.

2.5. Hydrogel Incubation

After gelation, the gels were transferred to six-well plates and incubated in chondrogenic differentiation medium (DMEM supplemented with 0.2 mM ascorbic acid 2-phosphate (Sigma), 0.4 mM proline (Sigma), 100 U/mL penicillin, and 100 µg/mL streptomycin (Invitrogen), 0.1 µM dexamethasone (Sigma), 100 µg/mL sodium pyruvate (Sigma), 50 µg/mL insulin–transferrin–selenite (ITS; Sigma), 10 ng/mL transforming growth factor β-3 (TGF-β3; R&D Systems)). The medium was refreshed three times every week, and the gels were harvested at time points day 0, 7, and 21.

2.6. Rheological Analysis

Rheological experiments were carried out using an MCR301 rheometer (Anton Paar, Oosterhout, Nederland) using parallel plates (8 mm diameter) at 20 °C under a 0.05 N normal force in the oscillatory mode with 0.5% strain and 1.0 Hz, which was in the LVE range according to the measured frequency and strain sweeps. The cylindrical hydrogels were prepared in 8 mm wide, 1.5 mm high molds and measured after equilibrating overnight in medium. At least three specimens were tested for each sample.

2.7. Hydrogel Swelling Ratio

The swelling ratio was based on the weight of the hydrogel samples:

$$Swelling\ ratio = \frac{w_{wet} - w_{dry}}{w_{dry}} \tag{1}$$

To assess swelling, the hydrogels were measured after equilibrating overnight in medium and compared to their dry weight. At least three specimens were tested for each composition.

2.8. Compression Tests

Compression testing was performed on the cylindrical gels as prepared and equilibrated for the rheological testing using a Texture Analyser TA–HD plus (StableMicro Systems Ltd., Surrey, UK) fitted with a 50 kg load cell. The hydrogels underwent three compression cycles with a maximum strain of 50% using a compression speed of 0.05 mm/s. The compression tests were conducted at room temperature, and at least three specimens were tested for each sample.

The following data were derived from the stress–strain curves. Maximum stress is the force needed to compress the sample until 50% strain is reached. The high strain compressive modulus was calculated from the stress–strain curves using a linear slope at a strain ranging from 40 to 49.5%. The percentage of energy dissipated during a compression–relaxation cycle was calculated by dividing the surface of the hysteresis loop by the surface under the compression trace.

2.9. Histology and Immunohistochemistry Staining

The samples were fixed in 10% formalin and then incubated in OCT (Thermo-scientific, Waltham, MA, USA) overnight at 4 °C. The samples embedded in OCT were then snap frozen using liquid nitrogen. Cryosections of 10 μm were cut using cryotome (Leica, Wetzlar, Germany, CM1100) and stained for sulfated glycosaminoglycan (GAG) with Alcian blue and Safranin O staining. For immunohistochemistry, cryosections were incubated with 0.3% H_2O_2 and blocked in 5% bovine serum albumin. Slides were subsequently incubated overnight at 4 °C with a rabbit polyclonal antibody against COL II (Abcam, Cambridge, UK). The sections were then incubated with a polyclonal goat–anti-rabbit HRP-conjugated secondary antibody (Dako, Glostrup, Denmark), followed by development with the DAB Substrate kit (Abcam). Counterstaining was performed with hematoxylin. Non-immune controls underwent the same procedure without primary antibody incubation. Both histology and immunohistochemistry stained slides were scanned with the NanoZoomer 2.0-RS slide scanner (Hamamatsu, Sendai City, Japan).

2.10. Live-Dead Staining

The effect of the hydrogel's composition on cell viability was studied using a live–dead assay. At day 0 and 21, the hydrogel constructs were rinsed with PBS and stained with calcein AM/ethidium homodimer using the live–dead assay Kit (Invitrogen), according to the manufacturer's instructions. Hydrogel/cell constructs were visualized using fluorescence microscopy (Leica DM IRB) and different areas were randomly selected. As a result, living cells fluoresce green and the nuclei of dead cells red. Image J software was used for cell counting. The cell viability was calculated by the percentage of live cells (green) in the total cells (green + red) from each area. Values represent the mean +/− standard deviation of at least 3 biological replicates.

2.11. RNA Isolation and Quantitative Polymerase Chain Reaction

The 5% *w/v* hydrogels were prepared for species-specific quantitative polymerase chain reaction (qPCR) analysis. At day 0 and 21, hydrogel samples were first homogenized by gentleMACS Dissociator according to the manufacturer's instructions (Miltenyi Biotec). Total RNA was then isolated using the TRIzol Reagent (Ambion) according to the manufacturer's protocol and reverse-transcribed into cDNA using the iScript cDNA Synthesis kit (Bio-Rad). A qPCR test was performed on cDNA samples by using the SensiMix SYBR& Fluorescein Kit (Bio-Rad). PCR reactions were carried out on CFX Connect™ Real-Time PCR Detection System (Bio-Rad). The cDNA was denatured at 95 °C for 10 min followed by 40 cycles. Each cycle consisted of following conditions: 15 s at 95 °C, 15 s at 60 °C, and 30 s at 72 °C. The sequence of primers for qPCR are listed in Table 1. The expression level of aggrecan (ACAN), collagen type I, II and IX (COL1, COL2, and COL9) and Osteopontin (OPN) were investigated.

Table 1. Primers used for quantitative RT-PCR.

Gene Name	Primer Sequence	Product Size (bp)
Bovine specific GAPDH	F: 5′ GCCATCACTGCCACCCAGAA 3′ R: 5′ GCGGCAGGTCAGATCCACAA 3′	207
Bovine specific Aggrecan	F: 5′ GACCAGAAGCTGTGCGAGGA 3′ R: 5′ GCCAGATCATCACCACACAG 3′	319
Bovine specific Collagen II	F: 5′ ATCAACGGTGGCTTCCACT 3′ R: 5′ TTCGTGCAGCCATCCTTCAG 3′	263
Bovine specific Collagen IX	F: 5′ GGACTCAACACGGGTCCACA 3′ R: 5′ ACAGGTCCAGCAGGGCTTTG 3′	102
Bovine specific Collagen I	F: 5′ GCGGCTACGACTTGAGCTTC 3′ R: 5′ CACGGTCACGGACCACATTG 3′	102
Bovine specific Osteopontin	F: 5′ ACTGGACTCTTCTCGCCGCC 3′ R: 5′ CGGAGGCAATGCCCAAGAGGC 3′	90

2.12. Statistical Analysis

Data were presented as mean $\pm$ standard deviation. Statistical significance between the two groups was analyzed using a Student's t-test. For three or more groups, a statistical comparison was done using the one-way Analysis of Variance (ANOVA) with Tukey's post hoc analysis. A p-value of < 0.05 was considered statistically significant.

3. Results

3.1. Hydrogel Formation and Morphology

In this study, the hydrogels were formed by dissolving the functionalized polymers (Dex–TA DS 10% and HA–TA DS 10%) and HRP in PBS (mixed with cells where desired) and adding minute quantities of H_2O_2 as an oxidizing agent. Upon the addition of H_2O_2, the crosslinking of tyramine was initiated. Precooling the solutions on ice and using a final concentration of 4.0 U/mL, HRP gave us a working time (i.e., gelation time) of about 30 s, after which the solution gelated in the mold. The quantity of H_2O_2 was adapted to the molar amount of tyramine groups, ensuring complete consumption of the oxidizing agent in all conditions within the different weight concentrations. The overview figures of the hydrogels (shown in Supplementary Figure S3) reveal that the size of the hydrogels decreased after the dextran concentration was increased from 0 to 100%. Both the hydrogels with and without cells showed the same trend; however, the addition of cells increased the gel size.

3.2. Cell Viability of Chondrocytes in Different Hydrogels

Cell viability of the bovine chondrocytes in the hydrogels was evaluated using a live–dead assay, in which living cells were stained green and dead cells stained red (Figure 1a). We counted the amount of live and dead cells and calculated the percentage of live cells (Figure 1b,c). The results showed that the chondrocytes were distributed homogeneously inside hydrogels, and over 90% of the cells remained viable in most conditions at day 0. After 21 days of culturing in the chondrogenic medium, as shown in the figure, chondrocytes maintained their characteristic round shapes, while cell viability decreased in all conditions over time. Groups B and C (75:25 and 50:50—HA–TA and Dex–TA, respectively) showed greater cell viability than under other conditions in both 5 and 10% w/v hydrogels. Meanwhile, groups D and E, which had a higher dextran concentration, presented lower cell viability in 5% w/v conditions compared to that in 10% w/v, especially in pure dextran hydrogels (group E). On the other hand, the other groups showed similar cell viability between 5 and 10% w/v hydrogels. However, after 21 days, there was still around 70 to 90% cell survival, indicating that these biomimetic hydrogels could provide a supportive environment for chondrocyte proliferation and differentiation as well as matrix deposition.

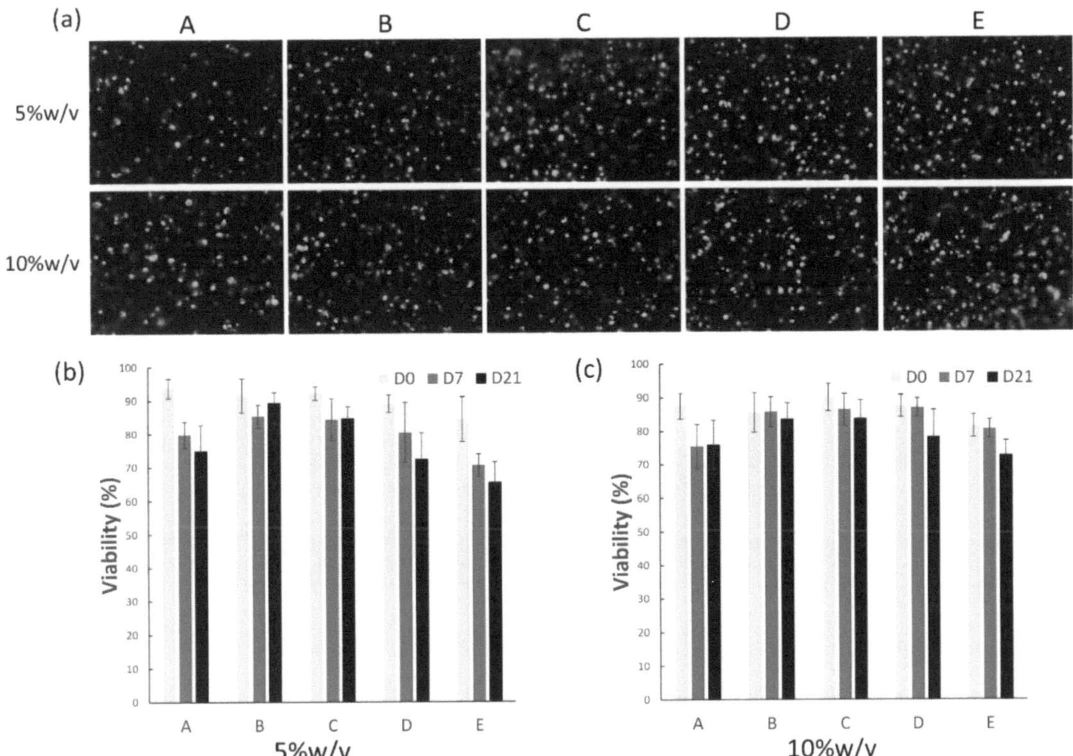

Figure 1. Representative figures show live–dead staining of 5 and 10% *w/v* hydrogels encapsulated with chondrocytes after culturing for 21 days in the chondrogenic medium (**a**). Conditions A to E represent different mix ratios of HA and Dex (100:0, 75:25, 50:50, 25:75, 0:100). Cell viability was quantified based on the live–dead staining figures of 5% *w/v* (**b**) and 10% *w/v* (**c**) hydrogels at day 0, 7 and 21.

3.3. Mechanical Properties

The rheological properties and equilibrium swelling of the hydrogels on day 0 are shown in Figure 2. As expected, increasing the polymer concentration of the hydrogels increased the storage modulus. Similarly, the equilibrium swelling ratio of the constructs with a lower polymer concentration was higher than for those with higher polymer concentrations. The inclusion of cells in our hydrogel constructs generally decreased the storage modulus and increased the swelling of the hydrogels, suggesting the decrease of crosslink density for these constructs. The average decrease in crosslink density was calculated to be 40% upon inclusion of cells. The calculations were based on the classical rubber elasticity theory and are provided in the Supplementary information (Table S1 and Equation (S1)).

The mechanical properties of the hydrogels were measured after 7 and 21 days of incubation in the chondrogenic differentiation medium. Although the hydrogels with cells were weaker, upon culturing they became stronger and more elastic. After 21 days, we saw an increase in the storage modulus measured by rheology, which indicated that the hydrogel could store more deformation energy in an elastic manner that it could on day 0 and day 7. It would indicate an increase in network density, which is related to the deposition of cartilage ECM proteins as confirmed by histology. This increase in the storage modulus was most prominent for 5% *w/v* hydrogels, indicating that the weaker, more open structure of the hydrogel was preferred for the deposition of the cartilaginous matrix (Figure 3a1–a4). This observation was confirmed by texture analysis showing that the maximum pressure needed to compress the hydrogels to 50% strain (Figure 3b1–b4) and the E-modulus under high strain (Figure 3c1–c4) were increased in the 5% w/v hydrogels. Next to that, the elasticity of the hydrogels was also increased after 21 days, which we derived by the hysteresis in the stress-strain curves recorded (Figure 3d1–d4).

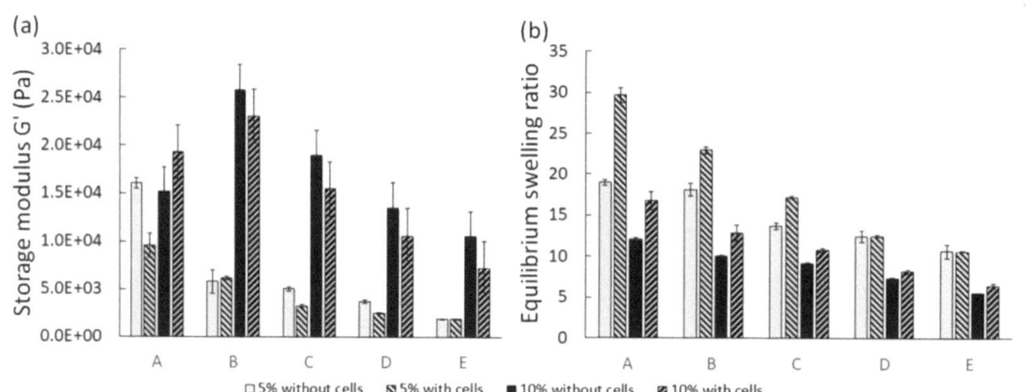

Figure 2. Rheological properties (**a**) and equilibrium swelling (**b**) of the hydrogels on day 0. A to E represent different HA-to-Dex ratios (100:0, 75:25, 50:50, 25:75, 0:100).

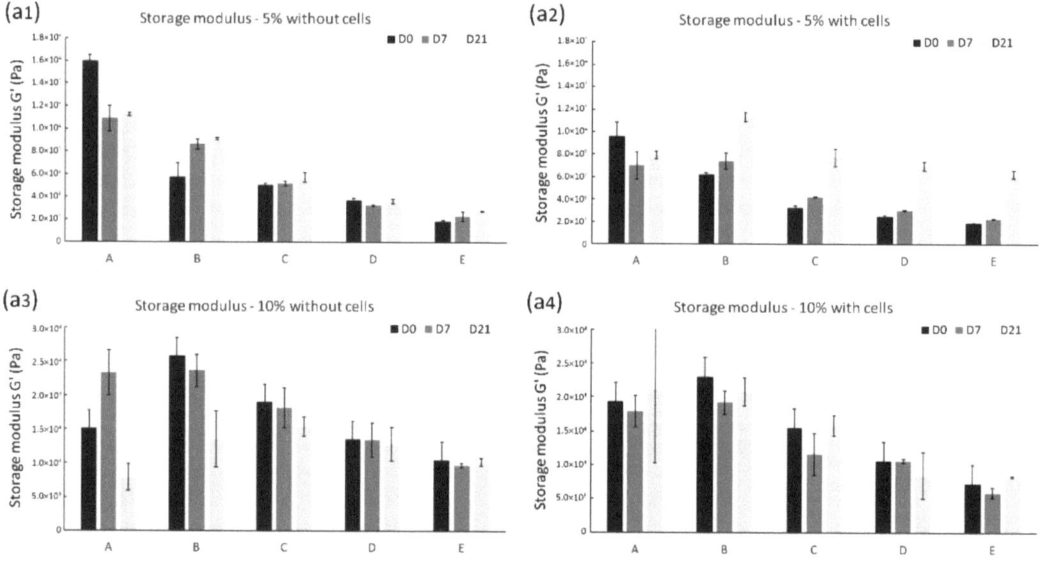

Figure 3. *Cont.*

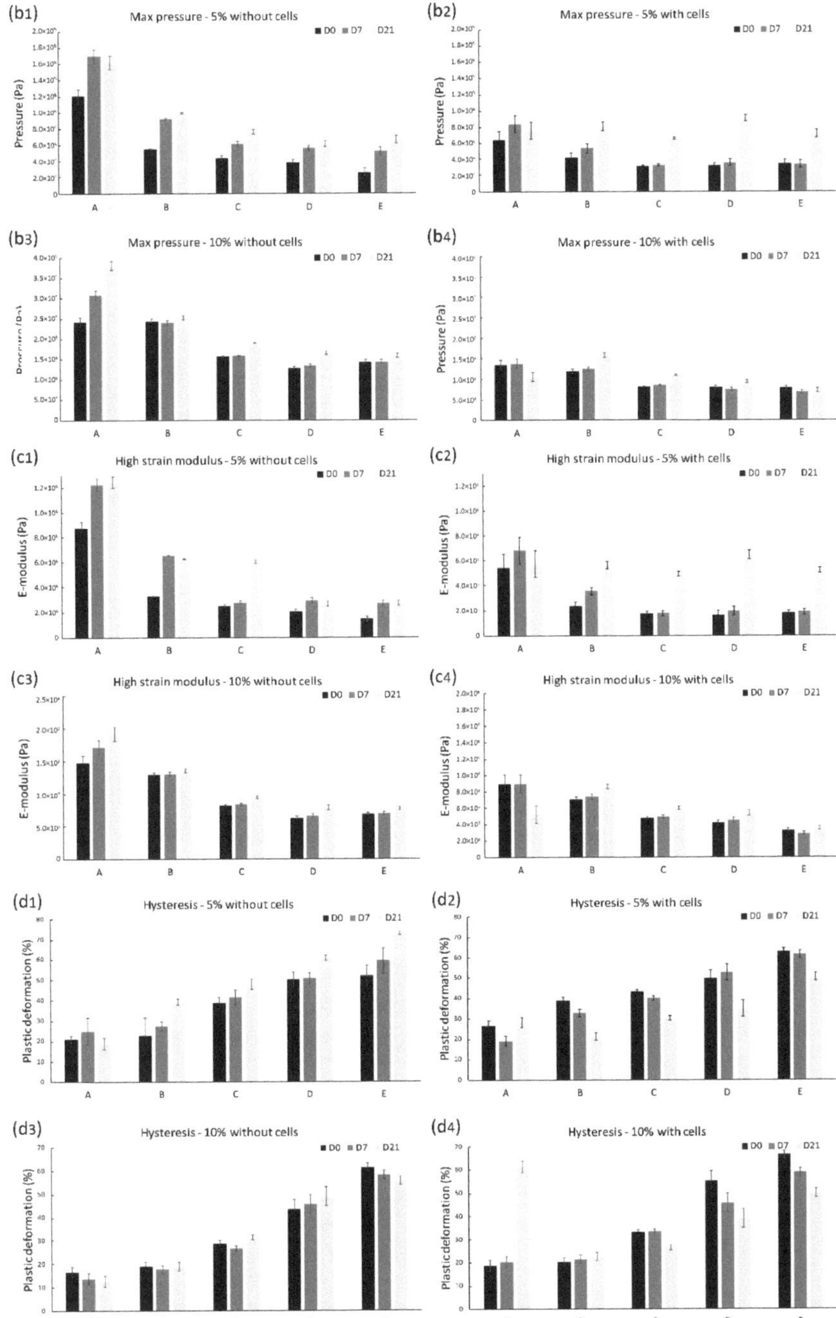

Figure 3. Mechanical properties of the hydrogels after 0, 7, and 21 days of culture. Conditions A to E represent different ratios of HA and Dex (100:0, 75:25, 50:50, 25:75, 0:100). (**a1–a4**) the storage modulus of the hydrogels, (**b1–b4**) the pressure needed for 50% deformation of the hydrogels, (**c1–c4**) the high strain Young's modulus of the hydrogels, (**d1–d4**) the elasticity of hydrogels.

3.4. Higher Dextran Concentration of Hydrogels Enhanced Matrix Deposition

GAG production by chondrocytes in all conditions was examined by histology using Safranin O staining (Supplementary Figure S4, but only Day 21 Safranin O staining results are shown). The staining results of HA hydrogel without cells showed that pure HA gel is stained by the dyes as expected (data not shown), which indicate that histology staining is not specific enough to distinguish the matrix production in these hydrogels. Nonetheless, histological staining confirmed that the chondrocytes incorporated in these hydrogels produced abundant ECM rich in GAGs after 21 days, as confirmed by the dense GAG staining in these gels. These results demonstrated that the incorporation of HA would improve the performance of Dex–TA gels in cartilage tissue engineering and that 5% *w/v* hydrogels showed better matrix production than did 10% *w/v* hydrogels.

Collagen type II is the primary type of collagen present in articular cartilage [20]. Consequently, we performed immunohistochemistry staining to detect the specific production of collagen type II (Figure 4 and Supplementary Figure S5). After 21 days, the cartilage matrix was deposited inside different hydrogel conditions. Moreover, it clearly showed that 5% *w/v* hybrid hydrogels encapsulated with chondrocytes exhibited greater deposition of type II collagen than the 10% *w/v* groups did. Besides, the condition that was composed of 25% HA and 75% dextran (condition D) displayed the most intense staining of all the 5% *w/v* hybrid hydrogels. The histochemical analysis also revealed that the cartilage matrix formation was more dominant at the periphery of the hydrogels, especially in hydrogels of the 10% *w/v* groups.

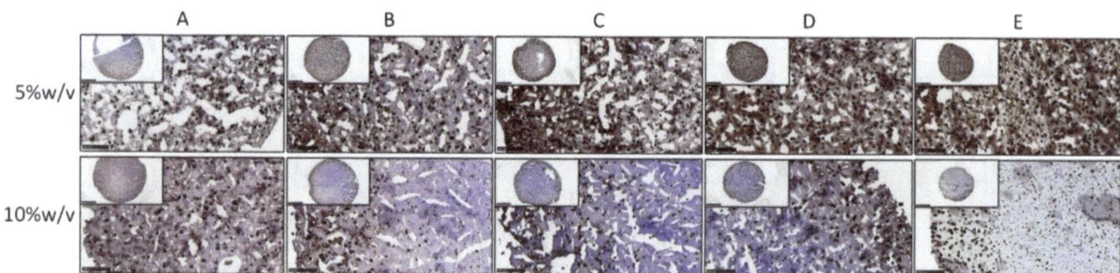

Figure 4. Immunohistochemistry staining of collagen type II for 5% *w/v* and 10% *w/v* hydrogels encapsulated with chondrocytes after culturing for 21 days in chondrogenic medium. From condition (**A–E**) represent the different conditions with different mix ratio of HA and Dex (100:0, 75:25, 50:50, 25:75, 0:100). Inserts indicate the overview of each hydrogel; scale bar = 2.5 mm. Pictures show the magnified view of each hydrogel; scale bar = 250 μm.

Next, we performed qPCR to study the mRNA expression of chondrogenic genes in 5% *w/v* hydrogels combined with bCHs that were cultured in the chondrogenic differentiation medium for 21 days. The histochemical results were corroborated by gene expression analysis (Figure 5). The relative fold expression of chondrogenic related genes such as ACAN, COL2A1, and COL9A1 was up-regulated in hydrogels containing higher dextran concentrations (Figure 5a–c). The expression of SOX9 was also measured in this study, which showed a similar trend as these genes (data not shown). However, the overall mRNA expression level of SOX9 remained low in all conditions. Additionally, a decrease in the expression of COL1A1 and OPN in conditions with a higher dextran content was observed compared to the pure HA group (Figure 5d,e).

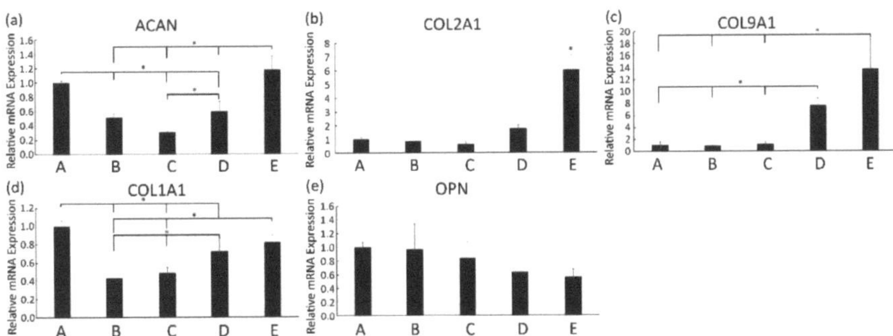

Figure 5. Relative mRNA expression levels for ACAN (**a**), COL2A1 (**b**), COL9A1 (**c**), COL1A1 (**d**), and OPN (**e**), expressed by bovine chondrocytes incorporated into 5% *w/v* hydrogels, after 21 days in culturing in the chondrogenic medium. Assignments A to E represent the different conditions with different HA-to-Dex ratio (100:0, 75:25, 50:50, 25:75, 0:100). Error bars reflect SD and * represents $p < 0.05$ compared to other indicated conditions.

4. Discussion

In this work, different ratios of HA and dextran-based hybrid hydrogels at both 10% *w/v* and 5% *w/v* were prepared with a mold to determine physical property changes and their effects on the cellular behavior and cartilage matrix formation. Our results indicated that the incorporation of chondrocytes in the hydrogels introduced soft pockets in the hydrogel matrix, which decreased their mechanical properties on a macroscopic level. Interestingly, the rheological and compression analysis indicated that 5% *w/v* hydrogels laden with cells showed a significant increase in mechanical properties after culturing for 21 days. Moreover, compared to 10% *w/v* hydrogels, the 5% *w/v* hybrid hydrogels showed enhanced deposition of cartilage matrix, especially in the constructs with higher Dex–TA concentrations.

The chondrocytes encapsulated inside hydrogels retained a round shape at 21 days in culture. However, compared to previously reported HA-g-Dex–TA hydrogels, hydrogels in this study showed decreased cell viability [12], which can be firstly explained by the difference in measured time points. Initially, over 90% of the incorporated cells were alive, demonstrating the cytocompatibility of the hybrid hydrogels. Decreased cell viability after 21 days could also be explained by the procedure of hydrogel formation. To make sure all components were incorporated homogeneously, the solution was mixed by vortexing after the addition of H_2O_2. The shear forces during vortexing could damage the incorporated cells. Further studies need to address the force effect on the cells and determine the proper vortexing speed. Moreover, the relatively low cell viability observed for the Dex–TA hydrogel may be attributed to the increasing crosslinking density of the constructs. The limited exchange of nutrients and waste products to the surrounding culture media can reduce cell viability [21].

With the help of the mold we designed, highly controlled cylindrical hydrogels were formed by dissolving the polymers and HRP in PBS and adding H_2O_2 as an oxidizing agent. Compared to the pure hydrogels without cells, the inclusion of cells increased hydrogel swelling, which could have indicated a decrease in crosslink density for these constructs. The increased size of hydrogels with higher HA concentration can be explained by an increase in water uptake resulting from the electrostatic repulsion of negatively charged HA chains at pH 7.4 [12]. This swelling behavior also suggested a decrease in network density as a result of degradation [22]. HA is an essential component of the ECM in cartilage tissue, which is biodegradable via enzymatic hydrolysis using hyaluronidase (HAse) [23,24]. HA degraded by the HAse expressed by the incorporated chondrocytes in the hybrid hydrogels could have been another reason for this behavior [25].

In the design of hydrogels as scaffolds for cartilage repair, adequate mechanical support is a critical requirement. The scaffold should be mechanically stable in order to protect the seeded cells and the developing tissue and to withstand the physiological load [26]. On a cellular scale, the mechanical properties of a scaffold are potent regulators of cell migration and their phenotypes [27]. The mechanical properties of the studied hydrogels were adjusted by varying the ratio of dextran and hyaluronic acid and the polymer concentration. The evaluation of these properties is an essential parameter in predicting the possibility of tissue production and construct quality. Rheological studies on the constructs were performed to determine storage and loss moduli, which are values for elasticity and viscosity, respectively [28]. Compression of cylindrical hydrogel samples between two plates yields a stress–strain curve, from which the elastic modulus and other mechanical properties can be derived [29]. Therefore, physical properties were determined by rheology and texture analysis at different time points to investigate how gel composition and mechanical properties could influence cell behavior and how the cells consequently would influence hydrogel characteristics.

In line with the previous report, increasing the polymer concentration in the hydrogels increased the storage modulus because hydrogels prepared at a concentration of 10% w/v showed a higher storage modulus than the 5% w/v hydrogels. Furthermore, by encapsulating the chondrocytes, the corresponding storage modulus G′ values decreased, suggesting a decrease of crosslink density of these constructs.

However, upon culturing in a differentiation medium for 21 days, these gels became stronger and more elastic. Compared to pure gels without cells, chondrocytes laden constructs showed enhanced storage moduli after 21 days. Especially in cell laden 5% w/v hydrogels, a significantly increased storage modulus was observed in rheological measurements. This improvement was most evident in constructs with a higher Dex–TA concentration. However, the 10% w/v hydrogels showed only moderate changes compared to the initial values. This observation was confirmed by rheological analysis. Previous reports indicated that the compressive modulus for articular cartilage is 0.24 to 0.85 MPa [30] which is substantially higher than the compressive modulus, which was obtained after three weeks of culture of the hydrogel constructs. The compressive modulus of the latter constructs approached the mechanical properties of the chondron, which has a modulus of around 70 kPa [31]. This suggested that in 21 days of being cultured the chondrocytes created an environment that resembled at least some of the properties of the native chondron. These results also suggested that the increased network density in the hydrogels was related to ECM deposition and that the 5% w/v hydrogels, which are weaker and have a more open structure, were preferred for the deposition of cartilaginous matrix. Furthermore, a higher Dex–TA concentration was shown to promote this.

This hypothesis was confirmed by the cartilage matrix-related staining. Compared to day 0 constructs, abundant deposition of cartilage matrix was observed inside different hydrogel conditions after 21 days, which can partly explain the increased stiffness and elasticity of the cell-laden hydrogels. Interestingly, it was clearly shown that 5% w/v hybrid hydrogels laden with chondrocytes exhibited a denser deposition of cartilage matrix compared to the 10% w/v constructs. These data were consistent with the significantly increased mechanical properties in 5% w/v hydrogels from rheology and compression analysis. It was likely caused by a greater diffusion of nutrients and growth factors in the 5% w/v hydrogels than in the 10% w/v hydrogels since 10% w/v hydrogels showed higher mechanical strength, which is known to compromise diffusion ability [32]. In addition, a higher polymer concentration also resulted in a decreased accumulation of matrix components such as proteoglycans and collagen type II [33]. Indeed, in 10% w/v hydrogels, the formation of cartilage matrix was predominantly observed on the periphery of the hydrogels.

Moreover, 5% w/v hybrid hydrogels with a higher Dex–TA content produced an abundant, homogeneously distributed cartilage matrix, which was corroborated by the up-regulated expression of chondrogenic related genes. Considering that HA is present in native cartilage and plays a role in influencing the cell phenotype and matrix production [34,35], incorporation of HA in a hybrid hydrogel would improve its performance in cartilage tissue engineering. In conclusion, these results indicated that 5% w/v hydrogels showed better matrix production than 10% w/v hydrogels and that a combination of 25% HA and 75% Dex is probably the optimal hybrid condition for cell growth and matrix formation.

5. Conclusions

We prepared hybrid hydrogels of different ratios of HA and dextran of controlled size and shape at both 10% w/v and 5% w/v using a designed mold. The behavior of chondrocytes incorporated in the hybrid hydrogels demonstrated that the gel systems had proper biocompatibility. Our data demonstrated that the presence of chondrocytes decreased the hydrogels' initial mechanical properties. Nonetheless, chondrocyte-laden constructs showed an enhanced storage modulus after 21 days. Rheological and compression analysis indicated that 5% w/v hydrogels laden with cells particularly showed a significant increase in mechanical properties. Moreover, compared to 10% w/v hydrogels, the 5% w/v hybrid hydrogels induced enhanced matrix deposition (increased glycosaminoglycan and collagen production). This observation was most evident in constructs with a higher Dex–TA concentration. Altogether, these data suggest that a 5% w/v hybrid hydrogel with 25% HA and 75% Dex is a promising construct for cartilage repair approaches.

Supplementary Materials: The following are available online at https://www.mdpi.com/article/10.3390/polym13091526/s1, synthetic methods, Figure S1: 1H NMR spectra of dextran and hyaluronic acid tyramine conjugates (Dex–TA and HA-TA), Figure S2: Sketches of the prepared hydrogel molds, Figure S3: Overview morphology of 5% w/v and 10% w/v hydrogels, Figure S4: Safranin O staining of 5% w/v and 10% w/v hydrogels, Figure S5: Immunohistochemistry staining of collagen type II for 5% w/v hydrogels. Table S1: Effective crosslink density, v_e (mM). Equation (S1): Crosslinking density.

Author Contributions: Conceptualization, S.B. and M.K.; methodology, Y.F. and B.Z.; formal analysis, Y.F.; investigation, Y.F. and B.Z.; data curation, Y.F. and B.Z; writing—original draft preparation, Y.F. and B.Z.; writing—review and editing, all authors; visualization, Y.F. and B.Z.; supervision, S.B., P.J.D. and M.K.; funding acquisition, M.K. All authors have read and agreed to the published version of the manuscript.

Funding: This project is supported by the European Commission H2020 Research and Innovation Staff Exchange (RISE) Project "Future Formulations (FutForm)", Hy2Care, and the China Scholarship Council. The authors would like to thank Netherlands Organisation for Scientific Research (NWO) P15-23 (Project 1) "Activating resident stem cells" for providing financial support to this project.

Institutional Review Board Statement: Not applicable.

Informed Consent Statement: Not applicable.

Data Availability Statement: The data presented in this study are available on request from the corresponding author.

Conflicts of Interest: The authors declare no conflict of interest.

References

1. Hayes, W.C.; Mockros, L.F. Viscoelastic properties of human articular cartilage. *J. Appl. Physiol.* **1971**, *31*, 562–568. [CrossRef] [PubMed]
2. Pearle, A.D.; Warren, R.F.; Rodeo, S.A. Basic science of articular cartilage and osteoarthritis. *Clin. Sports Med.* **2005**, *24*, 1–12. [CrossRef] [PubMed]
3. Barrère, F.; Mahmood, T.A.; de Groot, K.; van Blitterswijk, C.A. Advanced biomaterials for skeletal tissue regeneration: Instructive and smart functions. *Mater. Sci. Eng. R Rep.* **2008**, *59*, 38–71. [CrossRef]
4. Akkiraju, H.; Nohe, A. Role of Chondrocytes in Cartilage Formation, Progression of Osteoarthritis and Cartilage Regeneration. *J. Dev. Biol.* **2015**, *3*, 177–192. [CrossRef]
5. Newman, A.P. Articular cartilage repair. *Am. J. Sports Med.* **1998**, *26*, 309–324. [CrossRef]
6. Zhang, W.; Ouyang, H.; Dass, C.R.; Xu, J. Current research on pharmacologic and regenerative therapies for osteoarthritis. *Bone Res.* **2016**, *4*, 15040. [CrossRef]
7. Drury, J.L.; Mooney, D.J. Hydrogels for tissue engineering: Scaffold design variables and applications. *Biomaterials* **2003**, *24*, 4337–4351. [CrossRef]

8. Qiu, Y.S.; Shahgaldi, B.F.; Revell, W.J.; Heatley, F.W. Observations of subchondral plate advancement during osteochondral repair: A histomorphometric and mechanical study in the rabbit femoral condyle. *Osteoarthr. Cartil.* **2003**, *11*, 810–820. [CrossRef]

9. Van Tomme, S.R.; Storm, G.; Hennink, W.E. In situ gelling hydrogels for pharmaceutical and biomedical applications. *Int. J. Pharm.* **2008**, *355*, 1–18. [CrossRef]

10. Jeong, B.; Kim, S.W.; Bae, Y.H. Thermosensitive sol-gel reversible hydrogels. *Adv. Drug Deliv. Rev.* **2002**, *54*, 37–51. [CrossRef]

11. Yu, L.; Ding, J. Injectable hydrogels as unique biomedical materials. *Chem. Soc. Rev.* **2008**, *37*, 1473–1481. [CrossRef]

12. Jin, R.; Teixeira, L.M.; Dijkstra, P.J.; Van Blitterswijk, C.A.; Karperien, M.; Feijen, J. Enzymatically-crosslinked injectable hydrogels based on biomimetic dextran-hyaluronic acid conjugates for cartilage tissue engineering. *Biomaterials* **2010**, *31*, 3103–3113. [CrossRef]

13. Jin, R.; Moreira Teixeira, L.S.; Dijkstra, P.J.; Zhong, Z.; van Blitterswijk, C.A.; Karperien, M.; Feijen, J. Enzymatically crosslinked dextran-tyramine hydrogels as injectable scaffolds for cartilage tissue engineering. *Tissue Eng. Part A* **2010**, *16*, 2429–2440. [CrossRef]

14. Jin, R.; Hiemstra, C.; Zhong, Z.; Feijen, J. Enzyme-mediated fast in situ formation of hydrogels from dextran-tyramine conjugates. *Biomaterials* **2007**, *28*, 2791–2800. [CrossRef]

15. Wang, R.; Leber, N.; Buhl, C.; Verdonschot, N.; Dijkstra, P.J.; Karperien, M. Cartilage adhesive and mechanical properties of enzymatically crosslinked polysaccharide tyramine conjugate hydrogels. *Polym. Adv. Technol.* **2014**, *25*, 568–574. [CrossRef]

16. Ramirez, J.C.; Sánchez-Chaves, M.; Arranz, F. Dextran functionalized by 4-nitrophenyl carbonate groups. Aminolysis reactions. *Die Angew. Makromol. Chemie* **1995**, *225*, 123–130. [CrossRef]

17. Rydergren, S. Chemical Modifications of Hyaluronan using DMTMM-Activated Amidation. In *Synthetical Organic Chemistry*; Uppsala University: Uppsala, Sweden, 2013; Available online: https://www.diva-portal.org/smash/get/diva2:640661/FULLTEXT02.pdf (accessed on 1 April 2021).

18. D'Este, M.; Eglin, D.; Alini, M. A systematic analysis of DMTMM vs EDC/NHS for ligation of amines to hyaluronan in water. *Carbohydr. Polym.* **2014**, *108*, 239–246. [CrossRef]

19. Hendriks, J.; Riesle, J.; van Blitterswijk, C.A. Effect of stratified culture compared to confluent culture in monolayer on proliferation and differentiation of human articular chondrocytes. *Tissue Eng.* **2006**, *12*, 2397–2405. [CrossRef]

20. Aigner, T.; Stöve, J. Collagens—Major component of the physiological cartilage matrix, major target of cartilage degeneration, major tool in cartilage repair. *Adv. Drug Deliv. Rev.* **2003**, *55*, 1569–1593. [CrossRef]

21. Burdick, J.A.; Chung, C.; Jia, X.; Randolph, M.A.; Langer, R. Controlled degradation and mechanical behavior of photopolymerized hyaluronic acid networks. *Biomacromolecules* **2005**, *6*, 386–391. [CrossRef]

22. Lee, F.; Chung, J.E.; Kurisawa, M. An injectable enzymatically crosslinked hyaluronic acid–tyramine hydrogel system with independent tuning of mechanical strength and gelation rate. *Soft Matter* **2008**, *4*, 880–887. [CrossRef]

23. Akmal, M.; Singh, A.; Anand, A.; Kesani, A.; Aslam, N.; Goodship, A.; Bentley, G. The effects of hyaluronic acid on articular chondrocytes. *J. Bone Joint Surg Br.* **2005**, *87*, 1143–1149. [CrossRef]

24. Menzel, E.J.; Farr, C. Hyaluronidase and its substrate hyaluronan: Biochemistry, biological activities and therapeutic uses. *Cancer Lett.* **1998**, *131*, 3–11. [CrossRef]

25. Tanimoto, K.; Suzuki, A.; Ohno, S.; Honda, K.; Tanaka, N.; Doi, T.; Nakahara-Ohno, M.; Yoneno, K.; Nakatani, Y.; Ueki, M.; et al. Hyaluronidase expression in cultured growth plate chondrocytes during differentiation. *Cell Tissue Res.* **2004**, *318*, 335–342. [CrossRef]

26. Lu, L.; Zhu, X.; Valenzuela, R.G.; Currier, B.L.; Yaszemski, M.J. Biodegradable polymer scaffolds for cartilage tissue engineering. *Clin. Orthop. Relat. Res.* **2001**, *391*, S251–S270. [CrossRef]

27. Breuls, R.G.; Jiya, T.U.; Smit, T.H. Scaffold stiffness influences cell behavior: Opportunities for skeletal tissue engineering. *Open Orthop. J.* **2008**, *2*, 103–109. [CrossRef]

28. Malda, J.; Visser, J.; Melchels, F.P.; Jüngst, T.; Hennink, W.E.; Dhert, W.J.; Groll, J.; Hutmacher, D.W. 25th anniversary article: Engineering hydrogels for biofabrication. *Adv. Mater.* **2013**, *25*, 5011–5028. [CrossRef]

29. Wong, M.; Carter, D.R. Articular cartilage functional histomorphology and mechanobiology: A research perspective. *Bone* **2003**, *33*, 1–13. [CrossRef]

30. Little, C.J.; Bawolin, N.K.; Chen, X. Mechanical properties of natural cartilage and tissue-engineered constructs. *Tissue Eng. Part B Rev.* **2011**, *17*, 213–227. [CrossRef]

31. Alexopoulos, L.G.; Haider, M.A.; Vail, T.P.; Guilak, F. Alterations in the mechanical properties of the human chondrocyte pericellular matrix with osteoarthritis. *J. Biomech. Eng.* **2003**, *125*, 323–333. [CrossRef]

32. Nicodemus, G.D.; Bryant, S.J. Cell encapsulation in biodegradable hydrogels for tissue engineering applications. *Tissue Eng. Part B Rev.* **2008**, *14*, 149–165. [CrossRef] [PubMed]

33. Sontjens, S.H.; Nettles, D.L.; Carnahan, M.A.; Setton, L.A.; Grinstaff, M.W. Biodendrimer-based hydrogel scaffolds for cartilage tissue repair. *Biomacromolecules* **2006**, *7*, 310–316. [CrossRef] [PubMed]

34. Chung, C.; Burdick, J.A. Influence of three-dimensional hyaluronic acid microenvironments on mesenchymal stem cell chondrogenesis. *Tissue Eng. Part A* **2009**, *15*, 243–254. [CrossRef]

35. Liao, E.; Yaszemski, M.; Krebsbach, P.; Hollister, S. Tissue-engineered cartilage constructs using composite hyaluronic acid/collagen I hydrogels and designed poly(propylene fumarate) scaffolds. *Tissue Eng.* **2007**, *13*, 537–550. [CrossRef] [PubMed]

Article

Cellulose-Based Scaffolds: A Comparative Study for Potential Application in Articular Cartilage

Rachel Cordeiro [1,2], Rui D. Alvites [2,3,4], Ana C. Sousa [2,3,4], Bruna Lopes [2,3,4], Patrícia Sousa [2,3,4], Ana C. Maurício [2,3,4], Nuno Alves [1,5,*] and Carla Moura [1,5,6,*]

1   Centre for Rapid and Sustainable Product Development, Polytechnic of Leiria, 2430-028 Marinha Grande, Portugal
2   Veterinary Clinics Department, Abel Salazar Biomedical Sciences Institute (ICBAS), University of Porto (UP), Rua de Jorge Viterbo Ferreira, No. 228, 4050-313 Porto, Portugal
3   Animal Science Studies Centre (CECA), Agroenvironment, Technologies and Sciences Institute (ICETA), University of Porto, Rua D. Manuel II, Apartado 55142, 4051-401 Porto, Portugal
4   Associate Laboratory for Animal and Veterinary Science (AL4AnimalS), 1300-477 Lisbon, Portugal
5   Associate Laboratory for Advanced Production and Intelligent Systems (ARISE), 4050-313 Porto, Portugal
6   Applied Research Institute (i2A), Polytechnic Institute of Coimbra, Rua da Misericórdia, Lagar dos Cortiços–S. Martinho do Bispo, 3045-093 Coimbra, Portugal
*   Correspondence: nuno.alves@ipleiria.pt (N.A.); carla.moura@ipleiria.pt (C.M.); Tel.: +351-244569441 (C.M.)

Abstract: Osteoarthritis is a highly prevalent disease worldwide that leads to cartilage loss. Tissue engineering, involving scaffolds, cells, and stimuli, has shown to be a promising strategy for its repair. Thus, this study aims to manufacture and characterise different scaffolds with poly(ε-caprolactone) (PCL) with commercial cellulose (microcrystalline (McC) and methyl cellulose (MC) or cellulose from agro-industrial residues (corncob (CcC)) and at different percentages, 1%, 2%, and 3%. PCL scaffolds were used as a control. Morphologically, the produced scaffolds presented porosities within the desired for cell incorporation (57% to 65%). When submitted to mechanical tests, the incorporation of cellulose affects the compression resistance of the majority of scaffolds. Regarding tensile strength, McC2% showed the highest values. It was proven that all manufactured scaffolds suffered degradation after 7 days of testing because of enzymatic reactions. This degradation may be due to the dissolution of PCL in the organic solvent. Biological tests revealed that PCL, CcC1%, and McC3% are the best materials to combine with human dental pulp stem/stromal cells. Overall, results suggest that cellulose incorporation in PCL scaffolds promotes cellular adhesion/proliferation. Methyl cellulose scaffolds demonstrated some advantageous compressive properties (closer to native cartilaginous tissue) to proceed to further studies for application in cartilage repair.

Keywords: cartilage repair; PCL; cellulose; scaffold; tissue engineering

Citation: Cordeiro, R.; Alvites, R.D.; Sousa, A.C.; Lopes, B.; Sousa, P.; Maurício, A.C.; Alves, N.; Moura, C. Cellulose-Based Scaffolds: A Comparative Study for Potential Application in Articular Cartilage. *Polymers* **2023**, *15*, 781. https://doi.org/10.3390/polym15030781

Academic Editor: Chie Kojima

Received: 4 November 2022
Revised: 1 February 2023
Accepted: 1 February 2023
Published: 3 February 2023

1. Introduction

Osteoarthritis (OA) is a very complex process characterised by inflammation, synovial distension, articular cartilage (AC) loss, and bone hyperplasia [1]. A 2019 study indicates that OA is the 17th most prevalent disease worldwide, considering 369 possible diseases [2]. The most recent statistics indicate that about 250 million people are affected by this disease [3]. Furthermore, the number of people older than 60 with OA is expected to be about 2 billion in 2050 [4]. Thus, the field of tissue engineering (TE) has generated interest in the development of a long-term and effective treatment for AC repair [5].

TE uses mainly scaffolds, cells, and biochemical and/or biomechanical stimuli [6]. Scaffolds are temporary implants fundamental in the development of new tissue since they support the growth of tissues and cells [7]. To provide this support, the material that composes it is a factor to take into account [8]. Cells are also an important factor since they are fundamental for tissue maintenance. On the other hand, biochemical stimulation, i.e., growth factors, are used to support cell differentiation. Biomechanical stimulation in TE is used to improve the properties of the new tissue to be formed since this stimulation leads to increased extracellular matrix production [9].

Poly(ε-caprolactone) (PCL) is a semicrystalline polyester with melting temperatures around 55–60 °C, which makes it the perfect polymer for 3D extrusion [10,11]. It is the most used polymer in the TE field [12]. It is approved by the Food and Drug Administration [13], exhibits good mechanical properties, is biodegradable, has a slow degradation rate, and improves the cell differentiation and proliferation of chondrocytes [10,11,14]. However, their hydrophobicity makes cell attachment difficult [15–17]. Some studies blend this with natural materials to improve the biological activity in PCL scaffolds [18].

Cellulose is a natural polymer, the biopolymer most abundant in nature. The principal sources of cellulose were plants, bacteria, algae, and marine animals. In addition, it is renewable, biocompatible, and biodegradable. The use of the term biodegradable for cellulose, when applied to the human body, is from a perspective of comparison with other polymers since the human body does not have cellulases to degrade it [19]. The structure of cellulose is a repetition of glucose molecules joined by a $\beta(1{\rightarrow}4)$ bond with a high possibility of obtaining derivates since the OH groups can be replaced by other chemical groups [20,21].

The use of cellulose in scaffold manufacturing for application in different tissues (skin, bone, vessels, ocular tissue, and cartilage) has been growing [22]. This growth is due to the several advantages cellulose offers, such as the diversity of sources, different properties depending on the source, good cell adhesion, its crystalline and amorphous structure, and the consequence of the several arrangements in the chains, and its mechanical properties [23]. There are various polymorphs of cellulose, such as cellulose I, II, III, and IV. However, only cellulose II is completely biodegradable and appropriate for tissue repair, as it has no crystalline phase [24]. The interest in this polymer for cartilage TE is based on its chemical similarity to AC collagen fibres [25].

There are studies in the literature referring to the combination of PCL with cellulose with different scaffold manufacturing techniques [4,14,26]. Alemán-Domínguez (2018) developed scaffolds for bone TE with PCL and microcrystalline cellulose (McC) [4]. Four cellulose concentrations were evaluated, 0, 2, 5, and 10% (w/w), to verify the influence of its addition to PCL. The results showed that the scaffolds produced with 2% McC demonstrated the most suitable properties for the bone application. In the previous study, the same authors studied the influence of cellulose derivative, carboxymethyl cellulose, on PCL scaffolds [14]. Three concentrations were tested, 0, 2, and 5% (w/w), and, once again, 2% of carboxymethyl cellulose on PCL scaffolds proved to be the concentration with the highest potential for the same application. This cellulose was previously studied as its biomineralisation capacity had been reported [27]. Methyl cellulose (MC) is also a cellulose derivative. Its non-toxic, biocompatible, and gelling properties lead us to believe that it is a biomaterial with great potential to be included in scaffolds for cartilage repair. Roushangar (2018) included MC in scaffolds for cartilage repair and demonstrated that these had a high efficiency of living chondrocytes. In particular, MC increased the viscosity leading to the desired pore size [28]. Moreover, MC is considered a bioink since it can be used as a gel [29].

Kotcharat (2021) also studied the influence of bacterial cellulose in PCL implants for wound dressing [26]. Results proved that bacterial cellulose improves the swelling properties of the scaffold.

A study by our group comparing scaffolds with McC and corncob cellulose (CcC) indicated that CcC could be a sustainable alternative to wood cellulose. Furthermore, it was proven that cellulose promotes cell proliferation [30]. However, it has not been established which concentration is ideal for replacing commercial wood cellulose. In this study, PCL and McC, CcC and MC scaffolds were manufactured by fused deposition modelling technique. The scaffolds were manufactured with three different concentrations of each cellulose (1%, 2%, and 3%). They were characterised and compared with pure PCL in terms of morphology, chemical and mechanical properties, enzymatic degradation, and cell viability.

2. Materials and Methods

2.1. Materials

Poly(ε-caprolactone) (PCL) (Perstorp, Warrington, UK), with 6500 g/mol molecular weight, was used as a base-polymer. The different commercial celluloses used were microcrystalline cellulose (Avicel PH102, FMC Biopolymer, Braine l'Alleud, Belgium) and methyl cellulose (viscosity 15cP, Acros Organic, Tokyo, Japan). The corncob cellulose was kindly provided by the College of Agriculture of the Polytechnic Institute of Coimbra (ESAC-IPC) (Coimbra, Portugal). For the preparation of the polymeric mixture and for the enzymatic degradation analysis of scaffolds, dimethylformamide (DMF) (Chem-Lab, Zedelgem, Belgium), acetone (HoneyWell, Offenbach, Germany), lysozyme (Sigma-Aldrich, Oakville, ON, Canada), and phosphate-buffered saline (PBS) (Sigma Aldrich, Saint Louis, MO, USA) were used.

2.2. Preparation of Polymeric Mixtures

To manufacture the scaffolds, it is necessary to develop a composite, binding the PCL and cellulose. This mixture was prepared by solvent casting technique. Firstly, 17% (w/w) of cellulose solution in acetone:DMF (2:1) (w/w) was prepared at 50 °C under constant stirring (150 rpm) for 30 min. Then, PCL and DMF (1:3) (w/v) were added to the cellulose mixture. The final solution was placed under constant stirring (150 rpm) at 50 °C for 48 h. The mixtures were produced in 3 different concentrations, 1%, 2%, and 3% of cellulose:PCL (w/w). PCL: (acetone:DMF) (1:3 (w/v)) was used as a control sample.

Solutions were deposited in Petri dishes to dry at room temperature (RT) for 7 days to obtain the mixtures in the solid phase (membranes). In total, 10 mixtures were made, classified depending on the type and the proportion of cellulose used (Table 1).

Table 1. Nomenclature used for the different PCL and cellulose (microcrystalline, methyl-, and corncob) samples, with respective percentage and type of cellulose.

Base-Polymer	Type/Origin of Cellulose	Percentage of Cellulose	Nomenclature
	-	-	PCL
	Microcrystalline	1%	McC1%
		2%	McC2%
		3%	McC3%
PCL	Methyl	1%	MC1%
		2%	MC2%
		3%	MC3%
	Corncob	1%	CcC1%
		2%	CcC2%
		3%	CcC3%

2.3. Infrared Characterisation

The ATR–FTIR spectra were recorded to confirm that all DMF had been evaporated from the dried samples. A Bruker Alpha-P FTIR spectrometer (Billerica, MA, USA) was used in the absorbance mode, with ATR platinum–diamond coupling. The samples were analysed at RT, 2 cm^{-1} spectral resolution, at 64 scans per sample, and in the range 4000–400 cm^{-1}. The analysis was performed in triplicate, and the sample portion for analysis was randomly selected.

2.4. Scaffolds Manufacturing

Scaffolds were obtained by extrusion using the Fused Deposition Modelling (FDM) technique in Bioextruder (Portuguese Patent N. 104247, 2010), equipment developed by CDRSP, with a 0.4 mm nozzle needle. The membranes previously prepared were cut into small pieces and placed inside the extruder tank. The 3D structures were designed as cylinders, so their cross-section had a diameter of 10 mm, a fibre diameter of 300 μm, composed of 10 layers, and a $0°/90°$ lay-down pattern. The equipment and the extrusion parameters varied from sample to sample (Table 2) to obtain the desired scaffold features.

Table 2. Parameters of the equipment used in the extrusion of each PCL–cellulose mixture.

Sample	Crosshead Speed (mm/s)	Deposition Spindle Speed (rpm)	Nozzle Temperature ($°$C)
PCL	8	15.33	76
McC1%	8	12.33	84
McC2%	8	12.33	84
McC3%	7	13	85
MC1%	7	13.67	86
MC2%	7	14	85
MC3%	7	14.17	86
CcC1%	7	12.83	86
CcC2%	7	13.83	86
CcC3%	7	13.33	85

2.5. Morphology Analysis

The morphology of the scaffolds was evaluated by microscopy (Leica DM750M Microsystems, Northbrook, IL, USA) at a magnification of $5\times$, and by micro-computed tomography (MicroCT), using a SkyScan 1174TM (Brucker, Kontich, Belgium). The filament and pore dimensions of each scaffold were analysed using microscopy. For this, a scaffold was randomly selected and 5 images were captured at different points. The measurements were performed in 5 different pores and filaments in each image through the ImageJ v.1.53t software. For the MicroCT, the parameters used for the digitalisation of the samples were: a step of 0.6 degrees around the medio-lateral axis, an acceleration voltage of 50 kV; a beam current of 800 μA; an exposure time of 4000 ms; a pixel size of the image of 14 μm; and no filters were used. The 3D reconstruction was performed in NRecon v.1.7.3.1 software. The morphological parameters, porosity, and pore interconnectivity ($N = 3$) were calculated with CT-Analyser v.1.17.7.2 software.

2.6. Mechanical Tests

Resistance at compression and tensile was carried out to evaluate the effect of different celluloses and percentages in scaffolds and filaments produced by extrusion. According to ASTM standards, a compression test was performed using universal testing equipment (Instron 5544, Canton, MA, USA). The extension rate was 1 mm/min. During the test, samples were compressed to at least 50% of their height to assess the material's behaviour upon compressive forces to guarantee that the yield strength is not achieved, i.e., applied forces are within the elastic region. All mechanical testing were realised using dried scaffolds. The compression modulus was calculated by the slope of the linear region in the stress-strain curve (elastic region).

Regarding the tensile test, it was performed using a texturometer (TA.xTplusC, Stable Micro Systems, Godalming, UK). Tensile testing was carried out using a 50 Kg ($\approx$490 N) load cell at a speed of 10 mm/s with an initial spacing of 20 mm between strings. Filaments tested were cut with 30 mm of height and presented a slight variation in diameter (Table 3). The test was performed in triplicate, and all filaments were subjected to force until broken.

Table 3. Filament diameter (mean ± SD) manufactured for mechanical tensile test.

Sample	Filament Diameter (µm)
PCL	0.499 ± 0.016
McC1%	0.533 ± 0.005
McC2%	0.569 ± 0.019
McC3%	0.529 ± 0.007
MC1%	0.502 ± 0.008
MC2%	0.520 ± 0.010
MC3%	0.549 ± 0.041
CcC1%	0.506 ± 0.006
CcC2%	0.549 ± 0.013
CcC3%	0.559 ± 0.012

2.7. Enzymatic Degradation Assay

An enzymatic analysis using 1.5 mg/mL lysozyme in PBS (pH = 7.4) was performed to evaluate the degradation of scaffolds produced. The medium with lysozyme was stirred for 10 min at 150 rpm, added to each well, and incubated at 37 °C in an orbital shaker at 100 rpm. The enzymatic medium was refreshed every 3 days. On the 0th, 1st, 7th, 28th, and 42nd days, scaffolds were removed from the wells and dried for 3 days at RT. All scaffolds were weight before and after degradation. As a control, scaffolds were incubated under the same conditions without enzymes. Tests were performed in triplicate.

2.8. Cellular Population and In Vitro Culture

Human Dental Pulp stem/stromal cells (hDPSCs) previously characterised by J. M. Campos et al. (2019) [31], were acquired from AllCells, LLC (Cat. DP0037F, Lot N°DPSC090411-01) and cultured in Mem α, GlutaMAX™ Supplement, no nucleosides (Gibco, 32561029), supplemented with 10 mM HEPES buffer solution (Gibco, 15630122), 100 IU/mL penicillin, 0.1 mg/mL streptomycin (Gibco, 15140122), 2.05 µm/mL amphotericin B (Gibco, 15290026), and 10% (v/v) fetal bovine serum (FBS) (Gibco, A3160802). Cells were maintained under standard conditions, namely at 37 °C, with 80% humidity and 5% CO_2, as described in Branquinho et al. (2021) [32].

2.9. Cytocompatibility Evaluation

The PrestoBlue™ viability assay, as previously described by Alvites et al. (2021) [33], was used to determine the cytocompatibility between the hDPSCs and the different scaffolds. Before the assay, the scaffolds were sterilised using exposure to ultraviolet radiation in 20-min cycles on each surface of the biomaterials. Then, the scaffolds were placed in 24-well plates, and hDPSCs at P5 were seeded in a low-volume suspension on top of the devices at a density of 7000 cells per well/scaffold. The plates were kept under standard conditions for 1 h to maximise cell adhesion to the scaffolds, and then wells were filled with a complete culture medium. The same procedure was performed for the control group, in which cells were seeded directly onto the plate wells without contacting any biomaterial. Blank wells, i.e., without seeded cells, were considered for both groups receiving scaffolds and the control group. The determination of cytocompatibility between cells and scaffolds was determined at 4 different timepoints after incubation: 24 h, 72 h, 120 h, and 168 h. For each determination, the culture medium was removed from the wells and replaced with fresh culture medium plus 10% v/v 10× PrestoBlue™ reagent (Invitrogen, A13262, Eugene, OR, USA). After 1 h incubation under standard conditions, 100 µL of the supernatant was collected from each well and transferred to the 96-well plate. Absorbances were read in triplicates at 570 and 595 nm with a Multiskan™ FC Microplate Photometer (Thermo ScientificTM, 51119000, Woodlands, Singapore). After the reading, each well of the 24-well plate was washed abundantly with Dulbecco's phosphate-buffered saline solution (DPBS, Gibco, 14190169) to remove residual PrestoBlue™ reagent and received a fresh medium, returning to incubation conditions until the next timepoint.

The absorbance values obtained at 595 nm (emission wavelength) were subtracted from those obtained at 570 nm (excitation wavelength) for each well (pre-calculated absorbances). The mean values obtained in the blank wells were further subtracted from the respective pre-calculated absorbances of each experimental well. Two blank wells and four experimental wells were considered for each experimental group. The percentage of inhibition of cell viability induced by each scaffold was subsequently determined.

2.10. Statistical Analysis

Statistical analysis was carried out in GraphPad Prism v.8 software (GraphPad Software, Inc., San Diego, CA, USA), using one-way ANOVA, with Sidak's test for the mechanical compression test, and Tukey's test, for the mechanical tensile test. A two-way ANOVA with Tukey's test was performed for the enzymatic analysis. For cell assays, results were presented as mean ± standard error of the mean (SEM). A two-way ANOVA analysis with Tukey's test was performed. The level of statistical significance was set as 99.9% (* $p < 0.05$, ** $p < 0.01$, *** $p < 0.001$).

3. Results

The PCL–cellulose composite developed in this study was combined through an organic solvent, DMF, that is expected to evaporate totality during the drying process. Therefore, an FTIR analysis was performed on all membranes to confirm that all DMF has evaporated and compared with pure DMF FTIR spectra (Figure 1), between 800 cm^{-1} and 1850 cm^{-1}. The most prominent band concerns the C=O stretching of DMF (1658 cm^{-1}), which does not appear in any membrane analysed, confirming the total evaporation of DMF.

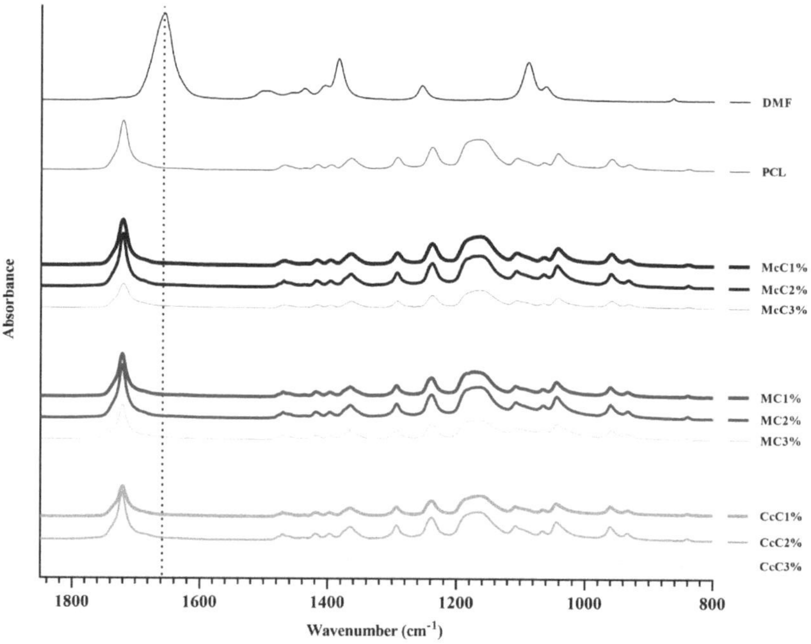

Figure 1. Fourier-transform infrared spectroscopy analysis of all the developed scaffolds: poly(ε-caprolactone) (PCL), microcrystalline cellulose at 1, 2, and 3% (McC1%, McC2%, and McC3%, respectively), methylcellulose at 1, 2, and 3% (MC1%, MC2%, and MC3%, respectively), and corncob cellulose at 1, 2, and 3% (CcC1%, CcC2%, and CcC3%, respectively) and pure DMF.

PCL–cellulose scaffolds were made with different cellulose and in different ratios. With this requirement, the fabrication conditions suffered some variations. However, they were optimised so that all samples had the same shape and similar dimensions. Morphological analyses were performed using microscopy and MicroCT (Figure 2) to obtain the average dimensions of each scaffold. Through optical microscopy, it is possible to obtain the filament and pore size, while through MicroCT, it is possible to obtain the porosity and interconnectivity of each scaffold.

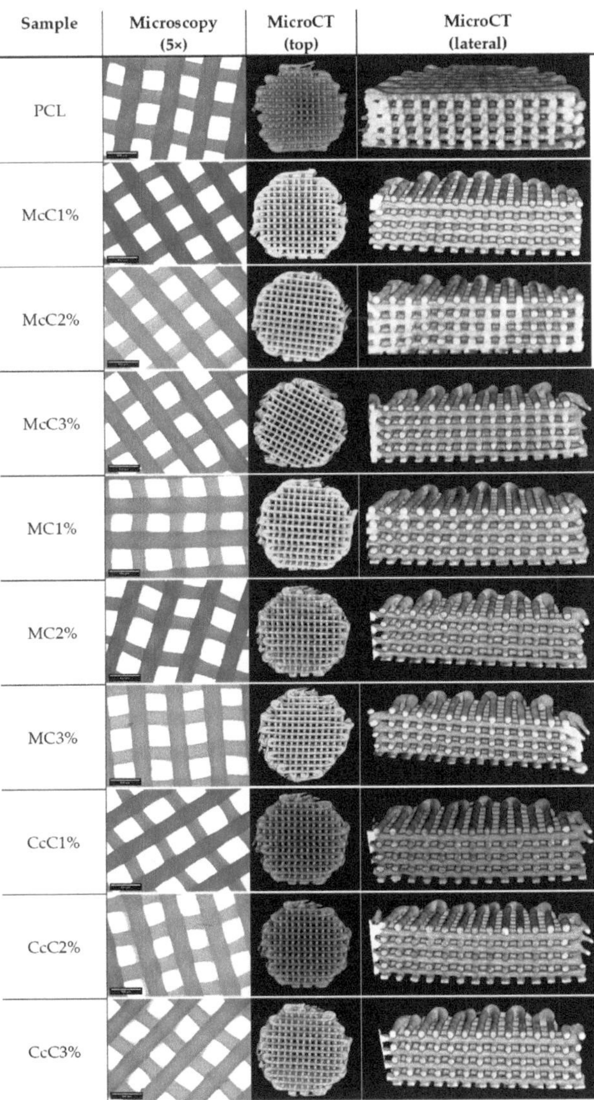

Figure 2. Morphological analysis of the scaffolds produced from each sample: poly(ε-caprolactone) (PCL), microcrystalline cellulose at 1, 2, and 3% (McC1%, McC2%, and McC3%, respectively), methyl cellulose at 1, 2, and 3% (MC1%, MC2%, and MC3%, respectively), and corncob cellulose at 1, 2, and 3% (CcC1%, CcC2%, and CcC3%, respectively), by microscopy, with a 5× resolution, and micro-tomography computed, observable at the top of the scaffold and in its medio-lateral region.

All scaffolds were designed to have approximately a 300 μm filament and 350 μm of pore size. The CcC1% scaffold presented the thinnest filament, with 270.14 ± 9.3005 μm, whereas the McC2% scaffold presented filaments with the largest dimension (326.45 ± 11.674 μm) (Table 4). On the other hand, the smallest pore corresponds to the CcC2% scaffold and the largest to the MC2% (329.72 ± 12.525 μm and 379.57 ± 24.389 μm, respectively). The porosity of the cartilaginous matrix varies from zone to zone between 92% and 70% [34]. Porosities of the scaffolds produced are below these values, between 59.40 ± 2.616% and 65.56 ± 1.663%, with the highest value corresponding to scaffold with MC2%. The interconnectivity between pores should be as close as possible to 100% so that cells can move freely in the scaffold. Except for the McC3% scaffold, all have interconnectivity higher than 99.99%, being 100% reached by the McC1% scaffold.

Table 4. Filament and pore dimensions, porosity, and interconnectivity of the scaffolds (mean ± SD) assessed through micro-computed tomography (MicroCT).

Sample	Filament (μm)	Pore (μm)	Porosity (%)	Interconnectivity (%)
PCL	324.31 ± 11.540	344.25 ± 16.391	57.40 ± 2.616	99.993 ± 0.00404
McC1%	309.40 ± 10.960	350.47 ± 21.683	59.10 ± 0.4051	100.000 ± 0.00058
McC2%	326.45 ± 11.674	344.27 ± 11.077	58.47 ± 0.7003	99.995 ± 0.00529
McC3%	303.92 ± 13.713	357.36 ± 16.376	62.36 ± 1.448 [a]	99.097 ± 1.5630
MC1%	305.89 ± 14.711	356.10 ± 21.137	57.76 ± 0.4972 [b]	99.994 ± 0.00872
MC2%	281.66 ± 14.610	379.57 ± 24.389	65.56 ± 1.663 [a,b,c]	99.999 ± 0.00153
MC3%	306.12 ± 8.6990	354.75 ± 11.359	59.07 ± 0.8932 [c]	99.999 ± 0.00058
CcC1%	270.14 ± 9.3005	369.53 ± 16.292	63.56 ± 1.501 [a]	99.997 ± 0.00379
CcC2%	292.11 ± 9.3061	329.72 ± 12.525	61.36 ± 2.510	99.998 ± 0.00321
CcC3%	286.46 ± 17.991	356.21 ± 19.100	62.65 ± 0.8638 [a]	99.998 ± 0.00100

a—statistical difference with control (PCL). b, c—statistical difference between same samples, different percentage.

Compression tests were also performed to test the mechanical strength of the scaffold produced. In the graph (Figure 3), it is visible that all scaffolds, except McC2% and MC1%, present a lower modulus than the scaffold-based (PCL). In the scaffold where commercial cellulose was incorporated, McC and MC, there is a tendency for the increase in cellulose, which leads to a decrease in strength. In the case of cellulose from residue, CcC, the resistance of the scaffold was not affected by cellulose incorporation.

Filament strength tests (tensile testing) were also performed. Figure 4 shows that there are also differences between commercial and waste cellulose. Fibres made from CcC show an equal resistance to the fibres only with PCL, while the fibres incorporated with McC and MC show higher resistance, except for MC1%. Further, visible changes in properties are visible within the same cellulose but in different concentrations. In McC, incorporating 2% of cellulose increases resistance, but at 3%, its resistance returns to values equal to 1%. Regarding MC, its incorporation in fibres only becomes significant at 3%, increasing its strength. As for CcC, the increase in concentration does not result in changes in its tensile strength.

An enzymatic degradation assay was also performed with lysozyme, the enzyme naturally present in cartilage. Figure 5 shows the graphs of mass gain for PCL, McC, MC and CcC scaffolds. In all scaffolds, it is possible to observe that after 1 day of testing, there was an abrupt increase in weight. By the 7th day, the weights decreased, remaining, in general, very close to a gain equal to 0. On days 28 and 42, scaffolds subjected to the enzyme presented weight loss for all types of cellulose, while scaffolds without lysozyme gained or maintained their weight. The cellulose with the highest variability in weight gain was McC, with different values among them.

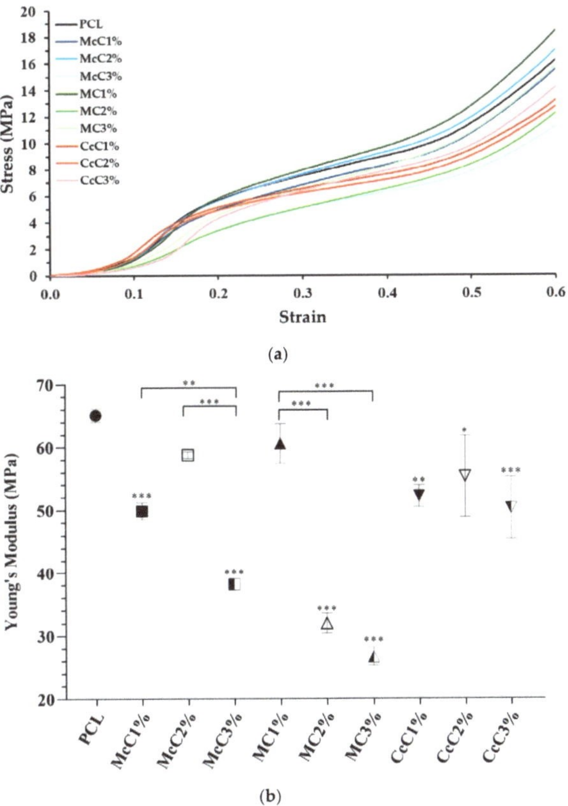

(a)

(b)

Figure 3. Mechanical performance of scaffolds upon compression: (**a**) Mean curves; (**b**) Compressive modulus (Mean ± SD, $n = 3$). Statistically significant differences are presented as: * $p < 0.05$; ** $p < 0.01$; *** $p < 0.001$.

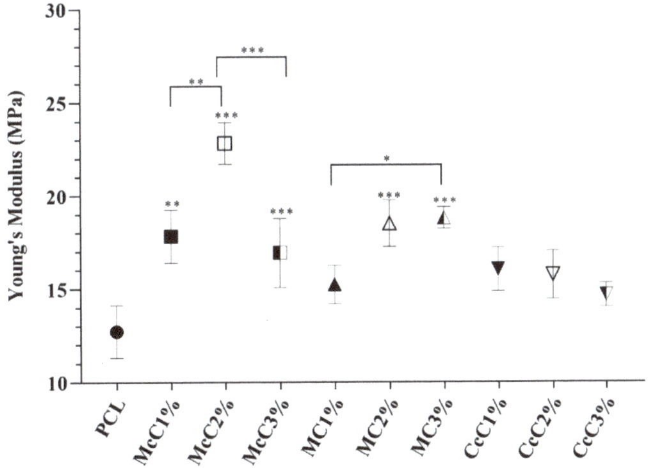

Figure 4. Modulus of tensile strength for filaments obtained by extrusion. (Mean ± SD, $n = 3$). Significant statistical differences are presented as: * $p < 0.05$; ** $p < 0.01$; *** $p < 0.001$.

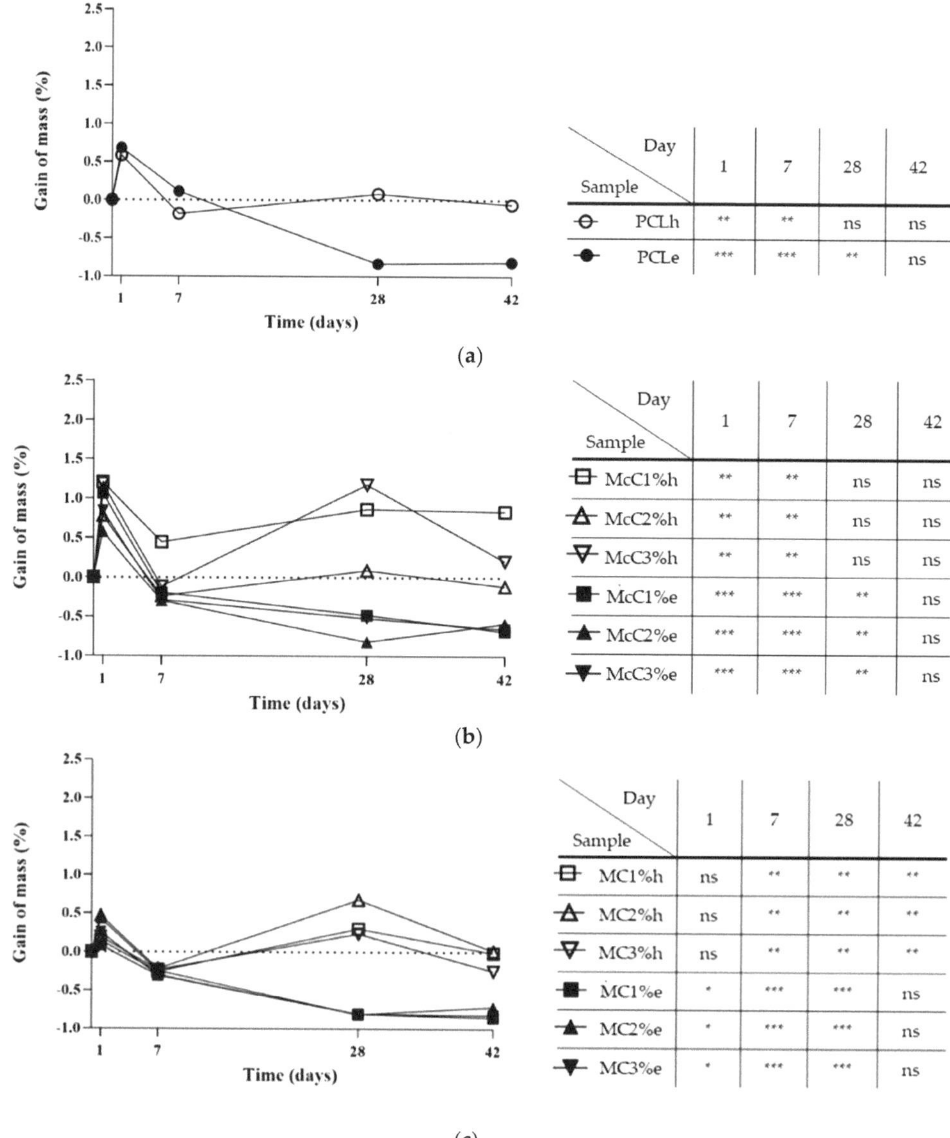

Figure 5. *Cont.*

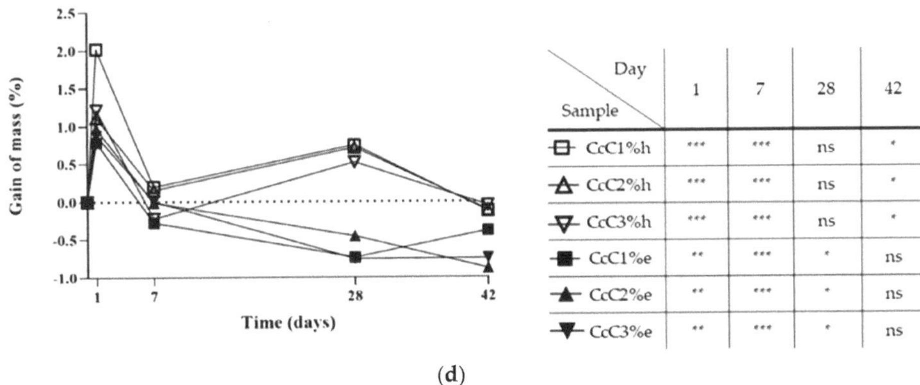

Sample \ Day	1	7	28	42
⊟ CcC1%h	***	***	ns	*
△ CcC2%h	***	***	ns	*
▽ CcC3%h	***	***	ns	*
■ CcC1%e	**	***	*	ns
▲ CcC2%e	**	***	*	ns
▼ CcC3%e	**	***	*	ns

(d)

Figure 5. Enzymatic and hydrolytic degradation assay of the scaffolds developed for 42 days: (**a**) poly(ε-caprolactone) (PCL); (**b**) microcrystalline cellulose at 1, 2, and 3% (McC1%, McC2%, and McC3%, respectively); (**c**) methyl cellulose at 1, 2, and 3% (MC1%, MC2%, and MC3%, respectively); (**d**) and corncob cellulose at 1, 2, and 3% (CcC1%, CcC2%, and CcC3%, respectively). Hydrolytic samples are identified with an "h" and are graphically represented with an unfilled symbol (○, □, △, ▽). Enzymatic samples are identified with an "e" and graphically represented with a symbol filled (●, ■, ▲, ▼). The assay was performed in triplicates, but only mean values are presented in the graphs. Statistically significant differences are presented as: * $p < 0.05$; ** $p < 0.01$; *** $p < 0.001$. ns refers to no statistically significant differences.

Cytocompatibility analysis of the scaffolds was performed, and the corrected absorbances determined at each timepoint can be seen in Figure 6. The corrected absorbance values at each time point can be consulted in Supplementary Material. Figure 7 demonstrates the statistical differences identified between experimental groups at each time point. The results reveal a general cytocompatibility of the studied biomaterials with hDPSCs, which are more evident at more advanced timepoints.

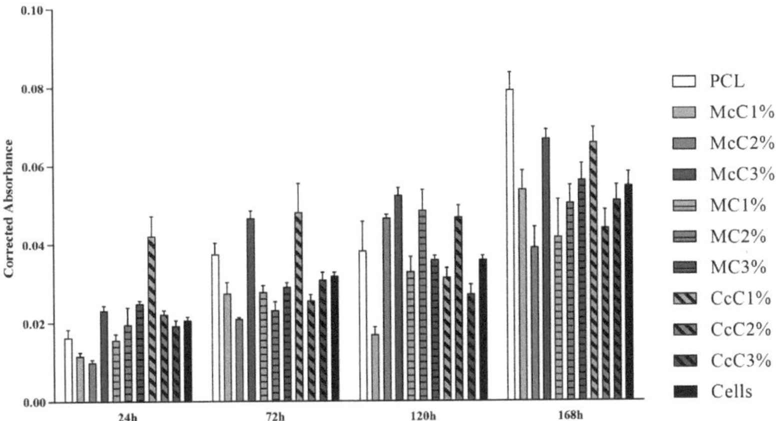

Figure 6. Cytocompatibility between scaffolds and hDPSCs determined at different timepoints using the PrestoBlueTM viability assay. (Mean ± Standard Error of the Mean (SEM)). Statistical differences at each timepoint can be consulted in Figure 7.

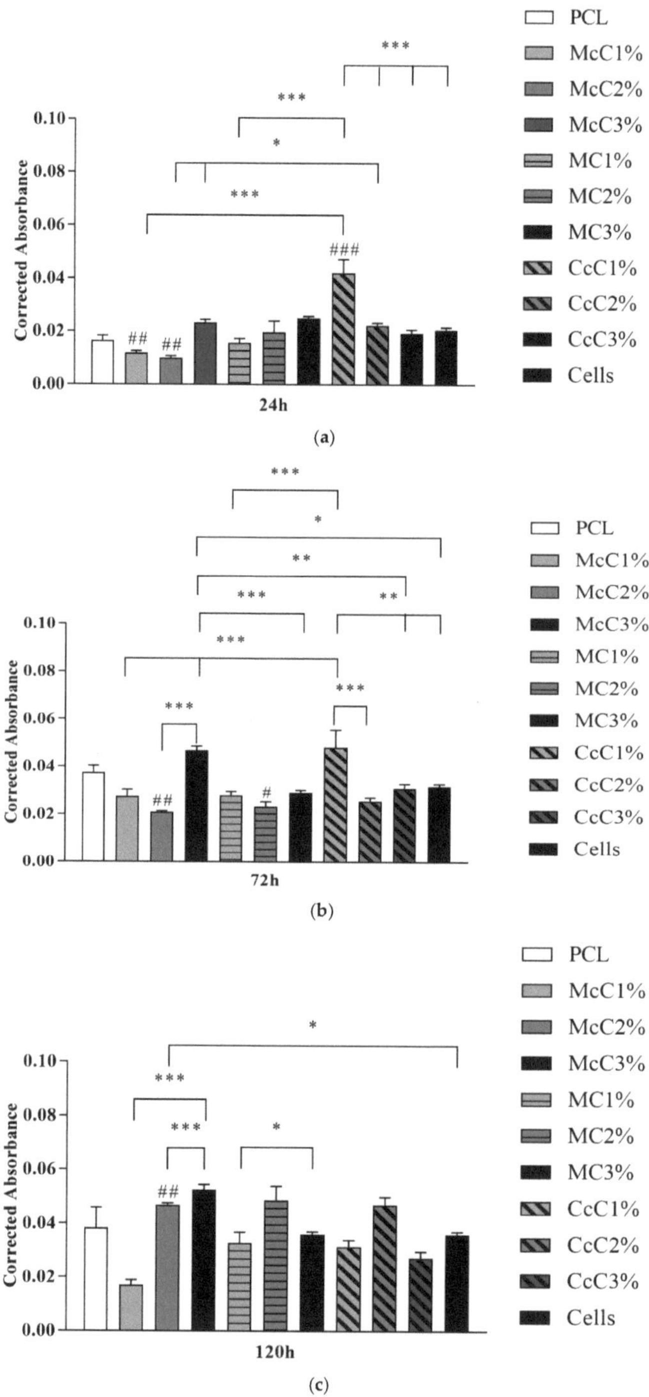

(a)

(b)

(c)

Figure 7. *Cont.*

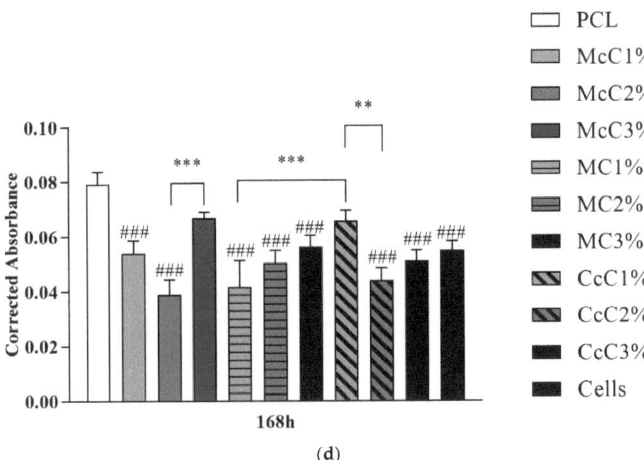

(d)

Figure 7. Cytocompatibility between scaffolds and hDPSCs using the PrestoBlue^{TM} viability assay, determined at: (**a**) 24 h; (**b**) 72 h; (**c**) 120 h; (**d**) 168 h. Results are expressed in Mean ± Standard Error of the Mean (SEM). Differences were considered statistically significant at $p < 0.05$. Results significances between different scaffolds were presented using the symbol (*). Differences between scaffolds and PCL were presented using the symbol (#). According to the p-value, *(#), **(##), and ***(###) corresponds to $p < 0.05$, $p < 0.01$, and $p < 0.001$, respectively.

The percentage of inhibition of cell viability is shown in Figure 8. The percentage values and the graph of each time point can be consulted in Supplementary Material. It is important to note that at some timepoints, cell viability inhibition percentage above 30% or on the borderline is observed. In these cases, biomaterials formulations are considered cytotoxic. However, this cytotoxicity is reversible in some cases, for example, at 24 h in the PCL (borderline values) and McC1% (cytotoxicity indicators). Afterwards, they indicate the absence of cytotoxicity and facilitation of cell proliferation. The McC2% scaffold can be considered the most cytotoxic.

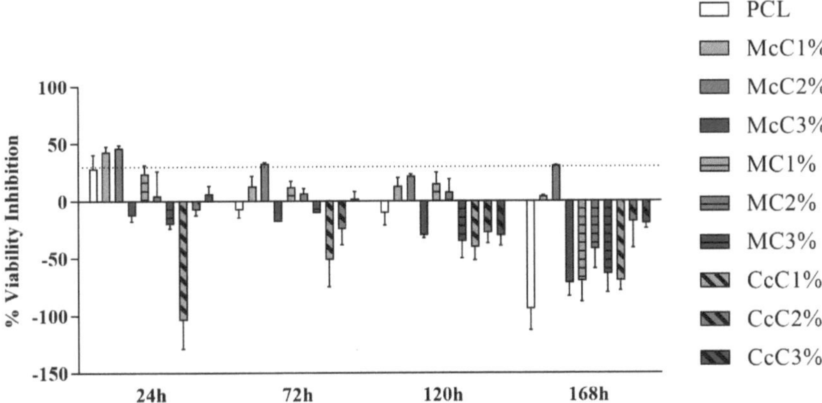

Figure 8. Percentage of inhibition of cell viability of each scaffold in contact with the hDPSCs. The dashed line represents the 30% inhibition limit, above which the effect is considered cytotoxic according to the ISO 10993-5:2009 guidelines.

4. Discussion

Previous work developed by our group [30] demonstrated that PCL–cellulose scaffolds could be a viable option to replace cartilage. In addition, CcC could be a potential substitute for McC [35]. Moreover, PCL and McC, CcC, and MC (at three different percentages in concentration) scaffolds were manufactured. MC was evaluated to assess its possibility as a viable alternative compared to the others.

In scaffold preparation and to develop the composites, it was necessary to use DMF. This organic solvent causes toxicity when in contact with cells [36] and also affects the mixing process [30]. To ensure that all DMF was removed from the membranes, an ATR-FTIR analysis was performed (Figure 1). The graph demonstrates that all DMF has been removed from membranes since the 1658 cm^{-1} band, concerning C=O stretching of DMF, does not appear in any of the membrane's spectra. The PCL spectrum was similar to the spectra of the different PCL–cellulose composites. The concentration of cellulose included in the membranes may not be enough to change the conformation of the spectrum of the base polymer.

Previous studies indicate that large pores, between 250 μm and 500 μm, are more beneficial to cell growth and new matrix production [37]. Oh, et al. (2010) revealed that pores between 300–320 μm and 370–400 μm are more conducive to gene expression of collagen type II and SOX-9, indicators of chondrogenesis, whereas smaller pores favour the production of collagen type I and X [38]. Our morphological results (Table 4), evaluated at 5× magnification, showed filaments with a diameter between 270.14 ± 9.3005 μm (CcC1%) and 326.45 ± 11.674 μm (McC2%) and pore size between 329.72 ± 12.525 μm (CcC2%) and 379.57 ± 24.389 μm (MC2%). Thus, the large pores obtained in our scaffold will be ideal for cell proliferation and infiltration and promote collagen II production, as previous studies have demonstrated. The pore sizes differences between samples are related to the production parameters and not directly to the composites, as these are manually adjusted.

Porosity is related to pore and filament sizes obtained in each scaffold. Although the native porosity values are around 70% to 92% [34], depending on the cartilage zone, the values obtained are a ratio between cell adhesion/proliferation capacity and mechanical properties. When compared with the control (PCL), the samples with higher porosity are McC3%, MC2%, CcC1%, and CcC3% (Table 4). These samples also correspond to the samples with the largest pore size. The remaining samples showed porosities similar to PCL. When comparing the porosity within the same cellulose at different percentages, the scaffold with MC is the only one revealing some variations. The incorporation of 2% of MC leads to an increase in porosity (65.56 ± 1.663%). However, the increase in the percentage leads to a decrease in porosity (59.07 ± 0.8932%). Some studies report increased porosity when MC is increased or no significant change, depending on the method of fabrication of the structure [29,39]. The MC1% and MC3% scaffolds presented similar porosity values in this study. In addition, in these scaffolds, similar pore sizes and filament diameters were obtained, leading to similar porosities.

Despite being lower than the reference values, a previous study of our group reveals that this low porosity does not interfere with cell viability and proliferation of fibroblasts [30]. Additionally, in MicroCT analysis is observable that no agglomerate of cellulose is formed. This result suggests that, although it is not soluble, all celluloses were well dispersed. Finally, regarding interconnectivity, almost all scaffolds presented approximately 100%. This value indicates that cells, nutrients, oxygen, residues, and other biomolecules can migrate freely throughout the whole scaffold [40].

The compression test results (Figure 3) show standard stress–strain thermoplastic curves [41]. The behaviour of the base thermoplastic, PCL, is not affected by the presence of cellulose. However, the mechanical properties of scaffolds are affected by the incorporation of the materials [42]. Compared with the PCL, all scaffolds, except for McC2% and MC1%, possess lower resistance. In general, it is concluded that incorporating cellulose decreases scaffolds' mechanical properties. Alemán-Dominguez (2018) achieved similar results when incorporating carboxymethyl cellulose into PCL scaffolds [14]. When evaluating the incorporation of cellulose, in McC, the incorporation of 2% did not result in a significant change in its mechanical properties. However, when this percentage is increased to 3%, a decrease in the scaffold properties is revealed. Alemán-Dominguez (2019) found the same effect when increasing the concentration of McC from 2% to 5% [4]. Our study shows that these properties decrease at a lower concentration. Regarding MC, the compression modulus at 2 and 3% were the lower of all samples (Figure 3). The incorporation of 2% immediately results in a decrease from 60.55 ± 3.185 MPa to 31.96 ± 1.565 MPa, almost half of the original properties. When incorporating 3% MC, its resistance is not affected compared to 2%. These values are closer to the desired for application in AC, since the compression modulus for human cartilage varies between 5.5 and 11.8 MPa [43]. CcC scaffolds were the only structure in which their resistance was not affected when CcC was incorporated. These results demonstrate that scaffolds with very similar porosities present similar mechanical properties since McC2% and MC1% present porosity values, $58.47 \pm 0.7003\%$ and $57.76 \pm 0.4972\%$, respectively, very close to PCL ($57.40 \pm 2.616\%$) (Table 4).

The AC is also subject to tensile forces [44]. For this reason, filaments were subjected to tensile tests (Figure 4). When compared with the PCL scaffold (12.74 ± 1.416 MPa), the MC1%, CcC1%, CcC2%, and CcC3% scaffolds demonstrated the same tensile resistance (15.24 ± 1.017 MPa, 16.05 ± 1.171 MPa, 15.74 ± 1.289 Mpa, and 14.67 ± 0.6385 Mpa, respectively). However, when incorporated McC in all concentrations, or MC at 2% and 3%, the tensile resistance is improved. The highest tensile strength corresponds to the McC2% scaffold (22.84 ± 1.424 MPa). When comparing the incorporation of cellulose, McC demonstrated that the incorporation of 2% improved the tensile force, but at 3%, this resistance decreased (16.94 ± 1.855 MPa). In MC, the incorporation of cellulose is only noticed when incorporated at 3%, with a slight increase in its tensile resistance (18.81 ± 0.5910 MPa). Once again, the increase in CcC does not result in changes in its resistance to this force.

Upon implantation in human/animal tissue, the scaffold is not just susceptible to hydrolytic degradation but also enzymatic. This way, scaffolds were submitted to a degradation test for 42 days (Figure 5). Regarding the base polymer (PCL), it was found that no hydrolytic (PCLh) degradation occurs (Figure 5a). The statistical differences detected in the gain of mass in the first 7 days are related to water absorption by the material. PCL is a hydrophobic polyester [45]. However, processing with DMF leads to an increase in its wettability, making it hydrophilic ($\theta = 88.41°$) [46]. Enzymatically the behaviour is similar in the first days, but the scaffold decreases its weight until day 28, stabilising after that. This result shows that, when in contact with lysozyme, PCL starts its degradation at day 7, with a reduction of 0.8% in weight. The scaffold should be degraded and absorbed upon new tissue formation and growth. Thus, the scaffold degradation rate must be inversely proportional to the rate of new tissue formation [47]. We believe that this degradation rate in lysozyme is too fast for cartilage regeneration and, for this reason, may not be a viable solution by itself to be applied in AC regeneration.

When the enzymatic degradation of the scaffolds with cellulose was observed, they presented the same response as the PCL scaffold (Figure 5). Thus, it is concluded that the degradation of the scaffold occurs by the degradation of PCL and not cellulose. At the end of the 42-day assay, McC1%, McC2%, and McC3% had lost about 0.6% of their weight (Figure 5b). CcC1%, CcC2%, and CcC3% lost 0.4%, 0.9%, and 0.7%, respectively (Figure 5d), while MC1%, MC2%, and MC3% lost about 0.8% (Figure 5c), similar to the PCL.

Hydrolytically, the behaviour of McC, MC, and CcC scaffolds is similar to PCL. Thus, once again, results suggest that the changes observed during the test are due to changes in the PCL structure and not in different celluloses, proving that it is not affected by the presence of water. The mass gain in the scaffolds on the first day of the test is due to the hydrolytic property of cellulose [48], enhancing the ability to absorb water. PCL had a gain of 0.6% (Figure 5a). On the other hand, McC2% and MC2% had a similar gain in mass, between 0.5% and 0.8%. The scaffolds with the highest mass gain were CcC1%

with 2%, McC1%, McC3%, and CcC3% with 1.2%, and CcC2% with 1.1%. These scaffolds have a greater ability to incorporate water in your matrix and benefit the nutrient change. Scaffolds made with 1 and 3% of MC had a lower water absorption capacity (0.1% and 0.2%, respectively).

On the seventh day of testing, the mass values reached results very similar to those of day 0, as they have reached hydrostatic equilibrium. The scaffold with the slightest mass variation was the McC1%, with a loss of 0.8% of its weight compared to day 1. On day 28 of testing, the mass gain results were not statistically significant for PCL, McC, and CcC. However, when MC was incorporated into the scaffolds, their weight tended to increase, being more noticeable at 2% of MC (about 0.7%). Natural MC presents itself as a gel when in contact with water. Thus, its gelation occurs in the manufacture of the membrane continuing in this state throughout the process. When placed in an aqueous medium (PBS), it tends to absorb part of the medium and, therefore, gain mass [49]. After 42 days of hydrolytic testing, samples that continued to lose weight were MC and CcC. Being a gel, MC is probably more susceptible to starting its degradation. At the time of its extraction, CcC undergoes very aggressive processes [35]. Its structure may no longer be completely intact, and for this reason, it may start its degradation earlier than the other tested celluloses.

A cytocompatibility test was performed using hDPSCs to validate the cellular response of the produced scaffolds (Figure 6). The cytocompatibility was determined at different timepoints using the PrestoBlue^{TM} method. The analysis of the results was carried out following an adaptation of ISO 10993-5:2009 "Biological evaluation of medical devices"—Part 5—"Test for in vitro cytotoxicity", which has already shown in previous works to guarantee an evaluative performance similar to the MTT assay standardise in the guidelines [32,33]. Unlike the MTT method, Presto Blue is performed on live cells, allowing them to be used over different timepoints, allowing for the use of fewer cells and better assessing cumulative toxicity [50]. hDPSCs were selected as the cell population to be tested due to the prior identification of their ability to induce cartilage tissue regeneration [51]. Following the guidelines of ISO 10993-5:2009, a control group was established using only seeded cells without contact with a scaffold. This group was considered a negative control, representing the cell culture of normal performance without the presence of any cytotoxic or cytoinhibitory effect. Direct contact cytocompatibility was considered, with cells being directly seeded onto the biomaterials and keeping in contact with them in the well. Absorbance values were obtained at each considered timepoint, and the data obtained were analysed according to the manufacturing instructions and presented as corrected absorbances. Subsequently, the same data were normalised considering the absence of toxicity in the control group (0% cell inhibition) and presented as % Viability Inhibition.

In general, the scaffolds proved not to be cytotoxic (Figure 8). However, at some timepoints, some scaffolds are considered cytotoxic. In some scaffolds, this cytotoxicity is reversible, for example, at 24 h in the PCL (borderline values) and McC1% (cytotoxicity indicators). Afterwards, they indicate the absence of cytotoxicity and facilitation of cell proliferation. The McC2% scaffold can be considered the most cytotoxic among those studied since, throughout the entire study period, it always induces inhibition of cell viability with values around 30%. Despite these occasional cases, the percentage of inhibition of cell viability is never particularly high, and it is, even in cases with less favourable results, around the borderline reference value to consider a material as cytotoxic. Most formulations of PCL–cellulose show no cytotoxicity and even values indicative of cell adhesion and proliferation induction. The CcC1% scaffold seems to present the best performance throughout the study, at 24 h showing a clear beneficial effect on cell adhesion and always maintaining negative values of percentage of inhibition of cell viability throughout the study period. Despite the indicative values of some toxicity in the first timepoint, the PCL presents good values in the subsequent phases. At 168 h, practically all formulations show no toxicity, except for McC2% with borderline values (Figure 8).

Accordingly, the absorbance values continuously increase over time for all groups, revealing an increase in cell proliferation while maintaining cell viability (Figure 6). Unexpectedly, at 24 h, the control group is not the one with the highest absorbance values, and others surpass this group over the study period. This result demonstrates that several scaffolds promote cell adhesion and proliferation superior to the ones observed when seeding cells directly into the wells of the plate. Previous studies proved that the presence of cellulose in scaffolds promotes cell adhesion and proliferation, corroborating these results [30].

Cell group was used as a control to determine the percentage of inhibition of cell viability in the remaining groups. So, biomaterials are considered cytotoxic, when absorbance values are noticeably different from the control. However, few statistical differences are observed between groups, and the values indicative of toxicity are always borderline. At 24 h, the only absorbance values different from PCL were McC1%, McC2%, and CcC1% (Figure 6). All other scaffolds behave similarly to PCL in terms of cell adhesion. However, at 168 h, the only samples with absorbance values similar to PCL were McC3% and CcC1%. For MC, the incorporation of cellulose does not translate into significant changes in the absorbance value at 24 h nor over the cell assay time. However, for McC, the absorbance values of McC3% increase from 24 h to 168 h. As for CcC, the absorbance values of CcC1% are always higher than the other percentages. Scaffolds with the best absorbances are the CcC1% and McC3% along the different timepoints, and the PCL at the latter. These values corroborate those observed in evaluating the percentage of inhibition of viability.

Once the compatibility of the manufactured scaffolds is confirmed and considering the results obtained in the different assays, the best PCL–cellulose combinations will be submitted to chondrogenic differentiation assays to determine the ability of hDPSCs to differentiate in the presence of these biomaterials. The ability of cells to adhere to scaffolds will be attested by scanning electron microscopy assays, and changes in gene expression associated with genes influencing chondrogenic differentiation will be determined by RT-PCR assays. After these phases and confirming favourable results, the manufactured scaffolds could be applied in vivo to demonstrate that they are suitable for promoting cartilaginous regeneration.

5. Conclusions and Future Work

PCL and PCL–cellulose scaffolds were developed to evaluate their potential for the regeneration of damaged cartilage. Morphologically, manufactured scaffolds are very similar to each other. However, the incorporation of cellulose leads to an increase in the porosity of the scaffolds. Mechanically, this increase in porosity makes its resistance decrease. However, obtained values are within the expected and required for potential application in cartilage regeneration. Regarding the elasticity of the scaffold filaments, these were revealed to be more elastic upon cellulose incorporation, being the McC2% scaffold the most elastic. When submitted to enzymatic and hydrolytic degradation, it was verified that the PCL scaffolds suffered degradation after 7 days. This degradation is also observed in scaffolds with different incorporated celluloses. When their adhesion capacity and cell proliferation were evaluated, the scaffolds with the best results were PCL, McC3%, and CcC1%. In general, regarding the properties assessed in this study, scaffolds manufactured with PCL-MC 2% and 3% prove to be more advantageous to proceed to further studies for cartilage repair application. These scaffolds present compressive performance more similar to native cartilage, and all biological studies assure good cell adhesion and proliferation. In enzymatic tests, it behaves as a hydrogel, soft material, corroborating the "bumper" behaviour of the cartilage tissue. Regarding the ideal percentage, further studies are still required.

The outcomes observed in this work are thus a starting point for future tests, where the main aim will be to apply the characterized biomaterials in the promotion of cartilage regeneration and repair. The results obtained here allow us to understand the degree of toxicity induced by direct contact between the hDPSCs and the scaffolds, but more specific tests will be necessary to effectively determine the permissibility of these biomaterials in terms of cell adhesion on their surface and the proliferation into the pores. This adhesion and proliferation must thus be confirmed by scanning electron microscopy after a new culture by direct contact. Likewise, the ability of the scaffolds to induce chondrogenic differentiation when in contact with the cells still needs to be determined, and this should be tested both qualitatively and quantitatively by measuring the production of proteoglycans and by identifying variations in gene expressions associated with this differentiation by RT-PCR. This whole set of new tests is planned to determine the best scaffold to be applied in in vivo tests, thus allowing to confirm its effectiveness in promoting successful cartilaginous regeneration.

Supplementary Materials: The following supporting information can be downloaded at: https://www.mdpi.com/article/10.3390/polym15030781/s1, Table S1: Cytocompatibility assessed by PrestoBlue^{TM} viability assay with hDPSCs. Corrected absorbance results are presented in Mean $\pm$ SEM: Table S2: Percentage of inhibition of cell viability of each scaffold in contact with the hDPSCs. After normalisation of the Control group (100%), results are presented as Mean $\pm$ SEM for the percentage of viability inhibition, comparing to the control group; Figure S1: Percentage of inhibition of cell viability of each scaffold in contact with the hDPSCs at 24 h. The dashed line represents the 30% inhibition limit, above which the effect is considered cytotoxic according to the ISO 10993-5:2009 guidelines; Figure S2: Percentage of inhibition of cell viability of each scaffold in contact with the hDPSCs at 72 h. The dashed line represents the 30% inhibition limit, above which the effect is considered cytotoxic according to the ISO 10993-5:2009 guidelines; Figure S3: Percentage of inhibition of cell viability of each scaffold in contact with the hDPSCs at 120 h. The dashed line represents the 30% inhibition limit, above which the effect is considered cytotoxic according to the ISO 10993-5:2009 guidelines; Figure S4: Percentage of inhibition of cell viability of each scaffold in contact with the hDPSCs at 168 h. The dashed line represents the 30% inhibition limit, above which the effect is considered cytotoxic according to the ISO 10993-5:2009 guidelines.

Author Contributions: Conceptualization, A.C.M., N.A. and C.M.; Methodology, R.C., A.C.M., N.A. and C.M.; Investigation, R.C.; Resources, A.C.M., N.A. and C.M.; Writing—original draft, R.C. and R.D.A.; Writing—review & editing, R.C., R.D.A., A.C.S., B.L., P.S., A.C.M., N.A. and C.M.; Supervision, A.C.M., N.A. and C.M. All authors have read and agreed to the published version of the manuscript.

Funding: This research was funded by the "Fundação para a Ciência e a Tecnologia" FCT/MCTES (PIDDAC) through the following Projects: UIDB/04044/2020; UIDP/04044/2020; 2022.09638.BD; Associate Laboratory ARISE LA/P/0112/2020; PAMI—ROTEIRO/0328/2013 (N° 022158) and BIODISCUS (CENTRE-01-0247-FEDER-039969). This research was also funded through the institutional scientific employment program-contract (CEECINST/00077/2021).

Institutional Review Board Statement: Not applicable.

Informed Consent Statement: Not applicable.

Data Availability Statement: Not applicable.

Acknowledgments: The authors acknowledge the administrative, technical, and material support received from Centre for Rapid and Sustainable Product Development of Polytechnic of Leiria and from Instituto de Ciências Biomédicas Abel Salazar of the University of Porto. Carla Moura thanks the National funding by FCT- Foundation for Science and Technology, P.I., through the institutional scientific employment program-contract (CEECINST/00077/2021).

Conflicts of Interest: The authors declare no conflict of interest.

References

1. He, Y.; Li, Z.; Alexander, P.G.; Ocasio-Nieves, B.D.; Yocum, L.; Lin, H.; Tuan, R.S. Pathogenesis of Osteoarthritis: Risk Factors, Regulatory Pathways in Chondrocytes, and Experimental Models. *Biology* **2020**, *9*, 194. [CrossRef] [PubMed]
2. Long, H.; Liu, Q.; Yin, H.; Wang, K.; Diao, N.; Zhang, Y.; Lin, J.; Guo, A. Prevalence Trends of Site-Specific Osteoarthritis from 1990 to 2019: Findings from the Global Burden of Disease Study 2019. *Arthritis Rheumatol.* **2022**, *74*, 1172–1183. [CrossRef] [PubMed]
3. Mora, J.C.; Przkora, R.; Cruz-Almeida, Y. Knee Osteoarthritis: Pathophysiology and Current Treatment Modalities. *J. Pain Res.* **2018**, *11*, 2189–2196. [CrossRef] [PubMed]

4. Alemán-Domínguez, M.E.; Giusto, E.; Ortega, Z.; Tamaddon, M.; Benítez, A.N.; Liu, C. Three-Dimensional Printed Polycaprolactone-Microcrystalline Cellulose Scaffolds. *J. Biomed. Mater. Res. Part B Appl. Biomater.* **2019**, *107*, 521–528. [CrossRef] [PubMed]

5. Hu, J.C.Y.; Athanasiou, K.A. Structure and Function of Articular Cartilage. In *Handbook of Histology Methods for Bone and Cartilage*; Springer: Totowa, NJ, USA, 2003; pp. 73–95.

6. Kwon, H.; Brown, W.E.; Lee, C.A.; Wang, D.; Paschos, N.; Hu, J.C.; Athanasiou, K.A. Surgical and Tissue Engineering Strategies for Articular Cartilage and Meniscus Repair. *Nat. Rev. Rheumatol.* **2019**, *15*, 550–570. [CrossRef]

7. Chinta, M.L.; Velidandi, A.; Pabbathi, N.P.P.; Dahariya, S.; Parcha, S.R. Assessment of Properties, Applications and Limitations of Scaffolds Based on Cellulose and Its Derivatives for Cartilage Tissue Engineering: A Review. *Int. J. Biol. Macromol.* **2021**, *175*, 495–515. [CrossRef]

8. Athanasiou, K.A.; Darling, E.M.; Hu, J.C. Articular Cartilage Tissue Engineering: 3 In Vitro Tissue Engineering of Hyaline Articular Cartilage. In *Synthesis Lectures on Tissue Engineering*; Athanasiou, K.A., Ed.; Morgan & Claypool Publishers, 2009; Volume 1, pp. 31–54.

9. Campos, Y.; Almirall, A.; Fuentes, G.; Bloem, H.L.; Kaijzel, E.L.; Cruz, L.J. Tissue Engineering: An Alternative to Repair Cartilage. *Tissue Eng. Part B Rev.* **2019**, *25*, 357–373. [CrossRef]

10. Yang, X.; Wang, Y.; Zhou, Y.; Chen, J.; Wan, Q. The Application of Polycaprolactone in Three-Dimensional Printing Scaffolds for Bone Tissue Engineering. *Polymers* **2021**, *13*, 2754. [CrossRef]

11. Mirmusavi, M.H.; Ahmadian, M.; Karbasi, S. Polycaprolactone-Chitosan/Multi-Walled Carbon Nanotube: A Highly Strengthened Electrospun Nanocomposite Scaffold for Cartilage Tissue Engineering. *Int. J. Biol. Macromol.* **2022**, *209*, 1801–1814. [CrossRef]

12. Arif, Z.U.; Khalid, M.Y.; Noroozi, R.; Sadeghianmaryan, A.; Jalalvand, M.; Hossain, M. Recent Advances in 3D-Printed Polylactide and Polycaprolactone-Based Biomaterials for Tissue Engineering Applications. *Int. J. Biol. Macromol.* **2022**, *218*, 930–968. [CrossRef]

13. Fu, L.; Yang, Z.; Gao, C.; Li, H.; Yuan, Z.; Wang, F.; Sui, X.; Liu, S.; Guo, Q. Advances and Prospects in Biomimetic Multilayered Scaffolds for Articular Cartilage Regeneration. *Regen. Biomater.* **2020**, *7*, 527–542. [CrossRef] [PubMed]

14. Alemán-Domínguez, M.E.; Ortega, Z.; Benítez, A.N.; Monzón, M.; Garzón, L.V.; Ajami, S.; Liu, C. Polycaprolactone–Carboxymethyl Cellulose Composites for Manufacturing Porous Scaffolds by Material Extrusion. *Bio Des. Manuf.* **2018**, *1*, 245–253. [CrossRef]

15. Zhao, X.; Hu, D.A.; Wu, D.; He, F.; Wang, H.; Huang, L.; Shi, D.; Liu, Q.; Ni, N.; Pakvasa, M.; et al. Applications of Biocompatible Scaffold Materials in Stem Cell-Based Cartilage Tissue Engineering. *Front. Bioeng. Biotechnol.* **2021**, *9*, 603444. [CrossRef] [PubMed]

16. Moura, C.S.; Silva, J.C.; Faria, S.; Fernandes, P.R.; da Silva, C.L.; Cabral, J.M.S.; Linhardt, R.; Bártolo, P.J.; Ferreira, F.C. Chondrogenic Differentiation of Mesenchymal Stem/Stromal Cells on 3D Porous Poly (ε-Caprolactone) Scaffolds: Effects of Material Alkaline Treatment and Chondroitin Sulfate Supplementation. *J. Biosci. Bioeng.* **2020**, *129*, 756–764. [CrossRef]

17. Silva, J.C.; Moura, C.S.; Alves, N.; Cabral, J.M.S.; Ferreira, F.C. Effects of Different Fibre Alignments and Bioactive Coatings on Mesenchymal Stem/Stromal Cell Adhesion and Proliferation in Poly (ε-Caprolactone) Scaffolds towards Cartilage Repair. *Procedia Manuf.* **2017**, *12*, 132–140. [CrossRef]

18. Wasyłeczko, M.; Sikorska, W.; Chwojnowski, A. Review of Synthetic and Hybrid Scaffolds in Cartilage Tissue Engineering. *Membranes* **2020**, *10*, 348. [CrossRef]

19. Hivechi, A.; Bahrami, S.H.; Siegel, R.A. Drug Release and Biodegradability of Electrospun Cellulose Nanocrystal Reinforced Polycaprolactone. *Mater. Sci. Eng. C* **2019**, *94*, 929–937. [CrossRef]

20. Homem, N.C.; Amorim, M.T.P. Synthesis of Cellulose Acetate Using as Raw Material Textile Wastes. *Mater. Today Proc.* **2020**, *31*, S315–S317. [CrossRef]

21. Alavi, M. Modifications of Microcrystalline Cellulose (MCC), Nanofibrillated Cellulose (NFC), and Nanocrystalline Cellulose (NCC) for Antimicrobial and Wound Healing Applications. *e-Polymers* **2019**, *19*, 103–119. [CrossRef]

22. Luo, H.; Cha, R.; Li, J.; Hao, W.; Zhang, Y.; Zhou, F. Advances in Tissue Engineering of Nanocellulose-Based Scaffolds: A Review. *Carbohydr. Polym.* **2019**, *224*, 115144. [CrossRef]

23. Hickey, R.J.; Pelling, A.E. Cellulose Biomaterials for Tissue Engineering. *Front. Bioeng. Biotechnol.* **2019**, *7*, 45. [CrossRef] [PubMed]

24. Bodea, I.M.; Cătunescu, G.M.; Stroe, T.F.; Dîrlea, S.A.; Beteg, F.I. Applications of Bacterial-Synthesized Cellulose in Veterinary Medicine—A Review. *Acta Vet. Brno* **2019**, *88*, 451–471. [CrossRef]

25. Nandgaonkar, A.G.; Krause, W.E.; Lucia, L.A. Fabrication of Cellulosic Composite Scaffolds for Cartilage Tissue Engineering. In *Nanocomposites for Musculoskeletal Tissue Regeneration*; Liu, H., Ed.; Woodhead Publishing: Sawston, UK, 2016; pp. 187–212. ISBN 9781782424529.

26. Kotcharat, P.; Chuysinuan, P.; Thanyacharoen, T.; Techasakul, S.; Ummartyotin, S. Development of Bacterial Cellulose and Polycaprolactone (PCL) Based Composite for Medical Material. *Sustain. Chem. Pharm.* **2021**, *20*, 100404. [CrossRef]

27. Pasqui, D.; Torricelli, P.; De Cagna, M.; Fini, M.; Barbucci, R. Carboxymethyl Cellulose-Hydroxyapatite Hybrid Hydrogel as a Composite Material for Bone Tissue Engineering Applications. *J. Biomed. Mater. Res. Part A* **2014**, *102*, 1568–1579. [CrossRef]

28. Roushangar Zineh, B.; Shabgard, M.R.; Roshangar, L. Mechanical and Biological Performance of Printed Alginate/Methylcellulose/Halloysite Nanotube/Polyvinylidene Fluoride Bio-Scaffolds. *Mater. Sci. Eng. C* **2018**, *92*, 779–789. [CrossRef] [PubMed]

29. Kanimozhi, K.; Khaleel Basha, S.; Sugantha Kumari, V. Processing and Characterization of Chitosan/PVA and Methylcellulose Porous Scaffolds for Tissue Engineering. *Mater. Sci. Eng. C* **2016**, *61*, 484–491. [CrossRef]

30. Cordeiro, R.; Henriques, M.; Silva, J.C.; Antunes, F.; Alves, N.; Moura, C. Corncob Cellulose Scaffolds: A New Sustainable Temporary Implant for Cartilage Replacement. *J. Funct. Biomater.* **2022**, *13*, 63. [CrossRef]
31. Campos, J.M.; Sousa, A.C.; Caseiro, A.R.; Pedrosa, S.S.; Pinto, P.O.; Branquinho, M.V.; Amorim, I.; Santos, J.D.; Pereira, T.; Mendonça, C.M.; et al. Dental Pulp Stem Cells and Bonelike® for Bone Regeneration in Ovine Model. *Regen. Biomater.* **2019**, *6*, 49–59. [CrossRef]
32. Branquinho, M.V.; Ferreira, S.O.; Alvites, R.D.; Magueta, A.F.; Ivanov, M.; Sousa, A.C.; Amorim, I.; Faria, F.; Fernandes, M.H.V.; Vilarinho, P.M.; et al. In Vitro and In Vivo Characterization of PLLA-316L Stainless Steel Electromechanical Devices for Bone Tissue Engineering—A Preliminary Study. *Int. J. Mol. Sci.* **2021**, *22*, 7655. [CrossRef]
33. Alvites, R.D.; Branquinho, M.V.; Sousa, A.C.; Amorim, I.; Magalhães, R.; João, F.; Almeida, D.; Amado, S.; Prada, J.; Pires, I.; et al. Combined Use of Chitosan and Olfactory Mucosa Mesenchymal Stem/Stromal Cells to Promote Peripheral Nerve Regeneration In Vivo. *Stem Cells Int.* **2021**, *2021*, 6613029. [CrossRef]
34. Liang, X.; Duan, P.; Gao, J.; Guo, R.; Qu, Z.; Li, X.; He, Y.; Yao, H.; Ding, J. Bilayered PLGA/PLGA-HAp Composite Scaffold for Osteochondral Tissue Engineering and Tissue Regeneration. *ACS Biomater. Sci. Eng.* **2018**, *4*, 3506–3521. [CrossRef] [PubMed]
35. Vallejo, M.; Cordeiro, R.; Dias, P.A.N.; Moura, C.; Henriques, M.; Seabra, I.J.; Malça, C.M.; Morouço, P. Recovery and Evaluation of Cellulose from Agroindustrial Residues of Corn, Grape, Pomegranate, Strawberry-Tree Fruit and Fava. *Bioresour. Bioprocess.* **2021**, *8*, 25. [CrossRef]
36. Zhang, J.; Zhou, D.; Zhang, L.; Lin, Q.; Ren, W.; Zhang, J.; Nadeem, L.; Xu, G. Dual Effects of N,N-Dimethylformamide on Cell Proliferation and Apoptosis in Breast Cancer. *Dose-Response* **2017**, *15*, 155932581774445. [CrossRef] [PubMed]
37. Li, Y.; Liu, Y.; Xun, X.; Zhang, W.; Xu, Y.; Gu, D. Three-Dimensional Porous Scaffolds with Biomimetic Microarchitecture and Bioactivity for Cartilage Tissue Engineering. *ACS Appl. Mater. Interfaces* **2019**, *11*, 36359–36370. [CrossRef] [PubMed]
38. Oh, S.H.; Kim, T.H.; Im, G.I.; Lee, J.H. Investigation of Pore Size Effect on Chondrogenic Differentiation of Adipose Stem Cells Using a Pore Size Gradient Scaffold. *Biomacromolecules* **2010**, *11*, 1948–1955. [CrossRef]
39. Valarmathi, N.; Sumathi, S. Biomimetic Hydroxyapatite/Silkfibre/Methylcellulose Composites for Bone Tissue Engineering Applications. *New J. Chem.* **2020**, *44*, 4647–4663. [CrossRef]
40. Lutzweiler, G.; Ndreu Halili, A.; Engin Vrana, N. The Overview of Porous, Bioactive Scaffolds as Instructive Biomaterials for Tissue Regeneration and Their Clinical Translation. *Pharmaceutics* **2020**, *12*, 602. [CrossRef]
41. Jin, M. Determination of Fracture Mechanics Behavior of Polyethylene Sheets. Master's Thesis, Linnaeus University, Växjö, Sweden, 2017.
42. Loh, Q.L.; Choong, C. Three-Dimensional Scaffolds for Tissue Engineering Applications: Role of Porosity and Pore Size. *Tissue Eng. Part B Rev.* **2013**, *19*, 485–502. [CrossRef]
43. Shepherd, D. The "Instantaneous" Compressive Modulus of Human Articular Cartilage in Joints of the Lower Limb. *Rheumatology* **1999**, *38*, 124–132. [CrossRef]
44. Sophia Fox, A.J.; Bedi, A.; Rodeo, S.A. The Basic Science of Articular Cartilage: Structure, Composition, and Function. *Sports Health* **2009**, *1*, 461–468. [CrossRef]
45. Piazza, R.D.; Brandt, J.V.; Gobo, G.G.; Tedesco, A.C.; Primo, F.L.; Marques, R.F.C.; Junior, M.J. MPEG-Co-PCL Nanoparticles: The Influence of Hydrophobic Segment on Methotrexate Drug Delivery. *Colloids Surf. A Physicochem. Eng. Asp.* **2018**, *555*, 142–149. [CrossRef]
46. Biscaia, S.; Silva, J.C.; Moura, C.; Viana, T.; Tojeira, A.; Mitchell, G.R.; Pascoal-Faria, P.; Ferreira, F.C.; Alves, N. Additive Manufactured Poly(ε-Caprolactone)-Graphene Scaffolds: Lamellar Crystal Orientation, Mechanical Properties and Biological Performance. *Polymers* **2022**, *14*, 1669. [CrossRef] [PubMed]
47. Hu, X.; Li, W.; Li, L.; Lu, Y.; Wang, Y.; Parungao, R.; Zheng, S.; Liu, T.; Nie, Y.; Wang, H.; et al. A Biomimetic Cartilage Gradient Hybrid Scaffold for Functional Tissue Engineering of Cartilage. *Tissue Cell* **2019**, *58*, 84–92. [CrossRef] [PubMed]
48. Wei, D.W.; Wei, H.; Gauthier, A.C.; Song, J.; Jin, Y.; Xiao, H. Superhydrophobic Modification of Cellulose and Cotton Textiles: Methodologies and Applications. *J. Bioresour. Bioprod.* **2020**, *5*, 1–15. [CrossRef]
49. Coughlin, M.L.; Liberman, L.; Ertem, S.P.; Edmund, J.; Bates, F.S.; Lodge, T.P. Methyl Cellulose Solutions and Gels: Fibril Formation and Gelation Properties. *Prog. Polym. Sci.* **2021**, *112*, 101324. [CrossRef]
50. Boncler, M.; Różalski, M.; Krajewska, U.; Podsędek, A.; Watala, C. Comparison of PrestoBlue and MTT Assays of Cellular Viability in the Assessment of Anti-Proliferative Effects of Plant Extracts on Human Endothelial Cells. *J. Pharmacol. Toxicol. Methods* **2014**, *69*, 9–16. [CrossRef]
51. Santanna, J.; Fernandez, T.L.; Bueno, D.F.; Pinheiro, C.C.; Hernandez, A.J. Cartilage Regeneration with Human Dental Pulp Stem Cells—A Systematic Review. *Cytotherapy* **2020**, *22*, S175–S178. [CrossRef]

Article

A Novel Approach for Design and Manufacturing of Curvature-Featuring Scaffolds for Osteochondral Repair

Pedro Marcelino [1,2,3], João Carlos Silva [1,2,3,*,†], Carla S. Moura [3,4,5], João Meneses [3], Rachel Cordeiro [3,6], Nuno Alves [3,4,7,†], Paula Pascoal-Faria [3,4,8,*,†] and Frederico Castelo Ferreira [1,2,*,†]

1. Department of Bioengineering and iBB-Institute for Bioengineering and Biosciences, Instituto Superior Técnico, Universidade de Lisboa, Av. Rovisco Pais, 1049-001 Lisboa, Portugal; pmama@tecnico.ulisboa.pt
2. Associate Laboratory i4HB—Institute for Health and Bioeconomy, Instituto Superior Técnico, Universidade de Lisboa, Av. Rovisco Pais, 1049-001 Lisboa, Portugal
3. CDRSP-Centre for Rapid and Sustainable Product Development, Polytechnic of Leiria, Rua de Portugal-Zona Industrial, 2430-028 Marinha Grande, Portugal; carla.moura@ipc.pt (C.S.M.); joao.p.meneses@ipleiria.pt (J.M.); rachel.s.cordeiro@ipleiria.pt (R.C.); nuno.alves@ipleiria.pt (N.A.)
4. Associate Laboratory for Advanced Production and Intelligent Systems (ARISE), 4050-313 Porto, Portugal
5. Polytechnic Institute of Coimbra, Applied Research Institute, Rua da Misericórdia, Lagar dos Cortiços—S. Martinho do Bispo, 3045-093 Coimbra, Portugal
6. Veterinary Clinics Department, Abel Salazar Biomedical Sciences Institute, University of Porto, Rua de Jorge Viterbo Ferreira 228, 4050-313 Porto, Portugal
7. Department of Mechanical Engineering, School of Technology and Management, Polytechnic of Leiria, Morro do Lena—Alto do Vieiro, Apartado 4163, 2411-901 Leiria, Portugal
8. Department of Mathematics, School of Technology and Management, Polytechnic of Leiria, Morro do Lena—Alto do Vieiro, Apartado 4163, 2411-901 Leiria, Portugal

* Correspondence: joao.f.da.silva@tecnico.ulisboa.pt (J.C.S.); paula.faria@ipleiria.pt (P.P.-F.); frederico.ferreira@tecnico.ulisboa.pt (F.C.F.)
† These authors jointly supervised this work.

Citation: Marcelino, P.; Silva, J.C.; Moura, C.S.; Meneses, J.; Cordeiro, R.; Alves, N.; Pascoal-Faria, P.; Ferreira, F.C. A Novel Approach for Design and Manufacturing of Curvature-Featuring Scaffolds for Osteochondral Repair. *Polymers* **2023**, *15*, 2129. https://doi.org/10.3390/polym15092129

Academic Editor: Ali Reza Zanjanijam

Received: 22 March 2023
Revised: 24 April 2023
Accepted: 26 April 2023
Published: 29 April 2023

Abstract: Osteochondral (OC) defects affect both articular cartilage and the underlying subchondral bone. Due to limitations in the cartilage tissue's self-healing capabilities, OC defects exhibit a degenerative progression to which current therapies have not yet found a suitable long-term solution. Tissue engineering (TE) strategies aim to fabricate tissue substitutes that recreate natural tissue features to offer better alternatives to the existing inefficient treatments. Scaffold design is a key element in providing appropriate structures for tissue growth and maturation. This study presents a novel method for designing scaffolds with a mathematically defined curvature, based on the geometry of a sphere, to obtain TE constructs mimicking native OC tissue shape. The lower the designed radius, the more curved the scaffold obtained. The printability of the scaffolds using fused filament fabrication (FFF) was evaluated. For the case-study scaffold size (20.1 mm × 20.1 mm projected dimensions), a limit sphere radius of 17.064 mm was determined to ensure printability feasibility, as confirmed by scanning electron microscopy (SEM) and micro-computed tomography (μ-CT) analysis. The FFF method proved suitable to reproduce the curved designs, showing good shape fidelity and replicating the expected variation in porosity. Additionally, the mechanical behavior was evaluated experimentally and by numerical modelling. Experimentally, curved scaffolds showed strength comparable to conventional orthogonal scaffolds, and finite element analysis was used to identify the scaffold regions more susceptible to higher loads.

Keywords: 3D printing; curvature-featuring scaffolds; finite element modelling; mechanical properties; osteochondral regeneration; tissue engineering

1. Introduction

The human joints are dynamic complex structures responsible for providing a near-frictionless interface between bones and allowing for constrained coordinated motion [1,2].

In diarthrodial joints (e.g., knee joint), articular cartilage and subchondral bone form a composite structure designated as the osteochondral (OC) unit, which comprises a complex and fine interplay between its components under physiological conditions [3].

Two types of OC defects can be described according to the phenotype. Focal lesions are generally well delineated and usually caused by trauma or illnesses, such as osteochondritis dissecans or osteonecrosis. Degenerative lesions are attributed to progressive deteriorating changes to the structures of the joints, resulting, for example, from ligament instability, meniscal tears, or OC unit diseases such as osteoarthritis (OA) [4]. These diseases create a very high socioeconomic burden on national healthcare systems. OA, the most common joint disease, is estimated to affect over 250 million people worldwide, and its prevalence is expected to continue to increase [5]. Among the consequences of OA are the loss in mobility and performance limitations in daily activities, with an associated cost, including direct medical expenses and indirect expenditures due to the loss in productivity, estimated to be between 1% and 2.5% of the Gross Domestic Product in high-income countries [6].

Regardless of the origin, damage in the OC unit leads to a cascade of events attempting to repair the injury. However, due to its avascular nature, these self-healing processes are very limited in articular cartilage, often leading to the formation of scar tissue of inferior quality. In the long run, the inadequacy of this scar tissue will likely lead to an increase in the severity of the defect, which will reach the subchondral bone [7]. Therefore, the repair of both components of the OC unit should be therapeutically addressed, since they are physiologically deeply interconnected. This presents a major challenge due to the intrinsically different properties and healing capacities of the articular cartilage and subchondral bone [4]. Therapeutic options for OC repair are selected based on the size and severity of the defect [8]. Currently available treatments range from non-operative conservative strategies, managing the symptoms and reducing risk factors for the less severe cases, to increasingly extensive surgical interventions such as arthroscopic lavage and debridement, abrasion arthroplasty, microfracture, OC autografting and allografting, whole-joint replacement surgeries, and also cell-based procedures such as autologous chondrocyte implantation (ACI) and matrix-induced autologous chondrocyte implantation (MACI), possibly combined with growth factors [9,10]. However, all of these methods still fail to fully restore tissue structure and properties, compromising long-term clinical outcomes [11,12].

Tissue engineering (TE) has been proposed as a promising therapeutic alternative for producing OC tissue substitutes. These strategies involve the proper combination of cells, biomaterial scaffolds, and external stimuli in the form of biochemical (e.g., growth factors and cytokines) and physical factors (e.g., mechanical or electrical stimulation) provided by bioreactor culture systems [13–16]. Among these components, scaffolds play an important role in supporting cell proliferation and differentiation and extracellular matrix production. In an ideal case, they should emulate structurally and functionally the native tissue, providing a biocompatible and biodegradable environment with a degradation rate compatible with the formation of newly regenerated tissue [17,18]. Requirements can also consider other important aspects related to the following: the scaffold structure, providing a hierarchical organization with proper porosity and interconnectivity that enables cell migration and nutrient and waste diffusion, tailored to the target tissue and cells; the scaffold functionality, giving the ability to interact with host cells and integrate with host tissue by incorporation of growth factors and biological cues and exhibiting native-like biomechanical properties; and the fabrication method, which should be precise, easily scalable and reproducible, and versatile to accommodate individualized "patient-tailored" variations in the construction [11,19–22].

The available scaffold fabrication technologies can be grouped into two main categories: conventional and additive manufacturing (AM) methods. Conventional techniques use mostly subtractive methods, in which parts of a material are removed to obtain the final conformation, while in AM technology, the final construction is obtained by successive deposition of overlaying layers [23]. Methodologies such as solvent casting/particle leaching, freeze-drying, gas foaming, phase separation, and electrospinning are included in the conventional techniques, while stereolithography, selective laser sintering, inkjet 3D printing, and extrusion bioprinting are defined as AM technologies [24]. Some advantages are recognized to the AM techniques when compared to the conventional methods, such as scalability and reproducibility; high versatility, as they can use a wide range of materials; higher control over scaffold geometry, predictable and consistent at the macro- and micro-scale; the possibility of fabricating more complex structures, more easily adaptable to create patient-specific scaffold designs; and avoiding the use of organic solvents, which are required in some conventional techniques and could compromise cell viability and the biological performance of the scaffolds [23,25–28]. For these reasons, AM methods have been increasingly used for manufacturing scaffolds for different TE applications [25].

A common AM-based method for scaffold fabrication is material extrusion, owing to its accessible cost and ability to work with a wide range of materials, which has found applications in the fabrication of blood vessel, bone, cartilage, neural, cardiac, skeletal muscle, liver, and skin TE constructs [29–33]. These systems incorporate fused filament fabrication (FFF) using thermoplastic polymers and composites, to which cells and biochemical factors may be added after scaffolds construction [34–37]. The used materials can be grouped into synthetic polymers (e.g., poly(ethylene glycol) (PEG), poly(lactic acid) (PLA), Poly(lactic-co-glycolic acid) (PLGA), Poly(ϵ-caprolactone) (PCL), Poly(propylene fumarate) (PPF)), bioceramics (e.g., hydroxyapatite (HAp), tricalcium phosphate (TCP)), natural polymers (e.g., hyaluronic acid, chondroitin sulphate, alginate, agarose, collagen, gelatin), and extracellular matrix (ECM)-based materials (e.g., decellularized ECM, pulverized ECM particles) [11,19,38–40]. A key factor, typically considered in choosing the scaffold materials for OC TE applications, is the purpose of having constructs that can approach the mechanical properties of articular cartilage and bone to achieve the natural load-bearing properties of the native OC tissue [41].

In the context of OC TE, scaffold design has seen a significant evolution in architectural complexity, aiming to achieve a greater resemblance to the structure of the native tissue. The field evolved from scaffolds consisting of monophasic constructs to biphasic constructs, recognizing the distinction between articular cartilage and the subchondral bone; to triphasic constructs, realizing the significance of the calcified cartilage between articular cartilage and the subchondral bone; and finally to multiphasic and gradient constructs, in an attempt to mimic the native tissue hierarchical gradient structure [11,42]. However, despite this evolution, most described scaffolds still exhibit a cuboid or cylindrical shape, which could lead to a mismatch in geometry at the interface between the tissue and the scaffold, resulting in stress concentrations. This may hamper the engineered host tissues' integration, accelerate weakening of the scaffold structure, or increase local stress in the surrounding tissue [42,43]. Furthermore, the filament placement strategy used in extrusion fabrication systems has not seen a significant change from the successive deposition of overlaying of layers, which has mostly been dictated by the commonly available AM fabrication algorithms [44,45]. Nevertheless, some exceptions have been reported using extrusion, reproducing anatomical shapes, such as the articular surface of a rabbit synovial joint [46], the shape of an ear [47], a calvarial reconstruction [48], and a human vertebral body [49], but these cases are still greatly outnumbered.

Several key aspects have been identified as of significant relevance for the successful culture of living cells within scaffolds and bioreactor devices for TE strategies. For example, the interactions of cells with the surrounding ECM and other cells, as well as the influence of environmental factors such as mechanical (e.g., flow-induced effects, compressive loading), electrical, and biochemical stimuli, have been shown to modulate cell proliferation, differentiation, and metabolic functions [50,51]. As a complement to the in vitro experimental research on the influence of these aspects, the use of detailed mathematical models as a virtual in silico representation of the tissues/scaffolds/bioreactors can be advantageous. These models are designated digital twins and are developed to provide a more profound characterization of the culture systems/scaffold features, optimizing stimulation parameters and predicting experimental outcomes while reducing the time and costs involved [4,15,52,53].

In this work, we present a novel methodology for the design and rapid manufacturing of scaffolds to mimic the native curvature of OC tissue. The scaffold curvature was defined mathematically, and an automated parametric design process was implemented for the construction of scaffolds with the desired curvatures using computer-aided design (CAD) software. The FFF technique was chosen to produce the scaffolds, and a procedure was devised to determine the highest curvature this method could produce. Moreover, a structural analysis by micro-computed tomography (μ-CT) and scanning electron microscopy (SEM) was conducted to assess the shape fidelity of the printed scaffolds. Finally, due to its relevance in the context of OC repair strategies, the mechanical properties of the produced curvature-featuring FFF-based scaffolds were evaluated by uniaxial compressive testing and studied in more detail using finite element analysis (FEA).

2. Materials and Methods

2.1. CAD Modelling, Characterization, and Fabrication of Scaffolds

Scaffold models were created with the CAD software Autodesk Fusion 360, up to version 2.0.11415, automatically updated online (Autodesk, Inc., San Francisco, CA, USA). One advantage of CAD packages is parametric design. Concerning scaffold modelling, this allowed us to create scaffolds using the built-in geometric operations, and by assigning distinct values to only one parameter, which controls the curvature, we were able to use the same workflow to automatically create as many distinct curved scaffolds as needed. The detailed description of this modelling workflow is provided in Section 3.1.1. Common to all scaffold models, a fibril (the term used in this manuscript to designate the tube-like strands in the CAD drawings) with a diameter of 300 μm and a layer height of 300 μm was defined.

For the CAD models, the porosity was obtained through the following Equation (1):

$$Porosity = \left(1 - \frac{V_{solid}}{V_{total}}\right) \times 100\% \tag{1}$$

where V_{solid} and V_{total} are, respectively, the volume occupied by the solid fraction and the total volume in a volume of interest.

To create objects suitable for 3D printing, the models were exported from Fusion 360 as STL files with high refinement. The 3D printing was executed by FFF with a Prusa i3 MK3S commercial 3D printer (Prusa Research, Praha, Czech Republic), to which a 0.25 mm brass nozzle (Prusa Research) was adapted, instead of the standard 0.4 mm nozzle. The prints were made from a 1.75 mm spooled PLA filament from Velleman (Velleman Group nv, Gavere, Belgium). Furthermore, 3D-printer-readable G-code files were created using PrusaSlicer 2.3.1 (Prusa Research). A printing layer height of 0.15 mm was defined for all layers (since the 300 μm scaffold fibril height would be a dimension too big to print in a single run with a 0.25 mm nozzle) and an extrusion width of 0.3 mm. The printing temperature was set to 210 °C and the bed temperature to 65 °C. Printing speed was adjusted according to the features of the construct being produced. For external and internal perimeters of parts, speeds of 20 mm/s and 30 mm/s were employed, respectively. For infill and travel movements, a speed of 45 mm/s was employed. Depending on the applicability to the features of the construct being printed, the additional parameters were set: 100% infill density; rectilinear infill pattern with alternating -45° and 45° raster orientations; and 3 perimeters.

2.2. Micro-Computed Tomography (μ-CT) Analysis

The microstructure of the scaffolds was evaluated by μ-CT with a SkyScan 1174v2 instrument, Bruker version 1.1 (Bruker, Billerica, MA, USA). Image reconstruction was performed using NRecon version 1.7.4.6 (Bruker), and CTVox version 3.3.1 (Bruker) and CTVol version 2.3.2.0 (Bruker) were employed to obtain realistic 3D visualizations of the scanned scaffold samples. CTAn version 1.20.0 (Bruker) was used for the reconstruction analysis. The following acquisition parameters were used: source voltage of 50 kV; source current of 800 mA; image pixel size of 30.11 μm; exposure time of 9000 ms; rotation step of 0.5°; frame averaging on (3); no filter.

2.3. Scanning Electron Microscopy (SEM) Imaging

The surface morphology of the printed curved scaffolds was evaluated by SEM analysis with a Hitachi S2400 SEM instrument (Hitachi, Ltd., Tokyo, Japan) operating at 20 kV acceleration voltage. Prior to scanning, scaffold samples were sputter-coated with a thin layer of gold/palladium by a Q150T ES sputter coater (Quorum Technologies, Laughton, East Sussex, UK).

2.4. Compressive Mechanical Testing

The experimental structural behavior of the parts was assessed, under compressive mechanical loading, using an Instron 4505 machine (Instron, Norwood, MA, USA) equipped with a 100 kN load cell and applying a constant displacement rate of 1 mm/min. Seven specimens were used for each scaffold condition analyzed ($n = 7$). The compressive modulus was determined from the slope of the initial linear regions of the stress–strain curves. Yield strength was calculated using the offset yield method, with the offset line parallel to the modulus line and a displaced strain of 0.2%. The yield strength was identified as the point of intersection between the offset line and the stress–strain curves.

Due to the curvature of the scaffolds created, an assembly of blocks joined with the scaffolds was designed and 3D-printed to allow the performance of the compressive tests. A description of the design approach is presented in Section 3.1.3. Supplementary Figure S1 shows the designs of the parts manufactured and tested mechanically under compressive loading.

2.5. Finite Element Analysis

Finite element analysis (FEA) was performed using the Solid Mechanics module from COMSOL Multiphysics 5.2a software (COMSOL Inc., Stockholm, Sweden). A stationary study was conducted to determine the stresses when scaffolds are subjected to compressive loadings.

Models were created in Fusion 360 and imported into COMSOL as STEP files. To emulate the experimental compression testing, the imported models had a geometry equivalent to the geometry of models described in Section 2.4. Taking advantage of the existence of two planes of symmetry, the imported models result from sectioning along these two planes to facilitate the numerical computer calculations (Supplementary Figure S2). These planes of symmetry can be defined in COMSOL as boundary conditions, and the simulation results correspond to the compression of the whole models. The scaffold and block assemblies were defined as single domains in COMSOL. The flat surface on one side was considered fixed, while a displacement of 0.4 mm was prescribed to the flat surface on the opposing side (Supplementary Figure S2). This displacement corresponds approximately to a strain obtained from the experimental compression results described in Section 2.4 when a transition is identified from elastic to plastic behavior. As a result, in the implemented COMSOL simulations, the material model contemplated just the isotropic linear elastic behavior. For all simulations, the assemblies were specified to be constituted by PLA. A range of values is reported in the literature for the density and Poisson's ratio for PLA samples. Accordingly, this study was performed considering a density of 1.24 g/cm^3 [54] and a Poisson's ratio of 0.3 [55]. Concerning the Young's modulus, the value determined experimentally in the compression of the 100% infill block, made of the same PLA used in scaffold fabrication, was chosen as the most appropriate for this material characteristic, as discussed in Section 4. The mesh elements for all models were created with a tetrahedral geometry and using the physics-controlled element size definition, in which the option "finer" for the element size parameter was chosen (element sizes in the range 0.1 mm to 1.38 mm). Considering all models, the skewness average element quality was always above 0.6627. The numerical model predictions were interpreted using the calculated von Mises stresses. Plots were obtained for the entire volume of the domains and along lines crossing the entire length of the scaffolds.

2.6. Statistical Analysis

The results are presented as mean values ± standard deviations (SD) when applicable. The statistical analysis was performed using GraphPad Prism 7.0 (GraphPad Software, San Diego, CA, USA). To assess statistically significant differences between independent samples, ANOVA tests were performed, followed by Tukey's multiple comparison test (for significant *p*-values, * denotes $p < 0.05$, ** denotes $p < 0.01$, *** denotes $p < 0.001$, **** denotes $p < 0.0001$).

3. Results

3.1. Scaffold Design Methodology and Assumptions

3.1.1. Curved Scaffold Design Procedure

The motivation for developing curved scaffolds was the nonplanar native structure of tissues in the human musculoskeletal system, particularly of the OC tissue in the knee joint, which can be affected by highly debilitating conditions such as OA. Figure 1a shows the distal femur and two spheres that approximate the native curvature of both its condyles. As a first approach, a sphere was chosen for simplicity, as it only requires a radius to be defined and modelled. Furthermore, from a manufacturing perspective, with the accessible FFF printing technology, it was convenient that the scaffold would have straight sides, therefore averting the use of supporting structures connected to the build plate during the printing process. Accordingly, Figure 1b represents the square-shaped section made on the surface of spheres that formed the template for the design of the scaffolds.

(a)   (b)

Figure 1. (a) Representation of matching the curvature of the femoral condyles to specific sphere sizes, as a strategy for designing curved scaffolds. (b) Illustration of fitting a square segment on the surface of a sphere, with the resulting section to be used as a template for the design of the curved scaffolds.

Using conventional slicing algorithms, each layer of the scaffold would be printed at a time, and each successive layer would be printed on top of the previously deposited material. This strategy also carried the advantage of having fewer interruptions in filament flow during printing, resulting in parts with improved structural stability and mechanical strength. To have scaffolds with a grid-like pattern and, therefore, a porous structure, the design would have fibrils patterned with two alternating orientations: (i) one with curved filaments, following the intersection of the square-shaped sections with planes at established layer heights; (ii) the other with straight filaments, alternating with the curved ones (Figure 2). Two strategies were applied to define the location of the curved fibrils. In the designated "constant radius" strategy, the square-shaped section was translated to a distance corresponding to the sum of fibril and pore widths the number of times necessary to have the desired thickness in the scaffold (Figure 2a). In the "concentric radius" strategy, a square-shaped section was made in a circle with a specified radius. Then, circles with the same center position but with increasingly larger radii were drawn, and sections of those circles defined the locations of the curved fibrils (Figure 2b). A distinctive consequence of modelling according to these two strategies can be visible towards the edges of the scaffolds. With the "constant radius" strategy, the porosity will decrease, while with the "concentric radius" strategy, the porosity will increase. Without an anticipated advantage of one over the other, both strategies were used in the subsequent stages of the curved scaffolds design.

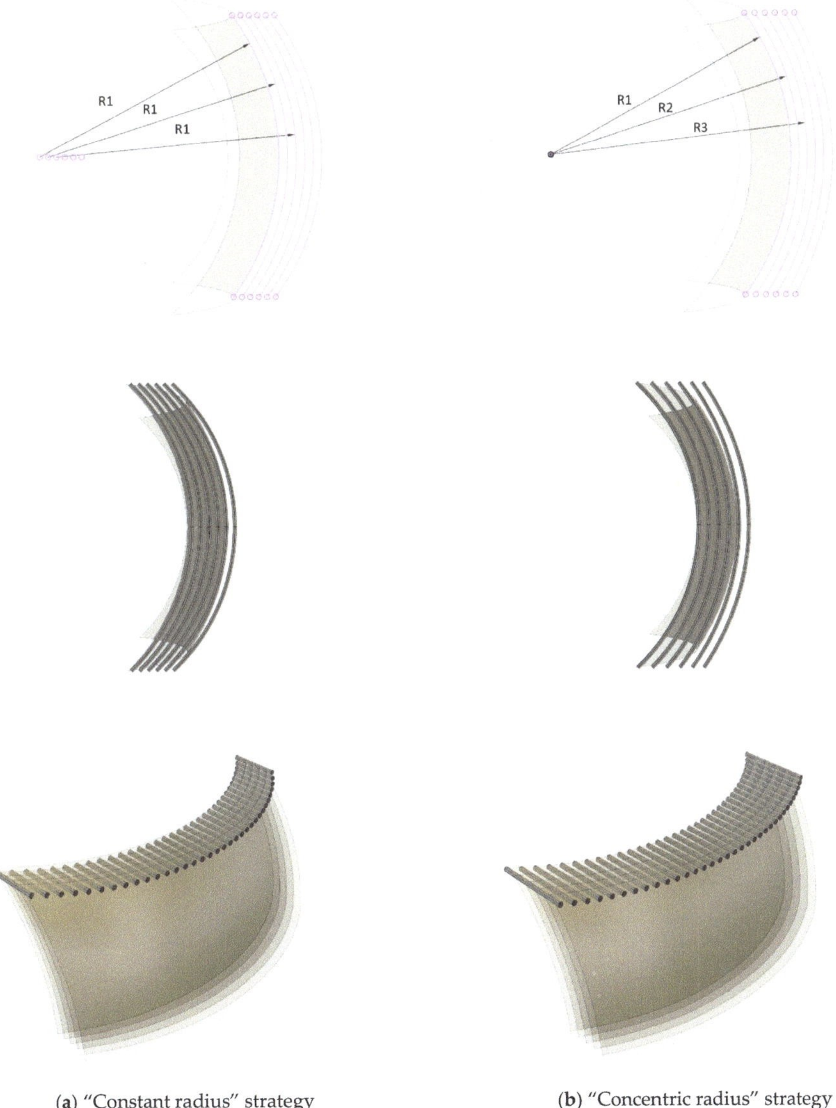

(a) "Constant radius" strategy   (b) "Concentric radius" strategy

Figure 2. Strategies for curved scaffold design: (top) Shaded square-shaped section of a spherical surface and arcs resulting from the intersection with a plane. (**a**) In the "constant radius" strategy, equal arcs are obtained from a plane intersecting translated sections (all radii have the same value R1); (**b**) in the "concentric radius" strategy, successively larger arcs are obtained from the plane intersecting concentric sections (R1 < R2 < R3). (center) Illustration, on the top layer, of the curved fibrils drawn following the arcs. (bottom) Illustration of the curved fibrils; the straight fibrils are designed to connect the two farthest-apart spherical sections.

For the effectiveness of printing with FFF, the material being deposited should be supported by the material already in place to have the parts match the desired geometry. More visible deformations can be seen when the deposited filament has to bridge across larger spans or the deposited unsupported sections are farther from being connected in a straight line. The decision to not use support structures or material was also made because this leads to parts with inferior surface quality and properties due to scarring resulting from the supports' removal. Therefore, a limit was imposed for each layer to be supported solely by already-deposited material. Due to the scaffold curvature and the path already determined for the filament deposition, the least supported locations would be noticeable in the first and last printing layers, where there is a steeper slope between the layers (Figure 3a). Therefore, the design objective became to calculate the model parameters, which led to the limit of tangential support between successive layers. This limit could not be exceeded for a print to be considered successful.

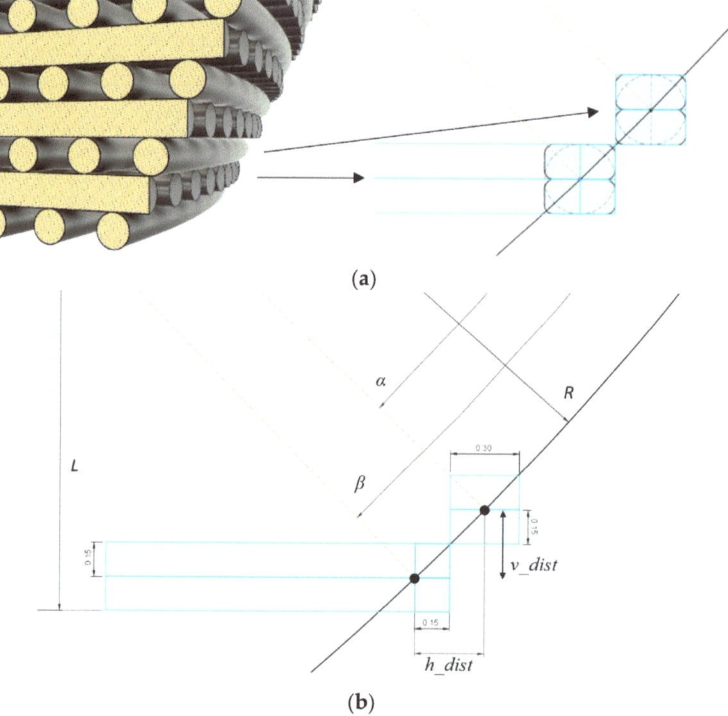

Figure 3. Printability of scaffolds: (**a**) Representation of the limit criteria assumed for the printability of a scaffold by FFF. There should be, at least, a tangential contact between successive layers. (**b**) CAD representation of a cut through the bottom layer of a scaffold, where the limit tangential contact will manifest. A system of equations was set up to relate the indicated parameters: the horizontal (*h_dist*) and vertical (*v_dist*) distances between the two black dots (the intersection of successive filament layers with the sphere section defining the curvature), the radius of the sphere section (*R*), the distance from the center of that sphere to the bottom of the scaffold (*L*), and the angles α and β measured from the center of the sphere.

The relevant parameters established to calculate the curvature limit are indicated in Figure 3b, which shows a cut through the bottom half of a sphere's square-shaped section and the fibrils deposited in the first four printed layers. The black dots denote the intersection of the section with the layer heights defining the position of the deposited filaments. All parameters are related in a system of equations (Equation (2)) that describes the following: (i) the difference in horizontal distance between the two dots (h_dist), (ii) the difference in vertical height between the two dots (v_dist), and (iii) the height of the designed scaffold, relative to its midplane (L). Here, R represents the radius of the sectioned sphere, and α and β the angles from the midplane to the filaments' positions, considering as vertex the sphere's center.

$$\begin{cases} h\_dist = R(\cos \alpha - \cos \beta) \\ v\_dist = R(\sin \beta - \sin \alpha) \\ L - 0.15 = R \sin \beta \end{cases} \tag{2}$$

Figure 3b and Equation (2) indicate some dimensions pertinent to the scaffold design and printing process. In TE scaffolds, pore size is one of the most relevant parameters influencing cell proliferation, differentiation, tissue formation, and vascularization, and it has been used as a means to elicit the desired biological response [56]. Therefore, to have dimensions compatible with OC TE constructs, both the fibril diameter and the pores were set to have a width of 300 µm, consistent with dimensions previously described in the literature [57,58]. Regarding the pores, this dimension was measured at the center of the scaffold. This choice considers the fact that for both design strategies, the pore size is the same at the center but varies towards the edges of the scaffolds, with the porosity increasing in the "concentric radius" design and decreasing in the "constant radius" design. As stated in Section 2.1, each filament was split into two printhead passages, represented by the blue line splitting the fibrils horizontally.

The system in Equation (2) can be solved for the desired variables, if sufficient parameters have been assigned. For example, given a height L for the scaffold and assuming a tangential contact between the first layers, the system can be solved for the radius R and the angles α and β. Conversely, if a radius is assigned to match the curvature of a tissue, and a dimension L is assigned to the scaffold, the system can be solved for h_dist. Thus, a conclusion could be reached about whether this scaffold architecture would be self-supported during the printing process. Given a scaffold height, the system in Equation (2) can be solved for the radius R. The lower the R value is, the more curved the scaffold obtained. However, there is a lower limit for its value, imposed by printability. This represents the situation, previously mentioned as limit of tangential support, in which scaffold integrity is compromised due to lack of support between successive layers, typically observed towards the sides of the scaffold.

Defining the curvature with a parameter independent of the scaffold height and width was more convenient for design and comparison purposes. For this reason, the radius through the center of the scaffold was calculated (Figure 4). Then, a scaffold can be designed with a specific curvature and later verified if it features unsupported fibrils. Thus, if the width or height of the scaffolds were to be modified, the overlapping sections would still have the same geometry. In Table 1, a summary of the scaffold dimensions and calculated parameters is presented.

Table 1. Summary of dimensions of curved scaffolds and solution to the limit printability condition.

Scaffold Dimension	Assigned Parameters	Calculated Limit Curvature
height = 20.1 mm width = 20.1 mm	L = 10.05 mm h_dist = 0.3 mm v_dist = 0.3 mm	$R_{side\ of\ scaffold}$ = 13.790 mm α = 44.12° β = 45.88° R = 17.064 mm

To compare and verify the supported material printability assumptions taken in model definition, scaffolds were designed with the limit-determined curvature and with greater and smaller curvatures, using both "constant radius" and "concentric radius" design strategies. The greater and smaller curvature radii were chosen so that both had a similar difference concerning the limit printability radius and, with the smaller radius, the square section would be close to intersecting the projection of the sphere. This resulted in modelling scaffolds with 14 mm and 20 mm curvature radii. The developed CAD models are represented in Figure 5.

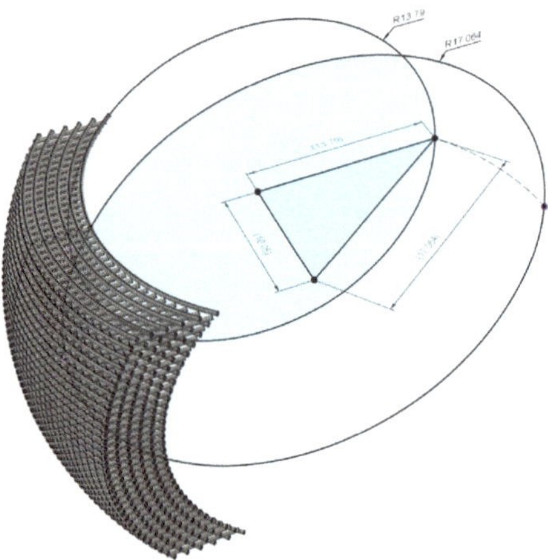

Figure 4. Determination of radius through the center of the scaffold. This radius is used to define the curvature in CAD, as a parameter set to define the curvature independent of the scaffold height and width.

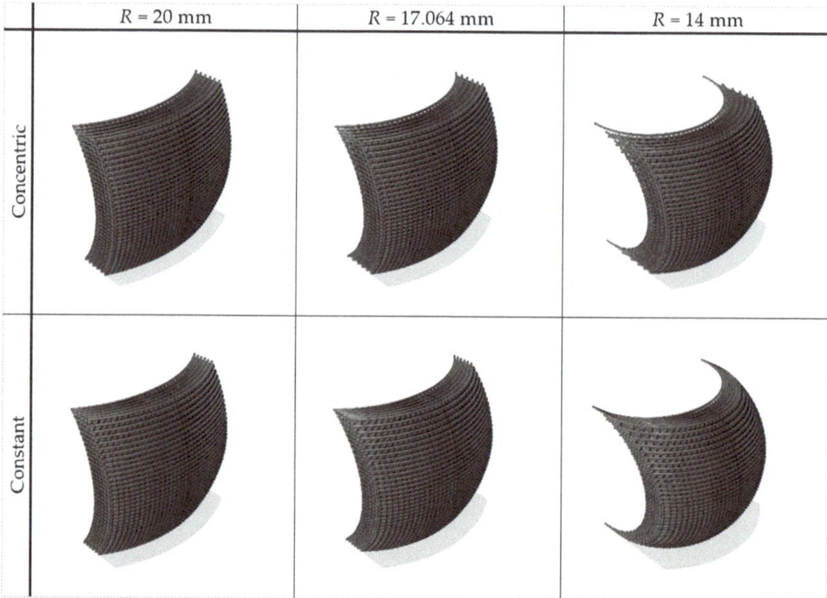

Figure 5. CAD models of the designed curved scaffolds. Designs for both strategies, "concentric radius" and "constant radius". The chosen radii (*R*) correspond to the determined limit of printability and values over (20 mm) and below (14 mm) that limit.

3.1.2. Orthogonal Scaffold Design

An orthogonal scaffold was also designed with the same top projected dimensions as the curved scaffolds, i.e., the same side lengths of the square used to achieve the sectioning of the sphere. The side on which it was printed was equivalent to those used to print the curved scaffolds. As for the curved scaffolds, pore size and fibril width were defined as 300 μm. A representation of this scaffold is shown in Supplementary Figure S3.

3.1.3. Block Design for Mechanical Testing under Compressive Loading

Since mechanical testing equipment compresses parts between two flat plates, the curved scaffolds could not be subjected directly to this test, given that the load would be transmitted only between the contact points of the curved scaffold with the plates instead of the whole surface of the scaffold. For this reason, solid blocks were modelled and added to all sides of the scaffolds so that they would, on one side, follow the curvature of the scaffolds and, on the other, have the proper flat surface necessary for the compression tests. The models included a slight overlap between the blocks and the scaffold, and in the 3D-printing process, they were printed as one single part to avoid slippage during testing. In all cases, the height of these assemblies was designed to be 25 mm, with the contact compression area corresponding to the square area used to achieve the sectioning of the sphere. The models created for each scaffold condition are shown in Supplementary Figure S1. Additionally, blocks integrating the orthogonal scaffold and solid blocks with the same global dimensions were also designed for comparison.

3.2. Scaffold Structural Characterization

Curved scaffolds were printed using PLA by FFF, with a curvature defined by the radius of a sphere and with projected dimensions of 20.1 mm × 20.1 mm (Figure 6). For structural and shape fidelity analyses, the scaffolds were assessed through SEM imaging (Figure 7) and by μ-CT analysis (Figure 8). Figure 7 specifically details the corners, where the more severe material detachment was predicted and observed. For the 20 mm curved scaffolds, obtained following both design strategies, the filaments were printed in the expected location, and no detachments were observed. The division of each fibril into two distinct layers is clearly visible. In the curved fibrils, a wavy pattern is seen in the bottom printed layer, but a connection to the straight fibrils is still clearly visible. For the scaffolds with a radius of 17.064 mm, the curvature of the top fibrils shows some deviation from a circular profile, suggesting a poor adhesion in relation to the fibrils printed underneath, with the filament being dragged out from the designed position. Some filaments are also clearly detached in some of the scaffold in the "constant radius" strategy. For the 14 mm scaffolds, the top curved layers are completely detached from the scaffolds for both design strategies. In particular, in the "concentric radius" strategy, one top curved fibril is even missing, since it had very limited contact points with the other parts of the scaffold. In the "constant radius" strategy, the change in position of discrete fibrils in a surface can be seen in the corners of the top layer. Due to the overlap, the slicing algorithm joined the extremities of the fibrils into the observed surface.

Overall, in the scaffold corners, an increase is observed in imprecisely deposited filament with the decrease in the radius, due to material deposition without support. Another printing artifact, also observed in Figure 7, is stringing, resulting from release of the residual pressure inside the nozzle between deposition movements. Both are artifacts that impact scaffold geometry and properties.

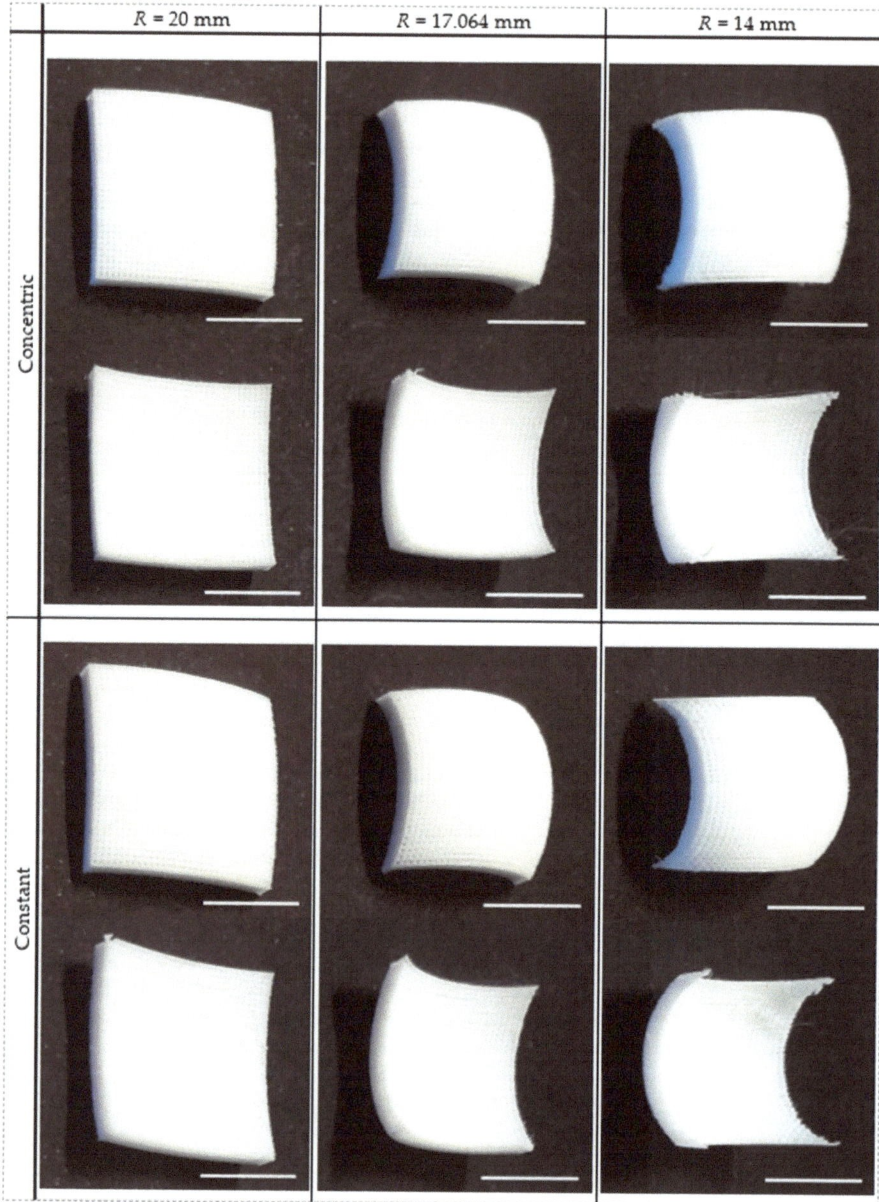

Figure 6. Examples of FFF-printed curved scaffolds based on the CAD models. Scaffolds were printed using both design strategies: "concentric radius" and "constant radius". The chosen radii (R) correspond to the determined limit of printability (17.064 mm) and values over (20 mm) and below (14 mm) that limit (scale bars: 1 cm).

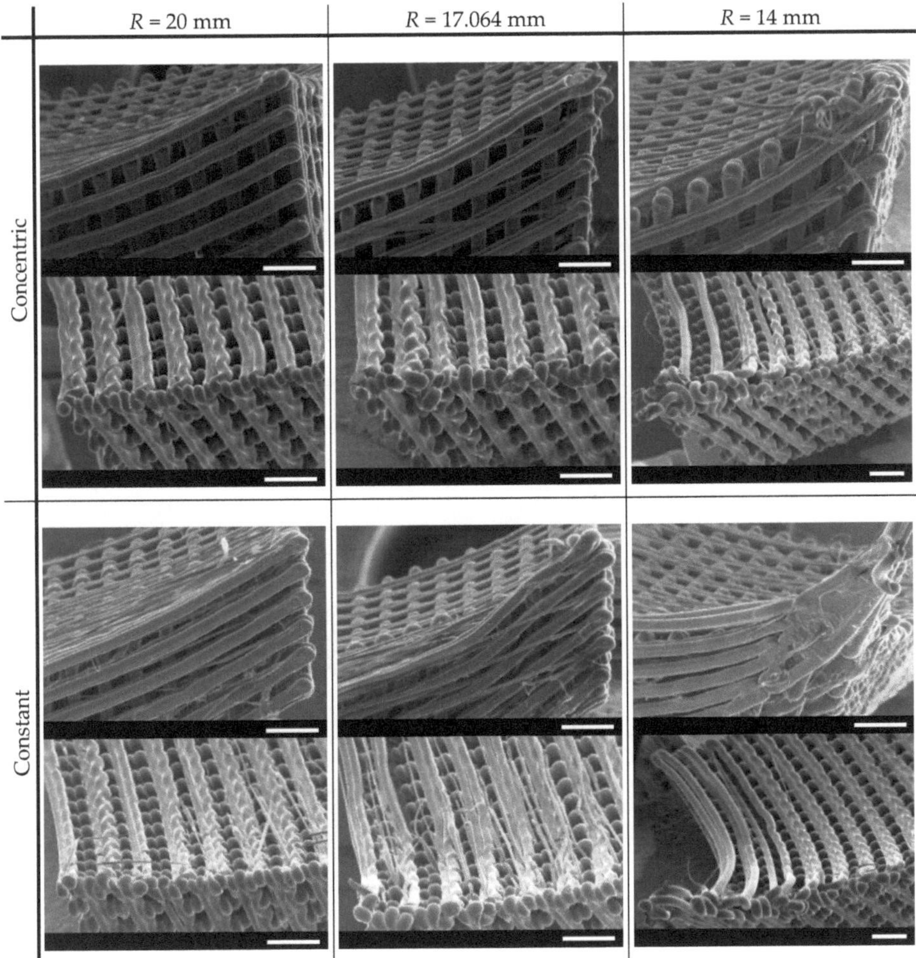

Figure 7. SEM imaging in the corners of the scaffolds using both design strategies, "concentric radius" and "constant radius". The chosen radii (*R*) correspond to the determined limit of printability (17.064 mm) and values over (20 mm) and below (14 mm) that limit (scale bars: 1000 μm).

In Figure 8, the difference between the two design strategies is clearly perceptible from the images shown. In the "concentric radius" strategy, scaffolds are thinner at the center and become wider towards the extremities. Additionally, by decreasing the radius, this enlargement in the extremities becomes even more noticeable. Conversely, in the "constant radius" strategy, a width reduction is observed towards the sides of the scaffolds, and such reduction is more perceptible with the decrease in the radius. For both strategies, no detachments were observed for the 20 mm radius, neither on the top nor on the bottom layers, the locations where misplacement of filament deposition during printing is more challenging. However, for the 17.064 mm and 14 mm radii, detachment of some fibrils is observed in the top and bottom printed layers. For the scaffolds made following the "concentric radius" strategy with radii of 17.064 mm and 14 mm, the fibrils were printed with the correct shape. This result suggests that the minimal contact with the previously deposited layer was achieved, since the fibril was not stretched out of position, although the minimal contact led to a later detachment. The detachment is also more pronounced towards the sides, where less contact between filaments had been predicted. In some sections of the scaffolds made using the "constant radius" strategy with radii of 17.064 mm and 14 mm, the filament being deposited became straight instead of maintaining the curvature, indicating no contact with the previously deposited material. For the 17.064 mm radius scaffolds, this problem is also more noticeable towards the sides. On the other hand, for the 14 mm radius scaffolds, due to the proximity of the fibrils being printed towards the sides, they were printed as a surface instead of individual filaments. Consequently, for those scaffolds, the detachment was seen not at the extremity but closer to the center of the scaffold.

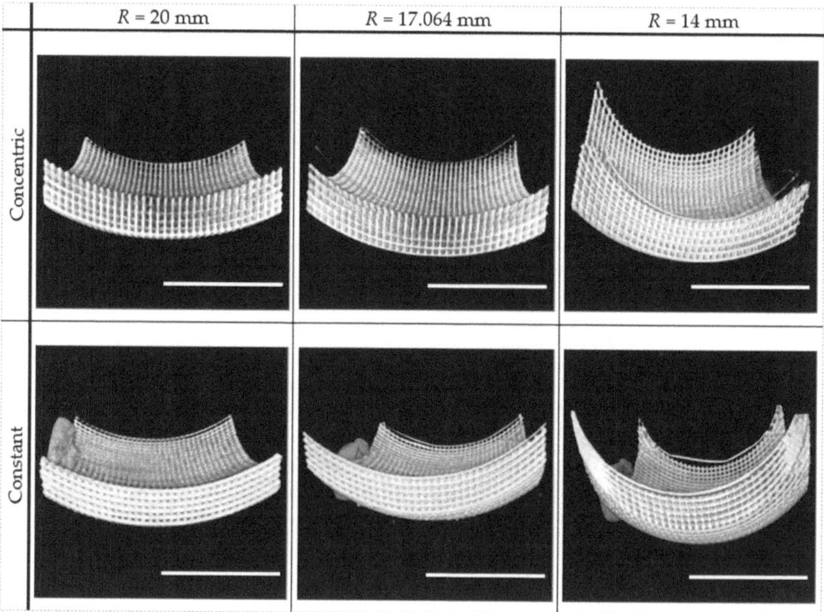

Figure 8. Representation of the µ-CT 3D reconstructions of the curved scaffolds using both design strategies, "concentric radius" and "constant radius". The chosen radii (*R*) correspond to the determined limit of printability (17.064 mm) and values over (20 mm) and below (14 mm) that limit (scale bars: 1 cm).

As represented in Figure 9, scaffold characterization focused on specific volumes of interest (VOIs) to analyze local differences in printed scaffolds due to the different scaffold structures resulting from the two design strategies and radius variation. Properties of regions in the CAD models, approximately corresponding to these sections, were calculated for comparison.

The scaffold properties were estimated from µ-CT reconstructions and are presented in Table 2 and Supplementary Table S1. Measurements of porosity (%), interconnectivity (%), and surface area/volume ratio were made for the sections of the scaffolds. The experimentally estimated values for porosity are frequently slightly higher than the ones determined from the CAD models (Table 2), which could indicate an under-extrusion of filament during printing.

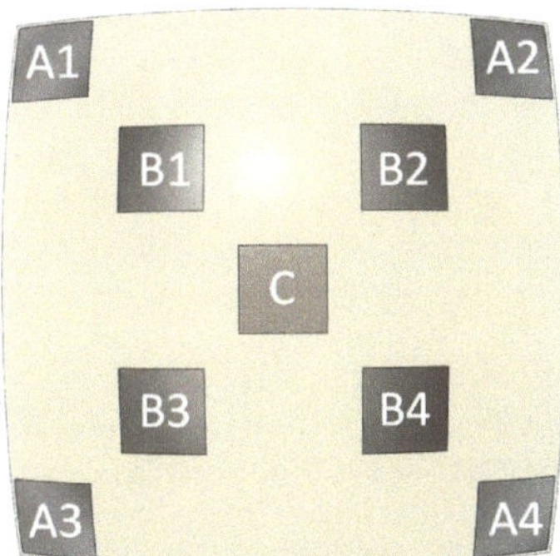

Figure 9. Scaffold sections (VOIs) identification of where a particular shape analysis is dedicated.

Table 2. Percentage (%) of porosity of the curved scaffold CAD models and of the printed scaffolds, obtained from the μ-CT analysis. The sections (VOIs) of the scaffolds are indicated in Figure 9.

VOI	Concentric						Constant						Orthogonal	
	14 mm		17.064 mm		20 mm		14 mm		17.064 mm		20 mm			
	CAD	μ-CT	CAD	μ-CT	CAD	μ-CT	CAD	μ-CT	CAD	μ-CT	CAD	μ-CT	CAD	μ-CT
A1		64.2		64.0		56.9		48.7		53.7		50.8		
A2	60.3	70.8	59.1	63.4	58.4	61.6	45.1	52.9	52.0	55.7	53.7	53.9		
A3		67.3		64.6		61.6		45.0		53.9		62.5		
A4		67.0		66.4		61.3		52.9		56.8		53.9		
C	56.3	62.1	56.3	65.9	56.3	64.7	56.3	65.0	56.3	61.0	56.3	60.3	56.3	62.3
B1		63.7		64.3		66.2		58.7		58.7		63.4		
B2	57.6	65.3	57.1	64.9	57.0	56.4	55.7	61.0	55.8	58.4	56.0	63.9		
B3		63.5		66.2		65.4		57.5		61.8		62.4		
B4		62.1		61.9		65.1		54.9		59.0		59.5		

Different results for the porosity variation, from the extremities of the scaffolds towards the center, are expected according to the design strategy. In the "concentric radius" strategy, due to the enlargement towards the extremities, porosity was generally higher at the extremities and lower towards the center, as expected. Additionally, the variation was greater in the scaffold with the smaller radius. On the contrary, in the "constant radius" strategy, porosity was lower in the corners and increased towards the center of the scaffolds. The variation was also more pronounced in the scaffolds with the smaller radius. These results are expected, as scaffolds with larger radius will approximate an orthogonal scaffold with equivalent dimensions.

Concerning the surface area/volume ratio (Supplementary Table S1), filaments are being deposited in single strands over each other, so no expressive variations should be expected between all sections of the scaffolds. The observed ratio ranged from 13.6 mm^{-1} to 16.6 mm^{-1}, which can be attributable to experimental variation and differences in the VOIs considered. Regarding interconnectivity, for every section considered, the value is approximately 100% (Supplementary Table S1), meaning there were no occlusions in the pores resulting from the manufacturing procedure.

3.3. Mechanical Behavior of Curved Scaffolds under Compressive Loading

As explained in Section 3.1.3, dedicated assemblies were designed and printed for mechanical assessment. Such assemblies integrate the different curved scaffolds designed into the middle of solid blocks placed on both sides of the scaffolds, allowing the compression of curved scaffolds between the two flat plates of the test machine in spite of their curvature. Specimen pictures of these assemblies, before and after compressive tests, are shown in Supplementary Figure S4. The obtained stress–strain curves and corresponding compressive moduli, calculated from the initial linear regions, are shown in Supplementary Figure S5 and Table S2, respectively. Since the solid block has a compact structure, it displays a higher modulus. Upon the insertion of the porous scaffolds into the blocks, the calculated modulus decreases, pointing to a reduction in resistance to deformation.

The calculation of the stresses in Supplementary Figure S5 was performed with the force being applied considering two distinct areas: (i) the cross-sectional area, corresponding to the area of contact of the machine with the blocks; (ii) the area of the curved surfaces that define the curvature of the scaffolds. Considering the conventional cross-sectional area, statistically significant differences in the compressive modulus were only found among the assemblies when compared to the concentric 20 mm insertion, but these differences can result from wider dispersion of the experimental results. The curved surface area can be considered as more representative of the stress distribution on the assemblies. In this case, the force is considered to be applied over a larger area, resulting in the calculation of lower moduli for scaffolds with greater curvatures but also increasing the significance differences between the estimated moduli.

Concerning the yield stress, calculated by the 0.2% offset method, a similar behavior to the compressive modulus was observed, with the insertion of a porous structure within the block leading to a significant reduction in the yield stress in comparison to the solid block (Supplementary Figure S5 and Table S2). Again, when considering the stress calculated in relation to the curved surface areas instead of the cross-sectional areas, the determined yield stress was reduced, and the differences between the stresses depending on the curvatures became more significant (Supplementary Figure S6).

Furthermore, two linear regions are observed in the stress–strain plots, suggesting different stages in the compression. In an initial stage, the strain would be mostly absorbed by the scaffold insertions, and due to its small width, the onset of plasticity is observed earlier. A second linear region can be observed, corresponding to the compression of a structure where the scaffold inserts have collapsed. This behavior shows some resemblance to the solid structure. Note that while the moduli determined with the assemblies should not correspond to the moduli of the scaffolds, since some deformation might also have occurred in the solid blocks, the results obtained nevertheless allow making a comparison between the assemblies. In Figure 10, the moduli and yield strength calculated considering the curved surface area of the scaffolds are shown.

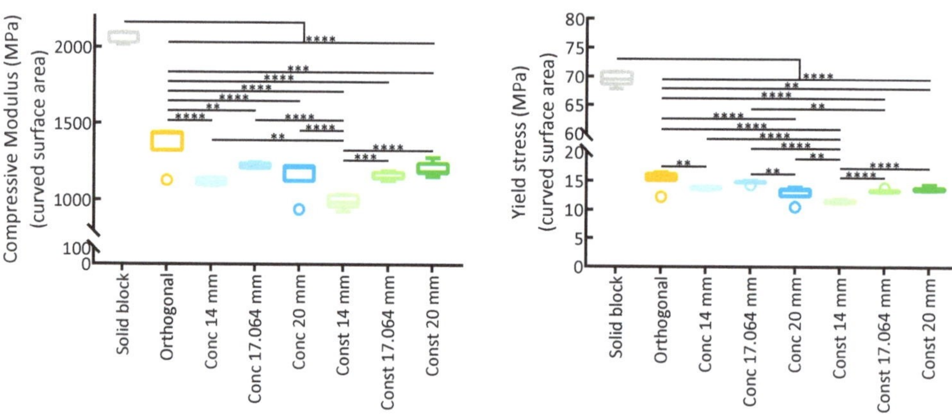

Figure 10. Compressive moduli (determined from initial linear regions) and yield stress (calculated with the 0.2% offset method) for the mechanical compression of the scaffold and block assemblies, considering the area of the curved surface of the scaffolds for the calculation of stresses ($n = 7$; ** $p < 0.01$, *** $p < 0.001$, **** $p < 0.0001$).

3.4. Finite Element Analysis

A deeper understanding of the compressive behavior of the assemblies can be gathered from finite element analysis. Supplementary Figure S7 shows the von Mises stresses calculated in the compression of the scaffold and block assemblies. The existence of two planes of symmetry allowed the simulation to be run in a fraction of the assembly. Only this fraction, corresponding to a quarter of the assembly, is represented in the figures.

For all assemblies, the simulations show higher von Mises stresses in the scaffold fibrils perpendicular to the direction of the applied load. On the other hand, the stresses in the curved fibrils, closer to a parallel orientation in relation to the applied load, are much smaller. It is understandable that higher stress is obtained upon a force being transmitted across smaller material areas, as occurs in the transition of the solid blocks to the scaffold sections in the assemblies. Additionally, higher stresses are associated with more extensive deformations. Therefore, when compressing the assemblies, a more extensive deformation is observed in the scaffold region, in comparison to the solid blocks of the assemblies. Still, the deformation is not limited to the scaffold region, and some deformation is also predicted in both solid blocks adjacent to the scaffolds, although not as noticeable as that in the scaffold sections (Supplementary Figure S8). Comparing the assemblies and the homogeneous solid block, the range of stress variation is smaller in the compression of the solid block. Since there is no sudden transition from blocks to scaffolds, the stress in the solid block does not reach values as high as in the assemblies. Furthermore, since in the assemblies the stress is mostly located within the scaffold region, the stress is not as high in the assembly block region as in the corresponding region of the solid block.

To assess the effect of the curvature and the design strategies on the estimated von Mises stresses, a stress plot was obtained through the assemblies and the solid block in five distinct locations, represented by the red segments in Figure 11a. Those locations are in the following regions: corner (A), top central (B), center (C), side middle (D), and midway between corner and center (E). Figure 11b shows the calculated von Mises stresses. An average of the von Mises stress was calculated on the scaffold region of the assemblies and in the corresponding section in the middle of the solid block (Supplementary Figure S9). Compared to the assemblies, the calculated average von Mises stress in the solid block is always inferior, reflecting the load distribution by a larger area. For the assemblies, a pattern is observed for the calculated stresses according to their location. Indeed, these stresses are higher in the center and become smaller towards the edges of the scaffold region of the assemblies. This reduction in estimated stresses is more evident towards the corners and the sides than towards the top positions. This is an observation that we hypothesize to be due to the scaffold fibrils' orientation on the corners and on the sides that may allow for a greater freedom for deformation and, therefore, a reduction of stress.

Additionally, with an increase in curvature, a greater freedom for deformation could also be anticipated, so an overall slight decrease in stress might be calculated. Moreover, the variation in stress in the different positions can also be analyzed relating to the curvature. On the orthogonal scaffold insert, the calculated stresses are smaller towards the edges. This observation is more marked for curved scaffolds, with smaller stresses estimated as the curvature increases. Further comparison of stresses in the same region of assemblies with different curvatures can provide further insights. In the center of the scaffolds, due to their geometric resemblance at these locations, the values of stresses estimated are very similar, regardless of their design. However, the reduction in stress values becomes more noticeable farther from the center, with the greater variations observed in the corners. The reduction is also more noticeable in the scaffolds manufactured with the "concentric radius" strategy than with the "constant radius" strategy, which could be a result of the structural widening towards the corners allowing a greater freedom for deformation in the former strategy (Figures 11b and S9).

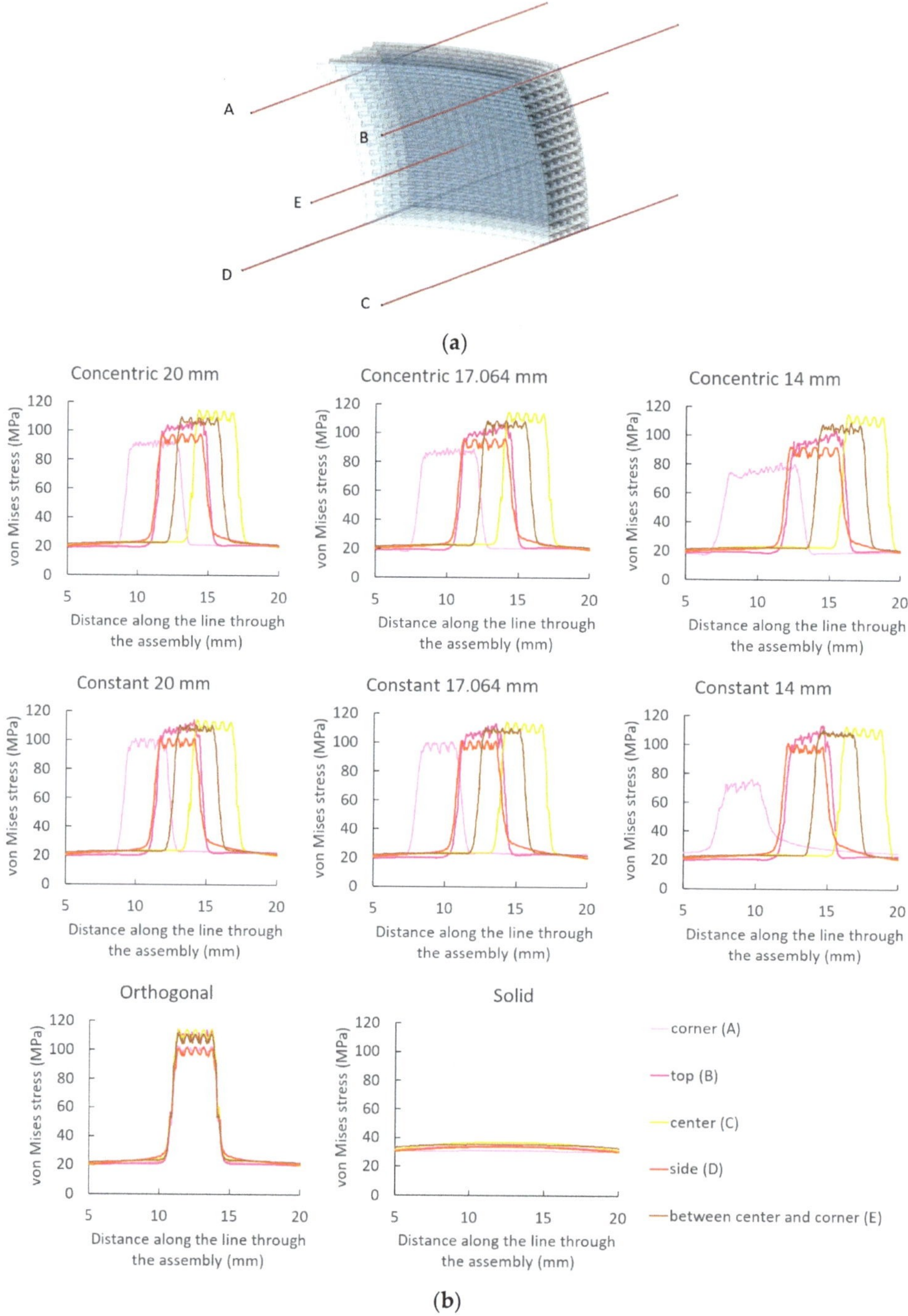

Figure 11. Von Mises stresses in the compression of scaffolds and block assemblies: (**a**) Position of the lines (red segments) along which stresses were determined. (**b**) Variation in stress along the lines represented in (**a**) corresponding to five different locations in the assemblies (corner (A), top (B), center (C), side (D) and between center and corner (E)).

4. Discussion

The ability to recreate the precise shape and heterogeneous architecture of tissues in the human body remains a challenge in TE strategies. The purpose could be, for instance, to create scaffolds fitting specific injured regions, inevitably variable by nature, and following the contours of surrounding tissue to achieve an appearance close to the native one. Additionally, at a smaller scale, recreating branching patterns of tissues, accounting for the possibility of vascularization, and gradient features would also be of great interest to produce more robust tissue substitutes [59,60]. In this sense, available medical imaging techniques, such as computed tomography and magnetic resonance imaging, have been combined with CAD software-assisted design and additive manufacturing methods to fabricate "patient-tailored" scaffolds able to fit perfectly into the defect [25]. Despite the recent shift in the TE field towards more complex scaffold designs, the produced constructs still fail to fully recreate the complexity of the native tissues and the curvature concept has been highly overlooked. In this work, we present a novel method to reproduce the curvature of the knee OC tissue, using a sphere as an approximation and as a template for the design and construction of scaffolds. Using an AM technique for scaffold manufacturing, the developed design presents the advantage of being highly reproducible. Furthermore, the use of parametric design allowed precise definition of scaffold parameters relating to curvature and overall dimensions, which we demonstrated could be altered to desired arbitrary values by modelling with three distinct curvatures. Thus, the purpose of having a scaffold that could fulfil specific curvature needs was achieved.

Fibers of material are designed at a specific distance from each other to manufacture, by extrusion, scaffolds exhibiting a porous structure. While more pronounced curvatures are expected for a lower design radius used, there is a minimal limit of the radius value that can be utilized without the use of supports. Still, the use of supports should be avoided due to constraints on printing fine details and potential scar formation upon support removal [61]. Therefore, with the introduction of a curvature, a procedure to evaluate the printability without using supports was devised, and the lower limit radius that allows printability without supports calculated by this procedure was 17.064 mm, considering a scaffold with 20.1 mm × 20.1 mm dimensions. Having in mind the objective of achieving an OC construct, it is worth highlighting that such a curvature is superior to the one found in the native tissue [62–64], and thus, these designs and manufacturing techniques can easily be adapted to the creation of clinically relevant scaffolds.

Considering the results obtained from the SEM and μ-CT imaging of the manufactured scaffolds (Figures 7 and 8), the printability limit is in accordance with what was expected. Clear deficiencies are observed when the scaffold radius of curvature is lower than this limit, and prints faithful to the design are observed when the radius is higher than it. Some faults observed in the prints made near the limiting radius of curvature can be considered reasonable to occur, as room for depositing new layers over the existing material is close to the physical limit. Being closer to the limit of printability also resulted in an increase in stringing printing artifacts. This suggests that the smaller support at the extremities of the scaffolds, where those artifacts were predominately observed, resulted from the release of the residual pressure at the nozzle, which compromised printability performance and potentiated appearance of those artifacts. A better tuning of the nozzle retraction settings could be attempted. However, it would not affect the lack of support at the steepest regions of the scaffolds with higher curvature. The fidelity potential of the selected manufacturing process can also be verified in the porosity measurements obtained from the μ-CT imaging (Table 2). These measurements confirm that, from the center of the scaffold towards the corners, the porosity decreases for the "constant radius" and increases for the "concentric radius" design strategies. Furthermore, as expected, this variation was more evident in the scaffolds with higher curvature. A valuable feature, quantified by μ-CT, was a pore interconnectivity of 100%, which is favorable for cell migration and colonization throughout the entire structure and also for widespread nutrient supply and waste removal [65]. Previous studies have shown that cell fate is highly affected by pore size [57,66]. Fibrils were designed with a width of 300 µm and, at the center of the scaffold, placed at 300 µm from each other. Such distance defines the pore size, which would increase or decrease towards the corners, whether built according to the "concentric radius" or "constant radius" strategies, respectively. Although the literature shows some variability, it has been suggested that larger pores (above 250–300 µm) favor a differentiation towards the osteogenic lineage by allowing vascularization to occur. Conversely, with smaller pores, blood vessel formation is hindered, and an environment closer to cartilage tissue is created, so chondrogenic differentiation is favored [67,68]. In the case of chondrogenesis, further differences in cell behavior have been reported, with smaller pores promoting type II collagen and aggrecan production, while larger pores seem more favorable to cell proliferation [67]. Since the design is automated in this work, it would be feasible to change parameters so that the differentiation would be favorable towards any specific lineage. Considering this, our future work will comprise the evaluation of the biological performance of these curved scaffolds through the study of their effects on the proliferation and differentiation of human mesenchymal stem/stromal cells, in comparison to standard planar orthogonal scaffolds.

To perform the compressive testing of the produced scaffolds, an assembly with solid blocks of material had to be conceived so that a load could be applied to the flat surfaces of the testing machine. However, being constituted by the same material as the scaffold, it is not possible to exclude the compression in the blocks. Still, a clear reduction in the yield strength is observed when scaffold inserts are introduced inside a solid block, as is a reduction in the compressive modulus, pointing to an initial compression exerted mainly in the scaffold region (Supplementary Figure S5). Compared to the standard orthogonal scaffold design, the introduction of curvature resulted in a small decrease in yield strength (considering the curved surface areas for the calculation), more noticeable with greater curvatures but still maintaining a similar magnitude (Supplementary Figure S6).

Considering only the orthogonal scaffold design, a compression test was also performed with just the scaffold insert (Supplementary Table S3 and Figure S10). Compared to the assembly with blocks, the yield strength is slightly lower (11.6 MPa vs. 15.2 MPa), which might indicate that some of the force being applied to assemblies is causing deformation of the blocks. The determined modulus for the scaffold alone was 215 MPa, which is in accordance with values reported by other works using the same PLA material [69]. Acknowledging the aims of TE, for optimal scaffold integration, their ideal properties should match the ones of the tissue they are intended to replace. Therefore, these scaffolds are closer to the strength and modulus of trabecular bone [70], rather than cartilage (lower modulus and strength) [47] or cortical bone (higher modulus and strength) [71]. In this regard, other materials and porosity variations could be envisaged to approximate the mechanical properties of those tissues. Moreover, the present study only evaluated the mechanical behavior of the scaffolds under specific conditions, and therefore, further work will include the analysis of their performance under more varied and realistic conditions (e.g., cyclic loading).

Taking advantage of the capabilities of numerical simulations, it was possible to better understand the assemblies' behavior during compression, improving on what was reached experimentally. The material was considered only as linear elastic, and the compression was restricted to the elasticity limit strain observed experimentally, so that the prescribed compression in the simulation would be within legitimate limits. The used modulus for the FEA simulations was obtained from the compression of the solid block (2062 MPa, Table S2). This value is within the range of the ones found in the literature for the compression of PLA specimens manufactured by FFF [72,73], and considering that the block was produced with the same manufacturing conditions as the assemblies prepared for the scaffolds mechanical tests, the choice of this value is in our opinion reasonable. Note that compression modulus value estimation depends on several factors, including testing mode (e.g., tension or compression), material specimens' preparation method (e.g., injection molding, FDM), and parameters used (temperature, extrusion speed, and infill pattern) [74,75]. Based on the results for von Mises stresses (Figure 11 and Supplementary Figure S8), it is possible to confirm that the load is mostly absorbed by the fibrils aligned with the applied force, yet some deformation is observed in the adjacent solid blocks. This result supports the hypothesis that the previously determined experimental modulus is not exclusively due to the scaffold compression. The simulation also reveals how the load is transmitted, which is a consequence of the scaffold design. The straight fibrils aligned with the applied load accommodate most of the load, and very little is transmitted to the curved fibrils. In order that the scaffolds could have a curved porous structure and be manufactured by FFF, the fibrils needed to be in the designed orientation. Therefore, the curved structure resulted in a limitation to the freedom of design, with implications for the transmission of loads. Additional designs can be proposed with the perspective of a more widespread distribution of loads, and numerical simulations can be employed for a fast evaluation of the mechanical behavior of such designs.

5. Conclusions

In the present study, we presented a novel strategy for designing and manufacturing curvature-featuring scaffolds using a mathematical approach to precisely define their structure. With the variation in design parameters, different curvature-featuring scaffolds could easily be obtained. In the context of scaffolds for OC TE, the ability to manufacture scaffolds that mimic the natural tissue and are adaptable to injury site structure variability has the potential to achieve better integration and successful therapeutic outcomes. A procedure was conceived to determine the manufacturability of the scaffolds, depending on their radius and curvature, which SEM and μ-CT experimental imaging corroborated. Mechanically, the curved scaffolds showed comparable properties to the common orthogonal scaffolds. A greater insight into their mechanical behavior was obtained from FEA, identifying the areas subjected to greater stresses. The insight provided by those models highlighted the potential of conjugating numerical modelling with experimental data towards the development of improved scaffolds for TE strategies.

Supplementary Materials: The following supporting information can be downloaded at https://www.mdpi.com/article/10.3390/polym15092129/s1. Figure S1: CAD models of the scaffold and block assemblies; Figure S2: Schematic representation of the scaffold and blocks assemblies imported into COMSOL software; Figure S3: Designed orthogonal scaffold with the same side lengths and top projected as the curved scaffolds; Figure S4: Representative images of the scaffold and blocks assemblies before (top) and after (bottom) the mechanical compression test; Figure S5: Stress-strain curves of the mechanical compression of the block and scaffold assemblies and respective compressive moduli values; Figure S6: Yield stress in the compression of the scaffold and blocks assemblies calculated with the 0.2% offset method from the stress-strain curves, in relation to the cross-sectional area (top) and the area of the curved surface of the scaffolds (bottom); Figure S7: Von Mises stresses in the numerical finite element analysis (FEA) of the mechanical behavior in compression of the block and scaffold assemblies; and a solid block with the same dimensions; Figure S8: Numerical simulation results of the rate of deformation of the scaffold and block assemblies along the direction of compression; Figure S9: Average von Mises stresses in the scaffold section of the assemblies and for an equivalent section of the solid block calculated from the plots in Figure 11; Figure S10: Stress-strain curves of the mechanical compression of the orthogonal scaffold without adjacent blocks; Table S1: Structural features (interconnectivity and surface area/volume ratio) of sections of the scaffolds measured from the μ-CT analysis; Table S2: Compressive modulus and yield stress (obtained with the 0.2% offset method) calculated for the solid block and for the scaffold and block assemblies; Table S3: Compressive modulus and yield stress (obtained with the 0.2% offset method) calculated for the orthogonal scaffold without adjacent solid blocks.

Author Contributions: Conceptualization, P.M., J.C.S., P.P.-F. and F.C.F.; Methodology, P.M. and J.M.; Software, P.M. and J.M.; Validation, P.M., J.C.S., P.P.-F., N.A. and C.S.M.; Investigation, P.M., J.C.S., J.M., R.C. and C.S.M.; Resources, P.P.-F., N.A. and F.C.F.; Data Curation, P.M., P.P.-F., N.A., and J.C.S.; Writing—Original Draft, P.M. and J.C.S.; Writing—Review and Editing, All Authors; Supervision,

Polymers **2023**, 15, 2129

J.C.S., P.P.-F., N.A., and F.C.F.; Project Administration, P.P.-F. and F.C.F.; Funding Acquisition, J.C.S., P.P.-F., N.A. and F.C.F. All authors have read and agreed to the published version of the manuscript.

Funding: The authors acknowledge funding from FCT—Portuguese Foundation for Science and Technology (FCT/MCTES), with dedicated funding from InSilico4OCReg (PTDC/EME-SIS/0838/2021) and OptiBioScaffold (PTDC/EME-SIS/4446/2020) project grants, as well as FCT/MCTES funding to C.M. (CEECINST/00077/2021), iBB (UIDB/04565/2020 and UIDP/04565/2020), CDRSP (UIDB/04044/2020 and UIDP/04044/2020), and i4HB (LA/P/0140/2020).

Institutional Review Board Statement: Not applicable.

Informed Consent Statement: Not applicable.

Data Availability Statement: The data that support the findings of this study are available from the corresponding authors upon proper request.

Acknowledgments: The authors thank Isabel Nogueira (MicroLab IST) for her assistance in obtaining the scaffold SEM images.

Conflicts of Interest: The authors declare no conflict of interest.

References

1. Goldring, S.R.; Goldring, M.B. Changes in the Osteochondral Unit during Osteoarthritis: Structure, Function and Cartilage–Bone Crosstalk. *Nat. Rev. Rheumatol.* **2016**, 12, 632–644. [CrossRef] [PubMed]
2. Carballo, C.B.; Nakagawa, Y.; Sekiya, I.; Rodeo, S.A. Basic Science of Articular Cartilage. *Clin. Sports Med.* **2017**, 36, 413–425. [CrossRef] [PubMed]
3. Di Luca, A.; Van Blitterswijk, C.; Moroni, L. The Osteochondral Interface as a Gradient Tissue: From Development to the Fabrication of Gradient Scaffolds for Regenerative Medicine. *Birth Defects Res. Part C—Embryo Today Rev.* **2015**, 105, 34–52. [CrossRef] [PubMed]
4. Gadjanski, I.; Vunjak-Novakovic, G. Challenges in Engineering Osteochondral Tissue Grafts with Hierarchical Structures. *Expert Opin. Biol. Ther.* **2015**, 15, 1583–1599. [CrossRef]
5. Long, H.; Liu, Q.; Yin, H.; Wang, K.; Diao, N.; Zhang, Y.; Lin, J.; Guo, A. Prevalence Trends of Site-Specific Osteoarthritis From 1990 to 2019: Findings from the Global Burden of Disease Study 2019. *Arthritis Rheumatol.* **2022**, 74, 1172–1183. [CrossRef]
6. Hunter, D.J.; Bierma-Zeinstra, S. Osteoarthritis. *Lancet* **2019**, 393, 1745–1759. [CrossRef]
7. Zhang, B.; Huang, J.; Narayan, R.J. Gradient Scaffolds for Osteochondral Tissue Engineering and Regeneration. *J. Mater. Chem. B* **2020**, 8, 8149–8170. [CrossRef]
8. Lepage, S.I.M.; Robson, N.; Gilmore, H.; Davis, O.; Hooper, A.; St. John, S.; Kamesan, V.; Gelis, P.; Carvajal, D.; Hurtig, M.; et al. Beyond Cartilage Repair: The Role of the Osteochondral Unit in Joint Health and Disease. *Tissue Eng. Part B Rev.* **2019**, 25, 114–125. [CrossRef]
9. Falah, M.; Nierenberg, G.; Soudry, M.; Hayden, M.; Volpin, G. Treatment of Articular Cartilage Lesions of the Knee. *Int. Orthop.* **2010**, 34, 621–630. [CrossRef]
10. Williams, G.M.; Chan, E.F.; Temple-Wong, M.M.; Bae, W.C.; Masuda, K.; Bugbee, W.D.; Sah, R.L. Shape, Loading, and Motion in the Bioengineering Design, Fabrication, and Testing of Personalized Synovial Joints. *J. Biomech.* **2010**, 43, 156–165. [CrossRef]
11. Wei, W.; Dai, H. Articular Cartilage and Osteochondral Tissue Engineering Techniques: Recent Advances and Challenges. *Bioact. Mater.* **2021**, 6, 4830–4855. [CrossRef] [PubMed]
12. Kwon, H.; Brown, W.E.; Lee, C.A.; Wang, D.; Paschos, N.; Hu, J.C.; Athanasiou, K.A. Surgical and Tissue Engineering Strategies for Articular Cartilage and Meniscus Repair. *Nat. Rev. Rheumatol.* **2019**, 15, 550–570. [CrossRef] [PubMed]
13. Nukavarapu, S.P.; Dorcemus, D.L. Osteochondral Tissue Engineering: Current Strategies and Challenges. *Biotechnol. Adv.* **2013**, 31, 706–721. [CrossRef]
14. Salgado, A.J.; Coutinho, O.P.; Reis, R.L. Bone Tissue Engineering: State of the Art and Future Trends. *Macromol. Biosci.* **2004**, 4, 743–765. [CrossRef]
15. Meneses, J.; Silva, J.C.; Fernandes, S.R.; Datta, A.; Ferreira, F.C.; Moura, C.; Amado, S.; Alves, N.; Pascoal-Faria, P. A Multimodal Stimulation Cell Culture Bioreactor for Tissue Engineering: A Numerical Modelling Approach. *Polymers* **2020**, 12, 940. [CrossRef]
16. Meneses, J.; Fernandes, S.; Alves, N.; Pascoal-Faria, P.; Miranda, P.C. How to Correctly Estimate the Electric Field in Capacitively Coupled Systems for Tissue Engineering: A Comparative Study. *Sci. Rep.* **2022**, 12, 11049. [CrossRef] [PubMed]
17. Seo, S.J.; Mahapatra, C.; Singh, R.K.; Knowles, J.C.; Kim, H.W. Strategies for Osteochondral Repair: Focus on Scaffolds. *J. Tissue Eng.* **2014**, 5, 204173141454185. [CrossRef] [PubMed]
18. Oliveira, J.M.; Ribeiro, V.P.; Reis, R.L. Advances on Gradient Scaffolds for Osteochondral Tissue Engineering. *Prog. Biomed. Eng.* **2021**, 3, 033001. [CrossRef]
19. Turnbull, G.; Clarke, J.; Picard, F.; Riches, P.; Jia, L.; Han, F.; Li, B.; Shu, W. 3D Bioactive Composite Scaffolds for Bone Tissue Engineering. *Bioact. Mater.* **2018**, 3, 278–314. [CrossRef]

20. Iulian, A.; Dan, L.; Camelia, T.; Claudia, M.; Sebastian, G. Synthetic Materials for Osteochondral Tissue Engineering. In *Osteochondral Tissue Engineering*; Oliveira, J.M., Pina, S., Reis, R.L., San Roman, J., Eds.; Springer International Publishing: Cham, Switzerland, 2018; pp. 31–52.
21. Armiento, A.R.; Stoddart, M.J.; Alini, M.; Eglin, D. Biomaterials for Articular Cartilage Tissue Engineering: Learning from Biology. *Acta Biomater.* **2018**, *65*, 1–20. [CrossRef]
22. Yang, J.; Zhang, Y.S.; Yue, K.; Khademhosseini, A. Cell-Laden Hydrogels for Osteochondral and Cartilage Tissue Engineering. *Acta Biomater.* **2017**, *57*, 1–25. [CrossRef] [PubMed]
23. Roseti, L.; Parisi, V.; Petretta, M.; Cavallo, C.; Desando, G.; Bartolotti, I.; Grigolo, B. Scaffolds for Bone Tissue Engineering: State of the Art and New Perspectives. *Mater. Sci. Eng. C* **2017**, *78*, 1246–1262. [CrossRef]
24. Mota, C.; Puppi, D.; Chiellini, F.; Chiellini, E. Additive Manufacturing Techniques for the Production of Tissue Engineering Constructs. *J. Tissue Eng. Regen. Med.* **2015**, *9*, 174–190. [CrossRef]
25. Koons, G.L.; Diba, M.; Mikos, A.G. Materials Design for Bone-Tissue Engineering. *Nat. Rev. Mater.* **2020**, *5*, 584–603. [CrossRef]
26. Leong, K.F.; Cheah, C.M.; Chua, C.K. Solid Freeform Fabrication of Three-Dimensional Scaffolds for Engineering Replacement Tissues and Organs. *Biomaterials* **2003**, *24*, 2363–2378. [CrossRef] [PubMed]
27. Gillispie, G.J.; Park, J.; Copus, J.S.; Pallickaveedu Rajan Asari, A.K.; Yoo, J.J.; Atala, A.; Lee, S.J. Three-Dimensional Tissue and Organ Printing in Regenerative Medicine. In *Principles of Regenerative Medicine*; Atala, A., Lanza, R., Mikos, A.G., Nerem, R., Eds.; Elsevier: Amsterdam, The Netherlands, 2019; pp. 831–852.
28. Biscaia, S.; Silva, J.C.; Moura, C.; Viana, T.; Tojeira, A.; Mitchell, G.R.; Pascoal-Faria, P.; Ferreira, F.C.; Alves, N. Additive Manufactured Poly(ε-Caprolactone)-Graphene Scaffolds: Lamellar Crystal Orientation, Mechanical Properties and Biological Performance. *Polymers* **2022**, *14*, 1669. [CrossRef]
29. Pedde, R.D.; Mirani, B.; Navaei, A.; Styan, T.; Wong, S.; Mehrali, M.; Thakur, A.; Mohtaram, N.K.; Bayati, A.; Dolatshahi-Pirouz, A.; et al. Emerging Biofabrication Strategies for Engineering Complex Tissue Constructs. *Adv. Mater.* **2017**, *29*, 1–27. [CrossRef]
30. Qu, H. Additive Manufacturing for Bone Tissue Engineering Scaffolds. *Mater. Today Commun.* **2020**, *24*, 101024. [CrossRef]
31. Mandrycky, C.; Wang, Z.; Kim, K.; Kim, D.H. 3D Bioprinting for Engineering Complex Tissues. *Biotechnol. Adv.* **2016**, *34*, 422–434. [CrossRef] [PubMed]
32. Cordeiro, R.; Henriques, M.; Silva, J.C.; Antunes, F.; Alves, N.; Moura, C. Corncob Cellulose Scaffolds: A New Sustainable Temporary Implant for Cartilage Replacement. *J. Funct. Biomater.* **2022**, *13*, 63. [CrossRef]
33. Silva, J.C.; Moura, C.S.; Borrecho, G.; de Matos, A.P.A.; da Silva, C.L.; Cabral, J.M.S.; Bártolo, P.J.; Linhardt, R.J.; Ferreira, F.C. Extruded Bioreactor Perfusion Culture Supports the Chondrogenic Differentiation of Human Mesenchymal Stem/Stromal Cells in 3D Porous Poly(ε-Caprolactone) Scaffolds. *Biotechnol. J.* **2020**, *15*, 1–14. [CrossRef] [PubMed]
34. De Mori, A.; Fernández, M.P.; Blunn, G.; Tozzi, G.; Roldo, M. 3D Printing and Electrospinning of Composite Hydrogels for Cartilage and Bone Tissue Engineering. *Polymers* **2018**, *10*, 285. [CrossRef] [PubMed]
35. Jammalamadaka, U.; Tappa, K. Recent Advances in Biomaterials for 3D Printing and Tissue Engineering. *J. Funct. Biomater.* **2018**, *9*, 22. [CrossRef]
36. Silva, J.C.; Moura, C.S.; Alves, N.; Cabral, J.M.S.; Ferreira, F.C. Effects of Different Fibre Alignments and Bioactive Coatings on Mesenchymal Stem/Stromal Cell Adhesion and Proliferation in Poly (ε-Caprolactone) Scaffolds towards Cartilage Repair. *Procedia Manuf.* **2017**, *12*, 132–140. [CrossRef]
37. Silva, J.C.; Moura, C.S.; Borrecho, G.; Alves de Matos, A.P.; Cabral, J.M.S.; Linhardt, R.J.; Ferreira, F.C. Effects of Glycosaminoglycan Supplementation in the Chondrogenic Differentiation of Bone Marrow- and Synovial- Derived Mesenchymal Stem/Stromal Cells on 3D-Extruded Poly (ε-Caprolactone) Scaffolds. *Int. J. Polym. Mater. Polym. Biomater.* **2021**, *70*, 207–222. [CrossRef]
38. Silva, J.C.; Carvalho, M.S.; Udangawa, R.N.; Moura, C.S.; Cabral, J.M.S.; da Silva, C.L.; Ferreira, F.C.; Vashishth, D.; Linhardt, R.J. Extracellular Matrix Decorated Polycaprolactone Scaffolds for Improved Mesenchymal Stem/Stromal Cell Osteogenesis towards a Patient-Tailored Bone Tissue Engineering Approach. *J. Biomed. Mater. Res.—Part B Appl. Biomater.* **2020**, *108*, 2153–2166. [CrossRef]
39. Kim, B.; Ventura, R.; Lee, B.T. Functionalization of Porous BCP Scaffold by Generating Cell-Derived Extracellular Matrix from Rat Bone Marrow Stem Cells Culture for Bone Tissue Engineering. *J. Tissue Eng. Regen. Med.* **2018**, *12*, e1256–e1267. [CrossRef]
40. Vernengo, A.J.; Grad, S.; Eglin, D.; Alini, M.; Li, Z. Bioprinting Tissue Analogues with Decellularized Extracellular Matrix Bioink for Regeneration and Tissue Models of Cartilage and Intervertebral Discs. *Adv. Funct. Mater.* **2020**, *30*, 1909044. [CrossRef]
41. Chung, J.J.; Im, H.; Kim, S.H.; Park, J.W.; Jung, Y. Toward Biomimetic Scaffolds for Tissue Engineering: 3D Printing Techniques in Regenerative Medicine. *Front. Bioeng. Biotechnol.* **2020**, *8*, 1–12. [CrossRef]
42. Pereira, D.R.; Reis, R.L.; Oliveira, J.M. Layered Scaffolds for Osteochondral Tissue Engineering. *Adv. Exp. Med. Biol.* **2018**, *1058*, 193–218. [CrossRef]
43. Leong, K.F.; Chua, C.K.; Sudarmadji, N.; Yeong, W.Y. Engineering Functionally Graded Tissue Engineering Scaffolds. *J. Mech. Behav. Biomed. Mater.* **2008**, *1*, 140–152. [CrossRef]
44. Zein, I.; Hutmacher, D.W.; Tan, K.C.; Teoh, S.H. Fused Deposition Modeling of Novel Scaffold Architectures for Tissue Engineering Applications. *Biomaterials* **2002**, *23*, 1169–1185. [CrossRef] [PubMed]
45. Gleadall, A.; Visscher, D.; Yang, J.; Thomas, D.; Segal, J. Review of Additive Manufactured Tissue Engineering Scaffolds: Relationship between Geometry and Performance. *Burn. Trauma* **2018**, *6*, 1–16. [CrossRef] [PubMed]

46. Lee, C.H.; Cook, J.L.; Mendelson, A.; Moioli, E.K.; Yao, H.; Mao, J.J. Regeneration of the Articular Surface of the Rabbit Synovial Joint by Cell Homing: A Proof of Concept Study. *Lancet* **2010**, *376*, 440–448. [CrossRef]

47. Lee, J.-S.; Hong, J.M.; Jung, J.W.; Shim, J.-H.; Oh, J.-H.; Cho, D.-W. 3D Printing of Composite Tissue with Complex Shape Applied to Ear Regeneration. *Biofabrication* **2014**, *6*, 024103. [CrossRef] [PubMed]

48. Probst, F.A.; Hutmacher, D.W.; Müller, D.F.; Machens, H.-G.; Schantz, J.-T. Calvarial Reconstruction by Customized Bioactive Implant. *Handchir. Mikrochir. Plast. Chir.* **2010**, *42*, 369–373. [CrossRef]

49. Daly, A.C.; Cunniffe, G.M.; Sathy, B.N.; Jeon, O.; Alsberg, E.; Kelly, D.J. 3D Bioprinting of Developmentally Inspired Templates for Whole Bone Organ Engineering. *Adv. Healthc. Mater.* **2016**, *5*, 2353–2362. [CrossRef]

50. Ding, S.; Kingshott, P.; Thissen, H.; Pera, M.; Wang, P.Y. Modulation of Human Mesenchymal and Pluripotent Stem Cell Behavior Using Biophysical and Biochemical Cues: A Review. *Biotechnol. Bioeng.* **2017**, *114*, 260–280. [CrossRef]

51. Saghati, S.; Nasrabadi, H.T.; Khoshfetrat, A.B.; Moharamzadeh, K.; Hassani, A.; Mohammadi, S.M.; Rahbarghazi, R.; Fathi Karkan, S. Tissue Engineering Strategies to Increase Osteochondral Regeneration of Stem Cells; a Close Look at Different Modalities. *Stem Cell Rev. Reports* **2021**, *17*, 1294–1311. [CrossRef]

52. Möller, J.; Pörtner, R. Digital Twins for Tissue Culture Techniques—Concepts, Expectations, and State of the Art. *Processes* **2021**, *9*, 447. [CrossRef]

53. Geris, L.; Lambrechts, T.; Carlier, A.; Papantoniou, I. The Future Is Digital: In Silico Tissue Engineering. *Curr. Opin. Biomed. Eng.* **2018**, *6*, 92–98. [CrossRef]

54. Farah, S.; Anderson, D.G.; Langer, R. Physical and Mechanical Properties of PLA, and Their Functions in Widespread Applications—A Comprehensive Review. *Adv. Drug Deliv. Rev.* **2016**, *107*, 367–392. [CrossRef] [PubMed]

55. Mirkhalaf, S.M.; Fagerström, M. The Mechanical Behavior of Polylactic Acid (PLA) Films: Fabrication, Experiments and Modelling. *Mech. Time-Dependent Mater.* **2021**, *25*, 119–131. [CrossRef]

56. Cheng, A.; Schwartz, Z.; Kahn, A.; Li, X.; Shao, Z.; Sun, M.; Ao, Y.; Boyan, B.D.; Chen, H. Advances in Porous Scaffold Design for Bone and Cartilage Tissue Engineering and Regeneration. *Tissue Eng.—Part B Rev.* **2019**, *25*, 14–29. [CrossRef]

57. Loh, Q.L.; Choong, C. Three-Dimensional Scaffolds for Tissue Engineering Applications: Role of Porosity and Pore Size. *Tissue Eng.—Part B Rev.* **2013**, *19*, 485–502. [CrossRef]

58. Moura, C.S.; Da Silva, C.L.; Bártolo, P.J.; Ferreira, F.C. Combination of 3D Extruded-Based Poly (ε-Caprolactone) Scaffolds with Mesenchymal Stem/Stromal Cells: Strategy Optimization. *Procedia Eng.* **2015**, *110*, 122–127. [CrossRef]

59. Kang, H.W.; Lee, S.J.; Ko, I.K.; Kengla, C.; Yoo, J.J.; Atala, A. A 3D Bioprinting System to Produce Human-Scale Tissue Constructs with Structural Integrity. *Nat. Biotechnol.* **2016**, *34*, 312–319. [CrossRef] [PubMed]

60. Murphy, S.V.; Atala, A. 3D Bioprinting of Tissues and Organs. *Nat. Biotechnol.* **2014**, *32*, 773–785. [CrossRef]

61. Zhang, X.; Le, X.; Panotopoulou, A.; Whiting, E.; Wang, C.C.L. Perceptual Models of Preference in 3D Printing Direction. *ACM Trans. Graph.* **2015**, *34*, 1–12. [CrossRef]

62. Mahfouz, M.; Abdel Fatah, E.E.; Bowers, L.S.; Scuderi, G. Three-Dimensional Morphology of the Knee Reveals Ethnic Differences. *Clin. Orthop. Relat. Res.* **2012**, *470*, 172–185. [CrossRef]

63. Irie, T.; Orías, A.A.E.; Irie, T.Y.; Nho, S.J.; Takahashi, D.; Iwasaki, N.; Inoue, N. Three-Dimensional Curvature Mismatch of the Acetabular Radius to the Femoral Head Radius Is Increased in Borderline Dysplastic Hips. *PLoS ONE* **2020**, *15*, e0231001. [CrossRef]

64. Li, K.; Tashman, S.; Fu, F.; Harner, C.; Zhang, X. Automating Analyses of the Distal Femur Articular Geometry Based on Three-Dimensional Surface Data. *Ann. Biomed. Eng.* **2010**, *38*, 2928–2936. [CrossRef] [PubMed]

65. Moura, C.S.; Silva, J.C.; Faria, S.; Fernandes, P.R.; da Silva, C.L.; Cabral, J.M.S.; Linhardt, R.; Bártolo, P.J.; Ferreira, F.C. Chondrogenic Differentiation of Mesenchymal Stem/Stromal Cells on 3D Porous Poly (ε-Caprolactone) Scaffolds: Effects of Material Alkaline Treatment and Chondroitin Sulfate Supplementation. *J. Biosci. Bioeng.* **2020**, *129*, 756–764. [CrossRef] [PubMed]

66. Xing, F.; Li, L.; Zhou, C.; Long, C.; Wu, L.; Lei, H.; Kong, Q.; Fan, Y.; Xiang, Z.; Zhang, X. Regulation and Directing Stem Cell Fate by Tissue Engineering Functional Microenvironments: Scaffold Physical and Chemical Cues. *Stem Cells Int.* **2019**, *2019*, 1–16. [CrossRef]

67. Ferlin, K.M.; Prendergast, M.E.; Miller, M.L.; Kaplan, D.S.; Fisher, J.P. Influence of 3D Printed Porous Architecture on Mesenchymal Stem Cell Enrichment and Differentiation. *Acta Biomater.* **2016**, *32*, 161–169. [CrossRef]

68. Gupte, M.J.; Swanson, W.B.; Hu, J.; Jin, X.; Ma, H.; Zhang, Z.; Liu, Z.; Feng, K.; Feng, G.; Xiao, G.; et al. Pore Size Directs Bone Marrow Stromal Cell Fate and Tissue Regeneration in Nanofibrous Macroporous Scaffolds by Mediating Vascularization. *Acta Biomater.* **2018**, *82*, 1–11. [CrossRef]

69. Alizadeh-Osgouei, M.; Li, Y.; Vahid, A.; Ataee, A.; Wen, C. High Strength Porous PLA Gyroid Scaffolds Manufactured via Fused Deposition Modeling for Tissue-Engineering Applications. *Smart Mater. Med.* **2021**, *2*, 15–25. [CrossRef]

70. Wu, D.; Isaksson, P.; Ferguson, S.J.; Persson, C. Young's Modulus of Trabecular Bone at the Tissue Level: A Review. *Acta Biomater.* **2018**, *78*, 1–12. [CrossRef]

71. Gerhardt, L.C.; Boccaccini, A.R. Bioactive Glass and Glass-Ceramic Scaffolds for Bone Tissue Engineering. *Materials* **2010**, *3*, 3867–3910. [CrossRef]

72. Vukasovic, T.; Vivanco, J.F.; Celentano, D.; García-Herrera, C. Characterization of the Mechanical Response of Thermoplastic Parts Fabricated with 3D Printing. *Int. J. Adv. Manuf. Technol.* **2019**, *104*, 4207–4218. [CrossRef]

73. Brischetto, S.; Torre, R. Tensile and Compressive Behavior in the Experimental Tests for Pla Specimens Produced via Fused Deposition Modelling Technique. *J. Compos. Sci.* **2020**, *4*, 140. [CrossRef]
74. Abeykoon, C.; Sri-Amphorn, P.; Fernando, A. Optimization of Fused Deposition Modeling Parameters for Improved PLA and ABS 3D Printed Structures. *Int. J. Light. Mater. Manuf.* **2020**, *3*, 284–297. [CrossRef]
75. Song, Y.; Li, Y.; Song, W.; Yee, K.; Lee, K.Y.; Tagarielli, V.L. Measurements of the Mechanical Response of Unidirectional 3D-Printed PLA. *Mater. Des.* **2017**, *123*, 154–164. [CrossRef]

MDPI AG
Grosspeteranlage 5
4052 Basel
Switzerland
Tel.: +41 61 683 77 34

Polymers Editorial Office
E-mail: polymers@mdpi.com
www.mdpi.com/journal/polymers

www.ingramcontent.com/pod-product-compliance
Lightning Source LLC
LaVergne TN
LVHW071930160726
843515LV00011B/2567